IMPROVEMENT STRATEGIES OF LEGUMINOSAE BIOTECHNOLOGY

FOCUS ON BIOTECHNOLOGY

Volume 10A

Series Editors
MARCEL HOFMAN
Centre for Veterinary and Agrochemical Research, Tervuren, Belgium

JOZEF ANNÉ
Rega Institute, University of Leuven, Belgium

Volume Editors
MARCEL HOFMAN
Société de Chimie Industrielle,
Centre for Veterinary and Agrochemical Research, Tervuren, Belgium

PHILIPPE THONART
Faculté Univ. des Sciences Agronomiques Gembloux,
Centre Wallon de Biologie Industrielle, Belgium

Colophon
Focus on Biotechnology is an open-ended series of reference volumes produced by Kluwer Academic Publishers BV in co-operation with the Branche Belge de la Société de Chimie Industrielle a.s.b.l.

The initiative has been taken in conjunction with the Ninth European Congress on Biotechnology. ECB9 has been supported by the Commission of the European Communities, the General Directorate for Technology, Research and Energy of the Wallonia Region, Belgium and J Chabert, Minister for Economy of the Brussels Capital Region.

The series is edited by Marcel Hofman, Centre for Veterinary and Agrochemical Research, Tervuren, and Jozef Anné, Rega Institute, University of Leuven, Belgium.

Improvement Strategies of Leguminosae Biotechnology

Edited by

PAWAN K. JAIWAL

and

RANA P. SINGH

Department of Biosciences, M.D. University
Rohtak – 124001, India

KLUWER ACADEMIC PUBLISHERS
DORDRECHT / BOSTON / LONDON

ISBN 1-4020-1405-8

Published by Kluwer Academic Publishers,
P.O. Box 17, 3300 AA Dordrecht, The Netherlands.

Sold and distributed in North, Central and South America
by Kluwer Academic Publishers,
101 Philip Drive, Norwell, MA 02061, U.S.A.

In all other countries, sold and distributed
by Kluwer Academic Publishers,
P.O. Box 322, 3300 AH Dordrecht, The Netherlands.

Printed on acid-free paper

Printed and bound in Great Britain by MPG Books Limited, Bodmin, Cornwall.

Dedicated to our beloved teacher
Professor H.S. Srivastava
(1946–2001)

CONTENTS

Foreword xvii

Preface xix

PART I: TECHNOLOGIES

Breeding Methodologies for the Improvement of Grain Legumes 3
Paolo Ranalli

Abstract 3
1. Introduction 4
2. Domestication of grain legumes 4
   2.1. Changed Growth Form 4
   2.2. Changed Life Form 4
   2.3. Suppression of Seed Dispersal Mechanisms 4
   2.4. Biochemical Changes 5
   2.5. Physiological Change 5
3. Synteny in grain legumes 6
   3.1. The Potential of *M. Truncatula* as a Model Plant in Grain Legume Crops 6
      3.1.1. Symbiosis establishment 6
      3.1.2. Nitrogen fixation efficiency, and C and N fluxes 6
      3.1.3. Seed composition 7
      3.1.4. Abiotic stress 7
      3.1.5. Biotic stress 7
4. Design of a complete breeding program 7
   4.1. Parental Selection and Hybridization 7
   4.2. Selection of Desirable Recombinants 8
5. Marker assisted selection 9
6. Constraints 10
   6.1. Restricted Genetic Base 10
   6.2. Qualitative Aspects 10
   6.3. Plant Architecture 11
   6.4. Biotic Stress 11
   6.5. Abiotic Stress 12
7. Improvement of yield 12
   7.1. Breeding Pyramid 13
      7.1.1. The apex of the breeding pyramid 13
      7.1.2. The intermediate level of the breeding pyramid 14
      7.1.3. The base of the breeding pyramid 15

8. Improvement of legume N_2 fixation 15
8.1. Strategies for Breeding Legumes with Enhanced N_2 Fixation 15
9. Improvement of disease resistance 16
9.1. Strategies and Methods of Screening for Resistance 17
10. Variety release 17
11. Conclusions and future prospects 18
References 19

***In Vitro* Morphogenesis in Grain Legumes: An Overview** 23
M.L. Mohan And K.V. Krishnamurthy

Abstract 23
1. Introduction 23
1.1. Economic Importance 23
1.2. Nutritional Value 24
1.3. Importance of *In Vitro* Techniques in Legume Improvement 24
2. *In vitro* studies in grain legumes 25
2.1. Organogenesis in Grain Legumes 26
2.1.1. Organogenesis in chickpea (*Cicer arietinum* L.) 28
2.1.2. Organogenesis in pigeonpea (*Cajanus cajan* (L.) Millsp.) 29
2.1.3. Organogenesis in mothbean (*Vigna aconitifolia* (Jacq) Marechal.) 32
2.2. Somatic Embryogenesis in Grain Legumes 33
2.2.1. Somatic embryogenesis in chickpea 33
2.2.2. Somatic embryogenesis in pigeonpea 38
2.2.3. Somatic embryogenesis in mothbean 39
2.3. Protoplast Regeneration in Grain Legumes 39
2.3.1. Protoplast regeneration in chickpea 40
2.3.2. Protoplast regeneration in pigeonpea 41
2.3.3. Protoplast regeneration in mothbean 41
2.4. Somaclonal Variation 41
2.5. Somatic Hybridization 42
2.6. Embryo Rescue 42
2.6.1. Embryo rescue in pigeonpea 43
3. Genetic transformation studies 43
4. Conclusions and future prospects 45
References 45

***In Vitro* Organogenesis** 65
A. Ganapathi, V.R. Anbazhagan, S. Amutha and R. Prem Anand

Abstract 65
1. Introduction 65
2. Organogenesis 66
3. Factors affecting organogenesis 68
3.1. Physiological Status of Explant 68

3.2. Size and Type of Explant ... 73
3.3. Genotype ... 73
3.4. Orientation of Explant on Medium ... 74
3.5. Medium Composition ... 74
4. *Vigna* species ... 75
5. *Phaseolus* species ... 75
6. *Glycine* species ... 76
7. *Arachis* species ... 77
8. *Cajanus* species ... 77
9. *Lupinus* species ... 78
10. *Pisum sativum* ... 78
11. *Cicer arietinum* ... 78
12. Importance of organogenesis ... 78
13. Conclusions and future perspectives ... 79
References ... 79

Somatic Embryogenesis ... 87
P. Venkatachalam, N. Geetha, P. Priya, N. Jayabalan and G. Lakshmi Sita

Abstract ... 87
1. Introduction ... 88
2. Types of somatic embryogenesis ... 94
3. Initiation and development of somatic embryos ... 95
4. Proliferation and maturation of somatic embryos ... 102
5. Embryo germination and conversion into plantlets ... 104
6. Effect of explants and genotypes on somatic embryogenesis ... 107
7. Effect of plant growth regulators on somatic embryogenesis ... 111
8. Effect of carbohydrates on somatic embryogenesis ... 113
9. Repetitive or secondary somatic embryogenesis ... 114
10. Artificial or synthetic seed technology ... 116
11. Somaclonal variation and evaluation of genetic stability of somatic embryo-derived plants ... 117
12. Biochemical and molecular studies on somatic embryogenesis ... 119
13. Conclusions and future prospects ... 121
References ... 122

Protoplasts of Grain and Forage Legumes: Their Exploitation in Genetic Manipulation, Physiological Investigations and Plant-Pathogen Interactions ... 133
M.R. Davey, R. Marchant and J.B. Power

Abstract ... 133
1. Introduction ... 133
2. Culture of protoplasts of grain and forage legumes ... 134
2.1. Grain Legumes ... 134
2.2. Forage Legumes ... 136

3. Isolated protoplasts for the genetic manipulation of legumes 137
3.1. Transformation 137
3.2. Somatic Hybridisation 138
3.2.1. Grain legumes 138
3.2.2. Forage legumes 141
4. Physiological investigations with legume protoplasts 143
4.1. Studies of Cell Membranes 144
4.2. Use of Protoplasts to Monitor Osmotic Changes 144
4.3. Studies of Gene Expression and Enzyme Activity 145
4.4. Investigations of the Effects of Growth Regulators on Cell Metabolism 146
4.5. Investigations of Salinity on Legume Cell Physiology 146
4.6. Studies of Photosynthesis and Cell Respiration 147
4.7. Effects of Herbicides on Cell Metabolism 148
5. Protoplast-based studies of plant-pathogen interactions 149
6. Conclusions and future prospects 149
References 150

Wide Hybridization in Important Food Legumes 155
Nalini Mallikarjuna

Abstract 155
1. Introduction 155
2. Groundnut 156
2.1. Interspecific Hybridization 158
3. Chickpea 160
3.1. Interspecific Hybridization 161
4. Pigeonpea 164
4.1. Interspecific Hybridization 165
5. Marker assisted gene introgression from wild species 168
6. Critical gaps in wild species utilization 168
References 170

PART II: NITROGEN METABOLISM

Perception and Processing of Nod Factor Signals 175
Hubert H. Felle

Abstract 175
1. Introduction 175
2. Early events 176
3. Nod factor perception 176
3.1. Functional Perception 177
3.2. Different Receptors for Entry and for Signalling? 178

3.3. Specificity of Nod Factor Recognition: Are the Fast Responses Required for Nodulation? ... 178
4. Processing of the signal ... 179
4.1. G-Protein Activation/Phosphoinositide Pathway ... 180
4.2. Protein Kinases/Map-Kinases ... 180
4.3. Calcium ... 180
4.4. Extracellular And Intracellular pH ... 182
4.5. The Nod O Protein ... 183
5. Intermediate responses ... 183
5.1. Cytoskeleton Rearrangements and Root Hair Deformation ... 183
5.2. Early Nodulin Gene Expression ... 184
6. Late responses: root cortical cell divisions and nodule formation ... 185
6.1. Imbalance of Phytohormones or Nod Factor Signal ... 185
6.2. Phytohormone Signalling ... 185
7. Conclusions and future prospects ... 186
References ... 187

Nitrate Assimilation ... 191
Niharika Shankar and H.S. Srivastava

Abstract ... 191
1. Introduction ... 191
2. Nitrate reduction in legumes ... 192
3. The characterization on nitrate reductase ... 193
4. Molecular aspects of nitrate reductase and transgenics ... 194
5. Improvement of nitrate assimilation ... 194
5.1. Carbohydrate Metabolism ... 194
5.2. Nitrate Supply ... 195
5.3. Growth Regulators ... 196
5.4. Environmental Factors ... 196
6. Conclusions and future prospects ... 196
References ... 197

Improvement of Nitrogen Utilization ... 201
Bertrand Hirel, Judith Harrison and Anis Limami

Abstract ... 201
1. Introduction ... 201
2. Modification of ammonia assimilation in legumes by genetic engineering: impact on plant physiology and development ... 204
2.1. In Alfalfa ... 204
2.2. In *Lotus* Species ... 206
3. Conventional plant breeding and mutagenesis: another way of improving nitrogen use efficiency in legumes ... 209

4. Conclusions and future prospects 215
References 217

PART III: ABIOTIC AND BIOTIC STRESS TOLERANCE

Genetic Engineering for Enhancing Abiotic Stress Tolerance 223
Rana P. Singh, Usha, S.M.H. Rizvi, Sonia and Pawan K. Jaiwal

Abstract 223
1. Introduction 224
2. Perception and transduction of stress signal 225
3. Stress responsive genes, promoters and transcription factors 228
4. Abiotic stress inducible proteins in legumes 230
5. Compatible solutes in the abiotic stress tolerance 231
6. Antioxidative defense system 233
7. Conclusions and future prospects 237
Acknowledgements 237
References 237

Genetic Engineering for Insect Resistance 245
Rajani Jaiswal and P. Anand Kumar

Abstract 245
1. Introduction 245
2. Insect pests of legume crops 246
3. Biotechnology and insect pest management 246
    3.1. Insecticidal Proteins of *Bacillus Thuringiensis* 247
    3.2. Protease Inhibitors 248
    3.3. Amylase Inhibitors 249
    3.4. Lectins 249
4. Bt-transgenic legume crops 250
5. Resistance management 250
6. Conclusions and future prospects 251
References 251

Genetic Engineering for Virus Resistance 255
R. Usha and K. Veluthambi

Abstract 255
1. Introduction 255
2. RNA viruses infecting legumes 256
    2.1. Alfalfa 256
        2.1.1. Alfalfa mosaic virus 256
    2.2. Pea 259
        2.2.1. Pea enation mosaic enamovirus 259
        2.2.2. Pea seed-borne mosaic potyvirus 260

2.3. Peanut ... 260
2.3.1. Tomato spotted wilt virus ... 260
2.4. Soybean ... 260
2.4.1. Bean pod mottle virus ... 260
2.5. Pigeonpea ... 261
2.5.1. Pigeonpea sterility mosaic virus ... 261
2.5.2. Pigeonpea isolate of Pothos latent virus ... 261
2.6. Legume-Infecting RNA Viruses for which Studies towards Pathogen-Derived Resistance has been Conducted in Tobacco ... 261
2.6.1. Cucumber mosaic cucumovirus ... 261
2.6.2. Cowpea mosaic comovirus ... 261
2.6.3. Pea early browning tobravirus ... 262
2.6.4. Peanut stripe potyvirus ... 262
3. DNA viruses infecting legumes ... 263
3.1. Geminiviridae ... 263
3.1.1. Begomovirus ... 263
4. Nanovirus ... 265
5. Caulimovirus ... 265
6. Genetic engineering of legumes for developing resistance against DNA viruses ... 265
References ... 266

Molecular Strategies for Fungal and Nematode Resistance ... 269
N.P. Singh, S. Datta and Bansa Singh

Abstract ... 269
1. Introduction ... 269
2. Molecular tools and techniques ... 270
2.1. DNA Molecular Marker ... 270
2.1.1. Restriction Fragment Length Polymorphism (RFLP) ... 270
2.1.2. Random Amplified Polymorphic DNA (RAPD) ... 270
2.1.3. Amplified Fragment Length Polymorphism (AFLP) ... 271
2.2. Recombinant DNA Technology ... 271
3. Molecular strategies for fungal and nematode management in legumes ... 272
3.1. Transgenic Approach ... 273
3.2. Molecular Marker Approach ... 275
4. Conclusions and future prospects ... 276
References ... 276

PART IV: PRODUCT QUALITY IMPROVEMENT

Improvement of Methionine-Deficient Legumes through Genetic Engineering ... 281
Lucilia Helena Marcellino and Eugen Silvano Gander

Abstract ... 281
1. Introduction ... 281

2. Biotechnological approaches for protein quality and quantity improvement .. 283
 2.1. Introduction of Heterologous High-Methionine Genes 283
 2.2. Introduction of *In Vitro* Engineered Transgenes 285
 2.3. Engineering of the Biosynthesis Pathways of Methionine 287
3. Conclusions and further considerations 287
 3.1. Food from Transgenic Crops – the Allergen Problem 288
Acknowledgements 289
References 289

Development of Regeneration Protocols to Exploit Somaclonal Variations in *Lathyrus Sativus* for Developing Toxin Free Cultivar 291
S.L. Mehta and I.M. Santha

Abstract 291
1. Introduction 291
2. Regeneration of *L. sativus* 292
3. Somatic embryogenesis 296
4. Conclusions and future prospects 299
References 299

PART V: BIODIVERSITY

Organisation and Genetic Mapping of the Chickpea Genome 303
Peter Winter, Christina Staginnus, Prakash C. Sharma and Günter Kahl

Abstract 303
1. Introduction 303
 1.1. Chickpea: The Third most Important Food Legume World-Wide 304
 1.2. The Genus *Cicer* 305
2. A closer look into chickpea chromosomes 307
 2.1. Chromosome Number 307
 2.2. Karyomorphology 307
3. Much of the genome is reiterated DNA 309
 3.1. Major Repeats in the Chickpea Genome: *CaSat1, CaSat2* and *CaRep* .. 310
 3.2. Repetitive Elements Support the Taxonomy of the Genus *Cicer* 311
 3.3. Less Abundant Repeat Families 312
 3.4. Microsatellites and their Distribution in the Chickpea Genome 312
4. Molecular markers for genetic mapping of the chickpea genome 313
 4.1. Some Repetitive Elements are useful: Microsatellite-Based Markers .. 314
 4.2. Locus-Specific Amplification of Microsatellites: STMS Markers 317
5. Genome mapping in chickpea 318
 5.1. Landmark Mapping with STMS and Extended Maps 320
6. Mapping of resistance loci: towards map-based cloning of resistance genes .. 322
 6.1. Genetics of *Ascochyta* Blight Resistance 323
 6.2. Mapping of *Ascochyta* Blight Resistance Genes 324

6.3. Genetics of *Fusarium* Wilt Resistance 326
6.4. Mapping of *Fusarium* Wilt Resistance Genes 327
7. Marker-assisted breeding: towards application of marker technology 330
7.1. Transferability of STMS Markers between Legume Species 331
8. Regeneration and transformation of chickpea: a step towards improved chickpea in the field 332
8.1. Transformation Systems 333
9. Legume genomics: a perspective 335
9.1. Model Legumes: *Medicago Truncatula* and *Lotus Japonicus* 336
10. Comparative genomics 337
10.1. Legume Comparative Genomics is Lagging Far Behind 338
11. Transcriptomics: EST collections and high-throughput transcript mapping .. 339
12. A deeper look into pathogenesis 339
13. Outlook 340
Acknowledgements 341
References 341

Biodiversity in Indian Underexploited/Tribal Pulses 353
K. Janardhanan, V. Vadivel and M. Pugalenthi

Abstract 353
1. Introduction 354
1.1. Biodiversity Concept 354
1.2. Reasons for Wide Prevalence of Protein-Energy-Malnutrition (pem) and its Implications 356
1.3. Traditionally Cultivated Pulse Crops in India 358
1.4. Importance of Indigenous Knowledge (IK) 358
1.5. Underexploited/Tribal Pulses 365
2. Tribal demography and geography of India 366
2.1. Collection and Documentation of Germplasm Seed Materials of Underexploited/Tribal Pulses in India 367
2.1.1. Definition of germplasm 367
2.1.2. Germplasm collection and conservation 368
2.1.3. Indian national plant genetic resource system 368
2.1.4. Underutilized/little known plant genetic resources (PGR) ... 369
2.1.5. Recorded genetic diversity in tribal pulse germplasm at the time of collection 379
3. Germplasm resource management of underexploited/tribal pulses 380
3.1. Evaluation of Agrobotanical Characters 380
3.2. Evaluation of Major Food Constituents 382
3.3. Traditional Medicinal Properties 388
3.3.1. *Abrus precatorius* 388
3.3.2. *Acacia catechu* 390
3.3.3. *Acacia nilotica* 390
3.3.4. *Alysicarpus rugosus* 390

3.3.5. *Bauhinia purpurea* 390
3.3.6. *Bauhinia vahlii* 390
3.3.7. *Cassia* species 390
3.3.8. *C. obtusifolia* 390
3.3.9. *C. obtusa* (= *C. obtusifolia* L.) 391
3.3.10. *Cassia tora* 391
3.3.11. *Canavalia ensiformis* 391
3.3.12. *Mucuna* species 391
3.3.13. *Mucuna monosperma* 391
3.3.14. *Mucuna pruriens* var. *pruriens* 391
3.3.15. *Mucuna pruriens* var. *utilis* 392
3.3.16. *Pongamia pinnata* (= *P. glabra*) 392
3.3.17. *Tamarindus indica* 392
3.3.18. *Uraria picta* 392
3.4. Identification of Elite Germplasm 393
3.5. Genetic Erosion and Conservation Strategies 393
4. Conclusions and future prospects 395
Acknowledgements 396
References 396

Subject Index 407

FOREWORD

Legumes include many very important crop plants that contribute critical protein to the diets of many around the world. Many important forages and green manure crops are legumes. Legumes are also large contributors to the vegetable oil and animal feed protein sectors. One characteristic of legumes that could become even more important as world energy sources decline and nitrogen fertilizer prices increase is nitrogen fixation, something few other plants can do. Thus legumes have a unique and important niche in agriculture.

While some of the small seeded forage legumes have been relatively easy to work with in tissue culture as far as culture initiation, plant regeneration and transformation are concerned, most large seeded legumes, like soybean, have been recalcitrant. Today, however, many laboratories are inserting genes into soybean and producing unique plants for both commercial and scientific uses. These advancements have taken a large amount of research effort and still require time and labour.

There are possible alternatives to the usual transformation methods using *Agrobacterium tumefaciens* cocultivation or particle bombardment of tissue cultures or direct DNA uptake by protoplasts that could circumvent the need for *in vitro* culture techniques. These include the pollen tube method which has been reported to be successful with several species but has not generally been thoroughly documented by molecular techniques or by reproducibility in other labs. Another method is the floral infiltration with *A. tumefaciens* method pioneered with *Arabidopsis thaliana*. This has been successful with some other species including *Medicago truncatula* so might be an approach amenable to other legumes.

Most of us are interested in our own favorite legume species but there could be greater importance in developing a "model" legume especially in the genomics era. Most plant phenomena can be studied with the most important plant model, *Arabidopsis*, but this species does not fix nitrogen so this critical event is ignored. The legumes *Lotus japonicus* and *M. truncatula* are being groomed as possible model systems. Large-scale efforts are under way to obtain expressed sequence tags (ESTs) from many legumes and in time some genomes will be sequenced. Transformation methods are crucial for functional genomic studies to determine what the genes do in the plant, so we need to continue to improve the methods so that any lab can transform their plant of interest.

Thus, as the chapters of this monograph demonstrate, progress is being made in legume biotechnology and this is being put into practice for crop improvement, both by improving breeding techniques and by increasing the number and types of genes that can be used by the breeders and other researchers. The initial commercial success of these steps is demonstrated by the fact that over half of the US soybean crop grown in 2001 will be Roundup herbicide resistant. This has allowed farmers to have very effective weed control for systems such as no-till that is environmentally friendly by minimizing both soil erosion and the energy needs for tillage.

There are many other crop improvements forthcoming that will be beneficial not only for the environment and farmers of the world but also to consumers in both developed and developing countries. Examples include the high vitamin A and high iron rice that clearly should benefit many. We also look forward to plants that can produce increasingly needed energy and chemical feedstocks to help decrease CO_2 in the atmosphere.

The chapters of this monograph were written by many experts from around the world and should be of great utility not only for researchers working in the area like plant physiologists, molecular biologists, geneticists and genomicists but also for students, technocrats and planners wishing to know the state of the science.

JACK M. WIDHOLM
Professor of Plant Physiology
University of Illinois
Department of Crop Sciences
ERML, 1201 W. Gregory
Urbana, IL 61801, USA
e-mail: widholm@uiuc.edu

PREFACE

The Leguminosae, comprising about 690 genera and 18,000 species is the most important family of dicotyledonous plants, and with an annual world market value of around 2 billion US $, is second only to cereals in world food production. Besides a less expensive protein source for vegetarian diet of growing population, these plants are used as fodder, forage and green manure, oil, fatty acids, timber, medicines, tannins and gums, etc. Leguminous plants fix atmospheric N_2 in symbiosis with rhizobia which improves soil fertility and overcomes the load of chemical nitrogen fertilizers. The productivity of leguminous crops has been limited due to their susceptibility to pathogens and pests, sensitivity of environmental stresses, poor pollination and low harvest index. Legumes are deficient in certain essential amino acids and contain many antinutrient metabolites. Improvement of these plants through conventional breeding has been of limited significance due to narrow genetic variability within germplasm and sexual incompatibility with wild and related species. An ideal ideotype for high yield in different agroclimatic conditions needs to be developed. The breeding methodologies for the improvement of grain legumes have been described in pages 3–21. Recent advances in *in vitro* culture and gene technologies have created unprecedented opportunities for the improvement of plants which are otherwise difficult through conventional breeding. A major bottleneck in use of these technologies is lack of efficient *in vitro* regeneration and transformation protocols due to their recalcitrant nature. Successful regeneration has been achieved in many of them via organogenesis and somatic embryogenesis (pages 23–63, 65–85 and 87–132). Both of these morphogenic pathways are dependent on genotype, age and physiological status of explants and hormonal and nutrient regime in the culture media. The efficient protocols for shoot organogenesis from truly dividing cells (callus) are limited. The *in vitro* rooting and establishment of plantlets in soil to get mature fertile plants is still problematic in many of them. The protocols of somatic embryogenesis in leguminous species suffers from low frequency induction, asynchronous and poor embryo maturation and conversion to plantlets. A better understanding of physiology and molecular biology of *in vitro* morphogenesis needs focal attention to reveal their recalcitrant nature. Legume protoplasts have been used in somatic hybridization and genetic transformation for genetic improvement, as an experimental material in physiological investigations and in studies of plant-microbe interaction (pages 133–153). However, regeneration from protoplasts still needs to be perfected in many of them. Wild and related species are the reservoirs of desirable agronomic traits. Interspecific hybridization has been limited due to post zygotic barriers leading to the abortion of hybrid embryo at a young stage. Rescue of hybrid embryo and gains in tissue culture techniques will determine the success of using incompatible wild species (pages 155–173).

Nitrogen in soil is one of the limiting nutritional factors which affects growth and productivity of plants. The biology of dinitrogen fixation and nitrate assimilation in leguminous plants has been discussed on pages 175–189 and 191–199, respectively. Ammonium produced by either of the above routes is incorporated into the organic cycle for the synthesis of amino acids and subsequently other vital nitrogenous macromolecules and metabolites by glutamine synthetase-glutamate synthase (GS-GOGAT) cycle. The manipulation of this pathway to improve nitrogen utilization efficiency of leguminous plants has been discussed on pages 201–220.

Environmental stresses cause a significant loss in productivity of leguminous crops. Many stress responses are induced in plants which lead to the enhanced stress tolerance. Success in improving tolerance of the stressed plants by the conventional breeding has been limited due to its multigenic nature. Genetic engineering paves a way for enhancement in the stress tolerance, though such studies on leguminous crops are very limited (pages 223–243). These crops are susceptible to pathogens and pests. Excessive use of synthetic organic insecticides to control these biotic agents has resulted in the degradation of the environment, adverse effect on human health and development of resistant biotypes of pathogens and pests. Genetic engineering has made it possible to develop stable resistance against insects (pages 245–253), viral (pages 255–268) and fungal pathogens and parasitic nematodes (pages 269–277).

Grain legumes are deficient in sulphur-containing amino acids and contain many antinutritional factors. Biotechnological tools have been applied to introduce and express genes coding for sulphur-rich proteins (pages 281–290) and to reduce a neurotoxin in *Lathyrus sativus* (pages 291–299).

Organization and genetic mapping of the plant genome is essential to understand evolution, structure and function of genes of various traits. Pages 303–351 analyze the progress made in genome structure and evolution, genetic mapping, mapping of resistance genes for important pathogens and synteny between related species and genera in chickpea. Knowledge on biodiversity in underexploited/tribal pulses in high biodiversity regions will be very useful for widening of the available gene pool to improve legumes for desired traits (pages 353–405).

We are grateful to Professors J. Finer, W.A. Parrott, I.K. Vasil, D.P.S. Verma, M. Chrispeels, Peggy Ozias Akins, P. Christou, H.J. Jacobsen, M. Ishimoto, S. Sopory and D. Pental and to Professor Marel Hofman, the series editor of "Focus on Biotechnology", for their kind suggestions and encouragement. We extend our warmest appreciation and thanks to Kluwer Academic Publishers for their keen interest in bringing out this title with quality work. We are thankful to our research scholars and family members for their understanding and patience during preparation of this title and appreciate the efforts made by Mr. Ashok Datta and Ms. Reema of *LaPrints* New Delhi, India for preparing camera ready version.

Rohtak, India
June, 2002

Pawan K. Jaiwal
Rana P. Singh

PART I
TECHNOLOGIES

BREEDING METHODOLOGIES FOR THE IMPROVEMENT OF GRAIN LEGUMES

PAOLO RANALLI
Istituto Sperimentale per le Colture Industriali,
Via di Corticella 133, I-40129 Bologna, Italy
e-mail: ranalli@bo.nettuno.it

Abstract

This review focuses on selection criteria and breeding methods used with grain legumes. The production of these crops has remained stagnant over time and research strategies which lead to the development of improved cultivars is urgently needed. The variables in the choice of breeding methods have been considered and selection targets have been pointed out. These include the improvement of plant architecture, nitrogen fixation (selection of plant host, inoculant strains and rhizobia by cultivar interaction), seed yield (quantitative, qualitative and stability) and resistance to biotic and abiotic stresses. Breeders must work within these specific constraints and a three-tiered pyramidal breeding strategy is proposed to facilitate yield improvement. Breeding of elite and agronomically valuable germplasm within the same market class is restricted to the apex of the pyramid. The intermediate level has fewer constraints and greater access to diverse germplasm. Interracial crosses within the same gene pool are utilized to exploit genetic differences within adapted material. Utilization of genetic diversity from unadapted sources, including wild germplasm, is conducted at the base of the pyramid. The objective of this breeding strategy is the movement towards the apex, using different breeding procedures to optimize improvement at each tier of the breeding pyramid. The implementation of conventional selection procedures with innovative methods (namely nucleic-acid based techniques) should considerably accelerate the breeding process (marker-assisted selection, MAS). The MAS would be more efficient provided there is tight linkage between the marker and the trait of interest and assuming selection for the marker is more convenient (faster, cheaper, reproducible, early expression). Selection based on the marker is widely used in soybean and *Phaseolus*; in the future these procedures will be able to penetrate plant breeding of other grain legumes. This approach would be of particular interest in the selection for disease resistance (disease diagnosis, pathogen identification, and screening methods), a basic prerequisite for improving and stabilizing the yield of grain legumes.

P.K. Jaiwal and R.P. Singh (eds.), Improvement Strategies for Leguminosae Biotechnology, 3–21.

1. Introduction

Globally, production of pulses remained static in the decade 1969–1971 to 1979–1981 at about 700 kg ha^{-1}, in 1989 was 828 kg ha^{-1} and in 1999 only 965 kg ha^{-1}. Among food legumes, soybean is the only crop where consistent improvements in yield have been achieved. Yields of soybean increased about 18.8 kg ha^{-1} $annum^{-1}$ in the period from 1900 to 1980, or more than 1.5 ton ha^{-1} over that 80 year period. This response comes as the result of a significant effort in research and breeding, especially in the USA and Brazil, and indicates that similar efforts will be needed to improve the yield of the other grain legumes. Fortunately, some of the stepping stones needed to achieve this progress are already in place. The germplasm resources needed for grain legume improvement have been collected, and are available through international and national centers of research; sources of resistance/tolerance to a number of the major pathogens and insects for these species have also been identified, and in many cases the methodologies needed for a more rapid incorporation and pyramiding of resistance traits in agronomically valuable cultivars have already been developed. In this chapter an overview on the basis for genetic improvement of the most prominent grain legumes is provided.

2. Domestication of grain legumes

The evolution of the domesticated forms of pulses was essentially based on loss of seed dormancy and suppression of active seed dispersal. In particular, the evolutionary sequences included:

2.1. CHANGED GROWTH FORM

A number of vining forms, largely members of the tribe *Phaseoleae*, in response to domestication and novel selection pressures have produced free-standing dwarf bean morphology. An interesting parallel evolutionary mode has been in the development of the tendril climber which led to a free standing dwarf herbaceous form culminating in the dwarf pea morphology.

2.2. CHANGED LIFE FORM

Many grain legumes appear to have been domesticated as short-lived perennials and to have become annuals as the consequences of a simple mutation which produced the change from indeterminate to determinate growth. In both *Phaseolus coccineus* and *P. lunatus* this has occurred. The actual form taken by growth in individual genotypes within determinate and indeterminate groupings is under strong background genotype influence, as can be seen in the range of variant growth forms of *P. vulgaris* recognized by CIAT (Smartt, 1990).

2.3. SUPPRESSION OF SEED DISPERSAL MECHANISMS

In wild legumes, most commonly, pods are dehiscent, often explosively so, and project mature seeds to some distance; in the pulses loss of pod dehiscence has been in a sense progressive. Some widely cultivated grain legumes, such as the soybean, are still somewhat prone to problems of pod dehiscence at maturity while at the other extreme pod dehiscence in the groundnut is totally suppressed even in the wild species. In *P. vulgaris* and *Pisum sativum* the complete range from explosively dehiscent to virtually indehiscent pods is found. In the french bean the selection of "stringless pods" has reduced the degree of lignification of the pod wall fibres so that the dehiscence mechanism is totally inoperative.

2.4. BIOCHEMICAL CHANGES

The range of biochemical changes established in the course of domestication is very considerable. Starting from the seed surface, one of the more obvious differences is very often a range of colours from black (or very dark) at one extreme to colourless (or white) at the other occurring in the cultivated materials. Light coloured (and especially white) testas lack the protection against fungi which is conferred by phenolic pigments.

The use of some legumes seeds as feed (for livestock) is precluded by the presence in the seed of a group of toxic materials collectively known as lectins. These are proteins in nature and have the property of agglutinating blood, hence the alternative name of phytohaemagglutinins. These are inactivated by heat and so seeds such as those of *Phaseolus* are wholesome when cooked but lethal when fed uncooked to rats. Selective breeding can reduce the content of lectins very sustantially (Osborn and Bliss, 1985).

Another group of compounds found in many legumes seeds are proteinase inhibitors. They inhibit the action of trypsin and chymotrypsin in the gut. The soybean has a high content of such compounds which are an impediment to the ready exploitation of the valuable protein reserves of its seed. These compounds can be inactivated by heat. The occurrence in some beans (e.g. *Phaseolus vulgaris*) of indigestible pentose polysaccharides leads to embarrassing problems of flatulence.

Biochemical variation within pulse species has been exploited by Gepts and coworkers in tracing the ultimate geographic origins of *P. vulgaris* found in the Americas, Africa and Europe to their primary centres of origin in Middle and South America (Gepts, 1988; Gepts and Bliss, 1988).

2.5. PHYSIOLOGICAL CHANGE

Physiological changes in development of legume crops of the greatest significance are probably those related to photoperiodic responses. Soybeans are very sensitive to photoperiod and there is a great range of photoperiod sensitivity within the species. While soybeans can be grown over a wide latitudinal range, from the tropics to the latitude of Southern Canada, individual cultivars have a very circumscribed latitudinal range indeed. In other grain legumes the situation appears to be less complex (it has certainly been studied in much less detail); a general division can be made into photoperiod sensitive and day-neutral classes. The lack of photoperiod sensitivity allows legumes such as *Phaseolus vulgaris* to be grown successfully from the tropics to Canada (as can the soybean) but without the complication of selecting for photoperiodic adaptation to specific areas.

3. Synteny in grain legumes

The gene content of related organisms is similar, but in addition for relatively close taxonomy groups gene order is also conserved. This conservation of gene order is referred to as syntheny, and segments of different genomes that have essentially the same genes in the same order are said to be syntenic. Cereal linkage maps are famously syntenic (Gale and Devos, 1998), similarly the tomato and potato maps can be aligned.

If the linkage maps of closely related organisms are similar, then we can combine, to a reasonable degree, the genetic information from these species. For example, if we know that a certain trait is associated with a segment of a genetic map defined by several molecular markers, then these markers are a good starting point for investigating the genetics of the same trait in a less well studied relative. For legumes there are two species being studied intensively at the molecular genetic level: *Medicago truncatula* and *Lotus japonicus*. Of the two model species, *M. truncatula* is more closely related to the major European grain legume crops.

The study of synteny among European grain legume crops is in its infancy, but data are beginning to emerge which suggest that studies of crop genetics will be able to exploit the detailed molecular genetic data emerging from model species (Ellis, 2000).

3.1. THE POTENTIAL OF *M. TRUNCATULA* AS A MODEL PLANT IN GRAIN LEGUME CROPS

Bridges between the genomics of different species will not be the only way to benefit from a legume model. Indeed, crop physiologists, pathologists, weed scientists and others can exploit it in their respective disciplines. However, bridges within genomics will appear first:

- through the exploitation of micro- and macro-syntheny to make progress in mapping, in understanding quantitative trait loci (QTL) and in the identification of candidate genes;
- through the exploitation of homologies in gene regulation;
- through the possibilities of discovering interesting genes in *M. truncatula* that may be regulated correctly and expressed when transferred into cultivated legumes.

By means of a model legume it is possible to address some of the following questions which are specific to pulses (Duc, 2000):

3.1.1. Symbiosis establishment

There is a need to increase scientific knowledge about the symbiotic mechanisms on this valuable feature of the legume family in order to enlarge the cultivated zones of legumes towards more difficult soils or climatic conditions.

3.1.2. Nitrogen fixation efficiency, and C and N fluxes

Nodule maintenance, N_2 fixing activity, and soil N assimilation are important factors for plant N accumulation that can be analysed in *M. truncatula*. In addition, fluxes of C and N between plant organs (nodules, roots, leaves and seeds), their direct transfer to the seed or their remobilisation and also the competition between organs are important mechanisms which need better understanding in legume crops. Such knowledge should help to improve the management of the N economy within a legume plant or after a legume crop. These are serious questions for legume performance and sustainable agriculture.

3.1.3. Seed composition

Using the *M. truncatula* model there may be valuable knowledge to acquire on protein distribution and accumulation in storage organelles, on oil bodies and starch granule formation, on cell-wall structure and on consequent seed germination capacity.

3.1.4. Abiotic stress

Adaptation of the legume plant together with its symbiosis to environmental constraints is an important demand from physiologists and breeders. Heat, drought and salt tolerance have been reported in *M. truncatula* and would be valuable for agronomic plants. The understanding of underlying mechanisms and the identification of candidate genes is a clear need and a study has started on the salt tolerance of *M. truncatula*. The water use efficiency by crop and adaptation to low irrigation systems is an attractive objective on which the model science may contribute.

3.1.5. Biotic stress

All breeders give high priority to incorporating genes of resistance to biotic constraints in cultivated legumes in order to stabilize yield and reduce chemical protection. Diseases or parasites generally appearing among legumes may be considered "model diseases" for which *M. truncatula* may bring solutions.

4. Design of a complete breeding program

The typical breeding program for grain legumes includes a broad scope of specific and non-specific objectives, each at various stages of planning, execution, revision, and completion.

4.1. PARENTAL SELECTION AND HYBRIDIZATION

The principal non-specific objective is to identify a wide range of useful genetic diversity for plant, pod, and seed characteristics, pest and disease resistance, environmental stress tolerance, physiologic capabilities, etc.

Sources of genetic diversity include the breeders'own collections, many accessible public and/or private breeders' collections, current and old cultivars, and collections of exotic or wild materials.

Sometimes the required parents first need to be developed via a preliminary or "prebreeding" program, i.e. if one of the required characteristics is only available in another species or in a wild line with negative traits.

Parental selection for each trait to be combined is based on thorough evaluation and reliable data obtained from contrasting environments. After evaluation and selection of parents, appropriate cross combinations are planned; the hybridization takes into account parental performance *per se*, their combining ability, the relationship between parents with regard to races and gene pools, their affinity for growth habit and seed. Multiple-parental or complex crosses are needed, especially when genes for each trait being improved are found in separate parents.

4.2. SELECTION OF DESIRABLE RECOMBINANTS

After the required parents are identified or developed, the breeding strategy is outlined on the basis of the knowledge of the inheritance of the available genetic diversity. The segregating populations from the hybridized parents are scheduled for appropriate screening to certain diseases or environmental resistance factors at specified generations. The selection for horticultural and agronomic characteristics is also scheduled for specific generations and usually in a defined environment and sequence (Ranalli and Cubero, 1997).

When the target combination of characteristics is finally assembled in advanced generation, the most promising genotypes are multiplied, evaluated under a wide range of conditions, and eventually released as a breeding line or a named cultivar. For desirable dominant and codominant alleles and possibly other segregant traits with high heritability (e.g. indeterminacy, climbing ability, and maturity), evaluation and selection commence in the F_1. But for seed yield and other quantitative traits requiring replications and evaluations across environments, selections have to be delayed until F_2 and F_3.

The F_1 hybrid seed resulting from each pair of plant-to-plant crosses is sown in separate hill plots. If desired, data on vigor, growth habit, flower color, maturity and reaction to diseases and insects are recorded. The F_2 seed obtained from each surviving F_1 plant is sown in a separate plant-to-progeny row in order to develop F_1-derived families for subsequent evaluation and identification of desirable recombinant genotypes. To maximize seed production, the F_1 is space-planted under good agronomic management.

For quantitative traits such as seed yield and tolerance to drought and diseases caused by variable pathogen populations (e.g. anthracnose), data obtained from replicated trials across contrasting environments are used. All uniformly susceptible F_1-derived F_2 families are discarded. Plants within each surviving or selected family are harvested in bulk. The F_3 seed of each family is used to prepare yield trials and separate complementary nurseries for each production constraint. Evaluation for production constraints and seed yield in replicated trials is repeated in the subsequent generation (F_4) if more reliable data are needed. After data analysis, only the high-yielding resistant families or those segregating for resistance and possessing all desirable traits are saved (Singh, 1994).

Theoretically, the proportion of families with all desirable alleles will be $(1/2)^n$, where n is the number of heterozygous alleles. Thus, if only four genes are segregating in the hybrid population, then 1/16 of all families should carry all four desirable alleles.

The plants within F_5 families are space-planted to maximize seed per plant and to maximize the number of single plant selections. All plants with undesirable seed characters are discarded. The F_6 progeny rows are checked for uniformity of flower colour, growth habit, maturity, and seed traits. Plants within selected and uniform rows are harvested in bulk. In F_7, separate complementary nurseries are used to evaluate each trait (similar to F_2, F_3, and F_4). Selected lines are then yield-tested in replicated trials under production constraints in representative environments (across locations and growing seasons) to identify new cultivars.

5. Marker assisted selection

Breeders of grain legumes, like other breeders, seek to exploit multiple plant breeding approaches to combine both traditional and modern techniques. More traits are still selected by conventional means in field sites where most important diseases, edaphic constraints and drought can be manipulated for the purpose of the selection.

Some types of molecular markers must be developed independently for each separate species under study; others, such as AFLP, can be used on any plant species with virtually no modification. A further advantage of AFLP is the high number of genetic loci that can be screened simultaneously on a single gel. Finally, the size of the products of the AFLP reaction are measured accurately on DNA sequencing gels, enabling the reliable fusion of data from different gels and from different years in order to accumulate a large genotype database over time.

Random amplified polymorphic DNA (RAPD) markers, obtained from PCR amplifications of genomic DNA with random primers, have been used successfully to tag a variety of genes in crop plants (Williams *et al.*, 1990; Michelmore *et al.*, 1991; Haley *et al.*, 1993; Yu and Pauls, 1994a, 1994b; Delourme *et al.*, 1994; Hormaza *et al.*, 1994). In particular, in grain legumes RAPD markers linked to major rust resistance genes (Haley *et al.*, 1993) and a dominant gene conferring resistance to anthracnose (Adam-Blondon *et al.*, 1994a) have been identified. In addition, RAPD markers linked to common bacterial blight resistant genes that may be useful for selecting disease resistant *Phaseolus vulgaris* beans carrying genes introduced from *Phaseolus acutifolius* have been identified (Bai *et al.*, 1997).

Putative quantitative trait loci (QTLs) for seed, pod, and leaf reactions to *Xanthomonas campestris* pv. *phaseoli* in common bean, which accounted for 18, 25, and 35% of the variation in common bacterial blight resistance, have been identified by Jung *et al.* (1994). Also, Nodari *et al.* (1993c) have reported four putative QTLs that accounted for 75% of the variation in common bacterial blight resistance in beans, using random fragment length polymorphism (RFLP) markers.

RFLP, AFLP, and microsatellites can also be applied to problems in taxonomic relatedness and to discriminate the selected parental genotypes. Well-spaced probes per linkage group of *P. vulgaris* could be chosen on the basis of the Bean Genes Data base (http://beangenes.cws.ndsu.nodak.edu) in order to detect polymorphism. The polymorphic probes can be applied to the individuals of mapping populations to follow the inheritance of the RFLPs. AFLP which detect a larger number of loci revealing greater polymorphism could be used for identification and for high resolution mapping (Jones *et al.*, 1997). Available microsatellites from CIAT which are more useful than RFLPs in self-pollinated species (i.e. *P. vulgaris*) for detecting polymorphisms within a gene pool could also be positioned. Nonetheless, RFLP probes can characterize several populations within the same species which is difficult with PCR-based techniques (Schneider *et al.*, 1997).

Molecular markers, such as RFLPs and RAPDs, are sufficiently polymorphic and numerous to allow the production of maps with extensive coverage of a genome using material produced by a single cross. All possible pairwise combinations between the loci detected by the selected probes can be estimated to build a reference map (MAPMAKER) and links between markers and the quantitative data are determined with MAPMAKER/QTL. Several linkage maps have been created in *Phaseolus* (Vallejos *et al.*, 1992; Nodari *et al.*, 1993b; Gepts *et al.*, 1993; Jung *et al.*, 1994) that identify regions of the genome associated with a number of important agronomic traits.

In lentil (*Lens* sp.) sources of resistance to key biotic and abiotic stresses are known in cultivated and wild germplasm and are being introduced into adapted lines. The availability of a detailed genetic linkage map would help to identify the location of genes controlling such traits to facilitate their directed manipulation in crop improvement. The current lentil maps consist of a small number of markers, mainly isozymes, covering a relatively small portion of the lentil genome. However, DNA molecular markers are being used to extend the linkage map. So, a genetic linkage map of *Lens* sp. was constructed with 177 markers (89 RAPD, 79 AFLP, 6 RFLP and three morphological markers) using 86 recombinant inbred lines. The map covered 1,073 cM of the lentil genome with an average distance of 6.0 cM between adjacent markers. The available map could be used to evaluate linkage between DNA markers and quantitative trait loci in this population (Eujayl *et al.*, 1997).

6. Constraints

Breeding for high yield in grain legumes must be conducted within the major constraints of growth habit, maturity (adaptation), and seed size (quality) preferences as well as disease resistance factors. Breeding for yield outside these constraints will result in germplasm with either little or no value in terms of commercial utility.

6.1. RESTRICTED GENETIC BASE

One of the constraints recognized by pulse breeders is the narrow genetic base available within cultivated forms; accordingly, breeders attempt to choose parents that are genetically diverse and which are identified by estimating the coefficient of parentage (CP) and genetic distance (GD) (Beebe *et al.*, 1995). Several molecular techniques are available for the study of genetic diversity of genetic resources. These include karyotype analysis, comparative immunological methods, isozyme electrophoresis, DNA-DNA hybridization, restriction fragment length polymorphism (RFLP) analysis, random amplification of polymorphic DNA (RAPD) and DNA sequencing.

6.2. QUALITATIVE ASPECTS

Legumes are nutritionally important because they are rich in proteins. Protein content across the major grain legumes varies from 18 to 48%, but it also depends upon soil type, climate, location, and management practices (Bahl, 1990). The major part of world faba bean production (more than 90%) is used for human consumption and the remainder, mainly in Europe, is used as a source of protein in animal feed.

Protein quality for monogastric feeding depends on three major features: (1) composition, especially of essential amino acids, (2) occurrence and content of protein anti-nutritional factors such as trypsin inhibitors and lectins, and (3) amino acid availability (Bahl, 1990).

Irrespective of the form of consumption of grain in human or in animal diets, the important nutritional characters which must be safeguarded, and if possible improved, are protein content, digestibility and freedom from anti-nutritional factors.

6.3. PLANT ARCHITECTURE

Growth habit. If high yield was chosen as the sole basis for selection in dry bean and growth habit was ignored, type IV growth habit would be indirectly selected. In one study, determinate genotypes were lower yielding than indeterminate genotypes, whereas another group of isogenic lines differing in growth habit showed no negative effect on yield due to growth habit (White *et al.*, 1992). A mere change in stem type from determinate to indeterminate did not increase the yield of large-seeded near-isogenic lines. Other traits, such as node and branch number, need to be increased to affect a change in yield. In an inter-gene pool cross between genotypes, lines with an indeterminate growth habit had significantly higher seed yield than lines with determinate growth habit (Welsh *et al.*, 1995). Determinate beans were found to be inherently less stable and lower yielding in trials conducted in the American midwest (Kelly *et al.*, 1987).

Sedgley *et al.* (1990) postulated that under a high input system with good weed control a communal chickpea plant type of erect habit with few branches, sown at high density, should give a slower rate of moisture depletion in winter and result in higher biological yield and harvest index.

Branch number of legume plants is highly variable, and is an important determinant of grain yield. Low plant densities can be compensated by branching (Hedley and Ambrose, 1981). Plant structure varies among cultivated grain legumes: pea and soybean branches appear mainly on the first vegetative nodes, whereas white lupin produces branches at the top of the main stem. Genotypes with greater branching ability generally have more determinate habits, probably because of the competition between branches in the canopy (Jeuffroy and Ney, 1997).

Leaf conformation. Leaflet and stipule mutations have been introduced into pea cultivars (Snoad *et al.*, 1974). The morphogenesis of leaflets and stipules is controlled by three major loci. Experimental evidence indicated that canopies of a leafless cultivar are as efficient in intercepting light energy, and therefore in accumulating dry matter, as normal phenotypes (McKerron and Thompson, 1983; Snoad, 1983); at the same time, this phenotype is less subject to lodging and thus is less susceptible to diseases, especially in wet climatic conditions.

6.4. BIOTIC STRESS

The most prominent disease and pest problems for breeders to address are the following:

- *Leaf fungal diseases. Mycosphaerella pinodes* in peas has one of the highest priorities and is considered a model diseases for many legumes. There is also a demand for resistance to powdery mildew, botrytis and rust.

- *Root fungal diseases*. Most of these diseases cannot be controlled by chemicals. The genetic control of *Aphanomyces* in pea is a major demand, *Fusarium* control is also requested.
- *Nematodes, aphids and sitona*. Aphids are serious pests of pea and faba bean. The stem nematode is prevalent in most temperate regions and it is a pest for many legumes.

6.5. ABIOTIC STRESS

Lack of water and unfavourable temperature are the most widespread stress conditions limiting grain legume yield. In pea, water stress and high temperature affect yield, reducing seed number when they occur before the final stage of seed development; the later the stress, the smaller the decrease in yield. Other stresses are more region-specific, and include salinity, waterlogging, mineral toxicity and unsuitable soil pH (Buddenhagen and Richards, 1988). In developing cultivars more resistant to soil moisture and temperature stresses, phenology must first be modified so that preanthesis growth and flowering avoid the most severe stress periods as far as possible. The above mentioned authors have advocated yield improvement in food legumes based on increasing early vigour and biomass accumulation under both optimum and stress conditions, as well as on partitioning of biomass into seed. For drought stress environments, characteristics such as denser and longer root system and small leaves are considered advantageous.

Selection for salt tolerance at germination was carried out in mungbean (Sekhar, 1994). Five cultivars of mungbean were germinated in increased levels of salinity (from 6 to 15 mmhos/cm) and seed germination decreased with salinity levels. The best genotypes exhibited 60% germination at the highest salinity compared to the least tolerant with 38% germination.

7. Improvement of yield

Most breeders aim at a particular ideotype, but this varies with environments and crop uses and with what are perceived as physiological constraints. One factor relevant to pod retention is the competition for assimilates between vegetative meristems and rapidly developing reproductive sinks, the young pods. This competition is most acute in rapidly growing plants with prolonged indeterminate vegetative growth. For this reason, breeding programmes have been carried out in *Vicia faba* to adjust the species towards determinate growth habit (Bond, 1977) and determinate types have been identified in many pulse crops. Also, in grain legumes, the yield components develop sequentially, with later-developing components under the control of earlier-developing ones (Thomas *et al.*, 1970). Assuming this causal relationship among variables, path analysis has been used extensively to partition phenotypic and genotypic correlation among yield components and to estimate the major direct contributors to seed yield.

In peas, the path coefficient analysis has shown the negligible effects of seeds per pod on final yield (Ranalli *et al.*, 1981). Thus, in a breeding program aimed at yield improvement the efforts have to be focused on the pods per plant and seed size which appear to be the main yield components.

In black-gram (*Vigna mungo*) a path analysis performed on six yield components measured in 30 genotypes indicated the yield positively and significantly correlated with 100 seed-weight, days to maturity, number of primary branches and plant height (Verma, 1992). In addition, seed yield/plant showed significant positive correlation with petiole length, total dry matter, height and primary leaf area (Renganayaki and Sreerengasamy, 1992).

Genetic aspects of harvest index in black-gram were also investigated. Low genetic advance and moderate heritability indicated a non-additive gene action and the influence of environment on this character (Anuradha and Murty, 1993). Moreover, studies on heterosis in mungbean in seven diverse genotypes and their F_1 hybrids showed the highest value for pods/plant (139.6% over better parent) followed by seed yield/plant (76.0%) and pod weight/plant (64.1%).

In chickpea, observations were recorded on seed yield and 14 other characters. Correlation and path analysis showed that biological yield and harvest index were the major direct contributors to seed yield. Therefore, these characters should receive the highest priority in selecting high yield plants in breeding programmes.

Once the yield character is dissected, the choice of parents, their crossing and then selection criteria of segregant populations must be defined to promote the recombination of the traits that affect the yield most.

In grain legumes, as in most crops, yield has a low heritability because of environmental variables. This has led a number of breeders to propose indirect selection using traits or trait combinations that are related to yield but more heritable (Ranalli, 1991).

7.1. BREEDING PYRAMID

In order to emphasize yield breeding in grain legumes, given the constraints mentioned earlier, a structured program based on a three-tiered breeding pyramid was proposed (Fig. 1) (Kelly *et al.*, 1998). The approach to yield breeding would be different at each level of the breeding pyramid, encompassing the variability within dry bean germplasm. Materials developed at the lower levels would be designed to move upwards sequentially or be maintained for additional improvement. The genetic diversity and the development time would be greatest at the lowest level. The exploitation and maitenance of this diversity would be optimized at the lower level while uniformity and yield performance would be sought at the apex of the pyramid. Since breeding approaches will differ at each level, they will be detailed separately, recognizing that a strong continuity among all three levels is necessary.

7.1.1. The apex of the breeding pyramid

This level would involve elite crosses. Crosses would be restricted within gene pools, within races, and within market classes. Selection for constraints other than yield is eliminated since genotypes at this level are relatively uniform for these traits. Breeding strategies in the elite group would utilize a limited number of crosses rapidly advanced using SSD to the F_5 generation after which preliminary yield trials (PYT) would be conducted (Ranalli, 1996). Selection intensity could vary from 20 to 35% depending on factors such

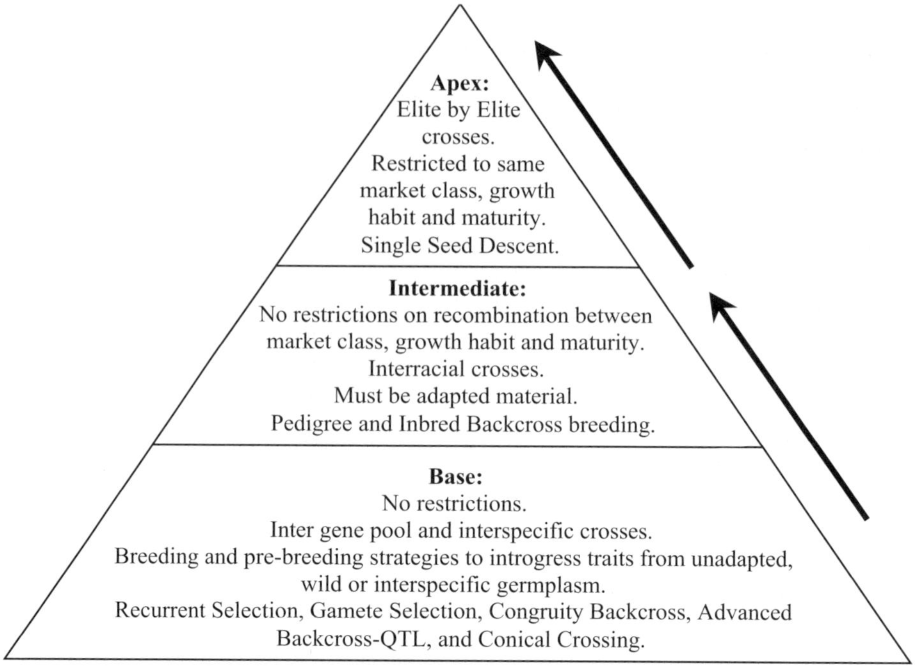

Figure 1. Breeding pyramid for yield improvement in grain legumes (after Kelley et al.*, 1998).*

as environment, resources, and optimization of subsequent advanced yield trial (AYT) nurseries. The genotypes selected from the PYT would be advanced to the AYT, where selection pressure would increase to 10–20%. Top performers from AYT would be considered for release as cultivars after extensive additional testing, consistent performance and consumer preference. Early generation testing could be used in the apex of the breeding pyramid provided resources are available, since many of the constraints (traits) listed above are fixed.

7.1.2. The intermediate level of the breeding pyramid

This would be structured to introgress greater levels of diversity into breeding materials, and to identify top performing intermediate lines to test and cross with elite line materials. Crosses in this level would be restricted within gene pools and between races differing in some agronomical traits (architectural types, maturity, etc.). A larger number of crosses would be made since no prior prediction of combining ability is known and larger F_2 populations would be developed. Single plant selection would be rigorously practiced for agronomical and seed traits (i.e. those demonstrating high heritability) which can be readily fixed in this generation. Selected individuals could be advanced rapidly to the F_5 generation by SSD and entered into PYT where selection would be based on performance compared to elite parental or other lines. Lines showing above-average performance for yield would enter the elite group at the apex to be tested and utilized accordingly. Use of three-way crosses, pedigree, modified inbred-backcross, and limited cyclic selection should all be considered as breeding methods in the intermediate level.

7.1.3. The base of the breeding pyramid

Increased diversity, through both intraspecific and interspecific hybridization, would be sought at the base level. Intraspecific hybridization would utilize a wide range of germplasm from within and between gene pools; interspecific hybridization leads to introgression of wild traits into cultivated species; once incorporated, efforts would be focused on bringing these traits into the intermediate level, and finally into the elite level for evaluation.

In order to improve the genetic value of the base population, the selection and advance of progeny from these wide crosses needs to be rigidly structured. Long-term selection and crossing programs need to be established to facilitate this objective. One example of such a program is a recurrent selection designed to improve the yield: high yielding progenies are recombined in additional cycles of recurrent selection (Ranalli *et al.*, 1996). The system demands patience and resources but offers a realistic way of effectively combining diverse germplasm sources without solely targeting single genes.

8. Improvement of legume N_2 fixation

The concept that legume N_2 fixation could be enhanced through selection and breeding has been around for decades (e.g. Lie, 1984; Nutman, 1949, 1952). The following research areas were emphasized:

- Examination of C and N allocation to various plant tissues in order to identify useful traits for N_2 fixation selection programs.
- Comparison of alternative selection strategies, including development of simple, rapid, non-destructive field-based techniques for assaying N_2 fixation.
- The development of genetic control system that could provide host specificity for superior rhizobia.
- The development of legumes that enhance the usefulness of N_2 fixation beyond that of biological yield *per se*, including those that can excrete N compounds and in which nodulation and N_2 fixation are not inhibited by soil N.

8.1. STRATEGIES FOR BREEDING LEGUMES WITH ENHANCED N_2 FIXATION

The general strategies for increasing legume N_2 fixation through breeding involve:

- maximizing legume (biomass) and seed yield within the constraints imposed by agronomic management and the environment. Because larger biomass requires more N, N_2 fixation will be increased as biomass yield is increased. This approach assumes a capacity for N_2 fixation sufficient to satisfy increased N demand of larger plants. This has particular application in lower-yielding species such as common bean, lentil, mung bean and chickpea, but less relevance for the larger, vigorously growing species like faba bean, pea and soybean (Herridge *et al.*, 1994).

- enhancing symbiotic nitrate tolerance, the ability of the legume to nodulate and fix N_2 in the presence of soil nitrate. This strategy may not result in greater yields (either biomass or grain).
- optimizing legume nodulation through specific nodulation traits (e.g. mass and duration) and, depending on the circumstances, for promiscuous or selective nodulation (Devine and Weber, 1977).

Most N_2 fixation breeding objectives can be met by selecting for N yield or related traits in a low N soil. The second option is to select directly for Pfix in a moderate to high N soil, i.e. select primarily for symbiotic nitrate tolerance. Elite material from either approach may have sufficient yield and suitable agronomic traits for immediate release. Alternatively, it may be used as a source of N_2 fixation genes for hybridization.

Over a period of 13 years, Bliss and co-workers, made substantial progress with common bean using selection for yield in low N soil. They released five high N_2 fixing lines in 1993 (Bliss *et al.*, 1989; Bliss, 1993).

Bliss and co-workers used a variety of methods with common bean, including plant yield, seed yield, seed N yield, ^{15}N methods, acetylene reduction assay and nodulation indices (Attewell and Bliss, 1985).

9. Improvement of disease resistance

All the major grain legumes including pea (*Pisum sativum* L.), chickpea (*Cicer arietinum* L.), faba bean (*Vicia faba* L.), lentil (*Lens culinaris* Medic.), common bean (*Phaseolus vulgaris* L.), and lupin (*Lupinus* spp.) are subject to significant yield loss following attack by pathogens. The diseases can be grouped into: root, vascular and foliar diseases caused by fungi and bacteria, virus diseases, and nematode diseases. Other diseases of some importance are caused by parasitic plants, as in the case of broom-rapes, agents of biotic stress in some species of grain legumes.

Fungi cause soilborne diseases including seed and seedling blights, root rots and wilts and important foliar diseases. *Pythium* spp. and *Rhizoctonia* spp. are the main agents of seed and seedling blights, while *Aphanomyces euteiches*, *Fusarium solani*, *Pythium* spp., *Macrophomina phaseolina* and *Phoma medicaginis* var. *pinodella* are the most important fungi causing root rots. Various formae speciales of *Fusarium oxysporum* are the agents of wilts. Soil-borne diseases occur worldwide with their severity affected by both plant physiological and genetic characters as well as by such environmental variables as excess or deficiency of water, type of soil and temperature.

Foliar diseases include down and powdery mildew, ascochyta blight, botrytis gray mould, rusts, anthracnose and bacterial blight.

Virus infection also causes substantial crop yield losses, and damages can vary greatly with host genotype and environmental conditions. Many viruses use insects and nematodes as vectors and several of them persist in various weeds or volunteers. Finally, it is worthy to mention that a large array of pathogens are seedborne. Several important diseases of grain legumes can be initiated from inoculum present on the seed.

9.1. STRATEGIES AND METHODS OF SCREENING FOR RESISTANCE

The rationale implied in screening for resistance in grain legumes includes a series of points:

(i) the plant material must be adequately challenged with a well identified single race or pathotype, at a realistic inoculum dose to allow disease development without obscuring minor differences required to identify partial resistance in the host (Porta-Puglia and Aragona, 1997). This objective could not be adequately achieved by mixtures of races or naturally infested debris (Parlevliet, 1979, 1983);
(ii) results obtained in laboratory tests, that are desirable for several practical reasons, must be interpreted with care and validate under natural environmental conditions.

The conventional procedures of screening for resistance are complex and time consuming, so the search for molecular markers linked to resistance and easy to select is gaining importance, and some results for grain legumes are already available. In pea, a large amount of work has been done on linkage of isozyme loci (Mahmoud *et al.*, 1984; Weeden and Marx, 1987). A close linkage has been demonstrated between Prx-3 peroxidase gene and the gene conferring resistance to pea seedborne mosaic virus, strain P_1. RFLPs and RAPDs are the molecular markers most used in marker-assisted selection (MAS). MAS is particularly suitable for traits which are highly variable in expression, as partial disease resistance. Moreover, it can permit the simultaneous selection of several genes and improve selection for recessive genes (Davies, 1990; Dirlewanger *et al.*, 1994; Kelly *et al.*, 1995; Yu *et al.*, 1995; Miklas *et al.*, 1996; Young and Kelly, 1996). RFLPs and RAPDs are likely to be far more useful than isozyme markers once appropriate molecular genetic maps are available.

Usually, host-plant resistance, alone or as a component of an integrated control strategy, appears to be the most practical and economic approach for managing these stresses. However, the task is challenging because several stresses often occur together, thus requiring the presence of multiple-stress resistance in suitable agronomic backgrounds. Also, the variations in pathotypes and the existence of biotype insect pests and parasitic weeds make durable resistance breeding difficult.

Conventional breeding strategies have shown some success but their effectiveness could be considerably increased with the application of suitable biotechnologies. Improvement in screening techniques, the use of molecular markers for identifying and utilizing economically important genes, the use of *in vitro* culture for transferring desirable genes from wild to cultivated species, and genetic engineering using alien genes, all offer opportunities for improving stress resistance in future cultivars of grain legumes (Saxena, 1993).

10. Variety release

A decision to release a line as a cultivar signals the end of the formal plant breeding process. In the United States, the first contacts are usually with a group of responsible elite seed growers, specialized growers who take the new variety through the seed build-up stages necessary before its mass availability to farmers. These contract growers are usually part of a seed growers' association and represent a facilitating process between the breeder, the agricultural experiment station, the private seed companies, and the farmer. Seed growers serve the public interest, but they also have personal interests at stake in the success or failure of a new variety. The result is a careful, scientific, and business-oriented scrutiny of all aspects of a potential new variety (Jensen, 1988).

The breeder's responsibilities lessen but do not end with the introduction of a variety. He is intimately involved in the two-to-three-year seed building and the monitoring of seed and crop purity and of production practices. The breeder serves as consultant to the specialized agencies which by law are responsible for these practices. For the life of the variety, the breeder remains a resource person to whom to turn for information and appropriate seed stocks.

11. Conclusions and future prospects

An unresolved feature of pulse crops is the great season-to-season variation in crop yield. It has been ascribed to poor pollination, water stress, competition from vegetative sinks and loss due to diseases. All these factors probably play some role at some sites, but they do not affect at all locations in every season. Thus, it is difficult to predict the ideal ideotype for maximum yield in favourable seasons and for yield stability under environmental stress conditions.

A better understanding at a physiological level of the interactions between vegetative and reproductive growth is essential as a basis for further improvements in yield. It is evident that none of the individual yield selection criteria discussed can be applied to all situations. Continuing research efforts are needed to find suitable combinations of criteria to improve efficiency. Resistance to biotic and abiotic stresses could provide the plant more stable performance in different environments. The use of genetic resources to overcome biotic and abiotic stresses and to design new plant ideotypes with enhanced biological nitrogen fixation has to be attempted. The proposed breeding pyramid would provide the approach needed to exploit wild germplasm as a source of desirable alleles for yield in the base population. Backcross populations and adequate linkage maps of wild species constructed within gene pools need to be developed to identify QTL associated with improved seed yield. The identification of QTL for yield would bypass the other limitations such as low heritability, and high GxE interactions that reduce progress in the improvement of complex traits such as yield. In this way it is possible to shift the paradigm from phenotypic to genotypic selection and the improvement of quantitative traits will be accelerated.

Although full attention will be dedicated to conventional breeding methods, the potential applications of new biotechnological techniques will be crucial in promoting further advances in the development of cultivars meeting farmers' needs.

Further advances in cell culture techniques and procedure of transformation will make gene transfer more routine for many legume grains. Such developments will also extend the use of non-traditional strategies such as molecular breeding and bio-engineering to legume improvement.

References

Adam-Blondon A F, Sevignac M, Bannerot H and Dron M (1994) SCAR, RAPD and RFLP markers linked to a dominant gene (*are*) conferring resistance to anthracnose in common bean. *Theor. Appl. Genet.*, **88**: 865–870.

Anuradha T and Murty B K (1993) Genetic of harvest index in blackgram (*Vigna mungo* (L.) Hepper). *Intl. J. Tropical Agric.*, **11**: 71–73.

Bahl P N (1990) The role of food legumes in the diet of populations of Mediterranean areas and associated nutritional factors. In: *The Role of Legumes in the Farming System of the Mediterranean Areas* (Eds Osmon A E, Ibrahim M H and Jones M A), Kluwer Academic Publishers, Dordrecht, The Netherlands, pp. 143–149.

Bai Y, Michaels T E and Pauls K P (1997) Identification of RAPD markers linked to common bacterial blight resistance genes in *Phaseolus vulgaris* L. *Genome*, **40**: 544–551.

Beebe S E, Ochoa I, Skroch P, Nienhuis J and Tivang J (1995) Genetic diversity among common bean breeding lines developed for Central America. *Crop Sci.*, **35**: 1178–1183.

Bliss F A (1993) Utilizing the potential for increased nitrogen fixation in common bean. *Plant Soil.*, **152**: 157–160.

Bliss F A, Pereira P A A, Araugo R S, Henson R A, Kiniecik K A, McFerson J R, Teixeira M G and da Silva C C (1989) Registration of five high nitrogen fixing common bean germplasm lines. *Crop Sci.*, **29**: 240–241.

Bond D A (1977) A breeder's approach to establishing production in field beans (*Vicia faba*). In: *Proceedings Symposium on the Production, Processing and Utilization of Field Bean* (*Vicia faba* L.) (Ed Thompson R), Bulletin No. 15, Scottish Horticultural Research Institute, Invergowrie, pp. 10–16.

Buddenhagen K W and Richards R A (1988) Breeding cool season food legumes for improved performance in stress environments. In: *World Crop: Cool season, Food Legumes*, Kluwer Academic Publishers, The Netherlands, pp. 81–95.

Davies D R (1990) The John Innes pea programme for the '90s. *Pisum Newslett.*, **35**: 214–223.

Delourme R, Bouchereau A, Hubert N and Renard M (1994) Identification of RAPD markers linked to a fertility restorer gene for the Ogura radish cytoplasmic male sterility of rapeseed. *Theor. Appl. Genet.*, **88**: 741–748.

Devine T E and Weber D F (1977) Genetic specificity of nodulation. *Euphytica*, **26**: 527–535.

Dirlewanger E, Isaac P G, Ranade S, Belajouza M, Cousin R and Devienne D (1994) Restriction fragment length polymorphism analysis of loci associated with disease resistance genes and developmental traits in *Pisum sativum* L. *Theor. Appl. Genet.*, **88**: 17–27.

Duc G (2000) The potential of *M. truncatula* as a model plant to hasten genomics in grain legumes crops. *Grain Legum.*, **28**: 18–19.

Ellis N T H (2000) Progress in genomics and synteny on grain legumes. *Grain Legum.*, **28**: 16–17.

Eujayl I, Baum M and Erskine W (1997) Current status of genetic mapping in lentils. *Grain Legumes*, **18**: 19.

Gepts P (1988) *Genetic Resources of Phaseolus Beans*. Dordrecht, Kluwer Academic Publishers, The Netherlands.

Gepts P and Bliss R A (1988) Dissemination pathways of common bean (*Phaseolus vulgaris*) deduced from phaseolin electrophoretic variability. II. Europe and Africa. *Economic Botany*, **42**: 86–104.

Gepts P, Nodari R, Tsai S M, Koinange E M K, Liaca V, Gilbertson R and Guzman P (1993) Linkage mapping in common bean. *Ann. Rep. Bean Improve. Cooperative*, **36**: 24–38.

Graham P H and Ranalli P (1997) Common bean (*Phaseolus vulgaris* L.). *Field Crop Res.*, **53**: 131–146.

Haley S D, Miklas P N, Stavely J R, Byrum J and Kelly J D (1993) Identification of RAPD markers linked to a major rust resistance gene block in common bean. *Theor. Appl. Genet.*, **86**: 505–512.

Hedley C L and Ambrose M J (1981) Designing "leafless" plants for improving yields of the dried pea crop. *Adv. Agron.*, **34**: 225–277.

Herridge D F, Rupela O P, Serraj R and Beck D P (1994) Screening techniques and improved biological nitrogen fixation in cool season food legumes. *Euphytica*, **73**: 95–108.

Hormaza J I, Dollo L and Polito V S (1994) Identification of a RAPD marker linked to sex determination in *Pistacia vera* using bulked segregant analysis. *Theor. Appl. Genet.*, **89**: 9–13.

Jensen N F (1988). *Plant Breeding Methodology*, John Wiley and Sons, New York, USA.

Jeuffroy M H and Ney B (1997) Crop physiology and productivity. *Field Crop Res.*, **53**: 3–16.

Jones N, Ougham H and Thomas H (1997) Markers and mapping: we are all geneticists now. *New Phytol.*, **137**: 165–177.

Jung G, Coyne D P, Skrock P W, Nienhuis J, Arnaud-Santana E, Bokosi J, Kaeppler S M and Steadman J R (1994) Construction of a genetic linkage map and locations of common blight, rust resistance and pubescence loci in *Phaseolus vulgaris* L. using random amplified polymorphic DNA (RAPD) markers. *Ann. Rep. Bean Improve. Cooperative*, **37**: 37–38.

Karp A, Ingram D and Isaac P G (1998) *Molecular Tools for Biodiversity: Plants and Animals*, Chapman and Hall, London, UK.

Kelly J D, Adams M W and Varner G V (1987) Yield stability of determinate and indeterminate dry bean cultivars. *Theor. Appl. Genet.*, **74**: 516–521.

Kelly J D, Afanador L and Haley S D (1995) Pyramid genes for resistance to bean common mosaic virus. *Euphytica*, **82**: 207–212.

Kelly J D, Kolkman J M and Schneider K (1998). Breeding for yield in dry bean (*Phaseolus vulgaris* L.). *Euphytica*, **102**: 343–356.

Knapp S J (1998) Marker assisted selection as a strategy for increasing the probability of selecting superior genotypes. *Crop Sci.*, **38**: 1164–1174.

Lander E S, Green P, Abrahamson J, Barlow A, Daly M J, Lincoln S E and Newberg L (1987) MAPMAKER. An interactive computer package for constructing primary genetic linkage maps of experimental and natural populations. *Genomics*, **1**: 174–181.

Lie T A (1984). Host genes in *Pisum sativum* L. conferring resistance to European *Rhizobium leguminosarum* strains. *Plant Soil.*, **82**: 415–425.

Mahmoud S H, Gatehouse J A and Boulter D (1984) Inheritance and mapping of isoenzymes in pea (*Pisum sativum* L.). *Theor. Appl. Genet.*, **68**: 559–566.

McKerron D K L and Thompson R (1983) Canopy performance in contrasting genotypes of pea (*Pisum sativum* L.). In: *Perspectives for Pea and Lupins as Protein Crops* (Eds Thompson R and Casey R), Martinus Nijhoff Publishers, Dordrecht, The Netherlands, pp. 139–145.

Michelmore R W, Paran I and Kesseli R V (1991) Identification of markers linked to disease resistance genes by bulked segregant analysis: A rapid method to detect markers in specific genomic regions using segregating populations. *Proc. Natl. Acad. Sci. USA*, **88**: 9828–9832.

Miklas P N, Afanador L and Kelly J D (1996) Recombination-facilitated RAPD marker-assisted selection for disease resistance in common bean. *Crop Sci.*, **36**: 86–90.

Nodari R O, Tsai S M, Gilbertson R L and Gepts P (1993b) Towards an integrated linkage map of common bean. II. Development of an RFLP-based linked map. *Theor. Appl. Genet.*, **85**: 513–520.

Nodari R O, Tsai S M, Guzman P, Gilbertson R L and Gepts P (1993c) Toward an integrated linkage map of common bean. III. Mapping genetic factors controlling host-bacteria interactions. *Genetics*, **134**: 341–350.

Nutman P S (1949) Nuclear and cytoplasmic inheritance of resistance to infection by nodule bacteria in red clover. *Heredity*, **3**: 263–291.

Nutman P S (1952) Studies on the physiology of nodule formation. III. Experiments on the excision of root-tips and nodules. *Ann. Bot.*, **16**: 80–102.

Osborn T C and Bliss F A (1985) Effects of genetically removing lectin seed protein on horticultural and seed characteristics of common bean. *J. Am. Soc. Hort. Sci.*, **110**: 484–488.

Parlevliet J E (1979) Components of resistance that reduce the rate of epidemic development. *Annu. Rev. Phytopathol.*, **17**: 203–222.

Parlevliet J E (1983) Can horizontal resistance be recognized in the presence of vertical resistance in plants exposed to a mixture of pathogen races? *Phytopathology*, **73**: 379.

Portapuglia A and Aragona M (1997) Improvement of grain legumes. General part: diseases. *Field Crops Res.*, **53**: 73–76.

Ranalli P (1996) Phenotypic recurrent selection in common bean (*Phaseolus vulgaris* L.) based on performance of S2 progenies. *Euphytica*, **87**: 127–132.

Ranalli P, Di Candilo M, Giordano I and Casarini B (1981) Correlation and path coefficient analysis in peas (*Pisum sativum* L.) for processing. *Z. Pflanzenzucht*, **86**: 81–86.

Ranalli P and Cubero J I (1997) Bases for genetic improvement of grain legumes. *Field Crops Res.*, **53**: 69–82.

Ranalli P, Ruaro G and Del Re P (1991) Response to selection for seed yield in bean (*Phaseolus vulgaris* L.). *Euphytica*, **57**: 117–123.

Ranalli P, Ruaro G, Del Re P and Faeti V (1996) Comparison of early generation yield testing and a single seed descent procedure in two bean (*Phaseolus vulgaris* L.) crosses. *J. Genet. Breed.*, **50**: 103–108.

Renganayaki K and Sreerengasamy S R (1992) Path coefficient analysis in blackgram. *Madras Agricult. J.*, **79**: 634–639.

Saxena M C (1993) The challenge of developing biotic and abiotic stress resistance in cool-season food legumes. In: *Breeding for Stress Tolerance in Cool-Season Food Legumes* (Eds Singh K B and Saxena M C), ICARDA-Wiley-Sayce, Chichester, pp. 3–14.

Schneider K A, Brothers M E and Kelly J D (1997) Marker-assisted selection to improve drought resistance in common bean. *Crop Sci.*, **37**: 51–60.

Sedgley R H, Siddique K H and Walton G H (1990) Chickpea ideotypes for Mediterranean environment. In: *Chickpea in the Nineties*. ICRISAT, Patancheru, India, 87–92.

Sekhar M R (1994). Salt tolerance of mungbean (*Vigna radiata* L. Wilczek) at germination stage. *Ann. Agric. Res.*, **15**: 90–91.

Singh S P (1994) Gamete selection for simultaneous improvement of multiple traits in common bean. *Crop Sci.*, **34**: 352–355.

Smartt J (1990) *Grain Legumes. Evolution and genetic resources*. Cambridge University Press, Cambridge, UK.

Snoad B (1983) Improving the pea crop. In: *Perspectives for Pea and Lupins as Protein Crops* (Eds Thompson R and Casey R), Martinus Nijhoff Publishers, Dordrecht, The Netherlands, pp. 101–112.

Snoad B, Payne A and Hobart J (1974) The yield potential of "leafless peas". *Annual Report (1973)*. John Innes Institute, Norwich, UK, pp. 23–24.

Soller M and Beckmann J S (1990). Marker based mapping of quantitative traits loci using replicated progenies. *Theor. Appl. Genet.*, **80**: 205–208.

Staub J E, Serquen F C and Gupta M (1996) Genetic markers, map construction and their application in plant breeding. *Hortscience*, **31**: 729–741.

Thomas R L, Grafius J E and Kahn S K (1970) Genetic analysis of correlated sequential characters. *Heredity*, **26**: 177–188.

Verma S (1992) Correlation and path analysis in blackgram (*Vigna mungo* (L.) Hepper). *Indian J. Pulses Res.*, **5**: 71–73.

Weeden N F and Marx G A (1987) Further genetic analysis and linkage relationships of isozyme loci in the pea: confirmation of the diploid nature of the genome. *J. Hered.*, **78**: 153–159.

Welsh W, Bushuk W, Roca W and Singh S P (1995) Characterization of agronomic traits and markers of recombinant inbred lines from intra- and interracial populations of *Phaseolus vulgaris* L. *Theor. Appl. Genet.*, **91**: 169–177.

White J W, Kornegay J, Castillo J, Cajiao C H and Tejeda G (1992) Effect of growth habitat on yield of large-seeded bush cultivars of common bean. *Field Crops Res.*, **29**: 151–161.

Williams J G K, Kubelik A R, Livak K J, Rafalksi J A and Tingey S V (1990) DNA poly-morphisms amplified by arbitrary primers are useful as genetic markers. *Nucleic Acids Res.*, **18**: 6531–6535.

Young R A and Kelly J D (1996) RAPD markers flanking the *are* gene for anthracnose resistance in common bean. *J. Am. Soc. Hort. Sci.*, **121**: 37–41.

Yu J, Giu W K, Provvidenti R and Weeden N F (1995) Identifying and mapping two DNA markers linked to the gene conferring resistance to pea enation mosaic virus. *J. Am. Soc. Hort. Sci.*, **120**: 730–733.

IN VITRO MORPHOGENESIS IN GRAIN LEGUMES: AN OVERVIEW

M.L. MOHAN AND K.V. KRISHNAMURTHY
Plant Tissue Culture Division, National Chemical Laboratory,
Pune – 411 008, India
e-mail: kvk@ems.ncl.res.in

Abstract

Legumes are one of the most significant groups of economically important crops and have therefore been the subject of efforts to improve desirable traits through *in vitro* manipulations. Efforts are directed for developing efficient regeneration system in order to facilitate development of a variety of technologies. Even though many legumes have undergone *de novo* regeneration, limited contributions to crop improvement have been achieved. Successful regeneration is, in many cases, genotype and explant specific. In recent years, focus has been on development of plant regeneration systems amenable for gene transfer technology. Organogenesis has been utilized in many cases for transformation, but somatic embryogenesis shows a great promise as prolific production system for gene transfer for those legumes which do not regenerate via organogenesis. Currently, sufficient knowledge is available only for designing and optimizing *de novo* regeneration systems for many legumes, but not for successful and hassle-free genetic transformation.

1. Introduction

Legumes are a group of plants which grow under diverse climatic conditions such as tropical jungles to temperate areas as annuals, perennials, bushes and trees and belong to one of the three largest families of flowering plants – the Leguminosae. The family Leguminosae encompasses 690 genera and 18,000 species which are characterized by keel-shaped flowers and pod-shaped fruits (Hulse, 1989).

1.1. ECONOMIC IMPORTANCE

The legumes are economically important as they form the third largest food crop (190 mt), following cereals (2054 mt) and root and tuber crops (625 mt) (Anonymous, 1999).

P.K. Jaiwal and R.P. Singh (eds.), Improvement Strategies for Leguminosae Biotechnology, 23–63.

The importance of legumes ranges from food to fodder, wood to spices and ornamentals. They also play a useful role in biological nitrogen fixation (Duke, 1981; Parrott *et al.*, 1992).

Grain legumes have been the main crop of Indian agriculture. These crops are generally included in rotation with other crops in most of the areas to keep the soil alive and productive by symbiotic nitrogen fixation with species specific *Rhizobia*. However, most of the grain legumes do not fix adequate amounts of nitrogen to support luxurious growth and development of plants to yield large quantities of protein-rich seeds.

1.2. NUTRITIONAL VALUE

Grain legumes (pulses) are an important source of dietary proteins, fibre and calories (Bliss, 1990; Muehlbauer, 1993). The protein concentration in grain legume seeds generally varies from 18 to 40% depending on the species and among cultivars within a species (Bliss, 1990). Due to its high protein content, grain legumes are important in the production of livestock and fish (Davey *et al.*, 1994). They contain approximately 70% of globulins forming the major storage proteins while the rest is made up by glutelins, albumins and free amino acids (Norton *et al.*, 1985). Proteins which are abundant in grain legumes are dietically inferior because of low content of sulphur amino acids mainly methionine and cysteine. Poor digestibility and presence of antinutritional factors also make them inferior (Bliss, 1990). However, the high level of lysine in grain legumes makes them a good supplement for cereals, which are deficient in this amino acid (Davey *et al.*, 1994). Limited success in improving the nutritional quality of seed proteins has been achieved by conventional breeding methods primarily because genes encoding seed storage proteins with high levels of essential amino acids do not normally exist in any given species (Kriz and Larkins, 1991). Modifications of genes coding for seed proteins thereby could be achieved using the tool of genetic engineering. The various approaches that could be advocated are protein sequence modification, development of synthetic genes, over expression of homologous genes and transfer and expression of heterologous genes (Sun and Larkins, 1993). A methionine-rich 2S albumin gene has been isolated from Brazil nut and transferred to *Vicia narbonensis* (Saalbach *et al.*, 1994).

Some of the grain legumes such as groundnut (*Arachis hypogaea* L.) and soybean (*Glycine max* L.) contain large amounts of edible oils (Christou *et al.*, 1993). Grain legumes are an excellent source of vitamins (thiamine and niacin), minerals (calcium and iron) and also contain about 60% carbohydrates, mainly as starch (Davey *et al.*, 1994).

1.3. IMPORTANCE OF *IN VITRO* TECHNIQUES IN LEGUME IMPROVEMENT

The yield of grain legumes has become static (Muehlbauer, 1993). Reduced yields are due to a range of factors, primarily abiotic (such as soil and inclement climatic conditions), biotic (such as pests and diseases) and drought. In addition, there are several constraints such as management, lack of improvement methods and inputs (Nene *et al.*, 1989). Grain legumes are susceptible to various fungal, bacterial and viral diseases and a host of insect and other pests.

Wild species may provide genetic diversity not present in cultivated species as they possess the traits for stress and disease resistance. However, these traits are often associated with undesirable characters such as seed shattering, hard seededness and indeterminate growth habit which are difficult to overcome in breeding programmes (Muehlbauer, 1993). Hence development of a molecular marker based system for transfer of specific segment of a genome is required (Muehlbauer, 1993).

The conventional breeding methods are time consuming and laborious and plant breeders take time releasing new genotypes due to the long process of crosses, back-crosses and progeny selection. This has led the plant breeders to explore the feasibility of using alternative biotechnological approaches of improvement of grain legumes, which include tissue culture techniques of plant differentiation, protoplast regeneration, somaclonal variation, somatic embryogenesis, somatic hybridization, embryo rescue and gene transfer with the help of *Agrobacterium*, biolistic gun and/or electroporation. However, for exploitation of aforementioned non-conventional methods of crop improvement, the following requirements have to be fulfilled:

(1) an efficient *in vitro* explant to plant regeneration system
(2) a method to deliver foreign DNA to plant tissues
(3) regeneration of plants from stably transformed tissues.

Biotechnology is now the cutting edge of plant science – offering new techniques, applications, and opportunities for crop improvement. Biotechnologists use a variety of techniques to identify genes that determine specific traits (for example, drought tolerance or disease resistance), make crosses between species previously believed to be incompatible, and produce improved genotypes much faster than was possible using conventional plant breeding. Application of biotechnological tools in crop improvement programmes can be effective in three different, complementary ways:

- speeding up the process of conventional breeding
- creating genetic variability through tissue culture and
- evolving novel genotypes through recombinant-DNA (r-DNA) technology.

Using various techniques of genetic engineering and tissue culture, it is now possible to introduce isolated genes derived from different organisms from bacteria to mammals, into plants without causing any additional change in the cultivar. These genetically modified plants can subsequently be incorporated into the conventional plant breeding programmes (Puonti-Kaerlas, 1993). Several methods have been developed for transfer of foreign DNA into plant cells.

2. *In vitro* studies in grain legumes

The *in vitro* culture methods, exploiting the regeneration capacity of plant cells, provided the opportunity to micropropagate elite plant clones by organogenesis or somatic embryogenesis. By manipulation of culture conditions, it is now possible to regenerate plants from intact organs, explants, callus and protoplasts (Davey *et al.*, 1994). Plant gene technology has catalyzed progress in plant breeding, but has not yet been applied to food legume improvement on a large scale (Kahl *et al.*, 1994). The advances made in the culture of grain legumes has been discussed in the following sections.

Legumes exhibit a diversity of responses when cultured *in vitro* (Parrott *et al.*, 1992). Until recently, grain legumes have been found to be less amenable (recalcitrant) to regeneration *in vitro* (Hammatt *et al.*, 1986; Puonti-Kaerlas, 1993). Plant regeneration from cultured explants occurs via somatic embryogenesis and/or organogenesis, either directly or indirectly via an intervening callus phase (Parrott *et al.*, 1992). Considerable progress has been made in the last ten years in the development of efficient plant regeneration systems for grain legumes. This review is written with special reference to chickpea, pigeonpea and mothbean.

2.1. ORGANOGENESIS IN GRAIN LEGUMES

Organogenesis, the process by which a cell or a group of cells, differentiates to form organs, may occur directly from the explanted tissue or from callus (Parrott *et al.*, 1992). Proliferation from pre-existing meristems could be exploited for microprojectile-mediated transformation, as it has several advantages over *de novo* organogenesis since it is less subject to somaclonal variation (Parrott *et al.*, 1992). The current status of *in vitro* regeneration of grain legumes by organogenesis is given in Table 1.

Table 1. In vitro *studies in grain legumes – organogenesis: current status*

Species	Reference
Arachis hypogaea L.	Illingworth, 1968; Martin and Rabechault, 1976; Mroginski and Fernandez, 1980; Bajaj *et al.*, 1980; Shyluk *et al.*, 1981; Kartha *et al.*, 1981; Bajaj *et al.*, 1981a, 1981b; Mroginski *et al.*, 1981; Sastri *et al.*, 1982, 1983a; Bajaj, 1983b; Pittman *et al.*, 1983; Oelck and Schieder, 1983; Narasimhulu and Reddy, 1983; Narasimhulu and Reddy, 1984; Atreya *et al.*, 1984; Mhatre *et al.*, 1985; Bhatia *et al.*, 1985; Seitz *et al.*, 1987; Nataraja and Patil, 1987; Banerjee *et al.*, 1988; McKently *et al.*, 1990; McKently *et al.*, 1991; Daimon and Mii, 1991; Cheng *et al.*, 1992; Kachonpadungkitti *et al.*, 1992; Eapen and George, 1993b; Li *et al.*, 1994; Kanyand *et al.*, 1994; Chengalrayan *et al.*, 1995; D'Silva and Podder, 1995; Ilahi *et al.*, 1995; Feng *et al.*, 1996; Venkatachalam *et al.*, 1996; Cheng and Yeh, 1997; Hu, 1997; Venkatachalam and Jayabalan, 1997; Kanyand *et al.*, 1997; Morris *et al.*, 1997; Ponsamuel *et al.*, 1998; Zharare *et al.*, 1998; Venkatachalam *et al.*, 1999b; Radhakrishnan *et al.*,1999; Pestana *et al.*, 1999; Victor *et al.*, 1999a; Gill and Ozias-Akins, 1999
Cajanus cajan L.	Mehta and Mohan Ram, 1980; Kumar *et al.*, 1983; Kumar *et al.*, 1984; Cheema and Bawa, 1991; Sarangi and Gleba, 1991; Eapen and George, 1993b; George and Eapen, 1994; Shiva Prakash *et al.*, 1994; Naidu *et al.*, 1995; Sreenivasan *et al.*, 1995; Mohan *et al.*, 1997; Mohan and Krishnamurthy, 1998; Eapen *et al.*, 1998; Franklin *et al.*, 1998; Geetha *et al.*, 1998
Cicer arietinum L.	Sharma *et al.*, 1979; Bajaj, 1979; Bajaj and Dhanju, 1979; Kartha *et al.*, 1981; Bajaj, 1983a; Khan and Ghosh, 1983; Altaf and Ahmad, 1985; Altaf and Ahmad, 1986; Rao and Chopra, 1987;

Table 1. Continued

Species	Reference
	Malik and Saxena, 1992c; Surya-Prakash *et al.*, 1992; Chandra *et al.*, 1993; Brandt and Hess, 1994; Barna and Wakhlu, 1994; Murthy *et al.*, 1996; Vani and Reddy, 1996; Polisetty *et al.*, 1997; Sagare, 1997; Chandra *et al.*, 1998; Nalini Mallikarjuna, 1999
Glycine canescens	Widholm and Rick, 1983; Grant, 1984
Glycine clandestina	Hammatt *et al.*, 1986; Hymowitz *et al.*, 1986
Glycine max L.	Oswald *et al.*, 1977; Cheng *et al.*, 1980; Kartha *et al.*, 1981; Kameya and Widholm, 1981; Wright *et al.*, 1986; Barwale *et al.*, 1986; Barwale and Widholm, 1987; Wright *et al.*, 1987; Freytag *et al.*, 1989; Yang *et al.*, 1990; Thome *et al.*, 1995; Kaneda *et al.*, 1997; Dan and Reichert, 1998
Glycine soja	Barwale *et al.*, 1986
Glycine spp.	Hammatt *et al.*, 1987b, 1987c
Lathyrus spp.	Malik *et al.*, 1993
Lens culinaris (L.) Medic	Polanco *et al.*, 1988; Malik and Saxena, 1992c; Ahmad *et al.*, 1996; Polanco and Ruiz, 1997
Lupinus spp.	Sroga, 1987; Mulin and Bellio Spataru, 2000
Phaseolus acutifolius L.	Dillen *et al.*, 1996; Zambre *et al.*, 1998
Phaseolus aureus L.	Bajaj and Singh, 1980
Phaseolus coccineus L.	Abou-Mandour and Hartung, 1980; Angelini and Allavena, 1989; Santalla *et al.*, 1998
Phaseolus vulgaris L.	McClean and Grafton, 1989; Franklin *et al.*, 1991; Malik and Saxena, 1992a; Fernandez Caso *et al.*, 1996; Benedicic *et al.*, 1997; Santalla *et al.*, 1998; Zambre *et al.*, 1998
Pisum sativum L.	Malmberg, 1979; Mroginski and Kartha, 1981; Hussey and Gunn, 1984; Rubluo *et al.*, 1984; Jackson and Hobbs, 1990; Nauerby *et al.*, 1991; Malik and Saxena, 1992c; Boehmer *et al.*, 1995; Sanago *et al.*, 1996; Popiers *et al.*, 1997; Madsen *et al.*, 1998
Vicia faba L.	Khalafall and Hattori, 1999
Vicia narbonensis L.	Tegeder *et al.*, 1996
Vigna aconitifolia (Jacq) Marechal	Bhargava and Chandra, 1983; Godbole *et al.*, 1984; Gill *et al.*, 1986; Krishnamurthy *et al.*, 1986; Sangeeta *et al.*, 1988; Jain and Chopra, 1988; Bhargava and Chandra, 1989; Gehlot *et al.*, 1989; Sankhla *et al.*, 1990; Sankhla *et al.*, 1991
Vigna mungo L. Hepper	Gosal and Bajaj, 1983; Geetha *et al.*, 1997a, 1997b; Sen and Guha Mukherjee, 1998; Das *et al.*, 1998; Ignacimuthu and Franklin, 1998; Saini and Jaiwal, 2002
Vigna radiata L. Wilczek	Gulati and Jaiwal, 1990, 1992, 1994; Chandra and Pal, 1995; Narciso *et al.*, 1996; Narciso *et al.*, 1997; Sen and Guha Mukherjee, 1998; Betal and Sen Raychaudhuri, 1999
Vigna sinensis L.	Pandey and Bansal, 1989
Vigna unguiculata (L.) Walp	Pellegrineschi, 1997; Brar *et al.*, 1997; Soh *et al.*, 1998; Brar *et al.*, 1999a, 1999b

2.1.1. Organogenesis in chickpea (Cicer arietinum *L.*)

In chickpea, reports on *de novo* organogenesis are scarce and are confined to the multiplication of pre-existing meristems (Table 2) (Bajaj and Dhanju, 1979; Kartha *et al.*, 1981; Altaf and Ahmad, 1985; Rao and Reddy, 1987; Sangvan *et al.*, 1989; Chandra *et al.*,

Table 2. In vitro *studies in chickpea: organogenesis*

Species	Explant	Mode of regeneration	Reference
Cicer arietinum L. *Cicer pinnatifidum* *Cicer judaicum* *Cicer reticulatum*	S, R	S→CA→S R→CA	Sharma *et al.*, 1979
Cicer arietinum L.	SA (FP)	SA→CP	Bajaj, 1979
Cicer arietinum L.	SA	SA→S	Bajaj and Dhanju, 1979
Cicer arietinum L.	SM	SM→AXB	Kartha *et al.*, 1981
Cicer arietinum L.	SM (FP)	SM→CP	Bajaj, 1983
Cicer arietinum L.	ANT	ANT→CA→R	Khan and Ghosh, 1983
Cicer arietinum L.	SA, CN	SA, CN→CP	Altaf and Ahmad, 1986
Cicer arietinum L.	CO, EP, HY, L, R, SM	CO, EP→CA HY, L, R, SM	Rao and Chopra, 1987
Cicer arietinum L.	IS	IS→MS	Malik and Saxena, 1992c
Cicer arietinum L.	SA, CO	SA→CA→S→PL CO	Surya-Parkash *et al.*, 1992
Cicer arietinum L.	SA, CN, HY	SA→MS CN→PL HY→CA	Chandra *et al.*, 1993
Cicer arietinum L.	CN, AM	CN, AM→PL	Brandt and Hess, 1994
Cicer arietinum L.	L, SA	L, SA→CA→PL	Barna and Wakhlu, 1994
Cicer arietinum L.	HY, EP	HY→CA→PL EP→S	Vani and Reddy, 1996
Cicer arietinum L.	SD	SD→MS→PL	Sagare, 1997
Cicer arietinum L.	SD	SD→MS→PL	Polisetty *et al.*, 1997
Cicer arietinum L.	AM, CN, HP	AM→MS→PL CN HP (Ethylene effects)	Chandra *et al.*, 1998
Cicer arietinum L.	Ovule, Embryo	Embryo rescue	Nalini Mallikarjuna, 1999

Abbreviations: AM – apical meristem, ANT – anther, AXB – axillary branching, CA – callus, CN – cotyledonary node, CO – cotyledons, CP – complete plants, EP – epicotyl, FP – freeze preserved, HP – hypocotyl, IEA – immature embryo axes, IS – intact seedlings, L – leaf, MS – multiple shoots, ORG – organogenesis, PL – plant, R – root, S – shoot, SA – shoot apex/shoot tip, SD – mature seeds, SM – shoot meristem.

1993; Brandt and Hess, 1994). Plant regeneration via callus has also been reported (Sharma *et al.*, 1979; Rao and Reddy, 1987; Schulz *et al.*, 1993; Surya-Prakash *et al.*, 1992; Barna and Wakhlu, 1994). However, histological evidence showing origin of shoots from truly dividing callus cultures and not from pre-existing meristems is lacking. Regeneration from actively dividing callus is eagerly awaited in chickpea (Altaf and Ahmad, 1990). Clonal multiplication from pre-existing meristems and/or *de novo* regeneration in chickpea may be useful for genetic transformation studies, selective multiplication of transgenic shoots and their possible induction of flowering *in vitro* to obtain pods without the need to transfer plants to isolated, controlled and environmentally safe conditions for the collection of transgenic seeds. In addition, the regeneration studies with pre-existing meristems may be utilized for the storage and maintenance of the germplasm (Bajaj and Dhanju, 1979) and clonally propagate F_1 hybrids to generate sufficient F_2 populations of difficult-to-achieve crosses in breeding programs (Rao and Chopra, 1989).

Multiple shoots could be produced from the cotyledonary node and shoot tip of intact seedlings of six cultivars when cultured on a N^6-benzylaminopurine (BAP, 0.5, 1.0, 2.0, 5.0 and 10 mg/L) containing medium. The number of multiple shoots obtained from intact seedlings could be increased significantly when they were raised in liquid medium containing 0.5 mg/L BAP and agitated on shaker for 20 days. These shoots were rooted in MS (Murashige and Skoog, 1962) medium supplemented with indole-3-butyric acid (IBA, 0.2 mg/L) or indole-3-acetic acid (IAA, 0.2 mg/L) in test tubes. Rooted plantlets from 0.2 mg/L IBA treatment were then transferred to soil, hardened and grown to maturity (Sagare, 1997).

In vitro flowering was observed in chickpea when shoot tips (1–1.5 cm in length) of 10-day-old seedlings of cultivar PG12 were cultured on MS medium supplemented with 0.2 mg/L IAA. Intact seeds of cultivar PG12 when cultured on MS medium without any growth regulator and shoot tips of *in vitro* grown seedlings when cultured on IAA containing medium also showed induction of flowers and pod formation. However, the percentage of shoots and seedlings that flowered was very low. Scanning electron micrograph of pollen grains obtained from *in vivo* flowers were found to be uniform in size when compared to the pollen grains from *in vitro* flowers which showed deformities (Sagare, 1997).

*2.1.2. Organogenesis in pigeonpea (*Cajanus cajan *(L.) Millsp.)*

Gosal and Bajaj (1979) obtained multiple shoots from *in vitro* cultured zygotic embryos. Multiple shoot production was observed when cotyledons were cultured on B_5 (Gamborg *et al.*, 1968) medium containing BAP at concentration of 10^{-5} M. Phenolic compounds were absorbed with polyvinylpyrrolidone (PVP) resulting in better growth of plantlets (Mohan Ram *et al.*, 1981). Mehta and Mohan Ram (1980) induced 5 to 35 shoots from cotyledons of young seedlings raised on B_5 medium supplemented with BAP. Shiva Prakash *et al.* (1994) obtained a mass of multiple shoot initials from the axillary bud region of the cotyledonary node explanted from the seedlings. Topical supplementation with IAA increased the formation of shoot initials. Shoot buds (2–3) were induced from cotyledon and hypocotyl explants when cultured on MS medium having BAP and naphthaleneacetic acid (NAA) (Cheema and Bawa, 1991). Eapen and George (1993b) observed *in vitro* shoot regeneration from leaf discs of pigeonpea on MS medium supplemented with BAP and IAA or IAA-amino acids conjugates. The frequency of shoot regeneration and the average number of shoot buds produced was dependent on the type of auxin present in the medium. The plant regeneration was obtained from leaves, mature cotyledons and whole seeds of cultivar ICPL-161 when cultured on MS + BAP + IAA (George and Eapen, 1994). The roots from seedlings germinated and maintained on BAP (15 mg/L) for 4 weeks gave rise to shoot buds on MS medium supplemented with BAP (1 mg/L) in combination with IAA (0.1 mg/L). Kumar *et al.* (1984) induced multiple shoots from epicotyl segments, excised cotyledons and shoot tips of pigeonpea and *Atylosia* on Blaydes' medium with BAP (Table 3).

Formation of multiple shoots was observed from mature embryo axes and mature cotyledons of seven genotypes when cultured on MS medium supplemented with BAP (1, 2, 3 and 5 mg/L) alone or in combination with kinetin (0.1, 0.5 and 1 mg/L) or IAA (0.1 and 1 mg/L) (Naidu *et al.*, 1995). Shoot buds were induced on epicotyl explant derived from 10–12 day old seedlings by culturing on MS medium containing BAP (1 mg/L) and IAA (0.1 mg/L) (Naidu *et al.*, 1995). The shoots were rooted individually on filter paper supports in half-strength MS medium (liquid) with 0.1 mg/L IBA within 14 days and the percentage of rooting was 80–85%. The rooted plantlets were transferred to soil : vermiculite mixture (1 : 1) for hardening at 25 ± 2°C for 20 days in pots. The plants flowered in pots as well as in the field and viable seeds were obtained after 160 days of transfer to soil.

The distal halves of cotyledons of the genotypes T-15-15 and GAUT-82-90 cultured on various basal media with cytokinins like BAP (5 mg/L), kinetin (0.5 mg/L) and adenine sulphate (AdS, 100 mg/L) swell and turn green after 2–3 weeks in culture, producing small, green, dome-like structures over the surface of the cotyledonary segment (Mohan and Krishnamurthy, 1998). After 3–4 weeks of culture these structures developed into shoot buds, without an intervening callus phase. The response of shoot bud formation varied from 25 to 95% and the average number of shoot buds per explant ranged from 13 to 41 with the maximum number of shoot buds on EC_6 basal medium in both genotypes. The shoot buds produced well developed shoots after two to three subcultures on half-strength MS medium supplemented with 1 mg/L gibberellic acid (GA_3). However, GA_3 was found to support only sporadic elongation of shoot buds into shoots and simultaneous elongation of all shoot buds into shoots could not be achieved. In fact, a large number of leafy shoots (leaves with petioles) were produced. The elongated shoots were rooted on half-strength MS medium with the growth regulator IBA (0.1 mg/L). The rooted plantlets were transferred to soil : vermiculite (1 : 1) mixture in pots for hardening at 25 ± 2°C.

The origin of shoot buds from cotyledonary segments was examined by histological preparations (Mohan and Krishnamurthy, 1998). Formation of a layer of non-meristematic compact mass of cells, which was due to the swelling and formation of small dome-like structures on the surface of cotyledonary segments, not easily separable from the explant, was observed prior to the development of meristematic pockets of small cells with dense cytoplasm and darkly stained nuclei. The anatomy of differentiated shoot bud along with leaf primordium originating from the compact mass of cells confirms the organogenetic pathway of morphogenesis.

Table 3. In vitro *studies in pigeonpea: organogenesis*

Cultivars	Explant	Mode of regeneration	Reference
Prabhat	CO	CO→MS→PL	Mehta and Mohan Ram, 1980
	CO	CO→CA→PL	Kumar *et al.*, 1983
ICP 6917, ICP 6974, ICP 7119, ICP 7263, Wild	CO	CO→MS→PL	Kumar *et al.*, 1984
	AM	AM→PL	
	EP	EP→SB→PL	
AL-15	AM	AM→PL	Cheema and Bawa, 1991
	HP	HP→SB→PL	
–	CO	CO→MS→PL	Sarangi and Gleba, 1991
	MEA	MEA→MS→PL	
ICPL 161	L	L→SB→PL	Eapen and George, 1993
ICPL 161	IEA	IEA→S	George and Eapen, 1994
	ICS	ICS→S	
	EP	EP→CA	
	SD	SD→SB	
	R	R→SB→PL	
	MCS	MCS→SB→PL,R	
	L	L→CA→SB→PL	
Gaut-89-8, Gaut-88-29, BP-86-34, SPMA-4	CN	CN→MS→PL	Shiva Prakash *et al.*, 1994
T-21, PT-22, T-Vishaka-1, ICPL 87, N-290-21	MEA	MEA→MS→PL	Naidu *et al.*, 1995
	CO	CO→MS→PL	
	EP	EP→SB→PL	
BDN-2	S	S→R	Sreenivasan *et al.*, 1995
T-15-15, Gaut-82-99	EP	EP→SB→PL	Mohan *et al.*, 1997
	MCS	MCS→SB→PL	
T-15-15, Gaut-82-99	MCS	MCS→SB→PL	Mohan and Krishnamurthy, 1998
ICPL 161, ICPL 88039, UPAS 120	L	L→SB→PL	Eapen *et al.*, 1998
Vamban-1	AM	AM→PL	Franklin *et al.*, 1998
	AB	AB→PL	
Hyderabad C	CN	CN→MS→PL	Geetha *et al.*, 1998
	EP	EP→MS→PL	
	HP	HP→MS→PL	
	CO	CO→MS→PL	
	L	L→MS→PL	

Abbreviations: AB – axillary bud, AM – apical meristem, CA – callus, CN – cotyledonary node, CO – cotyledons, EP – epicotyl, HP – hypocotyl, ICS – immature cotyledonary segments, IEA – immature embryo axes, IS – intact seedlings, L – leaf, MCS – mature cotyledonary segment, MEA – mature embryo axes, MS – multiple shoots, PL – plant, R – root, S – shoot, SB – shoot bud, SD – mature seeds.

Plant regeneration has been achieved using epicotyl region of mature embryo axes (ERMEA) and decapitated mature embryo axes (DCMEA) as explants (M.L. Mohan and K.V. Krishnamurthy, unpublished data). Shoots appear directly from explants of genotypes T-15-15, Gaut-82-99 and Gaup-9002 when cultured on EC_6 basal medium supplemented with BAP (1 mg/L) and IAA at various concentrations (0.1, 0.2, 0.3, 0.4 and 0.5 mg/L). The shoots elongated on half-strength MS medium fortified with 1 mg/L GA_3. Elongated shoots were rooted with 80–85% efficiency on half-strength MS medium with 0.1 mg/L IBA and plantlets were hardened. Survival of plants in the pots was 75–80%.

Basal part of the embryo and the part of the cotyledon adjacent to the embryo gave multiple shoots via callusing in MS medium containing 0.5 mg/L BAP. The distinct variation has been observed in the regenerated plants (Sarangi and Gleba, 1991). Kumar *et al.* (1983) obtained regeneration from cotyledonary callus and leaf callus cultures when incubated on Blaydes' medium fortified with BAP, NAA and GA_3. Shama Rao and Narayanaswamy (1975) reported that hypocotyl segments obtained only from γ-irradiated (5 kR) seeds produced abundant calli and shoot buds in 50 % of cultures.

2.1.3. Organogenesis in mothbean (Vigna aconitifolia *(Jacq) Marechal.)*

Callus cultures of two cultivars of mothbean (IPCMO-926, RDN-120) were raised and their growth and differentiation studied (Bhargava and Chandra, 1983). However, multiple shoot bud formation and differentiation was restricted to root and hypocotyl callus cultures. Further induction of rooting in the differentiated shoot buds and transfer of the resulting plantlets to soil has not been done. Plant regeneration from hypocotyl and stem explants (Gill *et al.*, 1986) has also been reported.

Incorporation of GA_3 in the culture medium enhanced the formation of roots from hypocotyl-derived callus cultures, but greatly diminished the formation of shoot buds *in vitro*. In contrast, abscissic acid promoted the formation of shoot buds, although development of roots was not significantly influenced by this growth regulator. The enhancement of root formation by GA_3 was effectively reversed by abscissic acid. Similarly, the promotive effect of abscissic acid on shoot bud formation was counteracted by increasing concentrations of GA_3 (Joshi *et al.*, 1988).

Sankhla *et al.* (1991) reported that the triazole plant growth regulators, paclobutrazol and uniconazole reduced *in vitro* growth of mothbean callus by 20–25% when added to the culture medium at 1 mg/L. The addition of 10 mg/L GA_3 to the culture medium in combination with the triazoles restored callus growth to a level equivalent to that of the untreated control. The addition of triazoles or GA_3 to the regeneration medium reduced the formation of green meristematic nodules, which are precursors of shoots in mothbean callus. Thus, exposure of mothbean callus tissue to micromolar concentrations of triazoles or GA_3 can significantly alter *in vitro* growth and differentiation.

Composition of the callusing medium strongly influenced the ability of *V. aconitifolia* callus to differentiate into shoot buds. The best regeneration response was obtained in callus produced on MS medium supplemented with BA (0.5 mg/L) and NAA (0.1 mg/L). The frequency of shoot regeneration from such callus depended both on the explant type and the relative concentrations of auxin and cytokinin in the differentiation medium. Highest frequency was obtained with callus of leaf origin at supplementation with kinetin (5 mg/L) and IAA (1 mg/L) (Jain and Chopra, 1988).

Leaf explants of *V. aconitifolia* taken from 8 to 10 day old seedlings were successfully regenerated on MS medium supplemented with various combinations of kinetin/BAP + IAA and transferred to soil (Bhargava and Chandra, 1989).

Incorporation of the plant growth regulator lactidichlor-Et (PPG-1721) in the culture medium at 1–10 mg/L decreased the growth of mothbean callus by 30%. The presence of PPG-1721 in a regeneration medium inhibited root formation but increased the percentage of cultures which formed green meristematic nodules which are precursors of shoots in mothbean callus (Sankhla *et al.*, 1990).

Gehlot *et al.* (1989) observed that addition of relatively low concentrations (1.7 to 6.8 mM) of paclobutrazol to the culture medium decreased the growth of mothbean cv. Jaadia callus. Paclobutrazol increased the content of sugars and total soluble protein in the callus. Paclobutrazol in the regeneration medium reduced the differentiation of roots and shoots *in vitro*.

The formation of complete plants and their transfer to field has been reported for the first time in literature for a cultivar of mothbean PLMO-130 (Godbole *et al.*, 1984). The plantlets have been obtained from callus derived from shoot apices, cotyledons and shoot apex of mothbean seedlings germinated on B_5 medium containing 2.25 mg/L BAP (B_5 + BAP) or modified White's medium (WMB). Shoot apices obtained from seedlings grown on B_5 + BAP produced multiples on MS and B_5 media without phytohormones, MS and B_5 media containing BAP alone (2.25 mg/L) or BAP (1 mg/L) and IAA (0.5 mg/L). However, shoot apices obtained from seedlings grown on WMB medium produced multiples only on B_5 medium supplemented with various growth regulators. Cotyledon explants derived from seedlings germinated on B_5 + BAP formed multiple shoots within 5–6 weeks on B_5 medium containing BAP (0.1 mg/L) and IBA, NAA or IAA at 0.05 mg/L level. The callus derived from explant shoot apices of seedlings grown on B_5 + BA and on WMB was transferred to MS or B_5 basal media without supplements, MS medium containing either coconut milk (CM, 10%) and NAA (1 mg/L) or BAP (1 mg/L) + IAA (0.5 mg/L). The response of callus on these different media was the same irrespective of the initial growth of seedlings on B5 + BAP or on WMB. The callus produced multiple shoot buds after seventh passage and elongated subsequently within 4 weeks. The multiple shoots obtained through callus differentiation, shoot apices and cotyledons were rooted with 60–70% survival and reared to maturity in the field. An analysis of the organogenetic and non-organogenetic calli of mothbean using scanning electron microscope has also been carried out (Krishnamurthy *et al.*, 1986) (Table 4).

2.2. SOMATIC EMBRYOGENESIS IN GRAIN LEGUMES

Somatic embryogenesis – a process whereby a cell or a group of cells from somatic tissues form an embryo (Parrott *et al.*, 1992) may be indirect with an intervening callus phase or direct – directly from the explants. Table 5 represents the studies on somatic embryogenesis in grain legumes.

2.2.1. Somatic embryogenesis in chickpea

In chickpea, somatic embryogenesis has been reported from immature embryo axes, immature cotyledonary segments, young leaflets and mature embryo axes (Table 6). The

Table 4. In vitro *studies in mothbean: organogenesis*

Cultivars	Explant	Mode of regeneration	Reference
IPCMO-926,	R	R→MS→PL	Bhargava and Chandra, 1983
RDN-120	HP	HP→MS→PL	
	HP	HP→PL	Gill *et al.*, 1986
	STEM	STEM→PL	
	HP	HP→CA→R	Sangeeta *et al.*, 1988
		HP→CA→S	
		CA	Sankhla *et al.*, 1991
	L	L→CA→PL	Jain and Chopra, 1988.
	L	L→PL	Bhargava and Chandra, 1989
		CA	Sankhla *et al.*, 1990
Jaadia		CA	Gehlot *et al.*, 1989
PLMO-130	SA	SA→MS→PL	Godbole *et al.*, 1984
	CO	CO→MS→PL	
	SA	SA→CA→PL	

Abbreviations: CA – callus, CO – cotyledons, HP – hypocotyl, L – leaf, MS – multiple shoots, PL – plant, R – root, S – shoot, SA – shoot apex/shoot tip, STEM – stem.

Table 5. In vitro *studies in grain legumes: somatic embryogenesis: current status*

Species	Reference
Arachis duranensis	Sabita Rani and Reddy, 1996
Arachis hypogaea L.	Pittman *et al.*, 1983; Banerjee *et al.*, 1988; Ozias-Akins, 1989; Hazra *et al.*, 1989; Sellars *et al.*, 1990; Daimon and Mii, 1991; McKently, 1991; Ozias-Akins *et al.*, 1992a; Ozias-Akins *et al.*, 1992b; Durham and Parrott, 1992; Gill and Saxena, 1992; Baker and Wetzstein, 1992; Saxena *et al.*, 1992; Wetzstein and Baker, 1993; Eapen and George, 1993a; Eapen *et al.*, 1993; Reddy and Reddy, 1993; Mhaske and Hazra, 1994; Baker and Wetzstein, 1994; Baker *et al.*, 1994; Chengalrayan *et al.*, 1994; Baker and Wetzstein, 1995; Baker *et al.*, 1995; Chengalrayan *et al.*, 1995; McKently, 1995; Murthy *et al.*, 1995; Sabita Rani and Reddy, 1996; Venkatachalam and Jayabalan, 1996; Venkatachalam *et al.*, 1997; Chengalrayan *et al.*, 1997; Chengalrayan *et al.*, 1998; Baker and Wetzstein, 1998; Venkatachalam *et al.*, 1998a; Venkatachalam *et al.*, 1999a; Venkatachalam *et al.*, 1999c; Victor *et al.*, 1999a; Victor *et al.*, 1999b; Murch *et al.*, 1999
Cajanus cajan L. (Millsp.)	Bajaj *et al.*, 1980; George and Eapen, 1994; Patel *et al.*, 1994; Nalini Mallikarjuna *et al.*, 1996; Sreenivasu *et al.*, 1998; Anbazhagan and Ganapathi, 1999
Cicer arietinum L.	Rao and Chopra, 1989; Shri and Davis, 1992; Sagare *et al.*, 1993; Barna and Wakhlu, 1993; Dineshkumar *et al.*, 1994; Suhasini *et al.*, 1994; Eapen and George, 1994a; Islam, 1994; Kumar *et al.*, 1995; Sagare *et al.*, 1995a, 1995b; Dineshkumar *et al.*, 1995; Adkins *et al.*, 1995; Suhasini *et al.*, 1996;

Table 5. Continued

Species	Reference
	Vani and Reddy, 1996; Ramana *et al.*, 1996; Murthy *et al.*, 1996; Suhasini *et al.*, 1997; Hita *et al.*, 1997; Guru *et al.*, 1999
Phaseolus acutifolius	Kumar *et al.*, 1988b
Phaseolus aureus L.	Bajaj and Singh, 1980
Phaseolus coccineus L.	Genga and Allavena, 1991; Malik and Saxena, 1992b
Glycine max L.	Christianson *et al.*, 1983; Gamborg *et al.*, 1983a; Lippmann and Lippmann, 1984; Lazzeri *et al.*, 1985; Li *et al.*, 1985; Ranch *et al.*, 1985; Kerns *et al.*, 1986; Ghazi *et al.*, 1986; Lazzeri *et al.*, 1987a; Buchheim *et al.*, 1989; Christou and Yang, 1989; Lazzeri *et al.*, 1987b; Komatsuda and Ohyama, 1988; Finer, 1988; Finer and Nagasawa, 1988; Hartweck *et al.*, 1988; Hepher *et al.*, 1988; Komatsuda *et al.*, 1992; Gill and Saxena, 1992; Liu *et al.*, 1992; Ranch, 1993; Lippmann and Lippmann, 1993; Gyulai *et al.*, 1993; Bailey *et al.*, 1993; Ma *et al.*, 1994; Li and Grabau, 1996; Santos *et al.*, 1997; Rajasekaran and Pellow, 1997; Santarem *et al.*, 1997; Samoylov *et al.*, 1998
Phaseolus vulgaris L.	Martin and Sondahl, 1984
Pisum sativum L.	Jacobsen and Kysely, 1984; Kysely *et al.*, 1987; Lehminger-Mertens and Jacobsen, 1989; Kysely and Jacobsen, 1990; Tetu *et al.*, 1990; Stejskal and Griga, 1992; Loiseau *et al.*, 1995;
Pisum sativum L.	Bencheikh and Gallais, 1996a; Bencheikh and Gallais, 1996b; Loiseau *et al.*, 1996; Loiseau *et al.*, 1998; Griga, 1998
Psophocarpus tetragonolobus (L.) DC	Ahmed *et al.*, 1996; Gupta *et al.*, 1997
Vicia faba L.	Tegeder *et al.*, 1995
Vicia narbonensis L.	Pickardt *et al.*, 1989; Albrecht and Kohlenbach, 1989; Tegeder *et al.*, 1996
Vigna aconitifolia (Jacq) Marechal	Kumar *et al.*, 1988a; Eapen and George, 1990
Vigna mungo (L.) Hepper	Eapen and George, 1990
Vigna radiata (L.) Wilczek	Eapen and George, 1990
Vigna sinensis	Li *et al.*, 1995a
Vigna unguiculata (L.) Walp	Kulothungan *et al.*, 1995

first successful report of induction of somatic embryos in chickpea (Rao and Chopra, 1989) described the formation of globular-stage embryos from leaflet derived callus on MS medium supplemented with 0.5 mg/L 2,4-dichlorophenoxyacetic acid (2,4-D) + 0.5 mg/L BAP. However, the further development of these embryos and their regeneration into plantlets was not observed. Shoots have been regenerated using immature cotyledons via cotyledon-like structures (CLS) (Shri and Davis, 1992). However, the mode of regeneration and bipolar nature of CLS was not studied and difficulty in rooting of these regenerated shoots was also reported. Regeneration of plants through somatic embryogenesis has been reported from immature cotyledons (Eapen and George, 1994a; Adkins *et al.*, 1995; Ramana *et al.*, 1996), immature leaflets (Barna and Wakhlu, 1993; Dineshkumar *et al.*, 1994; Dineshkumar *et al.*, 1995; Kumar *et al.*, 1995), de-embryonated cotyledons (Vani and Reddy, 1996) and mature seeds (Murthy *et al.*, 1996).

The successful regeneration of plants via somatic embryos was demonstrated using immature cotyledonary segments (Sagare *et al.*, 1993). Out of five genotypes evaluated for their embryogenic potential, somatic embryos were induced from immature cotyledons of genotypes PG12 and C235 and immature embryo axes of genotypes PG5, PG12 and C235.

Table 6. In vitro *studies in chickpea: somatic embryogenesis*

Species	Explant	Mode of regeneration	Reference
Cicer arietinum L.	L	L→CA→GSE	Rao and Chopra, 1989
Cicer arietinum L.	IC	IC→CLS→S	Shri and Davis, 1992
Cicer arietinum L.	IEA, ICS	ICS→SE→PL IEA→GSE	Sagare *et al.*, 1993
Cicer arietinum L.	L	L→CA→SE→PL	Barna and Wakhlu, 1993
Cicer arietinum L.	L	L→CA→SE→PL	Dineshkumar *et al.*, 1994
Cicer arietinum L.	MEA	MEA→DSE→PL	Suhasini *et al.*, 1994
Cicer arietinum L.	ICS	ICS→SE→PL	Eapen and George, 1994a
Cicer arietinum L.	IC	IC→SE→PL	Islam, 1994
Cicer arietinum L.	L	L→CA→SE→PL	Kumar *et al.*, 1995
Cicer arietinum L.	IEA, ICS, MEA, L	SE (Histology)	Sagare *et al.*, 1995
Cicer arietinum L.	L	L→CA→SE→PL	Dineshkumar *et al.*, 1995
Cicer arietinum L.	ICS, IEA	ICS, IEA→SE→PL	Adkins *et al.*, 1995
Cicer arietinum L.	MEA	SE (Histology)	Suhasini *et al.*, 1996
Cicer arietinum L.	CO	CO→CA→SE	Vani and Reddy, 1996
Cicer arietinum L.	ICS	ICS→DSE→PL	Ramana *et al.*, 1996
Cicer arietinum L.	SD	SD→DSE→PL	Murthy *et al.*, 1996
Cicer arietinum L.	ICS, MEA	Comparative study of SE and ZE	Suhasini *et al.*, 1997
Cicer arietinum L.	ICS	ICS→SE→PL	Hita *et al.*, 1997
Cicer arietinum L.	L	L→SE→PL (Ethylene effects)	Guru *et al.*, 1999

Abbreviations: CA – callus, CLS – cotyledon-like structures, CO – cotyledons, DSE – direct somatic embryogenesis, GSE – globular-stage somatic embryos, IC – immature cotyledons, ICS – immature cotyledonary segments, IEA – immature embryo axes, L – leaf, MEA – mature embryo axes, PL – plant, S – shoot, SD – mature seeds, SE – somatic embryos, ZE – zygotic embryo.

The disadvantages of non-availability of immature explants throughout the year and the difficulties to obtain the plant material at right developmental stage has been overcome by induction of somatic embryogenesis from mature embryo axes and the direct origin of somatic embryos was confirmed histologically (Suhasini *et al.*, 1994). Mature embryo axes were inoculated on MS medium supplemented with 3% sucrose and 3.0 mg/L 2,4,5-trichlorophenoxyacetic acid (2,4,5-T) (liquid T3 medium) supported with Whatman No. 1 filter paper bridges. Direct emergence of torpedo-stage somatic embryo with a prominent suspensor-like stalk and cotyledonary-stage somatic embryos was observed after 3–4 weeks of culture. This eliminated the subculture stage which is mandatory for the formation of cotyledonary structures from globular embryos when induced on solid T3 medium. This observation is very striking and significant in induction of somatic embryos in chickpea (Suhasini *et al.*, 1994). The time taken for the induction of somatic embryos was reduced from 12 weeks taken in the case of incubation on solid T3 medium to 4 weeks when incubated in liquid T3 medium. No effect of the culture vessel, position of tissues (floating or submerged) and physical state of the medium (semisolid or liquid, poorly or well aerated) on the induction of somatic embryogenesis was reported (Thorpe, 1988). However, Suhasini *et al.* (1994) have reported a significant reduction in the time taken for induction of somatic embryos by culturing the explants in liquid medium. The probable reason advocated for this has been that liquid medium facilitates better aeration and easy diffusion and flow of nutrients and growth regulators by capillary forces to the explants (Constabel and Shyluk, 1994).

Initiation of embryos from shoot tip portion of mature embryo axes and the description in similarity of their development to that of zygotic embryogenesis involving the globular, torpedo and cotyledonary-stage embryos has also been presented (Suhasini *et al.*, 1997).

Various concentrations of zeatin and BAP were evaluated for the efficient conversion of cotyledonary-stage somatic embryos. 1.0 and 2.0 mg/L zeatin supported the conversion of somatic embryo into plantlets whereas the other combinations of growth regulators lead to the callusing of the embryos. Plantlets with well developed roots and shoots were transferred to pots containing a mixture of peat, vermiculite and soil (1 : 1 : 1). After 15 days, the hardened plants were transferred to the greenhouse and grown to maturity. Mature plants after 90 days showed flowering and pod formation.

The mature seeds harvested from the somatic embryo regenerated plants were morphologically similar to the zygotic seeds. Comparison of the seeds of zygotic embryo raised plants and somatic embryo raised plants in terms of number of pods per plant and the average weight of seeds was carried out (A.P. Sagare and coworkers, unpublished data).

Different kinds of morphologically abnormal somatic embryos were observed when both immature and mature explants were cultured for induction of embryogenesis (Suhasini *et al.*, 1996) and were classified into six kinds based on the shape of the embryo and the number of cotyledons. Histological study of aberrant somatic embryos to ascertain the reasons for the lack of conversion of some of the aberrant somatic embryos revealed the lack of shoot apical meristem to be the reason for their non-conversion (Suhasini *et al.*, 1996).

Histology of somatic embryo initiation and development from four explants, viz. immature embryo axes, immature cotyledonary segments, mature embryo axes and young leaflet has been carried out (Sagare *et al.*, 1995a). It revealed that somatic embryos arose indirectly with an intervening callus phase from immature cotyledonary segments and immature embryo axes, directly from epidermal and sub-epidermal cells at apical meristem of mature embryo axes, both directly and indirectly from parenchymatous cells of young leaflets.

2.2.2. Somatic embryogenesis in pigeonpea

Patel *et al.* (1994) were successful in achieving somatic embryogenesis using distal halves of cotyledons in four genotypes. The somatic embryos were obtained on MS, B_5, Modified B_5, LS (Linsmaer and Skoog's medium), EC_6 and White's medium supplemented with three cytokinins BAP, kinetin and Adenine sulfate. Somatic embryogenesis was also obtained from immature leaflets, root discs, epicotyl, hypocotyl and mature cotyledons of genotype ICPL-87 on MS medium supplemented with NAA and BAP (Nalini Mallikarjuna *et al.*, 1996). Haploid embryogenesis and production of only globular embryos has been reported using anther culture technique (Bajaj *et al.*, 1980). Even though George and Eapen (1994) obtained somatic embryos from immature cotyledons and embryonal axes of variety ICPL-161, plantlet regeneration from somatic embryos capable of transfer to the field has not been reported. Thidiazuron induced somatic embryogenesis from leaf and cotyledon explants (Sreenivasu *et al.*, 1998) and cell suspension cultures (Anbazhagan and Ganapathi, 1999) has been reported (Table 7).

Somatic embryogenesis has been achieved from cotyledonary segments of mature seeds as explants (Mohan and Krishnamurthy, 2001). A large number of globular embryos was induced directly from cotyledons of genotypes T-15-15, GAUT-82-90 and GAUT-82-99 when cultured on EC_6 basal medium supplemented with BAP at 0.5–5 mg/L concentrations and thidiazuron at 0.1–2 mg/L concentrations. The cotyledonary embryos were

Table 7. In vitro *studies in pigeonpea: somatic embryogenesis*

Cultivars	Explant	Mode of regeneration	Reference
T-21	ANT	ANT→CA→SE	Bajaj *et al.*, 1980
ICPL 161	ICS IEA	ICS→SE→PL IEA→SE→PL	George and Eapen, 1994
T-15-15, Gaut-82-90, Bandapalera, NP(WR)15	MCS	MCS→SE→PL	Patel *et al.*, 1994
ICPL 87	L R	L→CA→SE→PL R→CA→SE→PL	Nalini Mallikarjuna *et al.*, 1996
Pusa-606, Pusa-609, Pusa-852, Pusa-855, Pusa-856, H-86-5	L CO	L→CA→SE→PL CO→CA→SE→PL	Sreenivasu *et al.*, 1998
Vamban-1	L	L→SC→SE→PL	Anbazhagan and Ganapathi, 1999

Abbreviations: ANT – anther, CA – callus, CO – cotyledons, ICS – immature cotyledonary segments, IEA – immature embryo axes, L – leaf, MCS – mature cotyledonary segment, PL – plant, R – root, SE – somatic embryos.

developed when the globular embryos were transferred to Murashige and Skoog's basal medium containing 1–5 mg/L GA_3. Maturation of somatic embryos was achieved on half-strength MS medium fortified with 0.1 mg/L ABA. The plantlets converted on MS medium supplemented with 0.1 mg/L BAP were hardened and transferred to the pots.

2.2.3. Somatic embryogenesis in mothbean

Plant regeneration has been achieved routinely from established cell suspension culture lines (Kumar *et al.*, 1988a) of 3-week-old leaf callus. Several media were tested for their effects on cell growth and the maximum growth was observed in L-6 medium containing 44.5 μM 2,4-D. After 6–8 weeks, the suspensions were filtered through 500, 250, 125, and 60 mm sieves, respectively, for four to five subcultures. An embryogenic cell line was obtained from the cell fraction collected below 250 mm and maintained on L-6 medium with 4.5 μM 2,4-D and 2.3 μM zeatin. Somatic embryogenesis was induced by transferring the cells to L-6 medium with 4.6 μM zeatin and the somatic embryos developed from most of the cell clusters when plated on L-6 agar medium with 2.3 μM BAP. Plantlets were obtained from the embryos on L-6 medium with 10.0 μM IBA. The regenerated plants were grown to maturity in the greenhouse (Kumar *et al.*, 1988a) (Table 8).

2.3. PROTOPLAST REGENERATION IN GRAIN LEGUMES

The first successful isolation of protoplasts from legumes was reported by Gamborg *et al.* (1968), from cell suspensions of *Glycine max*.

Many plant species have been successfully regenerated from protoplasts and it is now being routinely used for somatic hybridization and for direct gene transfer (Puite, 1992). However, regeneration of plants from protoplasts is difficult (Tegeder *et al.*, 1995) and limited to *Vigna aconitifolia*, *Pisum sativum* and *Glycine max*.

Kao *et al.* (1970) showed, for the first time, division in the protoplasts isolated from the cell suspensions of soybean. However, they failed to regenerate plants from protoplast-derived calli. Their pioneering work created interest in the isolation of protoplasts from cultured cells of most of the economically important legumes (Davey *et al.*, 1994). Protoplast technology is well established now for many plant species and is being routinely used for somatic hybridization and direct gene transfer (Puite, 1992). However, in grain legumes the regeneration of plants from single cells still remains difficult (Tegeder *et al.*, 1995). Few reports of successful recovery of plants from protoplasts are available for *Vigna aconitifolia, Psophocarpus tetragonolobus, Pisum sativum* and

Table 8. In vitro *studies in mothbean: somatic embryogenesis*

Cultivars	Explant	Mode of regeneration	Reference
	L	L→CA→SC→SE	Kumar *et al.*, 1988a

Abbreviations: CA – callus, L – leaf, SC – suspension culture, SE – somatic embryo.

Table 9. In vitro *studies in grain legumes – protoplast regeneration: current status*

Species	Reference
Arachis hypogaea L.	Oelck *et al.*, 1982; Mhatre *et al.*, 1985; Miyake and Taniguchi, 1994; Li *et al.*, 1995b; Venkatachalam and Jayabalan, 1996
Cajanus cajan L.	Kulkarni and Krishnamurthy, 1989; Sarangi *et al.*, 1992; Ramana Rao *et al.*, 1992
Cicer arietinum L.	Sagare and Krishnamurthy, 1991
Glycine max L.	Schwenk *et al.*, 1981; Lin, 1983; Gamborg *et al.*, 1983b; Tricoli *et al.*, 1986; Wei, 1988; Dhir *et al.*, 1991a, 1991b; Dhir *et al.*, 1992a, 1992b
Glycine spp.	Newell and Luu, 1985; Hammatt *et al.*, 1987a; Myers *et al.*, 1989; Hammat *et al.*, 1989
Phaseolus vulgaris L.	Crepy *et al.*, 1986
Pisum sativum L.	Gamborg *et al.*, 1975; von Arnold and Eriksson, 1976; Jia, 1982; Puonti-Kaerlas and Eriksson, 1988; Lehminger-Mertens and Jacobsen, 1989; Böhmer *et al.*, 1995; Gram *et al.*, 1996
Psophocarpus tetragonolobus L.	Wilson *et al.*, 1985
Vicia faba L.	Binding and Nehls, 1978; Roeper, 1980; Tegeder *et al.*, 1995
Vicia narbonensis L.	Donn, 1978; Tegeder *et al.*, 1996
Vigna aconitifolia	Shekhawat and Galston, 1983; Krishnamurthy *et al.*, 1984; Gill and Eapen, 1986; Kohler *et al.*, 1987; Eapen *et al.*, 1987

Glycine max (Table 9). Recent reports of thidiazuron-induced plant regeneration from protoplasts of *Vicia faba* (Tegeder *et al.*, 1995) and *Pisum sativum* (Boehmer *et al.*, 1995) have given a new ray of hope for the application of protoplast technology in the genetic modification of grain legumes of economic importance.

2.3.1. Protoplast regeneration in chickpea

Chickpea protoplasts were first isolated from leaves but sustained division of protoplasts was not observed. Factors affecting isolation of mesophyll protoplasts were studied but the division and regeneration of protoplasts was not observed. Protoplast division and regeneration up to callus stage was reported from hypocotyl explants of *in vitro* grown seedlings of cultivars PG5 and PG1 (Sagare and Krishnamurthy, 1991), while protoplasts of PG12 and Chafa failed to divide. Hypocotyl explants from 3-day-old seedlings cut into thin longitudinal sections, preplasmolyzed for 1 h and incubated in roller drum apparatus with the enzyme mixture 1.0% Cellulase R-10 and 0.5% Macerozyme R-10 for 6 or 8 h yielded a large number of protoplasts. Sustained division of hypocotyl protoplasts was observed on V-47 medium which lead to formation of microcalli which grew further when transferred on to fresh medium. Regeneration of plants from this callus has not been reported.

2.3.2. Protoplast regeneration in pigeonpea

Protoplast technology is well established now for many plant species and is being routinely used for somatic hybridization and direct gene transfer (Puite, 1992). Mesophyll protoplasts were isolated from pigeonpea cultivars UPAS 120 and Pant A2 and their wild relatives *Atylosia platycarpa*, *Rhynchosia aurea* and *Rhynchosia visicidia* (Ramana Rao *et al.*, 1992). Although protoclones were obtained successfully from mesophyll protoplasts, they did not differentiate into plantlets. The fusion of protoplasts of cultivars and wild species though was attempted, the fusion products viable only for several weeks (Ramana Rao *et al.*, 1992). Similarly the protocolonies from mesophyll protoplasts which proliferated to form green calli failed to regenerate into plants (Sarangi *et al.*, 1992) (Table 10).

Protoplasts were also isolated from primary leaves of 8–10 day old seedlings or first and second pairs of leaves from axenic cultures of 30–35 days of genotype T-21 and 148. High yields (5×10^7 per gram fresh weight of leaf tissue) of protoplasts were obtained which showed division after 4 days of culture on V-47 medium supplemented with NAA (3 mg/L) and BAP (0.8 mg/L) leading to the formation of calli. However, calli failed to regenerate into plants (Kulkarni and Krishnamurthy, 1989).

2.3.3. Protoplast regeneration in mothbean

In mothbean, regeneration from protoplasts has been reported from leaf mesophyll (Shekhawat and Galston, 1983), shoot apex derived callus (Krishnamurthy *et al.*, 1984) and hypocotyl (Gill and Eapen, 1986). PEG mediated transformation of protoplasts was demonstrated in mothbean (Kohler *et al.*, 1987).

A large number of viable protoplasts from shoot apex derived callus cultures of mothbean (cv. PLMO-130), their division leading to formation of cell clusters on modified MS medium, has been reported (Krishnamurthy *et al.*, 1986). The protoplast derived callus formed multiple shoot buds which elongated and gave rise to shoots (Krishnamurthy *et al.*, 1986). The plants were transferred to field, and these plants have undergone field trials up to F_5 progeny and the data have been analyzed for their growth, yield and other characteristics (Table 11).

2.4. SOMACLONAL VARIATION

Plant breeders are continually searching for new genetic variability that is potentially useful in cultivar improvement (Parrott *et al.*, 1992). Genetic variation which results as

Table 10. In vitro *studies in pigeonpea: protoplast regeneration*

Cultivars	Explant	Mode of regeneration	Reference
T-21, T-148	L	L→PRO→CA	Kulkarni and Krishnamurthy, 1989
ICPL 87, Rayagada Local, 82208 Kandula	L	L→PRO→CA	Sarangi *et al.*, 1992
UPAS 120, Pant A2	L	L→PRO→CA	Ramana Rao *et al.*, 1992

Abbreviations: CA – callus, L – leaf, PRO – protoplast.

Table 11. In vitro *studies in mothbean: protoplast regeneration*

Cultivars	Explant	Mode of regeneration	Reference
–	L	L→PRO→CA	Shekhawat and Galston, 1983
PLMO-130	SA	SA→CA→PRO→CA→PL	Krishnamurthy *et al.*, 1984
–	HP	HP→PRO→CA	Gill and Eapen, 1986
		PRO→PEG Transformation	Kohler *et al.*, 1987

Abbreviations: CA – callus, HP – hypocotyl, L – leaf, PEG – poly ethylene glycol, PL – plant, PRO – protoplast, SA – shoot apex.

a consequence of frequent culturing of plant cells (somaclonal variation) is the simplest form of genetic manipulation (Larkin and Scowcroft, 1981). It is rarely associated with gross cytological changes and is probably related to modifications in the nucleic acid composition. Although many somaclonal variants exhibit characteristics which are detrimental compared to those of parental plants, individuals have been produced which express superior traits (Davey *et al.*, 1994).

Somaclonal variations have been exploited recently to recover useful variants in soybean (Barwale and Widholm, 1990). Among the variants observed were chlorophyll deficiency, partial or complete sterility, numerous abnormalities in leaf morphology and number, twin seeds and multiple shoots. Changes in maturity date and increases in oil content among regenerants have also been reported (Barwale and Widholm, 1990). However, this variation was not inherited.

2.5. SOMATIC HYBRIDIZATION

Somatic hybridization has the potential for improving grain legumes, but the absence of protoplast to plant systems in many legumes has deterred the extension of the technique to grain legumes. In soybean, heterokaryons between *G. max* cv. HP-20-20 and perennial wild *Glycine* G1171 produced shoots, but these shoots failed to develop into plants (Hammat *et al.*, 1992).

2.6. EMBRYO RESCUE

Wild relatives of crop plants comprise an important germplasm resource for plant improvement (Davey *et al.*, 1994). Crosses between distantly related plants are generally unfruitful because of the abortion of embryos on the mother plant. These embryos can be precociously excised and cultured *in vitro* (Monnier, 1990).

By using this technique, a large number of hybrid plants have been obtained and several genetic characteristics have been transferred in grain legumes. Important among those are hybrids produced in *Arachis* (Bajaj, 1990) and *Glycine* (Grant, 1990). Recently, this technique has been successfully exploited to create a viable hybrid in *Glycine max* (Hu and Zanettini, 1995).

Table 12. In vitro *studies in pigeonpea: embryo rescue*

Hybrid	Reference
Cajanus lineata × *C. cajan*	Patel *et al.*, 1992
Cajanus platycarpus × *C. cajan*	Nalini Mallikarjuna and Moss, 1995

2.6.1. Embryo rescue in pigeonpea

Even though hybrid vigour is documented in crosses between pure lines, it is difficult to utilize them because of technical difficulties and the production of hybrid seed is expensive (Purseglove, 1988). The embryos from wide crosses could be rescued at early stages of development and cultured to get a viable hybrid. Patel *et al.* (1992) standardized a technique for embryo rescue of intervarietal and intergeneric hybrids by culturing on modified MS medium supplemented with IAA, kinetin and coconut water. The successes in intervarietal crosses were 80–90% but the cross of cultivars with wild *Atylosia lineata* was unsuccessful. Nalini Mallikarjuna and Moss (1995) developed an interspecific hybrid between *Cajanus platycarpus* and *Cajanus cajan* using an efficient embryo rescue technique to prevent the barrier to hybridization which is post-zygotic in nature. All earlier efforts to hybridize *C. platycarpus* with *C. cajan* were unsuccessful (Ariyanayagam and Spence, 1978; Dundas, 1985; Kumar, 1985; Pundir and Singh, 1987). The *C. platycarpus*, a wild species of pigeonpea, has many desirable characters important for the improvement of cultivated pigeonpea but is incompatible with the cultigen. The F_1 hybrids were found completely pollen sterile. Dundas (1985) recommended that the transfer of desirable genes from *C. platycarpus* may be possible by finding bridge-cross combinations between *C. platycarpus* and other compatible wild *Cajanus* species (Table 12).

3. Genetic transformation studies

Even though morphogenesis does not cover the scope of genetic transformation studies, in order to make a complete chapter on grain legumes liberty has been taken to include this area briefly. Grain legumes are one of the least amenable groups to transformation amongst dicotyledonous crops. Continued genetic improvement is a necessity for the development of crops with increased quality and yield. A well defined, simple shoot regeneration protocol is a prerequisite for the production of transgenic plants. In recent years, many of the shoot regeneration protocols have been developed specifically for the purpose of genetic manipulation (Boemer *et al.*, 1995). However, several requirements must be fulfilled in order to produce stably transformed plants. Initially, a suitable method is required to deliver foreign DNA to plant tissues, followed by appropriate procedure for culturing tissues prior to the regeneration of shoots leading to the recovery of transgenic plants. Subsequently, the newly introduced genes must be expressed in transgenic plants and finally, the foreign DNA must be heritable and expressed reproducibly in succeeding seed generations.

A range of somatic cell and molecular techniques are now available to supplement conventional plant breeding. Gene transfer (or DNA uptake) refers to a process which moves a specific piece of DNA (usually a foreign gene ligated to a bacterial plasmid) into protoplasts or cells (Jenes *et al.*, 1993). The introduction and expression of foreign DNA has been used to introduce commercially important characteristics such as herbicide and insect resistance, changes in oil and protein contents and virus tolerance (Walden, 1989). Several techniques for DNA delivery into plant cells are available, ranging from the *Agrobacterium*-mediated gene transfer, direct gene transfer through electroporation into protoplasts/intact tissues either by PEG method or by uptake of DNA into naked protoplasts by injection and the use of microprojectile bombardment-mediated transformation to introduce DNA into intact tissues. The current status of genetic transformation studies in grain legumes is given in Table 13.

Table 13. Genetic transformation studies in grain legumes: current status

Species	Reference
Arachis hypogaea L.	Lacorte *et al.*, 1991; Mansur *et al.*, 1993; Ozias-Akins *et al.*, 1993; Schnall and Weissinger, 1993; Eapen and George, 1994b; Brar *et al.*, 1994; McKently *et al.*, 1995; Schnall and Weissinger, 1995; Cheng *et al.*, 1996; Cheng *et al.*, 1997; de Freitas *et al.*, 1997; Lacorte *et al.*, 1997; Sarker *et al.*, 1997; Akasaka *et al.*, 1998; Wang *et al.*, 1998; Egnin *et al.*, 1998; Yang *et al.*, 1998; Venkatachalam *et al.*, 1998b; Livingstone and Birch, 1999; Kim *et al.*, 1999; Rohini and Rao, 2000
Cajanus cajan L.	Geetha *et al.*, 1999
Cicer arietinum L.	Srinivasan *et al.*, 1991; Mohapatra and Sharma, 1991; Fontana *et al.*, 1993; Islam *et al.*, 1994; Ramana *et al.*, 1996; Kar *et al.*, 1996; Krishnamurthy *et al.*, 2000
Cyamopsis tetragonoloba L.	Joersbo *et al.*, 1999
Glycine max L.	Owens and Cress, 1985; Baldes *et al.*, 1987; Christou *et al.*, 1987; Christou *et al.*, 1988; Hinchee *et al.*, 1988; Owens and Smigocki, 1988; Chee *et al.*, 1989; Parrott *et al.*, 1989; Christou, 1990; Christou *et al.*, 1990; Dhir *et al.*, 1991b, 1991c; Finer and McMullen, 1991; Sato *et al.*, 1993; Falco *et al.*, 1995; Di *et al.*, 1996; Bond *et al.*, 1996; Hadi *et al.*, 1996; Stewart *et al.*, 1996; Liu *et al.*, 1996; Santarem *et al.*, 1998; Trick and Finer, 1998; Maughan *et al.*, 1999; Santarem and Finer, 1999; Hu and Wang, 1999; Zhang *et al.*, 1999; Ponappa *et al.*, 1999
Lens culinaris (L.) Medic.	Warkentin and McHughen, 1991; Warkentin and McHughen, 1992
Lupinus augustifolius L.	Pigearie *et al.*, 1997
Phaseolus acutifolius (A.) Gray	Dillen *et al.*, 1997
Phaseolus vulgaris L.	Mariotti *et al.*, 1989; Genga *et al.*, 1991; Aragao *et al.*, 1992; Aragao *et al.*, 1993; Franklin *et al.*, 1993; Lewis and Bliss, 1994; Dillen *et al.*, 1995; Kim and Minamikawa, 1996; Kim and Minamikawa, 1997; Aragao *et al.*, 1996; Brasileiro *et al.*, 1996; Aragao and Rech,

Table 13. Continued

Species	Reference
	1997; Zhang *et al.*, 1997; Barros *et al.*, 1997; Giovinazzo *et al.*, 1997; Zhanyuan *et al.*, 1997; Aragao *et al.*, 1998
Pisum sativum L.	Hobbs *et al.*, 1989; Puonti-Kaerlas *et al.*, 1989; De Kathen and Jacobsen, 1990; Puonti-Kaerlas *et al.*, 1990; Schaerer and Pilet, 1991; Davies *et al.*, 1993; Jordan and Hobbs, 1993; Nicolaisen and Poulsen, 1993; Schroeder *et al.*, 1993; De Kathen and Jacobsen, 1995; Grant *et al.*, 1995; Nicoll *et al.*, 1995; Bean *et al.*, 1997; Grant *et al.*, 1998; Simonenko *et al.*, 1999
Vicia narbonensis L.	Pickardt *et al.*, 1991; Pickardt *et al.*, 1995; Saalbach *et al.*, 1995
Vigna aconitifolia (Jacq) Marechal	Eapen *et al.*, 1987; Kohler *et al.*, 1987; Bhargava and Smigocki, 1994
Vigna mungo (L.) Hepper	Pal *et al.*, 1991; Bhargava and Smigocki, 1994; Karthikeyan *et al.*, 1996; Saini *et al.*, 2002
Vigna radiata (L.) Wilczek	Bhargava and Smigocki, 1994; Jaiwal *et al.*, 1998, 2001
Vigna unguiculata (L.) Walp	Garcia *et al.*, 1987; Penza *et al.*, 1991; Akella and Lurquin, 1993; Muthukumar *et al.*, 1996
Kidney bean	Saker and Kuhne, 1998

4. Conclusions and future prospects

Eventhough legumes were considered to be recalcitrant with regards to regeneration *in vitro*, many species of grain legumes are now regenerated by tissue culture. Some species are amenable for regeneration by both organogenesis and embryogenesis but the reported regeneration systems are mostly genotype and explant specific. Media combinations will have to be standardized for each species, genotype and explant type. Plant regeneration systems for legumes have been and continue to be developed to meet the challenging demands of plant improvement and genetic engineering. The mode of regeneration required depends on specific goals, for example, for virus elimination and germplasm conservation only proliferation of shoot is all that needed, whereas *de novo* organogenesis is most suited for the purposes of genetic transformation. It is necessary to develop the regeneration systems suited for recovery of transformants since the gene transfer technology for crop improvement of legumes is the order of the day.

References

Abou-Mandour A A and Hartung W (1980) The effect of abscisic acid on growth and development of intact seedlings, root and callus cultures and stem and root segments of *Phaseolus coccineus*. *Z. Pflanzenphysiol.*, **100**: 25–33.

Adkins A L, Godwin I D and Adkins S W (1995) An efficient *in vitro* regeneration system for Australian-grown chickpea (*Cicer arietinum*). *Aust. J. Bot.*, **43**: 491–497.

Ahmad M, Fautrier A G, McNeil D L, Hill G D and Burritt D J (1996) *In vitro* propagation of *Lens* species and their F_1 interspecific hybrids. *Plant Cell Tiss. Org. Cult.*, **47**: 169–176.

Ahmed R, Gupta S D and De D N (1996) Somatic embryogenesis and plant regeneration from leaf derived callus of winged bean (*Psophocarpus tetragonolobus* (L.) DC). *Plant Cell Rep.*, **15**: 531–535.

Akasaka Y, Mii M and Daimon H (1998) Morphological alterations and root nodule formation in *Agrobacterium rhizogenesis*-mediated transgenic hairy roots of peanut (*Arachis hypogaea* L.). *Ann. Bot.*, **81**: 355–362.

Akella V and Lurquin P F (1993) Expression in cowpea seedlings of chimeric transgenes after electroporation into seed-derived embryos. *Plant Cell Rep.*, **12**: 110–117.

Albrecht C and Kohlenbach H W (1989) Induction of somatic embryogenesis in leaf-derived callus of *Vicia narbonensis* L. *Plant Cell Rep.*, **8**: 267–269.

Altaf N and Ahmad M S (1985) Plant regeneration and propagation of chickpea (*Cicer arietinum* L.) through tissue-culture techniques. In: *Proc. Symp. Nuclear Tech and In Vitro culture for plant improvement*, IAEA/FAO, Vienna, pp. 407–417.

Altaf N and Ahmad M S (1986) Chickpea (*Cicer arietinum*). In: *International Atomic Energy Authority*, Wein: IAEA, Wein, pp. 407–417.

Altaf N and Ahmad M S (1990) Chickpea (*Cicer arietinum*). In: *Biotechnology in Agriculture and Forestry, Vol. 10* (Ed Bajaj Y P S), Springer-Verlag, Berlin, pp. 101–113.

Anbazhagan V R and Ganapathi A (1999) Somatic embryogenesis in cell suspension cultures of pigeonpea. *Plant Cell Tiss. Org. Cult.*, **56**: 179–184.

Angelini R R and Allavena A (1989) Plant regeneration from immature cotyledon explant cultures of bean (*Phaseolus coccineus* L.). *Plant Cell Tiss. Org. Cult.*, **19**: 167–174.

Anonymous (1999) FAO Production Year Book, Rome.

Aragão F J L, Barros L M G, Brasileiro A C M, Ribeiro S G, Smith F D, Sanford J C, Faria J C and Rech E L (1996) Inheritance of foreign genes in transgenic bean (*Phaseolus vulgaris* L.) co-transformed via particle bombardment. *Theor. Appl. Genet.*, **93**: 142–150.

Aragão F J L, Grossi de Sa M F and Almeida E R (1992) Particle bombardment-mediated transient expression of a Brazil nut methionine-rich albumin in bean (*Phaseolus vulgaris* L.). *Plant Mol. Biol.*, **20**: 357–359.

Aragão F J L, Grossi de Sa M F, Davey M R, Brasileiro A C M, Faria J C and Rech E L (1993) Factors influencing transient gene expression in bean (*Phaseolus vulgaris* L.) using an electrical particle acceleration divice. *Plant Cell Rep.*, **12**: 483–490.

Aragão F J L and Rech E L (1997) Morphological factors influencing recovery of transgenic bean plants (*Phaseolus vulgaris* L.). of a Carioca cultivar. *Int. J. Plant Sci.*, **158**: 157–163.

Aragão F J L, Ribeiro S G, Barros L M G, Brasileiro A C M, Maxwell D P, Rech E L and Faria J C (1998) Transient beans (*Phaseolus vulgaris* L.) engineered to express viral antisense RNAs show delayed and attenuated symptoms to bean golden masaic geminivirus. *Mol. Breed.*, **4**: 491–499.

Ariyanayagam R P and Spence J A (1978) A possible gene source for early, day length neutral pigeonpea (*Cajanus cajan* (L.) Millsp). *Euphytica*, **27**: 505–509.

Atreya C D, Rao J P and Subrahmanyam N C (1984) *In vitro* regeneration of peanut (*Arachis hypogaea*) plantlets from embryo axis and cotyledon segments. *Plant Sci. Lett.*, **34**: 379–383.

Bailey M A, Boerma H R and Parrott W A (1993) Genotype-specific optimization of plant regeneration from somatic embryos of soybean. *Plant Sci.*, **93**: 117–120.

Bajaj Y P S (1979) Freeze preservation of meristems of *Arachis hypogaea* and *Cicer arietinum*. *Indian J. Exp. Biol.*, **17**: 1405–1407.

Bajaj Y P S (1983) Production of normal seeds from plants regenerated from the meristems of *Arachis hypogaea* and *Cicer arietinum* cryopreserved for 20 months. *Euphytica*, **32**: 425–430.

Bajaj Y P S (1990) Wide hybridization in legumes and oil seed crops through embryo, ovule and ovary culture. In: *Biotechnology in Agriculture and Forestry, Vol. 10* (Ed Bajaj Y P S), Springer-Verlag, Berlin, pp. 3–37.

Bajaj Y P S and Dhanju M S (1979) Regeneration of plants from apical meristem tips of some legumes. *Curr. Sci.*, **48**: 906–907.

Bajaj Y P S, Kumar P, Labana K S and Singh M M (1981b) Regeneration of plants from seedling explants and callus cultures of *Arachis hypogaea*. *Indian J. Exp. Biol.*, **19**: 1026–1029.

Bajaj Y P S, Ram A K, Labana K S and Singh H (1981a) Regeneration of genetically variable plants from the anther-derived callus of *Arachis hypogaea* and *Arachis villosa*. *Plant Sci. Lett.*, **23**: 35–39.

Bajaj Y P S and Singh H (1980) *In vitro* induction of androgenesis in mung bean (*Phaseolus aureus*). *Indian J. Exp. Biol.*, **18**: 1316–1318.

Bajaj Y P S, Singh H and Gosal S S (1980) Haploid embryogenesis in anther cultures of pigeonpea (*Cajanus cajan*). *Theort. Appl. Genet.*, **58**: 157–159.

Baker C M and Wetzstein H Y (1992) Somatic embryogenesis and plant regeneration from leaflets of peanut, *Arachis hypogaea* L.). *Plant Cell Rep.*, **11**: 71–75.

Baker C M and Wetzstein H Y (1994) Influence of auxin type and concentration on peanut somatic embryogenesis. *Plant Cell Tiss. Org. Cult.*, **36**: 361–368.

Baker C M and Wetzstein H Y (1995) Repetitive somatic embryogenesis in peanut cotyledon cultures by continual exposure to 2,4-D. *Plant Cell Tiss. Org. Cult.*, **40**: 249–254.

Baker C M and Wetzstein H Y (1998) Leaflet development, induction time and medium influence somatic embryogenesis in peanut (*Arachis hypogaea* L.). *Plant Cell Rep.*, **17**: 925–929.

Baker C M, Burns J A and Wetzstein H Y (1994) Influence of photoperiod and medium formulation on peanut somatic embryogenesis. *Plant Cell Rep.*, **13**: 159–163.

Baker C M, Durham R E, Burus J A, Parrott W A and Wetzstein H Y (1995) High frequency somatic embryogenesis in peanut (*Arachis hypogaea* L.) using mature, dry seed. *Plant Cell Rep.*, **15**: 38–42.

Baldes R, Moos M and Geider K (1987) Transformation of soybean protoplasts from permanent suspension cultures by cocultivation with cells of *Agrobacterium tumefaciens*. *Plant Mol. Biol.*, **9**: 135–145.

Banerjee S, Bandhopadhay S and Ghosh P D (1988) Cotyledonary node culture and multiple shoot formation in peanut: Evidences for somatic embryogenesis. *Curr. Sci.*, **57**: 252–255.

Barna K S and Wakhlu A K (1993) Somatic embryogenesis and plant regeneration from callus cultures of chickpea (*Cicer arietinum* L.). *Plant Cell Rep.*, **12**: 521–524.

Barna K S and Wakhlu A K (1994) Whole plant regeneration of *Cicer arietinum* from callus cultures via organogenesis. *Plant Cell Rep.*, **13**: 510–513.

Barros L M G, Gama M I C S, Gonsalves C H R D, Barreto C C, Santana E F and Carneiro V T D (1997) Bean tissue culture for introduction of foreign gene. *Pesquisa Agropecuaria Brasileira*, **32**: 267–275.

Barwale U B and Widholm J M (1987) Somaclonal variation in plants regenerated from cultures of soybean. *Plant Cell Rep.*, **6**: 365–368.

Barwale U B and Widholm J M (1990) Soybean: plant regeneration and somaclonal variation. In: *Biotechnology in Agriculture and Forestry, Vol. 10* (Ed Bajaj Y P S), Springer-Verlag, Berlin, pp. 114–133.

Barwale U B, Kerns H R and Widholm J M (1986) Plant regeneration from callus cultures of several soybean genoytpes via embryogenesis and organogenesis. *Planta*, **167**: 473–481.

Bean S J, Gooding P S, Mullineaux P M and Davies D R (1997) A simple system for pea transformation. *Plant Cell Rep.*, **16**: 513–519.

Bencheikh M and Gallais A (1996a) Study of the variation in the somatic embryogenesis ability of some pea lines (*Pisum sativum* L. and *Pisum arvense* L.). *Euphytica*, **90**: 251–256.

Bencheikh M and Gallais A (1996b) Somatic embryogenesis in pea (*Pisum sativum* L. and *Pisum arvense* L.): Diallel analysis and genetic control. *Euphytica*, **90**: 257–264.

Benedicic D, Ravnikar M and Gogala N (1997) The regeneration of bean plants from meristem culture. *Phyton-Ann. Rei. Bot.*, **37**: 151–160.

Betal S and Sen Raychaudhuri S (1999) Micropropagation of the endangered aromatic varieties of *Vigna radiata* (L.) Wilczek in West Bengal, India. In Vitro *Cell. Dev. Biol. Plant*, **35**: 76–78.

Bhargava S and Chandra N (1983) *In vitro* differentiation in callus cultures of mothbean *Vigna aconitifolia* (Jacq.) Marechal. *Plant Cell Rep.*, **2**: 47–50.

Bhargava S and Chandra N (1989) Factors affecting regeneration from leaf explants of mothbean *Vigna aconitifolia* (Jacq.) Marechal. *Indian J. Exp. Biol.*, **27**: 55–57.

Bhargava S C and Smigocki A C (1994) Transformation of tropical grain legumes using pariticle bombardment. *Curr. Sci.*, **66**: 439–442.

Bhatia C R, Murthy G S and Mathews V M (1985) Regeneration of plants from deembryonated peanut cotyledons cultured without nutrients and agar. *Z. Pflanzenzuchtg*, **94**: 149–155.

Binding H and Nehls R (1978) Regeneration of isolated protoplasts of *Vicia faba* L. *Z. Pflanzenphysiol.*, **88**: 327–332.

Bliss FA (1990) Genetic alteration of legume seed proteins. *HortSci.*, **25**: 1517–1520.

Boehmer P, Meyer B and Jocobsen H J (1995) Thidiazuron-induced high frequency of shoot induction and plant regeneration in protoplast derived pea callus. *Plant Cell Rep.*, **15**: 26–29.

Bond J E, McDonnell R E and Gresshoff P M (1996) Susceptibility of nodulation mutants of soybean (*Glycine max*) to *Agrobacterium tumefaciens. J. Plant Physiol.*, **148**: 684–692.

Brandt E B and Hess D (1994) *In vitro* regeneration and propagation of chickpea (*Cicer arietinum* L.) from meristem tips and cotyledonary nodes. In Vitro *Cell. Dev. Biol. Plant*, **30**: 75–80.

Brar M S, AlKhayri J M, Shamblin C E, McNew R W, Morelock T E and Anderson E J (1997) *In vitro* shoot tip multiplication of cowpea (*Vigna unguiculata* L.) Walp. In Vitro *Cell. Dev. Biol. Plant*, **33**: 114–118.

Brar M S, AlKhayri J M, Morelock T E and Anderson E J (1999a) Genotypic response of cowpea (*Vigna unguiculata* L.) Walp to *in vitro* regeneration from cotyledon explants. In Vitro *Cell. Dev. Biol. Plant*, **35**: 8–12.

Brar G S, Cohen B A, Vick C L and Johnson G W (1994) Recovery of transgenic peanut (*Arachis hypogaea* L.) plants from elite cultivars utilizing ACCELLâ technology. *Plant J.*, **5**: 745–753.

Brar M S, Moore M J, AlKhayri J M, Morelock T E and Anderson E J (1999b) Ethylene inhibitors promote *in vitro* regeneration of cowpea (*Vigna unguiculata* L.). In Vitro *Cell. Dev. Biol. Plant*, **35**: 222–225.

Brasileiro A C M, Aragao F J L, Rossi S, Dusi D M A, Barros L M G and Rech E L (1996) Susceptibility of common and tepary beans to *Agrobacterium* spp. strains and improvement of *Agrobacterium*-mediated transformation using microprojectile bombardment. *J. Am. Soc. Hort. Sci.*, **121**: 810–815.

Buchheim J A, Colburn S M and Ranch J P (1989) Maturation of soybean somatic embryos and the transition to plantlet growth. *Plant Physiol.*, **89**: 768–775.

Chandra M and Pal A (1995) Differential response of the two cotyledons of *Vigna radiata in vitro. Plant Cell Rep.*, **15**: 248–253.

Chandra R, Chatrath A, Polisetty R and Khetarpal S (1993) Differentiation of *in vitro* grown explants of chickpea (*Cicer arietinum* L.). *Indian J. Plant Physiol.*, **36**: 121–124.

Chandra R, Khetrapal S and Polisetty R (1998) Effect of plant growth regulators on evolution of ethylene and methane by different explants of chickpea. *Biol. Plant.*, **40**: 337–343.

Chee P P, Fober K A and Slightom J L (1989) Transformation of soybean (*Glycine max*) by infecting germinating seeds with *Agrobacterium tumefaciens. Plant Physiol.*, **91**: 1212–1218.

Cheema H K and Bawa J (1991) Clonal multiplication via multiple shoots in some legumes (*Vigna unguiculata* and *Cajanus cajan*). *Acta. Hortic.*, **289**: 93–94.

Cheng H-H and Yeh M-S (1997) Studies on immature embryo of peanut (*Arachis hypogaea* L.) III. *In vitro* organogenesis from immature embryo axes and cotyledons. *Guoli Zhongxing Daxue Nonglin Xuebao*, **46**: 39–51.

Cheng M, Hsi D C H and Phillips G C (1992) *In vitro* regeneration of Valencia-type peanut (*Arachis hypogaea* L.) from cultured petioles, epicotyl sections and other seedling explants. *Peanut Sci.*, **19**: 82–87.

Cheng M, Jarret R L, Li Z and Demski J W (1997) Expression and inheritance of foreign genes in transgenic peanut plants generated by *Agrobacterium* mediated transformation. *Plant Cell Rep.*, **16**: 541–544.

Cheng M, Jarret R L, Li Z, Xing A and Demski J W (1996) Production of fertile transgenic peanut (*Arachis hypogaea* L.) plants using *Agrobacterium tumefaciens. Plant Cell Rep.*, **15**: 653–657.

Cheng T-Y, Saka H and Voqui-Dinh T H (1980) Plant regeneration from soybean cotyledonary node segments in culture. *Plant Sci. Lett.*, **19**: 91–99.

Chengalrayan K, Mhaske V B and Hazra S (1995) *In vitro* regulation of morphogenesis in peanut (*Arachis hypogaea* L.). *Plant Sci.*, **110**: 259–268.

Chengalrayan K, Mhaske V B and Hazra S (1997) High-frequency conversion of abnormal peanut somatic embryos. *Plant Cell Rep.*, **16**: 783–786.

Chengalrayan K, Mhaske V B and Hazra S (1998) Genotypic control of peanut somatic embryogenesis. *Plant Cell Rep.*, **17**: 522–525.

Chengalrayan K, Sathaye S S and Hazra S (1994) Somatic embryogenesis from mature emryo-derived leaflets of peanut (*Arachis hypogaea* L.). *Plant Cell Rep.*, **13**: 578–581.

Christianson M L, Warnick D A and Carlson P S A (1983) A morphogenetically competent soybean suspension culture. *Science*, **222**: 632–634.

Christou P (1990) Morphological description of transgenic soybean chimeras created by the delivery, integration and expression of foreign DNA using electric discharge particle acceleration. *Ann. Bot.*, **66**: 379–386.

Christou P and Yang N-S (1989) Developmental aspects of soybean (*Glycine max*) somatic embryogenesis. *Ann. Bot.*, **64**: 225–234.

Christou P, McCabe D E and Swain W F (1988) Stable transformation of soybean callus by DNA-coated gold particles. *Plant Physiol.*, **87**: 671–674.

Christou P, McCabe D E, Swain W E and Russell D R (1993) *Legume Transformation*, CRC Press Inc, UK, pp. 547–564.

Christou P, Murphy J E and Swain W F (1987) Stable transformation of soybean by electroporation and root formation from transformed callus. *Proc. Natl. Acad. Sci. USA*, **84**: 3962–3966.

Christou P, Swain W F, Yang N and McCabe D E (1990) Cotransformation frequencies of foreign genes in soybean cell cultures. *Theort. Appl. Genet.*, **79**: 337–341.

Constabel F and Shyluk J P (1994) Initiation, nutrition and maintenance of plant cell and tissue cultures. In: *Plant Cell and Tissue Culture* (Eds Vasil I K and Thorpe T A), Kluwer Academic Publishers, Dordrecht, pp. 3–15.

Crepy L, Barros L M G and Valente V R N (1986) Callus production from leaf protoplasts of various cultivars of bean (*Phaseolus vulgaris* L.). *Plant Cell Rep.*, **5**: 124–126.

D'Silva I and Podder S K (1995) Peanut agglutinin from callus and cell suspension cultures of *Arachis hypogaea* L. *Plant Cell Rep.*, **14**: 50–54.

Daimon H and Mii M (1991) Plant formation of peanut by somatic embryogenesis and organogenesis. *Jpn. J. Breed.*, **41**: 461–466.

Dan Y H and Reichert N A (1998) Organogenic regeneration of soybean from hypocotyl explants. In Vitro *Cell. Devlop. Biol. Plant*, **34**: 14–21.

Das D K, Prakash N S and Bhalla-Sarin N (1998) An efficient regeneration system of black gram (*Vigna mungo* L.) through organogenesis. *Plant Sci.*, **134**: 199–206.

Davey M R, Kumar V and Hammatt N (1994) *In vitro* culture of legumes. In: *Plant Cell and Tissue Culture* (Eds Vasil I K and Thorpe T A), Kluwer Academic Publishers, Dordrecht, pp. 313–329.

Davies M R, Hammilton J and Mullineaux (1993) Transformation of peas. *Plant Cell Rep.*, **12**: 180–183.

de Freitas V G, Lacorte C, Sanchetto-Martins, Krul W R, de Oliveira D E, Neves L J and Mansur E (1997) Identification of competent cells for *Agrobacterium* transformation and *in vitro* regeneration in peanut leaf and cotyledon explant. *On Revo Bras. Fisiol. Veg.*, **9**: 157–167.

de Kathen A D and Jacobsen H J (1990) *Agrobacterium tumefaciens*-mediated transformation of *Pisum sativum* L. using binary and cointegrate vectors. *Plant Cell Rep.*, **9**: 276–279.

de Kathen A D and Jacobsen H J (1995) Cell competence for *Agrobacterium*-mediated DNA transfer in *Pisum sativum* L. *Transgenic Res.*, **4**: 184–191.

Dhir S K, Dhir S and Widholm J M (1991a) Plantlet regeneration from immature cotyledon protoplasts of soybean (*Glycine max* (L.) Merrill). *Plant Cell Rep.*, **10**: 39–43.

Dhir S K, Dhir S and Widholm J M (1992a) Regeneration of fertile plants from protoplasts of soybean (*Glycine max* (L.) Merrill): genotypic differences in culture response. *Plant Cell Rep.*, **11**: 285–289.

Dhir S K, Dhir S, Hepburn A and Widholm J M (1991b) Factors affecting transient gene expression in electroporated *Glycine max* protoplasts. *Plant Cell Rep.*, **10**: 106–110.

Dhir S K, Dhir S, Savka M A, Belanger F, Kriz A L, Farrand S K and Widholm J M (1992b) Regeneration of transgenic soybean (*Glycine max*) plants from electroporated protoplasts. *Plant Physiol.*, **99**: 81–88.

Dhir S K, Dhir S, Sturtevant A P and Widholm J M (1991c) Regeneration of transformed shoots from electroporated soybean (*Glycine max* (L.) Merrill) protoplasts. *Plant Cell Rep.*, **10**: 97–101.

Di R, Purcell V, Collins G B and Ghabrial S A (1996) Production of transgenic soybean lines expressing the bean pod mottle virus coat protein precursor gene. *Plant Cell Rep.*, **15**: 746–750.

Dillen W, DeClerq J, Goossens A, Montagu M V and Angenon G (1997) *Agrobacterium*-mediated transformation of *Phaseolus acutifolius* A Gray. *Theort. Appl. Genet.*, **94**: 151–158.

Dillen W, DeClerq J, Montagu M V and Angenon G (1996) Plant regeneration from callus in a range of *Phaseolus acutifolius* A. Gray genotypes. *Plant Sci.*, **118**: 81–88.

Dillen W, Engler G, Montagu MV and Angenon G (1995) Electroporation-mediated DNA delivery to seedling tissues of *Phaseolus vulgaris* L. (common bean). *Plant Cell Rep.*, **15**: 119–124.

Dineshkumar V, Kirti P B, Sachan J K S and Chopra V L (1994) Plant regeneration via somatic embryogenesis in chickpea (*Cicer arietinum* L.). *Plant Cell Rep.*, **13**: 468–472.

Dineshkumar V, Kirti P B, Sachan J K S and Chopra V L (1995) Picloram induced somatic embryogenesis in chickpea (*Cicer arietinum* L.). *Plant Sci.*, **109**: 207–213.

Donn G (1978) Cell division and callus regeneration from leaf protoplasts of *Vicia narbonensis. Z. Pflanzenphysiol.*, **86**: 65–75.

Duke J A (1981) *Handbook of Legumes of World Economic Importance*, Plenum Press, New York.

Dundas I S (1985) Studies into overcoming some crossability barriers involving pigeonpea (*Cajanus cajan* (L.) Millsp.) and related *Atylosia* species. Post-Doctoral Internship Report, Legumes Program, ICRISAT, Patancheru, India.

Durham R E and Parrott W A (1992) Repetitive somatic embryogenesis from peanut cultures in liquid medium. *Plant Cell Rep.*, **11**: 122–125.

Eapen S and George L (1990) Ontogeny of somatic embryos of *Vigna aconitifolia, V. mungo* and *V. radiata. Ann. Bot.*, **66**: 219–226.

Eapen S and George L (1993a) Somatic embryogenesis in peanut: Influence of growth regulators and sugars. *Plant Cell. Tiss. Org. Cult.*, **35**: 151–156.

Eapen S and George L (1993b) Plant regeneration from leaf discs of peanut and pigeonpea – influence of benzyladenine, indoleacetic acid and indoleacetic acid amino acid conjugates. *Plant Cell. Tiss. Org. Cult.*, **35**: 223–227.

Eapen S and George L (1994a) Somatic embryogenesis in *Cicer arietinum* L.: Influence of genotype and auxins. *Biol. Plant.*, **36**: 343–349.

Eapen S and George L (1994b) *Agrobacterium tumefaciens* mediated gene transfer in peanut (*Arachis hypogaea* L.). *Plant Cell Rep.*, **13**: 582–586.

Eapen S, George L and Rao P S (1993) Plant regeneration through somatic embryogenesis in peanut (*Arachis hypogaea* L.). *Biol. Plant*, **35**: 499–504.

Eapen S, Kohler F, Gerdemann M and Schieder O (1987) Cultivar dependence of transformation rates in mothbean after co-cultivation of protoplasts with *Agrobacterium tumefaciens. Theort. Appl. Genet.*, **75**: 207–210.

Eapen S, Tivarekar S and George L (1998) Thidiazuron-induced shoot regeneration in pigeonpea (*Cajanus cajan* L.). *Plant Cell Tiss. Org. Cult.*, **53**: 217–220.

Egnin M, Mora A and Prakash C S (1998) Factors enhancing *Agrobacterium*-mediated gene transfer in peanut (*Arachis hypogaea* L.). In Vitro *Cell. Dev. Biol. Plant*, **34**: 310–318.

Erhardt M, Herzog E, Lauber E, Fritsch C, Guilley H, Jonard G, Richards K and Bouzoubaa S (1999) Transgenic plants expressing the TGB1 protein of peanut clump virus complement movement of TGB1-defective peanut clump virus but not of TGB1-defective beet necrotic yellow vein virus. *Plant Cell Rep.*, **18**: 614–619.

Falco S C, Guida T, Mauvais J, Sanders C, Ward R T and Webber P (1995) Transgenic canola and soybean seed with increased lysine. *Bio/Technology*, **13**: 577–582.

Feng Q L, Stalker H T and Pattee H E (1996) Plant recovery of selfs and interspecific hybrids of *Arachis* by *in vitro* culture of peg tips. *Crop Sci.*, **36**: 1660–1666.

FernandezCaso M, Pelaez M I and Ruiz M L (1996) Onset of *in vitro* morphogenic response and protein pattern changes in *Phaseolus vulgaris* L. *J. Plant Physiol.*, **149**: 757–761.

Finer J J (1988) Apical proliferation of embryogenic tissue of soybean (*Glycine max* (L.) Merrill). *Plant Cell Rep.*, **7**: 238–241.

Finer J J and McMullen M D (1991) Transformation of soybean via particle bombardment of embryogenic suspension culture tissue. In Vitro *Cell. Dev. Biol. Plant*, **27**: 175–182.

Finer J J and Nagasawa A (1988) Development of embryogenic suspension culture of soybean (*Glycine max* Merrill). *Plant Cell Tiss. Org. Cult.*, **15**: 125–136.

Fontana G S, Santini L, Caretto S, Frugis G and Mriotii D (1993) Genetic transformation in the grain legume *Cicer arietinum* L. (chickpea). *Plant Cell Rep.*, **12**: 194–198.

Franklin C I, Trieu T N, Cassidy B G, Dixon R A and Nelson R S (1993) Genetic transformation of green bean callus via *Agrobacterium* mediated DNA transfer. *Plant Cell Rep.*, **12**: 74–79.

Franklin C I, Trieu T N, Gonzales R A and Dixon R A (1991) Plant regeneration from seedling explants of green bean (*Phaseolus vulgaris* L.) via organogenesis. *Plant Cell Tiss. Org. Cult.*, **24**: 199–206.

Franklin G, Jeyachandran R, Melchias G and Ignacimuthu S (1998) Multiple shoot induction and regeneration of pigeonpea (*Cajanus cajan* (L.) Millsp.) cv. Vamban 1 from apical and axillary meristem. *Curr. Sci.*, **74**: 936–937.

Freytag A H, Rao-Arelli A P, Anand S C, Wrather J A and Owens L D (1989) Somaclonal variation in soybean plants regenerated from tissue culture. *Plant Cell Rep.*, **8**: 199–202.

Gamborg O L, Davis B P and Stahlhut R W (1983a) Somatic embryogenesis in cell culture in *Glycine* species. *Plant Cell Rep.*, **2**: 209–212.

Gamborg O L, Davis B P and Stahlhut R W (1983b) Cell division and differentiation in protoplasts from cell cultures of *Glycine* species and leaf tissues of soybean. *Plant Cell Rep.*, **2**: 213–215.

Gamborg O L, Miller R A and Ojima K (1968) Nutrient requirements of suspension cultures of soybean root cultures. *Exp. Cell Res.*, **50**: 151–158.

Gamborg O L, Shyluk J and Kartha K K (1975) Factors affecting the isolation and callus formation in protoplasts from the shoot apices of *Pisum sativum* L. *Plant Sci. Lett.*, **4**: 285–292.

Garcia J A, Hille J, Vos P and Goldbach R (1987) Transformation of cowpea *Vigna unguiculata* with a full-length DNA copy of cowpea mosaic virus m-RNA. *Plant Sci.*, **48**: 89–98.

Geetha N, Venkatachalam P and Rao G R (1997a) Plant regeneration and propagation of black gram (*Vigna mungo* (L.) Hepper) through tissue culture. *Trop. Agric.*, **74**: 73–76.

Geetha N, Venkatachalam P and Rao G R (1997b) *In vitro* plant regeneration from different seedling explants of black gram (*Vigna mungo* (L.) Hepper) via organogenesis. *Breed. Sci.*, **47**: 311–315.

Geetha N, Venkatachalam P, Prakash V and Laxmi Sita G (1998) High frequency induction of multiple shoots and plant regeneration from seedling explants of pigeonpea (*Cajanus cajan* L.). *Curr. Sci.*, **75**: 1036–1041.

Geetha N, Venkatachalam P and Sita G L (1999) *Agrobacterium*-mediated genetic transformation of pigeonpea (*Cajanus cajan* L.) and development of transgenic plants via direct organogenesis. *Plant Biotech.*, **16**: 213–218.

Gehlot H S, Upadhyaya A, Davis T D and Sankhla N (1989) Growth and organogenesis in mothbean as affected by paclobutrazol. *Plant Cell Physiol.*, **30**: 933–936.

Genga A and Allavena A (1991) Factors affecting morphogenesis from immature cotyledons of *Phaseolus coccineus* L. *Plant Cell Tiss. Org. Cult.*, **27**: 189–196.

Genga A, Ceriott A, Bollini R, Bernacchia G and Allavena A (1991) Transient gene expression in bean tissues by high velocity microprojectiles bombardment. *J. Genet. Breed.*, **45**: 129–134.

George L and Eapen S (1994) Organogenesis and embryogenesis from diverse explants in pigeonpea (*Cajanus cajan* L.). *Plant Cell Rep.*, **13**: 417–420.

Ghazi T D, Cheema H V and Nabors M W (1986) Somatic embryogenesis and plant regeneration from embryogenic callus of soybean, *Glycine max* L. *Plant Cell Rep.*, **5**: 452–456.

Gill R and Eapen S (1986) Plant regeneration from hypocotyl protoplasts of mothbean (*Vigna aconitifolia*). *Curr. Sci.*, **55**: 100–102.

Gill R and Ozias-Akins P (1999) Thidiazuron-induced highly morphogenic callus and high frequency regeneration of fertile peanut (*Arachis hypogaea* L.) plants. In Vitro *Cell. Dev. Biol. Plant.*, **35**: 445–450.

Gill R and Saxena P K (1992) Direct somaic embryogenesis and regeneration of plants from seedling explants of peanut (*Arachis hypogaea*): Promotive role of thidiazuron. *Can. J. Bot.*, **70**: 1186–1192.

Gill R, Eapen S and Rao P S (1986) Tissue culture studies in mothbean-factors influencing plant regeneration from seedling explants of different cultivars. *Proc. Indian Acad. Sci. (Plant Sci.).*, **96**: 55–61.

Giovinazo G, Greco V, Vitale A and Bollini R (1997) Bean (*Phaseolus vulgaris* L.) protoplasts as a model system to study the expression and stability of recombinant seed proteins. *Plant Cell Rep.*, **16**: 705–709.

Godbole D A, Kunachgi M N, Potdar U A, Krishnamurthy K V and Mascarenhas A F (1984) Studies on a drought resistant legume: The mothbean *Vigna aconitifolia* (Jacq) Marechal II. Morphogenetic studies. *Plant Cell Rep.*, **3**: 75–78.

Gosal S S and Bajaj Y P S (1979) Establishment of callus-tissue cultures, and the induction of organogenesis in some grain legumes. *Crop Improvement*, **6**: 154–160.

Gosal S S and Bajaj Y P S (1983) *In vitro* hybridization in an incompatible cross – Blackgram × Greengram. *Curr. Sci.*, **52**: 556–557.

Gram T, Matson O and Joersbo M (1996) Division frequency of pea protoplasts in relation to starch accumulation. *Plant Cell Tiss. Org. Cult.*, **45**: 179–183.

Grant J E (1984) Regeneration from cotyledonary tissue of *Glycine canescenes*, a perenial relative of the soybean. *Plant Cell Tiss. Org. Cult.*, **3**: 169–173.

Grant J E (1990) Soybean: Wide *hybridization* through embryo culture. In: *Biotechnology in Agriculture and Forestry, Vol. 10* (Ed Bajaj Y P S), Springer-Verlag, Berlin, pp. 134–148.

Grant J E, Cooper P A, Gilpin B J, Hoglund S J, Reader J K, PitherJoyce M D and Timmerman-Vaughan G M (1998) Kanamycin is effective for selecting transformed peas. *Plant Sci.*, **139**: 159–164.

Grant J E, Cooper P A, McAra A E and Frew T J (1995) Transformation of peas (*Pisum sativum* L.) using immature cotyledons. *Plant Cell Rep.*, **15**: 254–258.

Griga M (1998) Direct somatic embryogenesis from shoot apical meristems of pea, and thidiazuron-induced high conversion rate of somatic embryos. *Biol. Plant.*, **41**: 481–495.

Gulati A and Jaiwal P K (1990) Culture conditions effecting plant regeneration from cotyledons of *Vigna radiata* (L.) Wilczek. *Plant Cell Tiss. Org. Cult.*, **23**: 1–7.

Gulati A and Jaiwal P K (1992) *In vitro* induction of multiple shoots and plant regeneration from shoot tips of mungbean (*Vigna radiata* (L.) Wilczek). *Plant Cell Tiss. Org. Cult.*, **29**: 199–205.

Gulati A and Jaiwal P K (1994) Plant regeneration from cotyledonary node explants of mungbean (*Vigna radiata* (L.) Wilczek). *Plant Cell Rep.*, **13**: 523–527.

Gupta S D, Ahmed R and De D N (1997) Direct somatic embryogenesis and plantlet regeneration from seedling leaves of winged bean, *Psophocarpus tetragonolobus* (L.) DC. *Plant Cell Rep.*, **16**: 628–631.

Guru S K, Chandra R, Raj A, Khetrapal S and Polisetty R (1999) Evolution of ethylene and methane in relation to somatic embryogenesis in chickpea. *Biol. Plant.*, **42**: 149–154.

Gyulai G, Kiss E, Csillag A and Hesky L E (1993) Developmental analysis of primary and secondary somatic embryogenesis in soybean tissue culture. *Acta Biol. Hung.*, **44**: 189–196.

Hadi M Z, McMullen M D and Finer J J (1996) Transformation of 12 different plasmids into soybean via particle bombardment. *Plant Cell Rep.*, **15**: 500–505.

Hammatt N, Ghose T K and Davey M R (1986) Regeneration in legumes. In: *Cell Culture and Somatic Cell Genetics of Plants, Vol. 3* (Ed Vasil I K), Academic Press, New York, pp. 67–95.

Hammatt N, Jones B and Davey M R (1989) Plant regeneration from seedling explants and cotyledon protoplasts of *Glycine argyrea* Tind. In Vitro *Cell Dev. Biol. Plant*, **25**: 669–672.

Hammatt N, Kim H-I, Davey M R, Nelson R S and Cocking E C (1987a)) Plant regeneration from cotyledon protoplasts of *Glycine canescenes* and *G. clandestina. Plant Sci.*, **48**: 129–135.

Hammat N, Lister A, Jones B, Cocking E C and Davey M R (1992) Shoot formation from somatic hybrid callus between soybean and a perennial wild relative. *Plant Sci.*, **85**: 215–222

Hammatt N, Nelson R S and Davey M R (1987b) Plant regeneration from seedling explants of perennial *Glycine* species. *Plant Cell Tiss. Org. Cult.*, **11**: 3–11.

Hammatt N, Nelson R S and Davey M R (1987c) Plant regeneration from seedling cotyledons, petioles, and leaves of *Glycine clandestina. Plant Physiol.*, **68**: 125–128.

Hartweck L M, Lazzeri P A, Cui D, Collins G B and Williams E G (1988) Auxin-orientation effects on somatic embryogenesis from immature soybean cotyledons. In Vitro *Cell Dev. Biol. Plant*, **24**: 821–828.

Hazra S, Sathaye S S and Mascarenhas A F (1989) Direct somatic embryogenesis in peanut (*Arachis hypogaea*). *Bio/Technology*, **7**: 949–951.

Hepher A, Boulter M E, Harris N, Nelson R S (1988) Development of a superficial meristem during somatic embryogenesis from immature cotyledons of soybean (*Glycine max* L.). *Ann. Bot.*, **62**: 513–519.

Hinchee M A W, Connor-Ward D V, Newell C A, McDonnell R E, Sato S J, Gasser C S, Fischoff D A, Re D B, Fraley R T and Horsch R B (1988) Production of transgenic soybean plants using *Agrobacterium*-mediated gene transfer. *Bio/Technology*, **6**: 915–922.

Hita O, Lafarga C and Guerra H (1997) Somatic embryogensis from chickpea (*C. arietinum* L.) immature cotyledons: The effect of zeatin, gibberellic acid and indole-3-butyric acid. *Acta Physiol. Planta.*, **19**: 333–338.

Hobbs S L A, Jackson J A and Mahon J D (1989) Specificity of strain and genotype in the susceptibility of pea to *Agrobacterium tumefaciens. Plant Cell Rep.*, **8**: 274–277.

Hu C Y and Wang L Z (1999) *In planta* soybean transformation technologies developed in China: Procedure, confirmation and field performance. In Vitro *Cell. Dev. Biol. Plant*, **35**: 417–420.

Hu C Y and Zanettini M H B (1995) Embryo culture and embryo rescue for wide cross hybrids In: *Plant Cell, Tissue and Organ Culture, Fundamental Methods* (Eds Gamborg O L and Phillips G C), Springer-Verlag, Berlin, pp. 129–141.

Hu T K (1997) *In vitro* culture of *Arachis hypogaea* peg tips as influenced by peg age, tip length and media. *Guoli Zhongxing Daxue Nonglin Xuebao*, **46**: 1–12.

Hulse J H (1989) Uses of tropical grain legumes. In: *Proceedings of a Consultants' Meeting*, 27–30 Mar 1989, ICRISAT Center, Patancheru, India.

Hussey G and Gunn H V (1984) Plant production in pea (*Pisum sativum* L. cvs Puget and Upton) from long-term callus with superficial meristems. *Plant Sci. Lett.*, **37**: 143–148.

Hymowitz T, Chalmers N L, Costanza S H and Saam M M (1986) Plant regeneration from leaf explants *Glycine clandestina* Wendl. *Plant Cell Rep.*, **3**: 192–194.

Iantcheva A, Vlahova M, Bakalova E, Kondorosi E, Elliott M C and Atanassov A (1999) Regeneration of diploid annual medics via direct somatic embryogenesis promoted by TDZ and BAP. *Plant Cell Rep.*, **18**: 904–910.

Ignacimuthu S and Franklin G (1998) Regeneration of plantlets from cotyledon and embryonal axis explants of *Vigna mungo* (L.) Hepper. *Plant Cell Tiss. Org. Cult.*, **55**: 75–78.

Ilahi I, Sami R and Jabeen M (1995) Plantlet regeneration from mature embryos of peanut (*Arachis hypogaea* L.). *Pakistan J. Bot.*, **27**: 405–409.

Illingworth J E (1968) Peanut plants from single deembryonated cotyledons. *Hort. Sci.*, **3**: 238–276.

Islam R (1994) Somatic embryogenesis from immature cotyledons of chickpea (*Cicer arietinum* L.). *Pakistan J. Bot.*, **26**: 197–199.

Islam R, Malik T, Hussain T and Riazuddin S (1994) Strain and cultivar specificity in the *Agrobacterium*-chickpea interaction. *Plant Cell Rep.*, **13**: 561–563.

Jackson J A and Hobbs S L A (1990) Rapid multiple shoot production from cotyledonary node explants of pea (*Pisum sativum* L.). In Vitro *Cell. Dev. Biol. Plant*, **26**: 835–838.

Jacobsen H J and Kysley W (1984) Induction of somatic embryos in pea, *Pisum sativum* L. *Plant Cell Tiss. Org. Cult.*, **3**: 319–324.

Jain J and Chopra V L (1988) Genotypic differences in response to regeneration of *in vitro* cultures of mothbean *Vigna aconitifolia* (Jacq.) Marechal. *Indian J. Exp. Biol.*, **26**: 654–656.

Jaiwal P K, Kumari R, Ignacimuthu S, Potrykus I and Sautter C (2001) *Agrobacterium*-mediated transformation of mungbean (*Vigna radiata* (L.) Wilczek), a recalcitrant grain legume. *Plant Sci.*, **161**: 239–247.

Jaiwal P K, Sautter C and Potrykus I (1998) *Agrobacterium rhizogenes*-mediated gene transfer in mungbean (*Vigna radiata* (L.) Wilczek). *Curr. Sci.*, **75**: 41–45.

Jenes B, Moore H, Cao J, Zhang W and Wu R (1993) Techniques for gene transfer. In: *Transgenic Plants, Vol. 1* (Eds Kung S D and Wu R), Academic Press, Inc, New York, pp. 125–146.

Jia S R (1982) Factors affecting the division frequency of pea mesophyll protoplasts. *Can. J. Bot.*, **60**: 2192–2196.

Joersbo M, Brunstedt J, Marcussen J and Okkels F T (1999) Transformation of the endospermous legume guar (*Cyamopsis tetragonoloba* L.) and analysis of transgene transmission. *Mol. Breed.*, **5**: 521–529.

Jordan M C and Hobbs L A (1993) Evaluation of a cotyledonary node regeneration system for *Agrobacterium*-mediated transformation of pea (*Pisum sativum* L.). In Vitro *Cell. Dev. Biol. Plant*, **29**: 77–82.

Joshi S, Upadhyaya A, Davis T D, Sankhla A K, Sankhla D and Sankhla N (1988) Interaction between abscisic acid and GA_3 on *in vitro* regeneration from mothbean callus. *Proc. Plant Growth Regul. Soc. Am.*, **15**: 35–38.

Kachonpadungkitti Y, Hisajima S and Arai Y (1992) Life cycle of peanut (*Arachis hypogaea* L.) plant *in vitro*. *Biosci. Biotech. Biochem.*, **56**: 543–546.

Kahl G, Kaemmer D, Weising K, Kost S, Weigand F and Saxena M C (1994) The potential of gene technology and genome analysis for cool season food legume crops – Theory and Practice. *Euphytica*, **73**: 177–189.

Kameya T and Widholm J (1981) Plant regeneration from hypocotyl sections *Glycine* species. *Plant Sci. Lett.*, **21**: 289–294.

Kaneda Y, Tabei Y, Nishimura S, Harada K, Akihama T and Kitamura K (1997) Combination of thidiazuron and basal media with low salt concentrations increases frequency of shoot organogenesis in soybeans (*Glycine max* (L.) Merrill). *Plant Cell Rep.*, **17**: 8–12.

Kanyand M, Desai A P and Prakash C S (1994) Thidiazuron promotes high frequency regeneration of peanut (*Arachis hypogaea*) plants *in vitro*. *Plant Cell Rep.*, **14**: 1–5.

Kanyand M, Peterson C M and Prakash C S (1997) The differentiation of emergences into adventitious shoots in peanut. *Plant Sci.*, **126**: 87–95.

Kao K N, Keller W A and Miller R A (1970) Isolation of protoplasts from cultured soybean root cells. *Exp. Cell Res.*, **62**: 338–340.

Kar S, Johnson T M, Nayak P and Sen S K (1996) Efficient plant regeneration through *Agrobacterium*-mediated transformation of chickpea (*Cicer arietinum* L.). *Plant Cell Rep.*, **16**: 32–37.

Kartha K K, Pahl K, Leung N L and Mroginski L A (1981) Plant regeneration from meristems of grain legumes: Soybean, cowpea, peanut, chickpea, and bean. *Can. J. Bot.*, **59**: 1671–1679.

Karthikeyan A S, Sarma K S and Veluthambi K (1996) *Agrobacterium tumefaciens*-mediated transformation of *Vigna mungo* (L.) Hepper. *Plant Cell Rep.*, **15**: 328–331.

Kerns H R, Barwale U B, Meyer Jr M M and Wildholm J M (1986) Correlation of cotyledonary node shoot proliferation and somatic embryo development in suspension cultures of soybean (*Glycine max*. Merrill). *Plant Cell Rep.*, **5**: 140–143.

Khalafalla M M and Hattori K (1999) A combination of thidiazuron and benzyladeninl promotes multiple shoot production from cotyledonary node explants of faba bean (*Vicia faba* L.). *Plant Growth Reg.*, **27**: 145–148.

Khan S K and Ghosh P D (1983) *In vitro* induction of androgenesis and organogenesis in *Cicer arietinum* L. *Curr. Sci.*, **52**: 891–893.

Kim J W and Minamikawa T (1996) Transformation and regeneration of French bean plants by the particle bombardment process. *Plant Sci.*, **117**: 131–138.

Kim J W and Minamikawa T (1997) Stable delivery of a canavalin promoter-beta-glucuronidase gene fusion into French bean by particle bombardment. *Plant Cell Physiol.*, **38**: 70–75.

Kim T, Chowdhury M K U and Wetzstein H Y (1999) A quantitative and histological comparison of *gus* expression with different promoter constructs used in microprojectile bombardment of peanut leaf tissue. In Vitro *Cell. Dev. Biol. Plant*, **35**: 51–56.

Kohler F, Golz C, Eapen S, Kohn H and Schieder O (1987) Influence of plant cultivar and plasmid-DNA on transformation rates in tobacco and mothbean. *Plant Sci.*, **53**: 87–91.

Komatsuda T and Ohyama K (1988) Genotypes of high competence for somatic embryogenesis and plant regeneration in *Glycine max*. *Theort. Appl. Genet.*, **75**: 695–700.

Komatsuda T, Lee W and Oka S (1992) Maturation and germination of somatic embryos as affected by sucrose and plant growth regulators in soybeans *Glycine gracilis* Skvortz and *G. max* (L.) Merrill. *Plant Cell Tiss. Org. Cult.*, **28**: 103–113.

Krishnamurthy K V, Godbole D A and Mascarenhas A F (1984) Studies on drought resistant legume: the mothbean, *Vigna aconitifolia* (Jacq.) Marechal. I. Protoplast culture and organogenesis. *Plant Cell Rep.*, **3**: 30–32.

Krishnamurthy K V, Godbole D A and Mascarenhas A F (1986) Scanning electron microscope analysis of organogenetic and non-organogenetic callus tissue of the mothbean, *Vigna aconitifolia* (Jacq.) Marechal: A comparative study. *Indian J. Exp. Biol.*, **24**: 408–413.

Krishnamurthy K V, Suhasini K, Sagare A P, Meixner M, de Kathen A, Pickardt T and Schieder O (2000) *Agrobacterium* mediated transformation of chickpea (*Cicer arietinum* L.) embryo axes. *Plant Cell Rep.*, **19**: 235–240.

Kriz A L and Larkins B A (1991) Biotechnology of seed crops: genetic engineering of seed storage proteins. *Hort. Sci.*, **26**: 1036–1041.

Kulkarni D D and Krishnamurthy K V (1989) Isolation and culture of protoplasts of pigeonpea, *Cajanus cajan* (L.) Millsp. *Indian J. Exp. Biol.*, **27**: 939–942.

Kulothungan S, Ganapathi A, Shajahan A and Kathiravan K (1995) Somatic embryogenesis in cell suspension culture of cowpea (*Vigna unguiculata* (L.) Walp). *Israel J. Plant Sci.*, **43**: 385–390.

Kumar A S, Gamborg O L and Nabors M W (1988a) Plant regeneration from cell suspension cultures of *Vigna aconitifolia*. *Plant Cell Rep.*, **7**: 138 141.

Kumar A S, Gamborg O L and Nabors M W (1988b) Regeneration from long-term cell suspension cultures of tepary bean (*Phaseolus acutifolius*). *Plant Cell Rep.*, **7**: 322–325.

Kumar A S, Reddy T P and Reddy G M (1983) Plantlet regeneration from different callus cultures of pigeonpea (*Cajanus cajan* L.). *Plant Sci. Lett.*, **32**: 271–278.

Kumar A S, Reddy T P and Reddy G M (1984) Multiple shoots from cultured explants of pigeonpea and *Atylosia* species. *SABRAO J.*, **16**: 101–105.

Kumar P A, Bisaria S, Pai R A and Sharma R P (1995) Comparative shoot regeneration in different genotypes of chickpea, *Cicer arietinum* L. *Indian J. Exp. Biol.*, **33**: 77–78.

Kumar P S (1985) Crossability genome relationships and inheritance studies in intergeneric hybrids of pigeonpea. Ph.D. thesis, University of Hyderabad, India.

Kysely W and Jacobsen H J (1990) Somatic embryogenesis from pea embryos and shoot apices. *Plant Cell. Tiss. Org. Cult.*, **20**: 7–14.

Kysely W, Myers J R, Lazzeri P A, Collins G B and Jacobsen H J (1987) Plant regeneration via somatic embryogenesis in pea (*Pisum sativum* L.). *Plant Cell Rep.*, **6**: 305–308.

Lacorte C, Aragão F J L, Almeida E R, Mansur E and Rech E L (1997) Transient expression of *Gus* and the 2S albumin gene from Brazil nut in peanut (*Arachis hypogaea* L.) seed explants using particle bombardment. *Plant Cell Rep.*, **16**: 619–623.

Lacorte C, Mansur E, Timmerman B and Cordeiro AR (1991) Gene transfer into peanut (*Arachis hypogaea* L.) by *Agrabacterium tumefaciens. Plant Cell Rep.*, **10**: 354–357.

Larkin P J and Scowcroft W R (1981) Somaclonal variation – a novel source of variability from cell culture for plant improvement. *Theort. Appl. Genet.*, **60**: 197–204.

Lazzeri P A, Hildebrand D F and Collins G B (1985) A procedure for plant regeneration from immature cotyledon tissue of soybean. *Plant Mol. Biol. Rep.*, **3**: 160–167.

Lazzeri P A, Hildebrand D F and Collins G B (1987a) Soybean somatic embryogenesis: Effects of hormones and culture manipulations. *Plant Cell Tiss. Org. Cult.*, **10**: 197–208.

Lazzeri P A, Hildebrand D F and Collins G B A (1987b) Soybean somatic embryogenesis: Effects of hormones and culture manipulations. *Plant Cell Tiss. Org. Cult.*, **10**: 209–220.

Lehminger-Mertens R and Jacobsen H-J (1989) Plant regeneration from pea protoplasts via somatic embryogenesis. *Plant Cell Rep.*, **8**: 379–382.

Lewis M E and Bliss F A (1994) Tumour formation and beta-glucuronidase expression in *Phaseolus vulgaris* inoculated with *Agrobacterium tumefaciens. J. Am. Soc. Hort. Sci.*, **119**: 361–366.

Li B J, Langridge W H R and Szalay A A (1985) Somatic embryogenesis and plantlet regeneration in the soybean, *Glycine max. Plant Cell Rep.*, **4**: 344–347.

Li J and Grabau A (1996) Comparison of somatic embryogenesis and embryo conversion in commercial soybean cultivars. *Plant Cell Tiss. Org. Cult.*, **44**: 87–89.

Li X B, Xu Z H and Wei Z M (1995a) Plant regeneration from protoplasts of immature *Vigna sinensis* cotyledons via somatic embryogenesis. *Plant Cell Rep.*, **15**: 282–286.

Li Z, Jarret, R L and Demski J W (1995b) Regeneration of plants from protoplasts of *Arachis* species. In: *Biotechnology in Agriculture and Forestry, Vol. 34* (Ed Bajaj Y P S), Springer-Verlag, Berlin, pp. 3–13.

Li Z, Robert L, Jarret L, Pittman R N and Demski J W (1994) Shoot organogenesis from cultured seed explants of peanut (*Arachis hypogaea* L.) using thidiazuron. In Vitro *Cell. Dev. Biol. Plant*, **30**: 187–191.

Lin W (1983) Isolation of mesophyll protoplasts from mature leaves of soybean. *Plant Physiol.*, **73**: 1067–1069.

Lippmann B and Lippmann G (1984) Induction of somatic embryos in cotyledonary tissue of soybean, *Glycine max* L. Merrill *Plant Cell Rep.*, **3**: 215–218.

Lippmann B and Lippmann G (1993) Soybean embryo culture: factors influencing plant recovery from isolated embryos. *Plant Cell Tiss. Org. Cult.*, **32**: 83–90.

Liu W, Moore P J and Collins G B (1992) Somatic embryogenesis in soybean via somatic embryo cycling. In Vitro *Cell. Dev. Biol. Plant*, **28**: 153–160.

Liu W N, Torisky R S, McAllister K P, Avdiushko S, Hildebrandt D and Collins G B (1996) Somatic embryo cycling: Evaluation of a novel transformation and assay system for seed-specific gene expression in soybean. *Plant Cell Tiss. Org. Cult.*, **47**: 33–42.

Livingstone D M and Birch R G (1999) Efficient transformation and regeneration of diverse cultivars of peanut (*Arachis hypogaea* L.) by particle bombardment into embryogenic callus produced from mature seeds. *Mol. Breed.*, **5**: 43–51.

Loiseau J, Marche C and Le Deunff Y (1995) Effects of auxins, cytokinins, carbohydrates and aminoacids on somatic embryogenesis induction from shoot apices of pea. *Plant Cell Tiss. Org. Cult.*, **41**: 267–275.

Loiseau J, Marche C and Le Deunff Y (1996) Variability of somatic embryogenic ability in the genus *Pisum* L.: Effects of genotype, explant source and culture medium. A*gronomie*, **16**: 299–308.

Loiseau J, Michaux-Ferriere N and Le Deunff Y (1998) Histology of somatic embryogenesis in pea. *Plant Physiol. Biochem.*, **36**: 683–687.

Ma H, McMullen M D and Finer J J (1994) Identification of a homeobox-containing gene with enhanced expression during soybean (*Glycine max* L.) somatic embryo development. *Plant Mol. Biol.*, **24**: 465–473.

Madsen M H, Nauerby B, Frederiksen C G and Wyndaele R (1998) Regeneration of pea (*Pisum sativum* L.) by the thin cell layer nodal system: Influence of explant culture media on rooting and plantlet formation. *Acta. Agric. Scand.*, **48**: 58–64.

Malik K A and Saxena P K (1992a) Regeneration in *Phaseolus vulgaris* L: High frequency induction of direct shoot formation in intact seedlings by N^6-benzylaminopurine and thidiazuron. *Planta*, **186**: 384–389.

Malik K A and Saxena P K (1992b) Somatic embryogenesis and shoot regeneration from intact seedlings of *Phaseolus acutifolius* A., *P. aureus* (L.) Wilczek, *P. coccineus* L., and *P. wrightii* L. *Plant Cell Rep.*, **11**: 163–168.

Malik K A and Saxena P K (1992c) Thidiazuron induces high-frequency shoot regeneration in intact seedlings of pea (*Pisum sativum*), chickpea (*Cicer arietinum*) and lentil (*Lens culinaris*). *Aust. J. Plant Physiol.*, **19**: 731–740.

Malik K A, Alikhan S T and Saxena P K (1993) High-frequency organogenesis from direct seed culture in *Lathyrus*. *Ann. Bot.*, **72**: 629–637.

Malmberg R L (1979) Regeneration of whole plants from callus cultures of diverse genetic lines of *Pisum sativum* L. *Planta*, **146**: 243–244.

Mansur E, Lacorte C, de Freitas V G, de Oliveira D E, Timmerman B and Cordeiro A R (1993) Regulation of transformation efficiency of peanut (*Arachis hypogaea* L.) explants by *Agrobacterium tumefaciens*. *Plant Sci.*, **89**: 93–99.

Marioti D, Fontana G S and Santini L (1989) Genetic transformation of grain legumes: *Phaseolus vulgaris* L. and *P. coccineus* L. *J. Genet. Breed.*, **43**: 77–82.

Martin I S and Sondahl M R (1984) Early stages of somatic embryo differentiation from callus cells of bean (*Phaseolus vulgaris* L.) growth in liquid medium. *J. Plant Physiol.*, **117**: 97–103.

Martin J P and Rabechault H (1976) The culture *in vitro* of groundnut stamens (*Arachis hypogaea* L.). – II. Establishment of tissue cultures and organogenesis (in Fr., summary in Eng.). *Oleagineux*, **31**: 19–25.

Maughan P J, Philip R, Cho M J, Widholm J M and Vodkin L O (1999) Biolistic transformation, expression, and inheritance of bovine beta-casein in soybean (*Glycine max*). In Vitro *Cell. Dev. Biol. Plant*, **35**: 344–349.

McClean P and Grafton K F (1989) Regeneration of dry bean (*Phaseolus vulgaris* L.) via organogenesis. *Plant Sci.*, **60**: 117–122.

McKently A H (1991) Direct somatic embryogenesis from axes of mature peanut embryos. In Vitro *Cell. Dev. Biol. Plant*, **27**: 197–200.

McKently A H (1995) Effect of genotype on somatic embryogenesis of axes of mature peanut embryos. *Plant Cell Tiss. Org. Cult.*, **42**: 251–254.

McKently A H, Moore G A and Gardener F P (1990) *In vitro* plant regeneration of peanut. *Crop Sci.*, **30**: 192–196.

McKently A H, Moore G A and Gardener F P (1991) Regeneration of peanut and perennial peanut from cultured leaf tissue. *Crop Sci.*, **31**: 833–837.

McKently A H, Moore G A, Doostar H and Niedz R P (1995) *Agrobacterium* mediated transformation of peanut (*Arachis hypogaea* L.) embryo axes and the development of transgenic plants. *Plant Cell Rep.*, **14**: 699–703.

Mehta U and Mohan Ram H Y (1980) Regeneration of plantlet from cotyledons of *Cajanus cajan* L. *Indian J. Exp. Biol.*, **18**: 800–802.

Mhaske V B and Hazra S (1994) Appearance of storage lipid (triglycerides) in somatic embryos of peanut (*Arachis hypogaea* L.). In Vitro *Cell. Dev. Biol. Plant*, **30**: 113–116.

Mhatre M, Bapat V A and Rao P S (1985) Micropropagation and protoplast culture in peanut (*Arachis hypogaea* L.). *Curr. Sci.*, **54**: 1052–1056.

Miyake H N and Taniguchi T (1994) An efficient method for callus formation of protoplasts from peanut (*Arachis hypogaea* L.) leaf mesophyll. *Jpn. J. Crop Sci.*, **63**: 362–367.

Mohan M L and Krishnamurthy K V (1998) Plant regeneration in pigeonpea (*Cajanus cajan* (L.) Millsp.) by organogenesis. *Plant Cell Rep.*, **17**: 705–710.

Mohan M L and Krishnamurthy K V (2001) Somatic embryogenesis and Plant regeneration in pigeonpea (*Cajanus cajan* (L.) Millsp.). *Biol. Plant.* (in press).

Mohan M L, Naidu R B, Kulkarni D D and Krishnamurthy K V (1997) Regeneration of plantlets in pigeonpea (*Cajanus cajan* (L.) Millsp.) by organogenesis. In: *Recent Advances in Biotechnological Applications of Plant Tissue and Cell Culture* (Eds Ravishankar G A and Venkataraman L A), Oxford and IBH Publishing Co. Pvt Ltd, New Delhi, pp. 151–155.

Mohan Ram H Y, Mehta U and Ramanuja Rao I V (1981) Tissue and protoplast culture and plant regeneration in legumes. In: *Proc. COSTED Symposium on Tissue Culture of Economically Important Plants* (Ed Rao A N), Singapore, pp. 66–69.

Mohan Ram H Y, Mehta U and Ramanuja Rao I V (1982) Haploid induction in legumes. In: *Plant Tissue Culture 1982* (Ed Fujiwara A). Japan Association of Plant Tissue Culture, Tokyo, pp. 541–542.

Mohapatra S T and Sharma R P (1991) *Agrobacterium* mediated genetic transformation of chickpea, *Cicer arietinum* L. *Plant Cell Tiss. Org. Cult.*, **28**: 45–51.

Monnier M (1990) Culture of zygotic embryos of higher plants. In: *Methods in Molecular Biology, Vol. 6, Plant Cell and Tissue Culture* (Eds Pollard J W and Walker J M), The Humana Press Clifton, New Jersey, pp. 129–139.

Morris J B, Dunn S, Pinnow D L, Hopkins M S and Pittman R N (1997) Meristem culture for virus elimination and peanut interspecific hybrid preservation. *Crop Sci.*, **37**: 591–594.

Mroginski L A and Fernandez A (1980) Obtained plantlets by *in vitro* culture of anthers of wild species of *Arachis* (Leguminosae) (in Sp., summary in Eng.). *Oleagineux*, **35**: 89–92.

Mroginski L A and Kartha K K (1981) Regeneration of pea (*Pisum sativum* L. cv. Century) plants by *in vitro* culture of immature leaflets. *Plant Cell Rep.*, **1**: 64–66.

Mroginski L A, Kartha K K and Shyluk J P (1981) Regeneration of peanut (*Arachis hypogaea*) plantlets by *in vitro* culture of immature leaves. *Can. J. Bot.*, **59**: 826–830.

Muehlbauer F J (1993) Use of wild species as a source of resistance in cool-season food legume crops. In: *Breeding for Stress Tolerance in Cool-season Food Legumes* (Eds Singh K B and Saxena M C), ICARDA, Wiley-Sayce Co Publications, pp. 359–372.

Mulin M and BellioSpataru A (2000) Organogenesis from hypocotyl thin cell layers of *Lupinus albus. Plant Growth Reg.*, **30**: 177–183.

Murch S J, Victor J M R, Krishnaraj S and Saxena P K (1999) The role of proline in thidiazuron-induced somatic embryogenesis of peanut. In Vitro *Cell. Dev. Biol. Plant*, **35**: 102–105.

Murthy B N S, Murch S J and Saxena P K (1995) Thidiazuron induced somatic embryogenesis in intact seedlings of peanut (*Arachis hypogaea*): Endogenous growth regulator levels and significance of cotyledons. *Physiol. Plant.*, **94**: 268–276.

Murthy B N S, Victor J, Singh R P, Fletcher R A and Saxena P K (1996) *In vitro* regeneration of chickpea (*Cicer arietinum* L.): Stimulation of direct organogenesis and somatic embryogenesis by thidiazuron. *Plant Growth Reg.*, **19**: 233–240.

Muthukumar B, Mariamma M, Veluthambi K and Gnanam A (1996) Genetic transformation of cotyledon explants of cowpea (*Vigna unguiculata* L. Walp) using *Agrobacterium tumefaciens. Plant Cell Rep.*, **15**: 980–985.

Myers J R, Lazzeri P A and Collins G B (1989) Plant regeneration of wild *Glycine* species from suspension culture-derived protoplasts. *Plant Cell Rep.*, **8**: 112–115.

Naidu R B, Kulkarni D D and Krishnamurthy K V (1995) Genotype dependent morphogenetic potentiality of various explants of a food legume, the pigeon pea (*Cajanus cajan* L.). In Vitro *Cell. Dev. Biol. Plant*, **31**: 26–30.

Nalini Mallikarjuna (1999) Ovule and embryo culture to obtain hybrids from interspecific incompatible pollinations in chickpea. *Euphytica*, **110**: 1–6.

Nalini Mallikarjuna and Moss J P (1995) Production of hybrids between *Cajanus platycarpus* and *Cajanus cajan. Euphytica*, **83**: 43–46.

Nalini Mallikarjuna, Reena M J T, Sastri D C and Moss J P (1996) Somatic embryogenesis in pigeonpea (*Cajanus cajan* L.). *Indian J. Exp. Biol.*, **34**: 282–284.

Narasimhulu S B and Reddy G M (1983) Plant regeneration from different callus cultures of *Arachis hypogaea. Plant Sci. Lett.*, **31**: 157–163.

Narasimhulu S B and Reddy G M (1984) *In vitro* flowering and pod formation from cotyledons of groundnut (*Arachis hypogaea* L.). *Theort. Appl. Genet.*, **69**: 87–91.

Narciso J O, Futsuhara Y, Hattori K and Wada T (1997) Effect of seed preconditioning and culture under different light intensities on callus induction and regeneration in excised cotyledon cultures of mungbean (*Vigna radiata* (L.) Wliczek). *Jpn. J. Crop Sci.*, **66**: 67–75.

Narciso J O, Hattori K and Wada T (1996) Histological observation of callus formation in mungbean (*Vigna vulgaris* L. cv. Goldstar). *Jpn. J. Crop Sci.*, **65**: 663–671.

Nataraja K and Patil S (1987) In: *Plant Cell and Tissue Culture of Economically Important Plants* (Ed G M Reddy), Osmania University, Hyderabad, 214–215.

Nauerby B, Madsen M, Chrystiansen J and Wyndaele R (1991) A rapid and efficient regeneration system for pea (*Pisum sativum*), suitable for transformation. *Plant Cell Rep.*, **9**: 676–679.

Nene Y L, Moss J P and Gowda C L L (1989) Problems and prospects for grain legume breeding. In: *The Role of Tissue Culture and Novel Genetic Technologies in Crop Improvement* (Eds Ketchum J L F and Gamborg O L), Proc. of the Third Conference of the International Plant Biotechnology Network (IPBNET), Nairobi, Kenya, TCCP, Colorado, 18–24.

Newell C A and Luu H T (1985) Protoplast culture and plant regeneration in *Glycine max. Plant Cell Tiss. Org. Cult.*, **4**: 145–149.

Nicolaisen M and Poulsen G B (1993) Optimization of polyethylene glycol mediated transient gene expression in pea protoplasts. *Plant Cell Tiss. Org. Cult.*, **35**: 93–97.

Nicoll S M, Brigham L A, Wen F and Hawes M C (1995) Expression of transferred genes during hairy root development in pea. *Plant Cell Tiss. Org. Cult.*, **42**: 57–66.

Norton F, Bliss F A and Bressani R (1985) Biochemical and nutritional attributes of grain legumes. In: *Grain Legume Crops* (Eds Summerfield R J and Roberts C H), Kluwer Academic Pub. Dordrecht, The Netherlands, pp. 73–114.

Oelck M M and Schieder O (1983) Genotype differences in some legume species affecting the redifferentiation ability from callus to plants. *Z. Pflanzenzuchtg.*, **91**: 312–321.

Oelck M M, Bapat V A and Schieder O (1982) Protoplast culture of three legumes: *Arachis hypogaea, Melilotus officinalis, Trifolium resupinatum. Z. Pflanzenphysiol.*, **106**: 173–177.

Oswald T H, Smith A E and Phillips D V (1977) Callus and plantlet regeneration from cell cultures of ladino clover and soybean. *Physiol. Plant.*, **39**: 129–134.

Owens L D and Cress D E (1985) Genotypic variability of soybean response to *Agrobacterium* strains harboring the Ti or Ri plasmids. *Plant Physiol.*, **77**: 87–94.

Owens L D and Smigocki C (1988) Transformation of soybean cells using mixed strains of *Agrobacterium tumefaciens* and phenolic compounds. *Plant Physiol.*, **88**: 570–573.

Ozias-Akins P (1989) Plant regeneration from immature embryos of peanut. *Plant Cell Rep.*, **8**: 217–218.

Ozias-Akins P, Anderson W F and Holbrook C C (1992a) Somatic embryogenesis in *Arachis hypogaea* L.: genotype comparison. *Plant Sci.*, **83**: 103–111.

Ozias-Akins P, Schnall J A, Anderson W F, Singsit C, Clemente T E, Adang M J and Weissinger A K (1993) Regeneration of transgenic peanut plants from stably transformed embryogenic callus. *Plant Sci.*, **93**: 185–194.

Ozias-Akins P, Singsit C and Branch W D (1992b) Interspecific hybrid inviability in crosses of *Arachis hypogaea* × *A. stenosperma* can be overcome by *in vitro* embryo maturation or somatic embryogenesis. *J. Plant Physiol.*, **140**: 207–212.

Pal M, Ghosh U, Chandra M, Pal A and Biswas B B (1991) Transformation and regeneration of mung bean (*Vigna radiata*). *Indian J. Biochem. Biophys.*, **28**: 449–455.

Pandey P and Bansal Y K (1989) Plantlet formation and callus cultures of cowpea (*Vigna sinensis* L.). *Curr. Sci.*, **58**: 394–396.

Parrott W A, Bailey M A, Durham R E and Mathews H V (1992) Tissue culture and regeneration in legumes. In: *Biotechnology and Crop Improvement in Asia* (Ed Moss J P), ICRISAT, Patancheru, India, pp. 115–148.

Parrott W A, Hoffman L M, Hildebrandt D F, Williams E G and Collins G B (1989) Recovery of primary transformants of soybean. *Plant Cell Rep.*, **7**: 615–617.

Patel D B, Barve D M, Nagar N and Mehta A R (1992) Regeneration of pigeonpea, *Cajanus cajan* through somatic embryogenesis. *Indian J. Exp. Biol.*, **30**: 871–873.

Patel D B, Barve D M, Nagar N and Mehta A R (1994) Regeneration of pigeonpea *Cajanus cajan*, through somatic embryogenesis. *Indian J. Exp. Biol.*, **32**: 740–744.

Pellegrineschi A (1997) *In vitro* plant regeneration via organogenesis of cowpea (*Vigna unguiculata* (L.) Walp). *Plant Cell Rep.*, **17**: 89–95.

Penza R, Lurquin P F and Filippone E (1991) Gene transfer by co-cultivation of mature embryos with *Agrobacterium tumefaciens*: Application to cowpea (*Vigna unguiculata* (L.) Walp). *J. Plant Physiol.*, **138**: 39–43.

Pestana M C, Lacorte C, deFreitas V G, deOliveira D E and Mansur E (1999) *In vitro* regeneration of peanut (*Arachis hypogaea* L.) through organogenesis: Effect of culture temperature and silver nitrate. In Vitro *Cell. Dev. Biol. Plant*, **35**: 214–216.

Pickardt T, Huancaruna Perales E and Schieder O (1989) Plant regeneration via somatic embryogenesis in *Vicia narbonensis*. *Protoplasma*, **149**: 5–10.

Pickardt T, Meixner M, Schade V and Schieder O (1991) Transformation of *Vicia narbonensis* via *Agrobacterium*-mediated gene transfer. *Plant Cell Rep.*, **9**: 535–538.

Pickardt T, Saalbach I, Waddell D, Meixner M, Müntz K and Schieder O (1995) Seed specific expression of the 2S albumin gene from Brazil nut (*Bertholletia excelsa*) in transgenic *Vicia narbonensis*. *Mol. Breed.*, **1**: 295–301.

Pigeaire A, Abernethy D, Smith P M, Simpson K, Fletcher N, Lu C Y, Adkins C A and Cornish E (1997) Transformation of a grain legume (*Lupinus angustifolius* L.) via *Agrobacterium tumefaciens*-mediated gene transfer to shoot apices. *Mol. Breed.*, **3**: 341–349.

Pittman R N, Banks D J, Kirby J S, Mitchell E D and Richardson P E (1983) *In vitro* culture of immature peanut (*Arachis* spp.) leaves: Morphogenesis and plantlet regeneration. *Peanut Sci.*, **10**: 21–25.

Polanco M C and Ruiz M L (1997) Effect of benzylaminopurine on *in vitro* and *in vivo* root development in lentil, *Lens culinaris* Medik. *Plant Cell Rep.*, **17**: 22–26.

Polanco M C, Pelaez M I and Riuz M L (1988) Factors affecting callus and shoot formation from *in vitro* cultures of *Lens culinaris* Medik. *Plant Cell Tiss. Org. Cult.*, **15**: 175–182.

Polisetty R, Paul V, Deveshwar J J, Khetrapal, Suresh K and Chandra R (1997) Multiple shoot induction by benzyladenine and complete plant regeneration from seed explants of chickpea (*Cicer arietinum* L.). *Plant Cell Rep.*, **16**: 565–571.

Ponappa T, Brzozowski A E and Finer J J (1999) Transient expresion and stable transformation of soybean using the jellyfish green fluorescent protein. *Plant Cell Rep.*, **19**: 6–12.

Ponsamuel J, Huhman D V, Cassidy B G and Post-Beittenmiller D (1998) *In vitro* regeneration via caulogenesis and brassin-induced shoot conversion of dormant buds from plumular explants of peanut (*Arachis hypogaea* L. cv. Okran). *Plant Cell Rep.*, **17**: 373–378.

Popiers D, Flandre F and Sangwan-Norreel (1997) Intensification of pea regeneration (*Pisum sativum* L.) using thidiazuron via the formation of organogenic cauline structures. *Can. J. Bot.*, **75**: 492–500.

Puite K J (1992) Progress in plant protoplast research. *Physiol. Plant.*, **85**: 403–410.

Pundir R P S and Singh R B (1987) Possibility of genetic improvement of pigeonpea (*Cajanus cajan* (L.) Millsp.) utilizing wild genes sources. *Euphytica*, **36**: 33–37.

Puonti-Kaerlas J (1993) Genetic engineering in pea crop improvement. *Acta Agric. Scand., Sect B, Soil Plant Sci.*, **43**: 65–73.

Puonti-Kaerlas J and Eriksson T (1988) Improved protoplast culture and regeneration of shoots in pea (*Pisum sativum* L.). *Plant Cell Rep.*, **7**: 242–245.

Puonti-Kaerlas J, Eriksson T and Engstrom P (1990) Production of transgenic pea (*Pisum sativum* L.) plants by *Agrobacterium tumefaciens*-mediated gene transfer. *Theort. Appl. Genet.*, **80**: 246–252.

Puonti-Kaerlas J, Stabel P and Eriksson T (1989) Transformation of pea (*Pisum sativum* L.) by *Agrobacterium tumefaciens*. *Plant Cell Rep.*, **8**: 321–324.

Purseglove J W (1988) *Tropical crops: Dicotyledons*, Longman Group Ltd, ELBS edition, London.

Radhakrishnan T, Murthy T G K, Desai S and Bandyopadhyay A (1999) Meristem culture of interspecific hybrids of groundnut. *Biol. Plant.*, **42**: 309–312.

Rajasekaran K and Pellow J W (1997) Somatic embryogenesis from cultured epicotyls and primary leaves of soybean (*Glycine max* (L.) Merrill). In Vitro *Cell. Dev. Biol.*, **33**: 88–91.

Ramana R V, Venu C, Jayasree T and Sadanadam A (1996) Direct somatic embryogenesis and transformation in *Cicer arietinum* L. *Indian J. Exp. Biol.*, **34**: 716–718.

Ramana Rao V V, Chary S N and Bhall J K (1992) Somatic embryogenesis from mesophyll protoplasts of wild and cultivated species of pigeonpea. In: *Biotechnology and Crop Improvement in Asia* (Ed Moss J P), ICRISAT, Patancheru, India, p. 161.

Ranch J P (1993) The potential for synthetic soybean seed. In: *Synseeds Applications of Synthetic Seeds to Crop Improvement* (Ed Redenbaugh K), CRS Press, Boca Raton/Ann Arbor/London/Tokyo, pp. 329–350.

Ranch J P, Oglesby L O and Zielinski A C (1985) Plant regeneration from embryo-derived cultures of soybean. In Vitro *Cell. Dev. Biol. Plant*, **21**: 653–658.

Rao B G and Chopra V L (1987) Genotypic and explant differences in establishment of chickpea (*Cicer arietinum* L.) callus. In: *Proc. Symp. Plant Cell and Tissue Culture of economically important plants* (Ed Reddy G M), Osmania University, Hyderabad, India.

Rao B G and Chopra V L (1989) Regeneration in chickpea (*Cicer arietinum* L.) through somatic embryogenesis. *J. Plant Physiol.*, **134**: 637–638.

Rao J D and Reddy G M (1987) *In Vitro* morphogenesis and multiple shooting in *Cicer arietinum* L. In: *Proc Symp Plant Cell and Tissue Culture of Economically Important Plants* (Ed Reddy G M), Osmania University, Hyderabad, India.

Reddy R L and Reddy G M (1993) Factors affecting direct somatic embryogenesis and plant regeneration in groundnut, *Arachis hypogaea* L. *Indian J. Exp. Biol.*, **31**: 57–60.

Roeper W (1980) Callus formation from protoplasts derived from cell suspension cultures of *Vicia faba* L. *Z. Pflanzenphysiol.*, **101**: 75–78.

Rohini V K and Rao K S (2000) Transformation of peanut (*Archis hypogaea* L.): a non-tissue culture based approach for generating transgenic plants. *Plant Sci.*, **150**: 41–49.

Rubluo A, Kartha K K, Mroginski L A and Dyck J (1984) Plant regeneration from pea leaflets cultured *in vitro* and genetic stability of regenerants. *J. Plant Physiol.*, **117**: 119–130.

Saalbach I, Pickardt T, Machemehl F, Saalbach G, Schieder O and Müntz K (1994) A chimeric gene encoding the methionine-rich 2S albumin of the Brazil nut (*Betrolletia excelsa* H.B.K.) is stably expressed and inherited in transgenic grain legumes. *Mol. Gen. Genet.*, **242**: 226–236.

Saalbach I, Waddell D, Pickardt T, Schieder O and Müntz K (1995) Stable expression of the sulphur-rich 2S albumin gene in transgenic *Vicia narbonensis* increases the methionine content of seeds. *J. Plant Physiol.*, **145**: 674–681.

Sabita Rani A and Reddy G M (1996) Induction of somatic embryogenesis from young leaflets of cultivated and wild species of groundnut. *Indian J. Exp. Biol.*, **34**: 569–571.

Sagare A P (1997) Tissue culture studies in grain legumes. Ph.D. thesis, University of Pune, India.

Sagare A P and Krishnamurthy K V (1991) Protoplast regeneration in chickpea, *Cicer arietinum* L. *Indian J. Exp. Biol.*, **29**: 930–932.

Sagare A P, Suhasini K and Krishnamurthy K V (1993) Plant regeneration via somatic embryogenesis in chickpea (*Cicer arietinum* L.). *Plant Cell Rep.*, **12**: 652–655.

Sagare A P, Suhasini K and Krishnamurthy K V (1995a) Histology of somatic embryo initiation and development in chickpea (*Cicer arietinum* L.). *Plant Sci.*, **109**: 87–93.

Sagare A P, Suhasini K and Krishnamurthy K V (1995b) Recent trends in improvement of chickpea (*Cicer arietinum* L.) through biotechnological methods. In: *Commemorative volume dedicated to Prof H Y Mohan Ram at the "National Seminar on Changing Scenario in Plant Sciences"*, Department of Botany, Banaras Hindu University, Varanasi, pp. 356–364.

Saini R and Jaiwal P K (2002) Age, orientation and position of explant in mother seedling determines the morphogenic response of epicotyl in *Vigna mungo* (L.) Hepper. *Plant Sci.*, **163**: 139–144.

Saker M M and Kuhne T (1998) Production of transgenic kidney bean shoots by electroporation of intact cells. *Biol. Plant.*, **40**: 507–514.

Samoylov V M, Tucker D M and Parrott W A (1998) Soybean (*Glycine max* (L.) Merrill) embryogenic cultures: The role of sucrose and total nitrogen content on proliferation. In Vitro *Cell. Dev. Biol. Plant*, **34**: 8–13.

Sanago M H M, Shattuck V I and Strommer J (1996) Rapid plant regeneration of pea using thidiazuron. *Plant Cell Tiss. Org. Cult.*, **45**: 165–168.

Sankhla A, Sankhla D, Upadhyaya A, Davis T D and Sankhla N (1990) Influence of lactidichlor-ethyl on growth, chemical composition, and regeneration of mothbean callus cultures. *Comp Physiol Ecol.*, **15**: 140–146.

Sankhla N, Davis T D, Gehlot H S, Upadhyaya A, Sankhla A and Sankhla D (1991) Growth and organogenesis in mothbean callus cultures as influenced by triazole growth regulators and gibberellic acid. *J. Plant Growth Reg.*, **10**: 41–45.

Santalla M, Power J B and Davey M R (1998) Efficient *in vitro* shoot regeneration responses of *Phaseolus vulgaris* and *P. coccineus*. *Euphytica*, **102**: 195–202.

Santarem E R and Finer J J (1999) Transformation of soybean (*Glycine max* (L.) Merrill). In Vitro *Cell. Dev. Biol. Plant*, **35**: 451–455.

Santarem E R, Pelissier B and Finer J J (1997) Effect of explant orientation, pH, solidifying agent and wounding on initiation of soybean somatic embryos. In Vitro *Cell. Dev. Biol. Plant*, **33**: 13–19.

Santarem E R, Trick H N, Essig J S and Finer J J (1998) Sonication assisted *Agrobacterium*-mediated transformation of soybean immature cotyledons: optimization of transient expression. *Plant Cell Rep.*, **17**: 752–759.

Santos K G B, Mundstock E and Bodanese-Zanettini M H (1997) Genotype-specific normalization of soybean somatic embryogenesis through the use of an ethylene inhibitor. *Plant Cell Rep.*, **16**: 859–864.

Sarangi B K and Gleba Y Y (1991) Direct multiple regeneration in *Cajanus cajan* (L.) Millsp. *Acta Hort.*, **289**: 149–150.

Sarangi B K, Kuchuk N and Gleba Y Y (1992) Isolation and culture of protoplasts of pigeonpea (*Cajanus cajan* L.). *Plant Cell Rep.*, **11**: 462–465.

Sarker R H, Chowdhury A P and Hoque M I (1997) Preliminary studies on *Agrobacterium*-mediated genetic transformation of peanut (*Arachis hypogaea* L.). *Bangladesh J. Bot.*, **26**: 155–162.

Sastri D C, Nalini M S and Moss J P (1982) Peanut. In: *Tissue Culture of Economically Important Plants* (Ed Rao A N), Costed and Anbs, Singapore, pp. 42–57.

Sato S, Newell C, Kolacz K, Tredo L, Finer JJ and Hinchee M (1993) Stable transformation via particle bombardment in two different soybean regeneration systems. *Plant Cell Rep.*, **12**: 408–413.

Saxena P K, Malik K A and Gill R (1992) Induction by thidiazuron of somatic embryogenesis in intact seedlings of peanut. *Planta*, **187**: 421–424.

Schaerer S and Pilet P E (1991) Roots, explants and protoplasts from pea transformed with strains. *Plant Sci.*, **78**: 247–258.

Schnall J A and Weissinger A K (1993) Culturing peanut (*Arachis hypogaea* L.) zygotic embryos for transformation via microprojectile bombardment. *Plant Cell Rep.*, **12**: 316–319.

Schnall J A and Weissinger A K (1995) Genetic transformation in *Arachis hypogaea* L. (peanut). In: *Biotechnology in Agriculture and Forestry, Vol. 34* (Ed Bajaj Y P S), Springer-Verlag, Berlin, pp. 135–144.

Schroeder H E, Schotz A H, Wardley-Richardson T, Spencer D and Higgins T J V (1993) Transformation and regeneration of two cultivars of pea (*Pisum sativum* L.). *Plant Physiol.*, **101**: 751–757.

Schulz A L, Adkins S W, Godwin I D and Dodd W A (1993) Focussed plant improvement (Towards responsible and sustainable agriculture). In: *Proc. 10th Australian Plant Breeding Conference*, Gold Coast, Australia, 187–188.

Schwenk F W, Pearson C A and Roth M R (1981) Soybean mesophyll protoplasts. *Plant Sci Lett.*, **23**: 153–155.

Seitz M H, Stalker H T and Green C (1987) Genetic variation for regenerative response in immature leaflet cultures of the cultivated peanut, *Arachis hypogaea. Plant Breed.*, **98**: 104–110.

Sellars R M, Southward G M and Phillips G C (1990) Adventitious somatic embryogenesis from cultured immature zygotic embryos of peanut using soybean as a model system. *Crop Sci.*, **30**: 408–414.

Sen J and Guha Mukherjee S (1998) *In vitro* induction of multiple shoots and plant regeneration in *Vigna. In Vitro Cell. Dev. Biol. Plant*, **34**: 276–280.

Shama Rao H K and Narayanaswamy S (1975) Effect of gamma irradiation on cell proliferation and regeneration in explanted tissues of pigeon pea, *Cajanus cajan* (L.) Millsp. *Radiat. Bot.*, **15**: 301–305.

Sharma D R, Kumari R and Chowdhury J B (1979) Plant regeneration in *Cicer* species through tissue culture. *Indian J. Exp. Biol.*, **17**: 607–609.

Shekhawat N S and Galston A W (1983) Isolation, culture, and regeneration of mothbean *Vigna aconitifolia* leaf protoplasts. *Plant Sci. Lett.*, **32**: 43–51.

Shiva Prakash N, Pental D and Bhalla-Sarin N (1994) Regeneration of pigeonpea (*Cajanus cajan*) from cotyledonary node via multiple shoot formation. *Plant Cell Rep.*, **13**: 623–627.

Shri P V and Davis T M (1992) Zeatin-induced shoot regeneration from immature chickpea (*Cicer arietinum* L.) cotyledons. *Plant Cell Tiss. Org. Cult.*, **28**: 45–51.

Shyluk J P, Kartha K K and Mroginski L A (1981) Plant regeneration from meristems of grain legume cowpea, chickpea, peanut. *Can. J. Bot.*, **59**: 1671–1679.

Simonenko Y V, Gleba Y Y and Kuchuk N V (1999) Double transformation: Producing transgenic phosphinothricin-resistant plants of commercial pea lines. *Russ. J. Plant Physiol.*, **46**: 804–807.

Soh W Y, Choi P S and Cho D Y (1998) Effects of cytokinin on adventitious root formation in callus cultures of *Vigna unguiculata* (L.) Walp. In Vitro *Cell. Dev. Biol. Plant*, **34**: 189–195.

Sreenivasan E, Pathak S S and Joshi R N (1995) Deproteinized leaf juice – a novel medium for rhizogenesis *in vitro. Curr. Sci.*, **68**: 586.

Sreenivasu K, Malik S K, Ananda Kumar P and Sharma R P (1998) Plant regeneration via somatic embryogenesis in pigeonpea (*Cajanus cajan* (L.) Millsp.). *Plant Cell Rep.*, **17**: 294–297.

Srinivasan M, Mohapatra S T and Sharma R P (1991) *Agrobacterium* mediated genetic transformation of chickpea, *Cicer arietinum* L. *Indian J. Exp. Biol.*, **29**: 758–761.

Sroga G E (1987) Plant regeneration of two *Lupinus* spp. from callus cultures via organogenesis. *Plant Sci.*, **51**: 245–249.

Stejskal J and Griga M (1992) Somatic embryogenesis and plant regeneration in *Pisum sativum* L. *Biol. Plant*, **34**: 15–22.

Stewart C N, Adang M J, All J N, Boerma H R, Cardineau G, Tucker D and Parrott W A (1996) Genetic transformation, recovery and characterization of fertile soybean transgenic for a synthetic *Bacillus thuringiensis* cry1Ac gene. *Plant Physiol.*, **112**: 121–129.

Suhasini K, Sagare A P and Krishnamurthy K V (1994) Direct somatic embryogenesis from mature embryo axes in chickpea (*Cicer arietinum* L.). *Plant Sci.*, **102**: 189–194.

Suhasini K, Sagare A P and Krishnamurthy K V (1996) Study of aberrant morphologies and lack of conversion of somatic embryos of chick pea (*Cicer arietinum* L.). In Vitro *Cell. Dev. Biol. Plant*, **32**: 6–10.

Suhasini K, Sagare A P, Sainkar S R and Krishnamurthy K V (1997) Comparative study of the development of zygotic and somatic embryos of chickpea (*Cicer arietinum* L.). *Plant Sci.*, **128**: 207–216.

Sun S S M and Larkins B A (1993) Transgenic plants for improving seed storage proteins. In: *Transgenic Plants, Vol. 1* (Eds Shain-dow K and Wu R), Academic Press Inc, New York, pp. 339–372.

Surya-Prakash, Chowdhury J B, Jain R K and Chowdhury V K (1992) Factors affecting plant regeneration in chickpea, *Cicer arietinum* L. *Indian J. Exp. Biol.*, **30**: 1149–1153.

Tegeder M, Gebhardt D, Schieder O and Pickardt T (1995) Thidiazuron-induced plant regeneration from protoplasts of *Vicia faba* cv. Mythos. *Plant Cell Rep.*, **15**: 164–169.

Tegeder M, Koehn H, Nibbe M, Schieder O and Pickardt T (1996) Plant regeneration from protoplasts of *Vicia narbonensis* via somatic embryogensis and shoot organogenesis. *Plant Cell Rep.*, **16**: 22–25.

Tetu T, Sangwan R S and Sangwan-Norreell B S (1990) Direct somatic embryogenesis and organogenesis in cultured immature zygotic embryos of *Pisum sativum* L. *J. Plant Physiol.*, **137**: 102–109.

Thome G C H, Santarem E R and Ferreira A G (1995) Adventitious bud induction and plant regeneration from soybean cotyledonary nodes. *Phyton-Int. J. Exp. Bot.*, **57**: 127–135.

Thorpe T A (1988) *In vitro* somatic embryogenesis. In: *ISI Atlas of Science, Animal and Plant Sciences*, pp. 81–88.

Trick HN and Finer J J (1998) Sonication assisted *Agrobacterium*-mediated transformation of soybean (*Glycine max* (L.) Merrill) embryogenic suspension culture tissue. *Plant Cell Rep.*, **17**: 482–488.

Tricoli D M, Hein M B and Carnes M G (1986) Culture of soybean mesophyll protoplasts in alginate beads. *Plant Cell Rep.*, **5**: 334–337.

Vani A K S and Reddy G M (1996) Morphogenesis from callus cultures of chickpea, *Cicer arietinum* L. *Indian J. Exp. Biol.*, **34**: 285–287.

Venkatachalam P and Jayabalan N (1996) Efficient callus induction and plant regeneration via somatic embryogenesis from immature leaf-derived protoplasts of groundnut (*Arachis hypogaea* L.). *Israel J. Plant Sci.*, **44**: 387–396.

Venkatachalam P and Jayabalan N (1997) Effect of auxins and cytokinins on efficient plant regeneration and multiple shoot formation from cotyledons and cotyledonary node explants of groundnut (*Arachis hypogaea* L.) by *in vitro* culture technology. *Appl. Biochem. Biotech.*, **67**: 237–247.

Venkatachalam P, Geetha N and Jayabalan N (1998a) Induction of somatic embryos and plantlet development in cell suspension cultures of *Arachis hypogaea* L. *Breed Sci.*, **48**: 231–236.

Venkatachalam P, Geetha N, Jayabalan N, Saravanababu and L Sita (1998b) *Agrobacterium*-mediated genetic transformation of groundnut (*Arachis hypogaea* L.): An assessment of factors affecting regeneration of transgenic plants. *J. Plant Res.*, **111**: 565–572.

Venkatachalam P, Geetha N, Khandelwal A, Shaila M S and Sita G L (1999a) Induction of direct somatic embryogenesis and plant regeneration from mature cotyledon explants of *Arachis hypogaea* L. *Curr. Sci.*, **77**: 269–273.

Venkatachalam P, Geetha N, Rao K S and Jayabalan N (1999b) Rapid and high-frequency *in vitro* plant regeneration from leaflet and petiole explants of groundnut (*Arachis hypogaea* L.). *Appl. Biochem. Biotech.*, **80**: 193–203.

Venkatachalam P, Kishor P B K, Geetha N, Thangavelu M and Jayabalan N (1999c) A rapid protocol for somatic embryogenesis from immature leaflets of groundnut (*Arachis hypogaea* L.). In Vitro *Cell Dev. Biol. Plant*, **35**: 409–412.

Venkatachalam P, Subrmaniampillai A and Jayabalan N (1996) *In vitro* callus culture and plant regeneration from different explants of groundnut (*Arachis hypogaea* L.). *Breed. Sci.*, **46**: 315–320.
Venkatachalam, P, Kavi Kishor P B and Jayabalan N (1997) High frequency somatic embryogenesis and efficient plant regeneration from hypocotyl explants of groundnut (*Arachis hypogaea* L.). *Curr. Sci.*, **72**: 271–275.
Victor J M R, Murch S J, Krishna Raj S and Saxena P K (1999a) Somatic embryogenesis and organogenesis in peanut: The role of thidiazuron and N^6-benzylaminopurine in the induction of plant morphogenesis. *Plant Growth Reg.*, **28**: 9–15.
Victor J M R, Murthy B N S, Murch S J, Krishna Raj S and Saxena P K (1999b) Role of endogenous purine metabolism in thidiazuron-induced somatic embryogenesis of peanut (*Arachis hypogaea* L.). *Plant Growth Reg.*, **28**: 41–47.
von Arnold S and Eriksson T (1976) Factors affecting the growth and division of pea mesophyll protoplasts. *Physiol. Plant.*, **36**: 193–196.
Walden R (1989) *Genetic Transformation of Plants.* Englewood Cliffs, NJ, USA.
Wang A M, Fan H L, Singsit C and Ozias-Akins P (1998) Transformation of peanut with a soybean vspB promoter-uidA chimeric gene. I. Optimization of a transformation system and analysis of *Gus* expression in primary transgenic tissues and plants. *Physiol. Plant.*, **102**: 38–48.
Warkentin T D and McHughen A (1992) *Agrobacterium tumefaciens*-mediated beta-glucuronidase (GUS) gene expression in lentil (*Lens culinaris* Medik.) tissues. *Plant Cell Rep.*, **11**: 274–278.
Warkentin T D and McHughen A (1991) Crown gall transformation of lentil (*Lens culinaris* Medik.) with virulent strains of *Agrobacterium tumefaciens. Plant Cell Rep.*, **10**: 489–493.
Wei Z (1988) Plant regeneration from protoplasts of soybean (*Glycine max* L.). *Plant Cell Rep.*, **7**: 348–351.
Wetzstein H Y and Baker C M (1993) The relationship between somatic embryo morphology and conversion in peanut *Arachis hypogaea* L. *Plant Sci.*, **92**: 81–89.
Widholm J M and Rick S (1983) Shoot regeneration from *Glycine canescens* tissue cultures. *Plant Cell Rep.*, **2**: 19–20.
Wilson V M, Haq N and Evans P K (1985) Protoplast isolation, culture and plant regeneration by organogenesis in *Glycine max. Plant Sci.*, **41**: 61–68.
Wright M S, Koehler S M, Hinchee M A and Carnes M G (1986) Plant regeneration by organogenesis in *Glycine max. Plant Cell Rep.*, **5**: 150–154.
Wright M S, Ward D V, Hinchee M A, Carnes M G and Kaufman R J (1987) Regeneration of soybean (*Glycine max* (L.) Merrill) from cultured primary leaf tissue. *Plant Cell Rep.*, **6**: 83–89.
Yang H, Singsit C, Wang A, Gonsalves D and Ozias-Akins P (1998) Transgenic peanut plants containing a nucleocapsid protein gene of tomato spotted wilt virus show divergent levels of gene expression. *Plant Cell Rep.*, **17**: 693–699.
Yang Y S, Wada K and Futsuhara (1990) Comparative studies of organogenesis and plant regeneration in various soybean explants. *Plant Sci.*, **72**: 101–108.
Zambre M A, DeClerq J, Vranova E, VanMontagu M, Angenon G and Dillen W (1998) Plant regeneration from embryo derived callus *Phaseolus vulgaris* L. (common bean) and *P. acutifolius* A Gray (Tepary bean). *Plant Cell Rep.*, **17**: 626–630.
Zhang Z Y, Coyne D P and Mitra A (1997) Factors affecting *Agrobacterium*-mediated transformation of common bean. *J. Am. Soc. Hort. Sci.*, **122**: 300–305.
Zhang Z Y, Xing A Q, Staswick P and Clemente TE (1999) The use of glufosinate as a selective agent in *Agrobacterium*-mediated transformation of soybean. *Plant Cell Tiss. Org. Cult.*, **56**: 37–46.
Zhanyuan Z, Coyne D P and Mitra A (1997) Factors affecting *Agrobacterium*-mediated transformation of common bean. *J. Am. Soc. Hort. Sci.*, **122**: 300–305.
Zharare G E, Blamey F P C and Asher C J (1998) Initiation and morphogenesis of groundnut (*Arachis hypogaea* L.) pods in solution culture. *Ann. Bot.*, **81**: 391–396.

IN VITRO ORGANOGENESIS

A. GANAPATHI, V.R. ANBAZHAGAN, S. AMUTHA AND R. PREM ANAND
Department of Biotechnology, School of Life Sciences,
Bharathidasan University, Tiruchirappalli – 620 024
e-mail: ganap@bdu.ernet.in

Abstract

Legumes are important sources of proteins for the growing population in many developing countries of the world. Their production is limited due to the crop's susceptibility to fungal, bacterial and viral diseases, insect pests and besides many other undesirable agronomic traits. Genetic improvement of legumes by classical breeding has met with limited success due to the lack of genetic variability within the germplasm. Strategies for increasing and stabilizing the production of legume crops depend on the development of varieties resistant to diseases, pests and with other desirable agronomic traits. Recent biotechnological advances have offered the opportunity to develop new germplasms. The development of such technology largely depends on the availability of efficient regeneration protocols. In the present review, regeneration via organogenesis in legumes is described. The advantages and limitations of this technique along with directions for future research are discussed.

1. Introduction

Plant tissue culture began as an intellectual curiosity. At present it has gained respect because of its significant contributions to our basic knowledge of plant biology. As early as 1902, Haberlandt, a German plant physiological anatomist stated that "every living plant cell is endowed with the potential to produce a whole plant". The concept of totipotency put forth by him provided impetus to the development of plant tissue culture.

The technique of plant tissue culture involves isolation of an organ, tissue, cell or protoplast from a plant under aseptic conditions and their growth in a well defined nutrient medium under a controlled environment. Like hydroponics – the art of growing plants without soil – tissue culture became an occupation for many scientists. A wide range of excised plant parts were grown on several nutrient formulations, supplements, extracts and juices such as coconut milk.

P.K. Jaiwal and R.P. Singh (eds.), Improvement Strategies for Leguminosae Biotechnology, 65–85.

Plant tissue culture has presently developed into a full-fledged technology with immense practical value. Its potential for large-scale applications in agriculture, horticulture and forestry has been visualized. This review reveals an important aspect of tissue culture, i.e. organogenesis which is currently used as one of the basic biotechnological tools for the improvement of crops in general and legumes in particular.

Legumes are one of the most significant groups of agriculturally important crops and have consequently been the subject of widespread efforts to improve desirable traits through *in vitro* manipulations. Legumes are next to the grasses in terms of diversity and economic importance, and are grown throughout the world as sources of food, feed, oil, forage, fuel, wood and even fiber. Legumes are also used as ornamentals, green manure and ground cover to control soil erosion. Many species within this family have been the subject of efforts towards non-conventional plant improvement using cell culture techniques. These efforts have resulted in the availability of a large number of *in vitro* protocols. Much effort has been expended to develop and optimize efficient regeneration systems in order to facilitate development of a variety of technologies. Despite the widely reported *in vitro* recalcitrance of legumes, at least 75 species from 25 genera have undergone *de novo* regeneration. To date, successful regeneration has been accomplished in species and specific determination of parameters critical to regeneration such as explant source, genotype and media constituents has been reported. Depending on several factors, regeneration occurs via organogenesis and/or embryogenesis, either directly from explanted tissue or indirectly after an intervening callus phase. While several species responded to either organogenesis or embryogenesis, only a few species regenerate via both the pathways. Methodologies leading to a diversity of *in vitro* responses are relatively a new accomplishment. With few exceptions, legumes are generally recalcitrant with respect to tissue culture. In recent years, attention has been focused on the development of regeneration systems amenable for gene transfer (Ramsay, 1993). Organogenic systems have been successfully employed in transformation of some legumes such as *Vigna unguiculata* (Muthukumar *et al.*, 1996; Prem Anand *et al.*, 2001). Presently, sufficient knowledge exists to design and optimize *de novo* regeneration systems for most legumes (Jaiwal and Gulati, 1995; Nagl *et al.*, 1997).

2. Organogenesis

Organogenesis is a process by which a cell, tissue, isolated protoplast or microspore differentiates to form adventitious organ or primordia in a well defined nutrient medium under controlled environmental conditions. Organogenesis refers to the formation of roots or shoots. The callus may remain in an undifferentiated condition regardless of the hormones and nutrients to which it is exposed. Organ formation generally follows cessation of unlimited proliferation of callus. Individual cells or groups of cells of smaller dimensions may form small nests of cells scattered throughout the callus tissue, the so-called meristemoids. These meristemoids become transformed into cyclic nodules from which shoot bud or root primordia may grow as shoots/roots. Shoot bud formation may decrease with age and subculture duration of the callus tissue but the capacity of rooting may persist for longer period. In some calli, rooting occurs more often than in other forms of organogenesis. During organogenesis, if the roots are first formed, then it is very difficult to induce adventitious shoot bud formation from the same callus tissue. But if the shoots are first formed, it may form roots later on or may remain in rootless condition unless and until the shoots are transferred to another medium or hormoneless medium or conditions that induce root formation. In certain cases, shoot and root formation may occur simultaneously. But organ connection (vascular connection) between two different organ primordia may or may not be established. The organ connection between shoot and root primordia is essential for the regeneration of complete plantlet from the culture. Shoot formation followed by rooting is the general feature of organogenesis.

The callus tissue may remain unchanged in colour during rhizogenesis or may develop yellow pigmentation. During shoot bud formation, the callus tissue generally develops green or pale-green pigmentation.

In many cases, the callus tissue shows a high potential for organogenesis when first initiated but gradually a decline sets in as subculture proceeds. The loss of potential for organogenesis may be due to a genetic or a physiological change induced on account of prolonged culture conditions or composition of culture media. Genetic variation in callus tissue occurs due to change in chromosome number such as aneuploidy, polyploidy and cryptic chromosomal rearrangements. Such chromosomal changes may lead to loss of totipotency in cells. During prolonged culture, totipotent cells of the callus tissue are gradually replaced by non-totipotent cells. It is generally believed that adventitious shoot bud formation takes place from the diploid cells of the callus tissue. At the early stage of culture, the callus tissue exhibits maximum number of diploid cells. However, at the later stage of culture, the cells of callus tissue become mixaploid due to alteration of chromosome number which ultimately inhibits organogenesis of callus tissue.

Occasionally, rooting occurs but shoot bud does not develop. But in some cases plantlets could also be regenerated from old subcultured callus tissue and the potential for organogenesis or embryogenesis could be enhanced in the later part of culture. An alternative physiological hypothesis has been put forward to explain the loss of organogenetic potential of the callus tissue due to prolonged sub-culture. According to this hypothesis, subculture often leads to loss of many endogenous factors or morphogens present at the critical stage of growth. Such factors present in the callus tissue at the initial stage may not be synthesized at all or synthesized only in insufficient quantities at later stages (Kalyan Kumar De, 1992). As a result, callus tissue fails to exhibit the potential for organogenesis or embryogenesis. However, if these factors are supplemented to the medium during subculture, the organogenic potential can be restored. It has been reported that addition of cytokinins could restore the decline in regenerative response in long term culture whereas, at the initial stages, no promotive effect of cytokinin was observed. The addition of cytokinin or any other additives is not always conducive for the regeneration. Therefore, it is plausible that both genetic as well as physiological processes are involved in the decline and loss of organogenic response during prolonged subculture.

The effects of chemical factors on organogenesis especially those of phytohormones, have been studied in explants from large number of species. The concept, as propounded by Skoog and Miller (1957), that induction of organogenesis would require, above all, the addition to culture medium of an appropriate balance of known phytohormones such as auxin and cytokinin has not proved to be so in many experimental materials. In few cultured tissues, the endogenous regulator complex can be adjusted to the required balance of phytohormones by an exogenous supply of auxin, cytokinin or gibberellin either separately or in combination. Generally high concentration of cytokinin brings about shoot bud initiation, whereas high levels of auxin favour rooting. Therefore, to obtain organogenesis, different permutations and combinations of hormones as well as various concentrations of hormones are supplemented in the culture medium.

Certain phenolic compounds are also used to induce shoot development. Phenolic compounds actually inactivate the auxins and consequently rise the physiological level of cytokinins which ultimately induce the initiation of buds. The use of an auxin inhibitor or auxin antagonist in culture medium also causes the induction of shoot bud. Addition of adenine in the culture medium induces shoot bud from the callus tissue of certain plants and addition of abscisic acid in place of cytokinin also induces shoot bud formation. Though the role of hormones and their quantitative interactions has been recognized, it is only recently that some efforts have been put to gain insight into the biochemistry of organ differentiation by hormonal interaction (Kalyan Kumar De, 1992). In the last few years, some indirect evidences have been obtained on organ forming tissues by estimating the level of structural and enzymatic proteins and the changes in isoenzyme patterns through gel electrophoresis during organogenesis.

Of different enzyme systems studied in plants, peroxidase has been shown to have a relation with organogenesis. One of the most important functions of peroxidase is its involvement in the metabolism of auxin. Plant tissue cultures need hormones like auxin and cytokinin for growth and differentiation *in vitro*. Hence, the study of peroxidase level by estimating the activity and the changes of isoperoxidase patterns during organogenesis is very important. Such isoperoxidases provide useful biochemical signals for morphogenic events (Kalyan Kumar De, 1992).

The activities of some isoenzymes in order to compare the regenerated plant with the mother plant during organogenesis have been studied in *Zizuphus mauritiana* (Gupta and Srivastava, 1996). Starch accumulation suggests high energy requirement for organogenic processes. A strong correlation has been found between the starch content of the callus, its rate of respiration and shoot formation. Gibberellic acid, which represses starch accumulation also inhibits shoot formation.

3. Factors affecting organogenesis

The expression of organogenic potential is influenced by a number of diverse factors, some of them related to the plant material such as physiological status of the plant, genotype and culture conditions (medium composition and physical nature of culture conditions).

3.1. PHYSIOLOGICAL STATUS OF EXPLANT

Cotyledons excised from two-day-old seedlings of *Vigna radiata* were more regenerative and those older than three days totally lacked this potential (Gulati and Jaiwal, 1990). The cotyledonary node with and without cotyledons excised from 4- and 6-day-old seedlings respectively were the most responsive for multiple shoot formation (Gulati and

Figure 1. a. Shoots from immature cotyledon calluses of cowpea. b. Direct regeneration from cotyledon segments of pigeonpea. c. In vitro *plant regeneration from callus induced from cotyledon of pigeonpea. d. Plant regeneration via adventitious bud formation from immature cotyledon explants of cowpea. e. Adventitious buds formed directly from mature cotyledon explants after 1 month. f. Shoot induction with simultaneous root formation in mungbean. C – cotyledon; S – shoots; Sb – shoot buds; R – roots.*

Table 1. Organogenesis in legumes

Species	Explant	Reference
Aeschynomene sensitiva	R	Nef-Campa *et al.*, 1996
Albizia lebbeck	H, R	Gharyal and Maheshwari, 1981
Albizia lebbeck	R	Tomar and Gupta, 1988; Miah *et al.*, 1996
Arachis hypogaea	CN, S	Daimon and Mii, 1991
Arachis hypogaea	L	Illingworth, 1974; Martin and Rabechault, 1976; Mroginski and Fernandez, 1980; Mroginski *et al.*, 1981c; Pittman *et al.*, 1983; Atreya *et al.*, 1984; Bhatia *et al.*, 1985; Narasimhulu and Reddy, 1985; Reddy and Narasimhulu, 1985; Rugman and Cocking, 1985; McKently *et al.*, 1990
Arachis pintoi	L	Burtnik and Mroginski, 1990
Arachis richardiana	H	Tomar and Gupta, 1988
Arachis villosa	A	Bajaj *et al.*, 1981b
Arachis villosulicarpa	L	Pittman *et al.*, 1983, 1984
Astragalus adsurgens	H	Luo and Jia, 1998
Cajanus cajan	C	Shiva Prakash *et al.*, 1994; Mohan and Krishnamurthy, 1998
Cajanus cajan	L	Kunjumon *et al.*, 1996
Cajanus cajan	Ca	Kumar *et al.*, 1983
Cicer arietinum	IM	Archana *et al.*, 1997
Cicer arietinum	S	Polisetty *et al.*, 1997
Cicer arietinum	CN	Murthy *et al.*, 1996
Crotalaria juncea	A, CP, L, S	Mohan Ram *et al.*, 1982
Crotalaria spp.	R,H,C	Sanghamitra *et al.*, 1998
Dalbergia sissoo	R	Mukhopadhyay and Mohan Ram, 1981
Glycine argyrea	C, CP, L, P	Hammatt *et al.*, 1989
Glycine clandestina	C, CP, H, L	Hymouwitz *et al.*, 1986; Hammatt *et al.*, 1987c; Myers *et al.*, 1989
Glycine falcata	C, L, P	Hammatt *et al.*, 1987a
Glycine latrobeana	C, L, P	Hammatt *et al.*, 1987a
Glycine max	CN	Cheng *et al.*, 1980; Thome *et al.*, 1995; Hammad, 1996
Glycine max	CP	Wei, 1988
Glycine max	C	Fu *et al.*, 1995
Glycine max	CN, H	Kandea *et al.*, 1997
Glycine soja	R, H, L, CN, E, P, L, ST	Sharma and Kothari, 1991
Glycine tomentella	H	Kameya and Widholm, 1981
Glycine tomentosa	H	Kameya and Widholm, 1981
Indigofera enneaphylla	C, H	Bharal and Rashid, 1979

Table 1. Continued

Species	Explant	Reference
Lathyrus sativus	L	Kunjumon *et al.*, 1996
Lens culinaris	N	Ahmad *et al.*, 1997
Lens culinaris	E, L	Williams and McHughen, 1986; Polanco *et al.*, 1988
Leucaena leucocephala	CN, N	Nangia *et al.*, 1996
Lupinus polyphyllus	A, H	Sator, 1990
Lupinus albus	CN	Harzic *et al.*, 1998
Lupinus angustifolius	H	Sator, 1990
Lupinus mutabilis	IS	Rahim *et al.*, 1999
Medicago borealis	P	Kuchuk *et al.*, 1990
Medicago arborea	R	Mariotti *et al.*, 1984
Medicago lupulina	Inf	Li and Demarly, 1995
Medicago sativa	ST	Cheyne and Dale, 1980
Medicago sativa	A	Dos Santos *et al.*, 1980; Bianchi *et al.*, 1988; Zagorska *et al.*, 1997
Medicago sativa	L, LP, O, P, S	Bianchi *et al.*, 1990;
Mimosa pudica	CN, ST	Gharyal and Maheshwari, 1982
Onobrychis viciaefolia	H, L	Arcioni and Mariotti, 1982
Onobrychis viciaefolia	IC, EA	Ozcan *et al.*, 1996
Onobrychis viciaefolia	L, P, S	Ozgen *et al.*, 1998
Phaseolus coccineus	ST	Ruiz *et al.*, 1986
Phaseolus coccineus	C	Santalla *et al.*, 1998
Phaseolus vulgaris	CN	McClean and Grafton, 1989
Phaseolus vulgaris	C	Santalla *et al.*, 1998
Phaseolus vulgaris	ST	Kartha *et al.*, 1981
Pisum sativum	Plu	Raruqui *et al.*, 1996
Pisum sativum	H	Sanago *et al.*, 1996
Pisum sativum	L	Mroginski and Kartha, 1981a
Pisum sativum	IM	Ozcan, 1995
Psophocarpus tetragonolobus	C, E, EP, I, L, P	Wilson *et al.*, 1985
Sesbania rostrata	C, H, ZE	Vlachova *et al.*, 1987
Stylosanthes guyanesis	H, R, L	Gregory *et al.*, 1980; Meijer and Broughton, 1981; Mroginski and Kartha, 1981b; Szabados and Roca, 1986
Trifolium nanum	C, H, N	Webb *et al.*, 1987
Trifolium nigrescens	E, H, R, S	Webb *et al.*, 1987
Trifolium pratense	E, H, R, S	Beach and Smith, 1979; Phillips and Collins, 1979b

Table 1. Continued

Species	Explant	Reference
Trifolium repens	C, Ca, CaP, H, LP, P, R	Oswald *et al.*, 1977; Gresshoff, 1980; Ahuja *et al.*, 1983a; White, 1984; Webb *et al.*, 1987; Yamada, 1989
Trifolium resupinatum	C	Webb *et al.*, 1987
Trifolium tridentatum	C, H	Webb *et al.*, 1987
Trifolium uniflorum	C, H	Webb *et al.*, 1987
Trifolium alpestre	ST	Parrott and Collins, 1983
Trifolium armenium	CN, ST	Webb *et al.*, 1987
Trifolium arnense	CN, ST	Webb *et al.*, 1987
Trifolium batmanicum	CN, ST	Webb *et al.*, 1987
Trifolium cherleri	CN, ST	Webb *et al.*, 1987
Trifolium curvisepalum	CN, ST	Webb *et al.*, 1987
Trifolium incarnatum	E, H, R, S	Beach and Smith, 1979
Trifolium pratense	E, FH, SN, ST	Phillips and Collins, 1979a; Cheyne and Dale, 1980; Collins and Phillips, 1982; Parrott and Collins, 1983; Oelck and Schieder, 1983; Broda, 1984; Campbell and Tomas, 1984; Webb *et al.*, 1987; Skucinska and Miszka, 1990
Trifolium repens	CN, SN, ST	Barnett *et al.*, 1975; Cheyne and Dale, 1980; Bhojwani, 1981; Parrott and Collins, 1983; Webb *et al.*, 1987
Trifolium rubens	ST	Parrott and Collins, 1983
Trifolium striatum	CN, ST	Webb *et al.*, 1987
Trifolium subterraneum	CN, ST	Parrot and Collins, 1983
Trifolium tomentosum	CN, ST	Webb *et al.*, 1987
Trifolium tridentatum	CN, ST	Webb *et al.*, 1987
Trifolium uniflorum	CN, ST	Webb *et al.*, 1987
Vicia faba	ST	Busse, 1986
Vigna mungo	ST	Bajaj and Dhanju, 1979
Vigna mungo	H, A, CN, L	Geetha *et al.*, 1997
Vigna mungo	CN, E, A	Ignacimuthu and Franklin, 1997, 1999
Vigna mungo	C	Geetha *et al.*, 1997
Vigna radiata	CN	Mathews, 1987; Gulati and Jaiwal, 1994
Vigna radiata	C	Mathews and Rao, 1984; Mathews *et al.*, 1986; Mathews, 1988; Gulati and Jaiwal, 1990; Chandra and Pal, 1995
Vigna radiata	ST	Bajaj and Dhanju, 1979; Gulati and Jaiwal, 1992
Vigna unguiculata	SI	Kulothungan *et al.*, 1993, 1997
Vigna aconitifolia	L	Kunjumon *et al.*, 1996

Table 1. Continued

Species	Explant	Reference
Vigna aconitifolia	P	Krishnamurthy *et al.*, 1984
Vigna glabrescens × *V. radiata*	ZE	Chen *et al.*, 1990
Vigna unguiculata	H, C	Pellegrineschi, 1997a, b
Vigna unguiculata	ST	Kartha *et al.*, 1981
Vigna unguiculata	C, H	Ladeinde and Soh, 1991; Muthukumar *et al.*, 1995; Amitha and Reddy, 1996a; Brar *et al.*, 1997, 1999; Prem Anand *et al.*, 2001

Abbreviations: A – anther; C – cotyledon; Ca – callus; CN – cotyledonary node; CaP – callus protoplast; E – epicotyl; EA – embryo axis; F – flower bud; FH – flower head; H – hypocotyl; I – internode; IC – immature cotyledon; Inf – inflorescence; L – leaf; LP – leaf protoplast; N – node; P – petiole; Plu – plumule; R – root; SN – shoot node; ST – shoot tip; ZE – zygotic embryo.

Jaiwal, 1994). Cotyledon explants excised after 15 days of self-pollination were effective and produced five to six buds at the cut surface (Prem Anand *et al.*, 2001).

3.2. SIZE AND TYPE OF EXPLANT

The morphogenic response in *in vitro* conditions varies with the size of the explant. The regenerative response of cotyledon, epicotyl and hypocotyl decreased when sliced into equal parts either longitudinally or transversely. The regeneration was more in longitudinal halves than proximal transverse halves. The distal transverse half of cotyledons did not show any differentiation (Gulati and Jaiwal, 1990; Kulothungan, 1997). A direct correlation between the size of meristem and the percentage of regeneration of plants of *V. radiata* (Bajaj and Dhanju, 1979; Gulati and Jaiwal, 1990) and *V. unguiculata* (Kulothungan, 1997) was observed. The growth regulator requirements of shoot-tips for induction of multiple shoots varied with explant size probably due to the differences in the levels of endogenous hormones in the explants (Gulati and Jaiwal, 1992).

The presence/absence of cotyledons and removal of a portion of one or both cotyledons from cotyledonary nodes affected shoot formation. More shoots were produced by explants when the proximal half of both cotyledons were present as compared to explants with one or two whole cotyledons removed. Complete removal of both the cotyledons not only delayed the shoot regeneration response of explants, but also reduced the regeneration frequency as well as the number of shoots per explant as compared to explants having both cotyledons (Gulati and Jaiwal, 1994). In *V. unguiculata* proximal end of the immature cotyledons produced large number of shoots when cultured in MS medium with 8.87 μM BAP and 0.46 μM zeatin (Prem Anand *et al.*, 2001).

3.3. GENOTYPE

Patel *et al.* (1991) and Mendoza *et al.* (1992) reported genotype specificity in efficient regenerability of cultivars of *V. radiata*. Similar results were reported by Kulothungan

(1997) and Prem Anand *et al.* (2001). In *V. radiata,* cotyledon, shoot-tip and cotyledonary node cultures showed that the frequency of shoot regeneration as well as the number of shoots per explant varied with cultivar (Gulati and Jaiwal, 1990, 1992, 1994). Different groundnut cultivars exhibited differential response to the regeneration regime in terms of multiple shoot bud production (Daimon and Mii, 1991; Narasimhulu and Reddy, 1983). Genotypic differences in the regeneration capacity of legumes have been reported (Rubluo *et al.*, 1984; Franklin *et al.*, 1991).

3.4. ORIENTATION OF EXPLANT ON MEDIUM

Kulothungan (1997) reported that cotyledon explants of *V. unguiculata* showed organogenic differentiation at proximal end of the cotyledon when distal ends were in contact with the medium. In cotyledon cultures of *V. radiata*, callusing and organogenic differentiation occurred only if the petiolar end of cotyledons was kept in contact with the medium (Gulati and Jaiwal, 1990).

3.5. MEDIUM COMPOSITION

MS and B_5 media are frequently used in legume tissue culture, but MS medium has been found to be the best for growth responses of different explants and calli (Mathews, 1987; Patel *et al.*, 1991; Prem Anand *et al.*, 2001). Reduction of MS inorganic salts to half has favoured induction of regenerable calli (Mendoza *et al.*, 1992). Ascorbic acid and citric acid (each 100 mg/L) have been used to prevent the browning of shoot-tip callus during subcultures (Mathews, 1987). Proline (1–2 mg/L) and yeast extract (50 mg/L) in combination with growth regulators have induced regenerable calli (Mendoza *et al.*, 1992). Various explants have shown differences in growth regulator requirements for caulogenesis. An auxin and a cytokinin are required for the induction of callus from various explants. Among cytokinins, 6 benzylaminopurine (BA), kinetin, zeatin, thidiazuron (TDZ) and among auxins 2,4-dichlorophenoxyacetic acid (2,4-D), naphthaleneacetic acid (NAA), indole-3-acetic acid (IAA) have been used to induce callusing. BAP in combination with auxin has supported callus growth better than kinetin. On the other hand, zeatin excels BAP in inducing shoot differentiation (Patel *et al.*, 1991).

Although considerable attention has been paid to the manipulation of culture medium to improve the *in vitro* plant regeneration response, the effects of physical conditions like temperature, photoperiod, etc., during culture have not been paid the attention they deserve. Irrespective of the nature of explants, cultures are generally grown at 25 ± 2°C (Mathews, 1987; Gulati and Jaiwal, 1990, 1992, 1994; Kulothungan, 1997). The age of the explants also plays a vital role in determining the organogenic response. Prem Anand *et al.* (2001) reported that the explants collected at different ages from the seedlings of *V. unguiculata* showed differences in organogenetic response.

Photoperiod also affects the regeneration frequency of explants or callus induction. Complete inhibition of multiple shoot formation was observed when shoot-tips (0.5–0.6 mm) were cultured on BAP enriched medium under dark for first two weeks (A. Gulati and P.K. Jaiwal, unpublished data).

It is clear from the aforesaid facts that regeneration from callus cultures is genotype specific and is restricted to one or two genotypes and that too with low efficiency. Moreover, effects of other factors (explant type, media composition, etc.) in controlling shoot differentiation efficiency are not clearly understood. Further studies are therefore required to define precisely these factors in order to improve the efficiency of regeneration in legumes. The age of the explant also influences the shoot bud regeneration. Differences in shoot bud regeneration ability is thus considered to be controlled by genetic factors (Bhojwani *et al.*, 1984; Ono *et al.*, 1994). The following are the details about the organogenesis of a few important legumes.

4. *Vigna* species

Direct regeneration of shoots without an intervening callus phase has been achieved from shoot tips (Bajaj and Dhanju, 1979; Mathews, 1987; Gulati and Jaiwal, 1992), cotyledons (Mathews, 1987; Gulati and Jaiwal, 1990; Patel *et al.*, 1991; Prem Anand *et al.*, 2001), cotyledonary nodes (Mathews, 1987; Gulati and Jaiwal, 1994) and leaf petioles (A. Gulati and P.K. Jaiwal, unpublished data). Shoot organogenesis has been reported from primary immature leaf derived callus (Mendoza *et al.*, 1992); and mature leaves of 7-day-old seedlings (Patel *et al.*, 1991). The callus derived from cotyledons, cotyledonary nodes, shoot-tips and root segments failed to regenerate shoot/root (Mathews, 1987; Patel *et al.*, 1991; Gulati and Jaiwal, 1990, 1992, 1994).

Gill *et al.* (1986) obtained *de novo* differentiation of shoot buds in *V. aconitifolia* on basal medium devoid of growth regulators. Gill *et al.* (1986) made morphogenetic studies on different seedling explants of *V. aconitifolia*. They reported that the frequency of shoot regeneration depended on both explant type and relative concentration of auxin and cytokinin. Highest frequency was obtained from leaf derived callus in presence of 5.71 μM IAA and 23.23 μM kinetin. Kunjumon *et al.* (1996) achieved regeneration in *V. aconitifolia* from leaf explants of 7-day-old seedlings on MS medium fortified with 30 μM IAA and 6 μM kinetin. Muthukumar *et al.* (1996) used primary leaf explants, including the intact petiole, to regenerate plants of *V. unguiculata*. When they cultured them on B_5 medium supplemented with 8×10^{-7} M 2,4,5-T, 1×10^{-2} M L-glutamine and 1×10^{-4} M adenine sulphate, callus developed at the petiolar ends. Shoot regeneration occurred, when this callus was subcultured on B_5 medium containing 5×10^{-5} M BAP. Pandey and Bansal (1989) reported plantlet formation from leaf callus on MS medium supplemented with IBA (100 μM) and kinetin (10 μM). Geetha *et al.* (1997) reported organogenesis from cotyledon explants of *Vigna mungo*. Pellegrineschi (1997a) reported organogenesis from cotyledon and hypocotyl explants of *Vigna unguiculata*. Ignacimuthu and Franklin (1999) reported organogenesis from cotyledon and epicotyl explants of *V. mungo*.

5. *Phaseolus* species

Franklin *et al.* (1991) induced callus from shoot tips of *P. vulgaris*, on modified MS medium supplemented with 10 μM kinetin, 5 μM 2,4-D and 10 μM IAA. Genotypic differences were found with respect to multiple shoot bud formation on MS medium supplemented with fractional combinations of BAP and NAA after 28 days of culture. McClean and Grafton (1989) regenerated sexually mature *P. vulgaris* plants from cotyledonary node tissue of 12–14-day-old seedlings on MS medium supplemented with 1–10 μM of BA or kinetin. Optimum results were obtained with 5 μM BA, whereas kinetin did not induce multiple shoots. Franklin *et al.* (1991) regenerated plants of *P. vulgaris* from three day old seedling explants via organogenesis. Malik and Saxena (1991) reported that intact seedling showed higher frequency of shoot regeneration than explant cultures. Mohamed *et al.* (1991) regenerated plants from embryonic axis explants of *P. acutifolius* and *P. vulgaris*. Mohamed *et al.* (1991) also obtained plantlets from four genotypes of *P. vulgaris* using cotyledonary explants grown on B_5 medium containing benzyl adenine. Santalla *et al.* (1998) reported organogenesis from embryonic axis with cotyledon in *Phaseolus vulgaris* and *P. coccineus*. Crocomo *et al.* (1976) reported plant regeneration from leaf explants on Veliky and Martins medium (1970) supplemented with IAA (11.42 μM), NAA (5.37 μM) and Kn (0.58 μM). Franklin *et al.* (1991) reported organogenesis from 3 day-old seedling explants of *Phaseolus vulgaris* on MS medium containing glutamine (10 μM), BAP (2.0 μM) and GA_3 (15.0 μM). Vaquero *et al.* (1993) described plant regeneration in *Phaseolus coccineus* from shoot-tip and cotyledonary node explants on MS medium supplemented with BAP (10 μM) and NAA (1.0 μM).

6. *Glycine* species

Myers *et al.* (1989), Hammatt *et al.* (1987b) and Hymowitz *et al.* (1986) reported organogenesis from cotyledon, hypocotyl and leaf explants of *Glycine clandestina*. Hammatt *et al.* (1987a) has reported organogenesis from cotyledon, petiole, hypocotyl and leaf explants of *Glycine falcata* and *G. latrobeana*. Wei (1988) reported organogenesis from cotyledon and petiole explants of *Glycine max*. Kameya and Widholm (1981) reported organogenesis from hypocotyl explants of *Glycine tomentella* on MS medium containing NAA (26.85 μM) and BA (44.38 μM). Hammatt *et al.* (1989) reported organogenesis from cotyledon, petiolated cotyledon, leaf and petiole explants of *Glycine argyra* on MS medium containing NAA and BAP. Cheng *et al.* (1980) reported organogenesis from cotyledon explants of *Glycine max*.

Hammad (1996) reported plant regeneration from cotyledonary node explants of *Glycine max* on MS + B_5 medium containing BAP and NAA. Kaneda *et al.* (1997) reported plant regeneration from hypocotyl and cotyledonary nodc callus of *G. max* on B_5 medium containing TDZ.

Adventitious bud and shoot formation from different seedling explants of *Glycine* species were obtained on MS medium supplemented with BAP (0.44–44.38 μM) and NAA (0.54–26.85 μM) (Kameya and Widholm, 1981). Widholm and Rick (1983) reported regeneration of plants from cotyledon and hypocotyl explants of *Glycine canescens* on MS and B_5 medium containing NAA (2.9 μM) and BAP (22.19 μM). Pandey and Bansal (1992) reported initiation and proliferation of shoots from leaf and hypocotyl explants of a perennial soybean (*Glycine wightii*) on MS medium containing NAA (0.54 μM) and BAP (22.19 μM). Sharma and Kothari (1991) reported plant regeneration from different seedling explants of *Glycine clandestina* on B_5 medium supplemented with 3% sucrose and different concentration of BAP and IBA. The regeneration of shoots was maximum from cotyledon explants.

7. *Arachis* species

Mroginski and Fernandez (1980), Bhatia *et al.* (1985), Atreya *et al.* (1984) and Illingworth (1974) reported plant regeneration from different explants such as axillary bud, cotyledon, embryonic axis, hypocotyl, leaf and zygotic embryos of *Arachis hypogaea*. Bajaj *et al.* (1981) reported organogenesis from *A. villosa*. Daimon and Mii (1991) reported organogenesis from cotyledonary node and shoot-tip explants of *A. hypogaea* on B_5 medium containing BAP (24.5 μM) and Kn (46.47 μM).

Vajranabhaiah *et al.* (1993) reported development of green compact callus on MS medium containing NAA (16.74 μM) and kinetin (0.93 μM) from hypocotyl segments of groundnut. Pittman *et al.* (1983) reported organogenesis from leaf callus of *Arachis chacoense* on MS medium supplemented with each of 4.44 μM BAP and NAA.

McKently *et al.* (1990) reported that the combination of NAA (5.37 μM) and BAP (8.87 μM) induced maximum frequency of callus from leaf explants within 21 days in peanut. He also reported that MS basal medium containing 22.19 μM BAP promoted shoot elongation and differentiation. Mroginski *et al.* (1981c) induced shoot regeneration from the callus of immature leaves of colorodo and Manfredi peanuts cultured on medium supplemented with 4.44 μM each of BAP and NAA. Narasimhulu and Reddy (1983) reported sporadic shoot development from peanut leaf callus cultured on medium supplemented with 4.65 μM kinetin, 2.85 μM IAA and 0.54 μM NAA.

Atreya *et al.* (1984) observed 60% of callus induction on MS medium supplemented with 10.74 μM NAA from embryo axes with cotyledon segments of peanut (*A. hypogaea*). Cheng *et al.* (1992) reported *in vitro* regeneration of valencia-type peanut *(A. hypogaea)* from cultured petioles on MS medium containing BAP (44.38 μM) and NAA (5.37 μM). Narasimhulu and Reddy (1983) reported that 38% of shooting occurred on a medium containing 4.44 μM BAP and 2.15 μM NAA.

8. *Cajanus* species

Mehta and Mohan Ram (1980) reported regeneration of plantlets from the cotyledons of *Cajanus cajan* on B_5 medium containing BAP (10^{-5} M). Organogenesis from leaf explants on MS medium containing IAA, 2,4-D, BA and kinetin has been reported by Kunjumon *et al.* (1996) in *Cajanus cajan*. Eapen and George (1993) reported organogenesis from leaf discs of pigeonpea and the effect of benzyladenine, indoleacetic acid and its amino acid conjugates. Rao and Narayanaswamy (1975) obtained plantlets from the hypocotyl derived callus on White's (1939) medium supplemented with kinetin (4.65 μM), IAA (0.57 μM) and casein hydrolysate (CH, 400 mg/L). Kumar *et al.* (1983) recorded organogenesis from different explants of *Cajanus cajan*. They reported callus induction and plant regeneration from different seedling explants cultured on

Blayde's (1966) medium supplemented with 2,4-D (9.05 μM), kinetin (2.32 μM) and BAP (9.11 μM). Patel *et al.* (1992) recovered complete plantlets from immature embryos on MS medium containing IAA, kinetin and coconut milk (CM).

9. *Lupinus* species

Multiple shoot regeneration was achieved via organogenesis from immature seeds of *Lupinus mutabilis* by Rahim *et al.* (1999). *In vitro* shoot formation of *Lupinus albus* from cotyledonary node by TDZ and NAA was achieved (Harzic *et al.*, 1998). The use of embryo culture for the regeneration of interspecific hybrids from *Lupinus mutabilis* × *L. haetwegii* were reported by Schaefer *et al.* (1988).

10. *Pisum sativum*

Tissue culture response in *Pisum* genotypes were reported by Raruqui *et al.* (1996). Rapid plant regeneration from hypocotyl explants of pea using TDZ was reported by Sanago *et al.* (1996). Ozcan (1995) induced *in vitro* micropropagation of pea (*Pisum sativum* L.) from immature embryos. Kubalakova *et al.* (1988) reported several factors affecting root formation in *in vitro* regenerated pea shoots. Hisajima (1985) reported micropropagation through multiple shoot formation from pea seeds.

11. *Cicer arietinum*

Altinkut *et al.* (1997) reported regeneration and hairy root formation in chickpea using callus derived plantlets and seedlings. Induction of callus and regeneration from immature embryos of chickpea has been reported (Singh *et al.*, 1997). Selection for herbicide resistance in tissue culture and isolation of phenotypic variants in cowpea and chickpea have been achieved by Rao and Patel (1997). Polisetty *et al.* (1997) achieved multiple shoot induction using benzyladenine and complete plant regeneration from seed explants of chickpea. Murthy *et al.* (1996) reported that TDZ stimulated direct organogenesis and somatic embryogenesis in chickpea. Morphogenesis from callus cultures of chickpea was reported by Vani and Reddy (1996). An efficient *in vitro* regeneration from immature cotyledon and embryonic axis of chickpea cultivars was also achicved (Adkins *et al.*, 1995).

12. Importance of organogenesis

A large number of plant species including economically important legume crops have been successfully regenerated from callus culture via organogenesis. Improvement of crop plants through manipulation at the cellular level such as the somatic hybridization by the fusion of isolated protoplasts of distantly related plants, the transfer of foreign genetic material to protoplasts, etc., is possible only if somatic cells are able to give rise to whole plant. Regeneration of whole plants through organogenesis is of special interest in mutagenic studies. Chemical mutagens and ionizing radiation can be used in bringing about genetic variation in plants and by means of cell/callus culture. The regeneration of whole plants from somatic mutant cell types results in the development of mutant plants through organogenesis. A new source of genetic variability is also available in plants regenerated from cell, tissue and organ culture. The regenerants from callus exhibit variation known as somaclonal variation. This somaclonal variation can be exploited to evolve new variants if regeneration via organogenesis is available. Direct DNA transfer to easily regenerable organised tissues has a good potential to engineer plant resistance to diseases and insect pests and with better nutritional quality. It may be noted that multiple shoot induction from cotyledon explant is a potential tool to transfer foreign genes through *Agrobacterium* infection. The lack of efficient regeneration systems will hamper the improvement of legumes through *in vitro* techniques and genetic transformation.

13. Conclusions and future perspectives

The development of procedures to regenerate plants from single cells and organized tissues and the discovery of methodology to transfer genes to plant cells provided the prerequisite for the practical use of genetic engineering to crop improvement. These techniques have given us the opportunity to create, characterize and select plant cultivars which could not be obtained by traditional breeding methods. Somaclonal variations can be developed via organogenesis resulting in the production of salt, drought and herbicide tolerant cultivars. Morphogenetic studies on regeneration of shoots via differentiation and dedifferentiation can also be studied.

General conclusions drawn from studies on legume organogenesis are

- Regeneration is genotype-specific.
- Most genera or species which have been seriously addressed contain genotypes capable of regeneration via one mode or another.
- Media compositions will probably need to be optimized for individual genotypes.
- Explant type and age can affect regeneration capability.
- Simple shoot proliferation from meristem culture is necessary to eliminate virus in germplasm conservation studies.
- Organogenic regeneration systems have traditionally been more effective for the purpose of genetic transformation since *de novo* regeneration occurs exclusively in several species.

References

Adkins A L, Godwin I D and Adkins S W (1995) An efficient *in vitro* regeneration system for Australian grown chickpea (*Cicer arietinum*) cultivars. *Aust. J. Bot.*, **43**: 491–497.

Ahmad M, Fautrier A G, McNeil D L, Hill G D and Burritt D J (1997) *In vitro* propagation of *Lens* species and their F_1 interspecific hybrids. *Plant Cell Tiss. Org. Cult.*, **47**: 169–176.

Ahuja P S, Lu D Y, Cocking E C and Davey M R (1983) An assessment of the cultural capabilities of *Trifolium repens* L. (white clover) and *Onobrychis viciaefolia* Scop. (sainfoin) mesophyll protoplasts. *Plant Cell Rep.*, **2**: 269–272.

Altinkut A, Gozukirmiz N, Bajrovic K and Altman A (1997) High percentage of regeneration and transformation in chickpea. *Acta Hort.*, **447**: 319–320.

Amitha K and Reddy T P (1996) Regeneration of plantlets from different explants and callus cultures of cowpea (*Vigna unguiculata* L.). *Phytomorphology*, **46**: 207–211.

Archana S, Singh N P, Astana A N and Singh A (1997) Callus induction and direct regeneration from immature embryo in chickpea. *Int. Chickpea Pigeonpea Newslett.*, **4**: 39–40.

Arcioni S and Mariotti D (1982) Tissue culture and plant regeneration in the legumes. *Onobrychis viciaefolia* Scop., *Coronilla varia* and *Lotus corniculatus* L. In: *Plant Tissue Culture* (Ed Fujiwara A), Japan Association for Plant Tissue Culture, Tokyo, pp. 707–708.

Atreya C D, Rao J and Subramanyam N C (1984) *In vitro* regeneration of peanut (*Arachis hypogaea* L.) plantlets from embryo axes and cotyledon segments. *Plant Sci. Lett.*, **34**: 379–383.

Bajaj Y P S and Dhanju M S (1979) Regeneration of plants from apical meristem tips of some legumes. *Curr. Sci.*, **48**: 906–907.

Bajaj Y P S, Ram A K, Labana K S and Singh H (1981) Regeneration of genetically variable plants from the anther-derived callus of *Arachis hypogaea* and *Arachis villosa*. *Plant Sci. Lett.*, **23**: 35–39.

Barnett O W, Gibson P B and Seo A (1975) A comparison of heat treatment, cold treatment and meristem tip-culture for obtaining virus-free plants of *Trifolium repens*. *Plant Disease Rep.*, **59**: 834–837.

Beach K H and Smith R R (1979) Plant regeneration from callus of red and Crimson clover. *Plant Sci. Lett.*, **16**: 231–237.

Bharal S and Rashid A (1979) Regeneration of plants from tissue cultures of the legume *Indigofera enneaphylla* Linn. *Pflanzenphysiologie*, **92**: 443–447.

Bhatia C R, Murty G S S and Mathews V H (1985) Regeneration of plants from "de-embryonated" peanut cotyledons cultured without nutrients and agar. *Pflanzenzuchtung*, **94**: 149–155.

Bhojwani S S (1981) A tissue culture method for propagation and low temperature storage of *Trifolium repens* genotypes. *Physiol. Plant.*, **52**: 187–190.

Bhojwani S S, Mullins K and Cohen D (1984) Intra-varietal variation for *in vitro* plant regeneration in the genus *Trifolium*. *Euphytica*, **33**: 915–921.

Bianchi S, Flament P and Dattee Y (1988) Somatic embryogenesis and organogenesis in alfalfa: Genotypic variation in regeneration ability. *Agronomie*, **8**: 121–126.

Brar M S, Al-Khayri J M, Morelock T E and Anderson E J (1999) Genotypic response of cowpea (*Vigna unguiculata* L.) to *in vitro* regeneration from cotyledon explants. In Vitro *Cell. Dev. Biol. Plant*, **35**: 8–12.

Brar M S, Al-Khayri J M, Shamblin C E, McNew R W, Morelock T E and Anderson E J (1997) *In vitro* shoot-tip multiplication of cowpea *Vigna unguiculata* (L.) Walp. In Vitro *Cell. Dev. Biol. Plant*, **33**: 114–118.

Broda Z (1984) Red clover (*Trifolium pratense* L.) vegetative propagation by tissue cultures with special consideration of genetic conditioning of the ability to regenerate from callus of *Wegetatywna propagacja* koniczyny czerwonej (*Trifolium pratense* L.). (in Pl., summary in Eng.). *Roczniki Ad-adeni Rolnicze, W Poznaniu Rozprawy Naukowe*, **140**: 5–41.

Burtnik O J and Mroginski L A (1990) Regeneration of *Arachis pintoi* (Leguminosae) plants throughout the *in vitro* culture of leaf tissue (in Sp. and Fr., summary in Eng.). *Oleagineux*, **40**: 609–612.

Busse G (1986) *In vitro* cultivation of *Vicia faba* and induction of morphogenesis. *Biologisches Zentralblatt*, **105**: 97–104.

Campbell C T and Tomes D T (1984) Establishment and multiplication of red clover plants by *in vitro* shoot-tip culture. *Plant Cell Tiss. Org. Cult.*, **3**: 49–57.

Chandra M and Pal A (1995) Differential response of the two cotyledons of *Vigna radiata in vitro*. *Plant Cell Rep.*, **15**: 3–4.

Chen H K, Mok M C and Mok D W S (1990) Somatic embryogenesis and shoot organogenesis from interspecific hybrid embryos of *Vigna glabrescens* and *V. radiata*. *Plant Cell Rep.*, **9**: 77–79.

Cheng M, His D C H and Philips G C (1992) *In vitro* regeneration of valencia type peanut (*Arachis hypogaea* L.) from cultured petioles, epicotyl sections and other seedling explants. *Peanut Sci.*, **19**: 82–87.

Cheng T Y, Saka H and Voqui-Dinh T H (1980) Plant regeneration from soybean cotyledonary node segments in culture. *Plant Sci. Lett.*, **19**: 91–99.

Cheyne V A and Dale P J (1980) Shoot-tip culture in forage legumes. *Plant Sci Lett.*, **19**: 303–309.

Collins G B and Phillips G C (1982) *In vitro* tissue culture and plant regeneration in *Trifolium pratense* L. In: *Regeneration from Cells and Tissue Culture* (Eds Earle E D and Demarly Y), Praeger Scientific Publishing, New York, USA, pp. 22–34.

Crocomo O J, Peters J E and Sharp W R (1976) Interactions of phytohormones on the control of growth and root morphogenesis in cultured *Phaseolus vulgaris* leaf explants. *Turriable*, **26**: 232–236.

Daimon H and Mii (1991) Multiple shoot formation and plantlet regeneration from cotyledonary node in peanut (*Arachis hypogaea* L.). *Jpn. J. Breed.*, **41**: 461–466.

Dos Santos A V P, Outka D E, Cocking E C and Davey M R (1980) Organogenesis and somatic embryogenesis in tissues derived from leaf protoplasts and leaf explants of *Medicago sativa. Pflanzenphysiologie*, **99**: 261–270.

Eapen S and George L (1993) Plant regeneration from leaf discs of peanut and pigeonpea: Influence of benzyladenine, indoleacetic acid and amino acid conjugates. *Plant Cell Tiss. Org. Cult.*, **35**: 223–227.

Franklin C I, Trieu T N, Gonzales R A and Dixon R A (1991) Plant regeneration from seedling explants of green bean (*Phaseolus vulgaris* L.) via organogenesis. *Plant Cell Tiss. Org. Cult.*, **24**: 199–206.

Fu Y Q, Lucchin M and Lupotto (1995) Rapid and efficient regeneration from cotyledonary explants of soybean cultivars (*Glycine max* L.). *J. Gen. Breed.*, **49**: 339–342.

Geetha N, Venkatachalam P and Rao GR (1997) *In vitro* selection and plant regeneration from polyethylene glycol adapted callus of blackgram. *Curr. Agri.*, **21**: 1–2.

Gharyal P K and Maheshwari S C (1981) *In vitro* differentiation of somatic embryos in a leguminous tree *Albizzia lebbeck* L. *Naturwissenshaten*, **68**: 379–380.

Gharyal P K and Maheshwari S C (1982) Plantlet formation in tissue cultures of the sensitive plant *Mimosa pudica* L. *Pflanzenphysiologie*, **105**: 179–182.

Gill R, Eapen S and Rao P S (1986) Tissue culture studies in mothbean – factors influencing plant regeneration from seeding explants of different cultivars. *Proc. Indian Acad. Sci.*, **96**: 55–61.

Gregory H M, Haq N and Evans P K (1980) Regeneration of plantlets from leaf callus of the winged bean *Psophocarpus tetragonolobus* (L.) DC. *Plant Sci. Lett.*, **18**: 395–400.

Gresshoff P M (1980) *In vitro* culture of white clover: callus, suspension, protoplast culture and plant regeneration. *Bot. Gaz.*, **141**: 157–164.

Gulati A and Jaiwal P K (1990) Culture conditions affecting plant regeneration from cotyledon of *Vigna radiata* (L.) Wilczek. *Plant Cell Tiss. Org. Cult.*, **23**: 1–7.

Gulati A and Jaiwal P K (1992) *In vitro* induction of multiple shoots and plant regeneration from shoot-tips of mungbean (*Vigna radiata* (L.) Wilczek). *Plant Cell Tiss. Org. Cult.*, **29**: 199–205.

Gulati A and Jaiwal P K (1994) Plant regeneration from cotyledonary node explants of mungbean (*Vigna radiata* (L.) Wilczek). *Plant Cell Rep.*, **13**: 523–527.

Gupta N and Srivastava P S (1996) *In vitro* regeneration and isozyme patterns in *Zizyphus mauritiana. J. Plant Biochem. Biotech.*, **5**: 87–90.

Hammad A H A (1996) Regeneration of some bean cultivars via *in vitro* organogenesis after irradiation by gamma rays. *Ann. Agri. Sci.*, **34**: 1117–1122.

Hammatt N, Nelson R S and Davey M R (1987a) Plant regeneration from seedlings explants of perennial *Glycine* species. *Plant Cell Tiss. Org. Cult.*, **11**: 3–11.

Hammatt N, Nelson R S and Davey M R (1987b) Plant regeneration from seedling cotyledons, petioles and leaves of *Glycine clandestina. Plant Physiol.*, **68**: 125–128.

Hammatt N, Nelson R S and Davey M R (1989) Plant regeneration from seedling explants and cotyledon protoplasts of *Glycine argyrea* Tind. In vitro *Cell. Dev. Biol. Plant*, **25**: 669–672.

Harzic N, Guilloteau A and Huyghe C (1998) *In vitro* shoot formation of *Lupinus albus* from cotyledonary node. In: *3rd European Conference on Grain Legumes. Opportunities for high quality healthy and added-value crops to meet European demands*. Valladolid, Spain, p. 369.

Hisajima S (1985) Micropropagation through multiple shoot formation from pea seeds (*Pisum sativum* L. var. arvense poir). *Jpn. J. Trop. Agri.*, **29**: 176–179.

Hymowitz T, Chalmers N L, Constanza S H and Saam M M (1986) Plant regeneration from leaf explants of *Glycine clandestina* Wendl. *Plant Cell Rep.*, **3**: 192–194.

Ignacimuthu S and Franklin G (1999) Regeneration of plantlets from cotyledon and embryonal axis explants of *Vigna mungo* L. Hepper. *Plant Cell Tiss. Org. Cult.*, **55**: 75–78.

Ignacimuthu S, Franklin G and Melchias G (1997) Multiple shoot formation and *in vitro* fruiting of *Vigna mungo* (L.) Hepper. *Curr. Sci.*, **73**: 733–735.

Illingworth J E (1974) Peanut plants from single de-embryonated cotyledons or cotyledonary fragments. *Hort. Sci.*, **9**: 462.

Jaiwal P K and Gulati A (1995) Current status and future strategies of *in vitro* culture techniques for genetic improvement of mungbean (*Vigna radiata* (L.) Wilczek). *Euphytica*, **86**: 167–181.

Kalyan Kumar De (1992) *An Introduction to Plant Tissue Culture*. New Central Book Agency, Calcutta, pp. 1–185.

Kameya T and Widholm J M (1981) Plant regeneration from hypocotyl sections of *Glycine* species. *Plant Sci. Lett.*, **21**: 289 –294.

Kaneda Y, Tabei Y, Nishimura S, Harada K, Akihama T and Kitamura K (1997) Combination of thidiazuron and basal media with low salt concentrations increases the frequency of shoot organogenesis in soybeans (*Glycine max* (L.) Merr). *Plant Cell Rep.*, **17**: 8–12.

Kartha K K, Pahl K, Leung N L and Mroginski L A (1981) Plant regeneration from meristems of grain legumes soybean, cowpea, peanut, chickpea and bean. *Can. J. Bot.*, **59**: 1671–1679.

Krishnamurthy K V, Godbole D A and Mascarenhas A F (1984) Studies on a drought resistant legume: The moth bean, *Vigna aconitifolia* (Jacq) Marechal. I. Protoplast culture and organogenesis. *Plant Cell Rep.*, **3**: 30–32.

Kubalakova M, Tejklova E and Griga M (1988) Some factors affecting root formation on *in vitro* regenerated pea shoots. *Biol. Plant.*, **30**: 179–184.

Kuchuk N, Komarnitski I, Shakhovsky A and Gleba Y (1990) Genetic transformation of *Medicago* species by *Agrobacterium tumefaciens* and electroporation of protoplasts. *Plant Cell Rep.*, **8**: 660–663.

Kulothungan S (1997) *In vitro culture studies on cowpea* (*Vigna unguiculata* (L.) Walp). Ph.D Thesis, Department of Biotechnology, Bharathidasan University, Tiruchirappalli, India.

Kulothungan S, Baskaran A, Kashinathan P, Shajahan A and Ganapathi A (1993) Morphogenetic studies on excised embryo culture of cowpea *Vigna unguiculata* (L.) Walp. *Legume Res.*, **16**: 71–74.

Kumar A S, Reddy T and Reddy G M (1983) Plantlet regeneration from different callus cultures of pigeonpea (*Cajanus cajan* L.). *Plant Sci. Lett.*, **32**: 271–278.

Kunjumon A, Kannan V R and Jasrai Y T (1996) Plant regeneration from leaf explants of three grain legumes on same medium. *J. Plant Biochem. Biotech.*, **5**: 27–29.

Ladeinde T A Q and Soh W H (1991) Effect of different growth regulators on organogenesis and total fresh weight gain in cultured leaf tissues of cowpea (*Vigna unguiculata* (L.) Walp). *Phytomorphology*, **41**: 99–107.

Li X Q and Demarly Y (1995) Characterization of factors affecting plant regeneration frequency of *Medicago lupulina* L. *Euphytica*, **86**:143–148.

Luo J P and Jia J F (1998) Callus induction and plant regeneration from hypocotyl explants of the forage legume *Astrogalus adsurgens*. *Plant Cell Rep.*, **17**: 6–7.

Malik K A and Saxena P K (1991) Regeneration in *Phaseolus vulgaris* L. promote role of N^6-benzylamino-purine in cultures from juvenile leaves. *Planta,* **814**: 148–150.

Mariotti D, Arcioni S and Pezzotti M (1984) Regeneration of *Medicago arborea* L. plants from tissue culture and protoplast cultures of different organ. *Plant Sci. Lett.*, **37**: 149–156.

Martin J P and Rabechault H (1976) The culture *in vitro* of groundnut stamens (*Arachis hypogaea* L.)–II. Establishment of tissue cultures and organogenesis (in Fr., summary in Eng.). *Oleagineux*, **31**: 19–25.

Mathews H (1987) Morphogenetic response from *in vitro* cultured seedling explants of mungbean (*Vigna radiata* (L.) Wilczek). *Plant Cell Tiss. Org. Cult.*, **11**: 233–246.

Mathews H (1988) *In vitro* responses of *Brassica juncea* and *Vigna radiata* to the antibiotic kanamycin. *Ann. Bot.*, **62**: 671–675.

Mathews H and Rao P S (1984) *In vitro* production of multiple seedlings from single seeds of mungbean (*Vigna radiata* L. Wilczek). *Pflanzen Physiologie.*, **113**: 325–329.

Mathews V H, Rao P S and Bhatia C R (1986) Somaclonal variation in cotyledonary plants of mungbean. *Pflanzenzuchtung*, **96**: 169–173.

McClean P and Grafton K F (1989) Regeneration of dry bean (*Phaseolus vulgaris* L.) via organogenesis. *Plant Sci.*, **60**: 117–122.

McKently A H, Moore G A and Gardner FP (1990). *In vitro* plant regeneration of peanut. *Crop Sci.*, **30**: 192–196.

Mehta V and Mohan Ram H Y (1980) Regeneration of plantlets from the cotyledons of *Cajanus cajan* L. *Indian J. Exp. Biol.*, **18**: 800–802.

Meijer E G M and Broughton W J (1981) Regeneration of whole plants from hypocotyl-, root-, and leaf-derived tissue cultures of the pasture legume *Stylosanthes guyanensis Physiol. Plant*, **52**: 280–284.

Mendoza A B, Hattori K, Nishimura T and Futsuhara Y (1992) Shoot regeneration from the callus of immature primary leaves in mungbean (*Vigna radiata* (L.) Wilczek). *Jpn. J. Breed.*, **42**: 134–149.

Miah S A R, Rao R and Rao R (1996) Regeneration of plantlets from excised roots of *Albizia lebbeck* Benth. *Indian J. Exp. Biol.*, **34**: 188–189.

Mohamed M F, Read P E and Coyne D P (1991) Plant regeneration *in vitro* from the embryonic axis of common and tepary beans. *Annu. Rep. Bean Improve. Co-op.*, **34**: 150–151.

Mohan M L and Krishnamurthy K V (1998) Plant regeneration in pigeonpea (*Cajanus cajan* (L.) Millsp.) by organogenesis. *Plant Cell Rep.*, **17**: 705–710.

Mohan Ram H Y, Mehta U, Ramanuja Rao I V and Narasimham M (1982) Haploid induction in legumes. In: *Plant Tissue Culture* (Ed Fujiwara A), Japanese Association for Plant Tissue Culture, Tokyo, Japan, pp. 541–542.

Mroginski L A and Fernandez A (1980) Obtain plantlets by *in vitro* culture of anthers of wild species of *Arachis* (Leguminosae) (in Sp., summary in Eng.), *Oleagineux*, **35**: 89–92.

Mroginski L A and Kartha K K (1981a) Regeneration of pea (*Pisum sativum* L. cv Century) plants by *in vitro* culture of immature leaflets. *Plant Cell Rep.*, **1**: 64–66.

Mroginski L A and Kartha K K (1981b) Regeneration of plants from callus tissue of the forage legume *Stylosanthes guianensis. Plant Sci. Lett.*, **23**: 245–251.

Mroginski L A, Kartha K K and Shyluk J P (1981c) Regeneration of peanut (*Arachis hypogaea*) plantlets by *in vitro* culture of immature leaves. *Can. J. Bot.*, **59**: 826–830.

Mukhopadhyay A and Mohan Ram H Y (1981) Regeneration of plantlets from excised roots of *Dalbergia sissoo. Indian J. Exp. Biol.*, **19**: 1113–1115.

Murthy B N S, Victor J, Singh R P, Fletcher R A and Saxena P K (1996) *In vitro* regeneration of chickpea (*Cicer arietinum* L.): Stimulation of direct organogenesis and somatic embryogenesis by thidiazuron. *Plant Growth Reg.*, **19**: 233–240.

Muthukumar B, Mariamma M and Gnanam A (1995) Regeneration of plants from primary leaves of cowpea. *Plant Cell Tissue Org. Cult.*, **42**: 153–155.

Muthukumar B, Mariamma M, Veluthambi K and Gnanam A (1996) Genetic transformation of cotyledon explants of cowpea (*Vigna unguiculata* (L.) Walp) using *Agrobacterium tumefaciens. Plant Cell Rep.*, **15**: 980–985.

Myers J R, Lazzeri P A and Collins G B (1989) Plant regeneration of wild *Glycine* species from suspension culture derived protoplasts. *Plant Cell Rep.*, **8**: 112–115.

Nagl W, Ignacimuthu S and Becker J (1997) Genetic engineering and regeneration of *Phaseolus* and *Vigna.* State of the art and new attempts. *J. Plant Physiol.*, **150**: 625–644.

Nangia S, Singh N and Singh N (1996) Micropropagation of *Leucaena leucocephala* (Lam.) de wit. *Ann. Biol.* (Ludhiana), **12**: 82–85.

Narasimhulu S B and Reddy G M (1983) Plant regeneration from different callus cultures of *Arachis hypogaea* L. *Plant Sci. Lett.*, **31**: 157–163.

Narasimhulu S B and Reddy G M (1985) Callus induction and morphogenesis in *Arachis hypogaea* L. In: *Proceedings of an International Workshop on Cytogenetics of Arachis*, 31 Oct.–2 Nov. 1983 (Eds Moss J P and Feakin S D), International Crops Research Institute for the Semi-Arid Tropics, Patancheru, India, pp. 159–163.

Nef-Campa C, Chaintreuil-Dongmo C and Dreyfus B L (1996) Regeneration of the tropical legume *Aeschynomene sensitiva* Sw from root explants. *Plant Cell Tiss. Org. Cult.*, **44**: 149–154.

Oelck M M and Schieder O (1983) Genotypic differences in some legume species affecting the redifferentiation ability from callus to plants. *Pflanzenzuchtung*, **91**: 312–321.

Ono Y, Takahata Y and Kaizuman N (1994) Effect of genotype on shoot regeneration from cotyledonary explants of rapeseed (*Brassica napus* L.). *Plant Cell Rep.*, **14**: 13–17.

Oswald T H, Smith A E and Phillips D V (1977) Callus and plantlet regeneration from cell cultures of ladino clover and soybean. *Physiol. Plant.*, **39**: 129–134.

Ozcan S (1995) *In vitro* micropropagation of pea (*Pisum sativum* L.) from immature embryos. *Turkish J. Bot.*, **19**: 427–429.

Ozcan S, Sevimary C S, Vildiz M, Sancak C and Ozgen M (1996) Prolific shoot regeneration from immature embryo explants of sainfoin (*Onobrychis viciaefolia* Scop.). *Plant Cell Rep.*, **16**: 3–4.

Ozgen M, Ozcan S, Sevimay C S, Sancak C and Yildiz M (1998) High frequency adventitious shoot regeneration in Sainfoin (*Onobrychis viciaefolia*). *Plant Cell Tiss. Org. Cult.*, **52**: 205–208.

Pandey P and Bansal Y K (1989) Plantlet formation from callus cultures of cowpea (*Vigna sinensis*). *Curr. Sci.*, **58**: 394–395.

Pandey P and Bansal Y K (1992). Plant regeneration from leaf and hypocotyl explants of *Glycine wightii* (Wand A) verdc. var longicauda. *Jpn. J. Breed.*, **42**: 1–5.

Parrott W A and Collins G B (1983) Callus and shoot-tip culture of eight *Trifolium* species *in vitro* with regeneration via somatic embryogenesis of *T. rubens. Plant Sci. Lett.*, **28**: 189–194.

Patel D B, Barve D M, Nagar N and Mehta A R (1992) *In vitro* development of immature and hybrid embryos of *Cajanus cajan* (L.) Mill sp. *Indian J. Exp. Biol.*, **30**: 871–873.

Patel M B, Bhardwaj B R and Joshi A (1991) Organogenesis in *Vigna radiata* (L.) Wilczek. *Indian J. Exp. Biol.*, **29**: 619–622.

Pellegrineschi A (1997a) *In vitro* plant regeneration via organogenesis of cowpea (*Vigna unguiculata* (L.) Walp). *Plant Cell Rep.*, **17**: 89–95.

Pellegrineschi A, Fatokun C A, Thottappilly G and Adepoju A A (1997b) Cowpea embryo rescue: 1. Influence of culture media composition on plant recovery from isolated immature embryos. *Plant Cell Rep.*, **17**: 133–138.

Phillips G C and Collins G B (1979a) Virus symptom-free plants of red clover using meristem culture. *Crop Sci.*, **19**: 213–216.

Phillips G C and Collins G B (1979b) *In vitro* tissue culture of selected legumes and plant regeneration from callus cultures of red clover. *Crop. Sci.*, **19**: 59–64.

Pittman R N, Banks D J, Kirby J S, Mitchell E D and Richardson P E (1983) *In vitro* culture of immature peanut (*Arachis* spp.) leaves: Morphogenesis and plantlet regeneration. *Peanut Sci.*, **10**: 21–25.

Pittman R N, Johnson B B and Banks D J (1984) *In vitro* differentiation of a wild peanut, *Arachis villosulicarpa* Hoehine. *Peanut Sci.*, **11**: 24–27.

Polanco M C, Pelaez M I and Ruiz M L (1988) Factors affecting callus and shoot formation from *in vitro* cultures of *Lens Culinaris* Medik. *Plant Cell Tiss. Org. Cult.*, **15**: 175–182.

Polisetty R, Paul V, Dereshwar J J, Khetarpal S, Suresh K and Chandra R (1997) Multiple shoot induction by benzyladenine and complete plant regeneration from seed explants of chickpea (*Cicer arietinum* L.). *Plant Cell Rep.*, **16**: 565–571.

Prem Anand R, Ganapathi A, Anbazhagan V R, Vengadesan G, Kolothungan S and Selvaraj N (2001) Plant regeneration from immature derived cotyledons of *Vigna unguiculata* (L.) Walp. *Curr. Sci.*, **80**: 671–674.

Rahim M A, Caligari P D S and Hill G D (1999) Multiple shoot regeneration in *Lupinus mutabilis* Sweet. Towards the 21st Century. In: *Proceedings of the 8th International Lupin Conference*, Asilomar, California, USA, pp. 264–266.

Ramsay G (1993) Regeneration in grain legume tissue culture. *Grain Legumes* (Paris) **2**: 16–17.

Rao S and Patel P (1997) *In vitro* selection of atrazine tolerant cell lines and regeneration of plantlets from the selected lines in *Cicer arietinum* (L.). *Phytomorphology*, **47**: 269–272.

Rao S H K and Narayanaswamy B (1975) Effect of gamma irradiation in cell proliferation and regeneration in explanted tissues of pigeonpea (*Cajanus cajan* (L.) Mill sp.). *Radiation Bot.*, **15**: 301–305.

Raruqui O R, Shahi V K, Harsh-Kumar and Kumar H (1996) Tissue culture response in *Pisum. Crop Improv.*, **23**: 39–43.

Reddy G M and Narasimhulu S B (1985) Plantlet regeneration and *in vitro* flowering in *Arachis hypogaea* L. *Genet Manipulat. Crops News Lett.*, **1**: 44–50.

Rubluo A, Kartha K K, Mroginski L A and Dyck J (1984) Plant regeneration from pea leaflets cultured *in vitro* and genetic stability of regenerants. *J. Plant Physiol.*, **117**: 119–130.

Rugman E E and Cocking E C (1985) The development of somatic hybridization techniques for groundnut improvement. In: *Proceeding of an International Workshop on Cytogenetics of Arachis*, 31 Oct.–2 Nov. 1983 (Eds Moss J P and Feakin S D), International Crops Research Institute for the Semi-Arid Tropics, Patancheru, India, pp. 167–174.

Ruiz M L, Pelaez Rueda J, Espino F J and Vasquesz A M (1986). A comparative study of callus formation and plant regeneration from different explants of *Phaseolus vulgaris* and *Phaseolus coccineus*. In: *Genetic*

Manipulation in Plant Breeding (Eds Horn W, Jenson DJ, Oden Bach W and Scheider O), Berlin, Germany, and New York, USA, pp. 495–497.

Sanago M H M, Shattuck V I and Strommer J (1996) Rapid plant regeneration of pea using thidiazuron. *Plant Cell Tiss. Org. Cult.*, **45**: 145–168.

Sanghamitra S, Rout G R, Premananda D, Samantaray S and Das P (1998) Plant regeneration from callus cultures of *Crotalaria* spp. *Bio. Bratislava*, **53**: 115–120.

Santalla M, Power J B and Davey M R (1998) Efficient *in vitro* shoot regeneration responses of *Phaseolus vulgaris* and *P. coccineus*. *Euphytica*, **102**: 195–202.

Sator C (1990) Lupinus (*Lupinus* spp.). In: *Biotechnology in Agriculture and Forestry, Vol. 10. Legumes and Oilseed Crops* I (Ed Bajaj Y P S), Springer-Verlag, Berlin, Germany, pp. 288–311.

Schaefer M A, Czerwinski T and Busmann A (1988). The use of embryo culture for the regeneration of interspecific hybrids from *Lupinus mutabilis* × *Lupinus hartwegii*. *Landbauforschung-Volkenrode* (Germany FR), **38**: 173–177.

Sharma V K and Kothari S C (1991) High frequency of plant regeneration in tissue culture of *Glycine clandestina*, a wild relative of soybean. *Phytomorphology*, **43**: 29–33.

Shiva Prakash N, Pental D and Sarin N B (1994) Regeneration for pigeonpea (*Cajanus cajan*) from cotyledonary node via multiple shoot formation. *Plant Cell Rep.*, **13**: 623–627.

Singh A, Singh N P, Asthana A N and Singh A (1997) Callus induction and direct regeneration from immature embryos in chickpea. *Int. Chickpea Pigeonpea News Lett.*, **4**: 39–40.

Skoog F and Miller C O (1957) Chemical regulation of growth and organ formation in plant tissue cultured *in vitro*. *Symp. Soc. Exp. Biol.*, **11**: 118–131.

Skucinska B and Miszke W (1980) *In vitro* propagation of red clover. *Pflanzenzuchtung*, **85**: 328–331.

Szabados L and Roca W M (1986) Regeneration of isolated mesophyll and cell suspension protoplasts to plants in *Stylosanthes guianesis*, a tropical legume. *Plant Cell Rep.*, **3**: 174–177.

Thome G C H, Santarem E R and Ferreira A G (1995) Adventitious bud induction and plant regeneration from soybean cotyledonary nodes. *Phyton*, (Buenos-Aires), **57**: 127–135.

Tomar U K and Gupta S C (1988) Somatic embryogenesis and organogenesis in callus cultures of a tree legume – *Albizzia lebbeck* King. *Plant Cell Rep.*, **7**: 70–73.

Vajranabhaiah S N, Purushotham M G, Reddy P C and Prakash A H (1993) Regeneration potential of hypocotyl derived long term callus cultures in groundnut (*Arachis hypogaea* L.) cv TMV-2. *Curr. Sci.*, **65**: 806–807.

Vani A K S and Reddy V D (1996) Morphogenesis from callus cultures of chickpea (*Cicer arietinum* L.). *Indian J. Exp. Biol.*, **34**: 285–287.

Vaquero F, Robles C and Riz M L (1993) A method for long term micropropagation of *Phaseolus coccineus* L. *Plant Cell Rep.*, **12**: 395–398.

Veliky I A and Martin S M (1970). A fermenter for plant cell suspension cultures. *Can. J. Microbiol.*, **16**: 223–226.

Vlachova M, Metz B A, Schell J and de Bruijn F J (1987) The tropical legume *Sesbania rostrata*: Tissue culture, plant regeneration and infection with *Agrobacterium tumefaciens* and *rhizogenes* strains. *Plant Sci.*, **50**: 213–223.

Webb K J, Fay M F and Dale P J (1987) An investigation of morphogenesis within the genus *Trifolium*. *Plant Cell Tiss. Org. Cult.*, **11**: 37–46.

Wei Z (1988) Plant regeneration from protoplasts of soybean (*Glycine max* L.). *Plant Cell Rep.*, **7**: 348–35.

White D W R (1984) Plant regeneration from long term suspension cultures of white clover. *Planta*, **162**: 1–7.

White P R (1939) Potentially unlimited growth of excised plant callus in an artificial medium. *Am. J. Bot.*, **26**: 59–64.

Widholm J M and Rick S (1993) Shoot regeneration from *Glycine canescens* tissue culture. *Plant Cell Rep.*, **2**: 19–20.

Williams D J and McHughen A (1986) Plant regeneration of the legume *Lens culinaris* Medik. (Lentil) *in vitro*. *Plant Cell Tiss. Org. Cult.*, **7**: 149–153.

Wilson V M, Haq N and Evans P K (1985) Protoplast isolation, culture and plant regeneration in the winged bean, *Psophocarpus tetragonolobus* (L.) DC. *Plant Sci.*, **41**: 61–68.

Yamada T (1989) Selection of a highly-regenerative genotype of white clover (*Trifolium repens* L.) and plant regeneration from protoplasts derived from this genotype. *Euphytica*, **44**: 181–186.

Zagorska N, Dimitrov B, Gadcva P and Robeva P (1997) Regeneration and characterization of plants obtained from anther cultures in *Medicago sativa* L. In Vitro *Cell. Dev. Biol. Plant*, **33**: 107–110.

SOMATIC EMBRYOGENESIS

P. VENKATACHALAM[2], N. GEETHA[3], P. PRIYA[2], N. JAYABALAN[4] AND G. LAKSHMI SITA[1]

1 Department of Microbiology and Cell Biology, Indian Institute of Science, Bangalore – 560 012, India

2 Biotechnology Division, 3 Plant Physiology Division, Rubber Research Institute of India, Kottayam – 686 009, India

e-mail: pchalam@rediffmail.com

4 Department of Plant Science, Bharathidasan University, Tiruchirappalli – 620 024, India

Abstract

Somatic embryogenesis (SE) has a great potential in the near future for mass multiplication and genetic improvement of several leguminous species. In the recent past, a significant progress has been made in the development of regeneration systems for various legumes, and many legumes are now amenable to regeneration via somatic embryogenesis. The purpose of this review is to focus on recent advances in the initiation and development of somatic embryos in various leguminous plants. Much emphasis has been given to identify the successful protocols for the establishment of an efficient somatic embryogenic system in legume species. It reveals that, in legumes, somatic embryogenesis can easily be achieved when young plant tissues are used as initial explants. 2,4-dichlorophenoxy-acetic acid (2,4-D) is widely used auxin for somatic embryogenesis in most of the leguminous species. Recently thidiazuron (TDZ) has been shown to induce high frequency direct somatic embryogenesis. Plant regeneration from somatic embryogenic tissue involves several steps. Firstly, the embryogenic tissue is developed from initial explant tissues, secondly, embryogenic tissue is allowed to proliferate into mature somatic embryos and thirdly somatic embryos germinate into plantlets. Although plants have been regenerated via somatic embryos from various legumes in the last two decades, low-frequency embryo induction, poor percentage of germination and conversion of somatic embryos into plants and somaclonal variations are the major obstacles limiting the exploitation of somatic embryogenic technology for biotechnological applications in legume improvement. The sequential and interactive action of many genes is clearly involved in the establishment of somatic embryos and plant regeneration. The precise role of these genes remains to be determined in most cases. Similarly, little is known of how regulatory genes function to specify major events. Many genes, e.g. calnexin-like homologues (early) and SNQ2-like

P.K. Jaiwal and R.P. Singh (eds.), Improvement Strategies for Leguminosae Biotechnology, 87–132.

genes (late), have been identified which express during somatic embryogenesis. In order to overcome these limitations, a systematic study of these patterns could help to identify various conditions such as the effect of carbohydrates, plant growth regulators, amino acids, media composition, explant age, biochemical markers and regulation of gene expression for fundamental understanding of somatic embryogenesis and the development of regeneration system for most legumes. It is our hope that as the technology develops further, it may become possible to produce large numbers of somatic embryos in bioreactors for use as artificial seeds to propagate elite or hybrid and genetically modified legume genotypes in the future.

1. Introduction

The phenomenon of inducing plant somatic embryogenesis (SE) in tissue culture was first discovered and reported four decades ago (Reinert, 1958; Steward *et al.*, 1958). Since then, this technique has been successfully applied to over three hundred diverse species of plants ranging from herbaceous annuals and perennials to trees (Bajaj, 1995). Somatic embryogenesis has been the subject of many excellent books and reviews. Numerous reviews during the past decade have examined the technical challenges associated with this technology. Most reviews are organized with tissue culture procedures, developmental aspects, molecular biology, physiological and biochemical aspects, etc. (Griga, 1999; Lakshmanan and Taji, 2000). Remarkable progress has been made in legume somatic embryogenesis in recent years (Table 1). This chapter is mainly focused on recent achievements in order to pinpoint the present approaches and problems associated with

Table 1. Somatic embryogenesis in leguminous species

Species name	Initial explants	Growth regulators used	Results	Reference
Acacia catechu	C	KIN, NAA	CD, SE, PD	Rout *et al.*, 1995a, b
Acacia koa	H, SA	2,4-D, KIN	SE	Skolmen, 1986
Acacia nilotica	C	2,4-D, BA	SE, PD	Garg *et al.*, 1996
Albizia amara	H	BA	SE	Tomar and Gupta, 1986
A. lebbeck	H	Nil	SE, PD	Gharyal and Maheshwari, 1981
A. lebbeck	C	2,4-D, IAA, KIN, BA	SE, PD	Gharyal *et al.*, 1983b
A. lucida	H	BA	SE	Tomar and Gupta, 1986
A. richardiana	H	BA	SE, PD	Tomar and Gupta, 1988
A. richardiana	H	BA	SE	Tomar and Gupta, 1986
Arachis hypogaea	IE	2,4-D	SE, PD	Hazra *et al.*, 1989
	IC	PIC	SE, PD	Ozias-Akins, 1989
	IE	PIC	SE, PD	Sellars *et al.*, 1990
	IC	2,4-D	SE, PD	Durham and Parrott, 1992; Wetzstein and Baker, 1993

Table 1. Continued

Species name	Initial explants	Growth regulators used	Results	Reference
	IS	TDZ	SE, PD	Saxena *et al.*, 1992; Murthy *et al.*, 1995
	IE, IC	2,4-D, DIC, 2,4,5-T, NAA, IPA	SE, PD	Eapen and George, 1993
	ME	2,4-D, 2,4,5-T, NAA, PIC	SE, PD	McKently, 1991, 1995
	L	2,4-D, KIN	CD, SE, PD	Baker and Wetzstein, 1992, 1998
	L	2,4-D	CD, SE, PD	Chengalrayan *et al.*, 1994, 1995, 1997, 1998
	L	2,4-D	CD, SE, PD	Mhaske *et al.*, 1998
	IC	NAA, 2,4-D	SE, PD	Baker *et al.*, 1994; Baker and Wetzstein, 1994
	S	2,4-D	SE, PD	Baker *et al.*, 1995
	H	2,4-D, NAA, BAP	CD, SE, PD	Venkatachalam *et al.*, 1996
	L	2,4-D, BAP	CD, SE, PD	Venkatachalam *et al.*, 1997
	IL	2,4-D, BAP, NAA	CD, SE, PD	Venkatachalam *et al.*, 1999a
	MC	NAA, BAP	SE, PD	Venkatachalam *et al.*, 1999b, 2000
	E	PIC, DIC, Cx	SE, PD	Little *et al.*, 2000
Arachis pintoi	L	PIC, BA or 2iP	CD, SE, PD	Rey *et al.*, 2000
A. hypogaea × *A. stenosperma*	IC	PIC	SE, PD	Ozias-Akins *et al.*, 1992
Astragalus adsurgens	PC	2,4-D, NAA, BA	CD, SE, PD	Luo and Jia, 1998
Cajanus cajan	A	IAA, KIN	C, E	Bajaj *et al.*, 1980
	IC, IE	PIC, 2,4-D, NAA, GA, ABA, Z	SE, PD	George and Eapen, 1994
	C	BAP, KIN, IBA, GA	SE, PD	Patel *et al.*, 1994
	C, L	TDZ, BA	SE, PD	Sreenivasu *et al.*, 1998
	L	2,4-D	CD, SE, PD	Anbazhagan and Ganapathy, 1999
Cassia fistula	A	IAA, KIN	SE	Bajaj and Dhanju, 1983
C. siamea	A	2,4-D, KIN	SE	Gharyal *et al.*, 1983a
Cercis canadensis	IE	2,4-D	SE, PD	Trigiano *et al.*, 1988, 1995; Buckley and Trigiano, 1994
	IE	2,4-D	SE, PD	Geneve and Kester, 1990; Geneve, 1990

Table 1. Continued

Species name	Initial explants	Growth regulators used	Results	Reference
Chamaecytisus austriacus	C	2,4-D, IAA, KIN	SE	Greinwald and Czygan, 1991
C. purpureus	C	IBA, BA	CD, SE, PD	Greinwald and Czygan, 1991
Cicer arietinum	IC	Z, IAA	SE, PD	Shri and Davis, 1992
	IL	2,4-D, NAA, KIN, BAP, GA	SE, PD	Barna and Wakhlu, 1993
	IC	2,4,5-T, 2,4-D, GA, Z, DIC	SE, PD	Sagare *et al.*, 1993, 1995
	IC, IE	2,4-D, PIC, NAA, DIC, GA	SE, PD	Eapen and George, 1994
	L	2,4-D, KIN, NAA, IBA, IAA	SE, PD	Kumar *et al.*, 1994
	L	PIC	SE, PD	Kumar *et al.*, 1995
	MS	TDZ	SE, PD	Murthy *et al.*, 1996
	IC, ME	2,4,5-T	SE, PD	Suhasini *et al.*, 1997
Cladastris lutea	IE	2,4-D	SE, PD	Weaver and Trigiano, 1991
Dalbergia latifolia	IE	2,4-D	SE, PD	Rao and Lakshmi Sita, 1992
D. sissoo	C	2,4-D, KIN	CD, SE	Das *et al.*, 1997
Genista monosperma	C	2,4-D, NAA, BA, Z	SE	Curir *et al.*, 1990
Glycine sp.	H	PIC, BA, 2,4-D	CD, SE, PD	Gamborg *et al.*, 1983
Glycine max	IE	2,4-D	CD, SE, PD	Christianson *et al.*, 1983
	IE	2,4-D	SE, PD	Lippmann and Lippmann, 1984
	IE	2,4-D, NAA	SE, PD	Lazzeri *et al.*, 1985
	IE	2,4-D	CD, SE, PD	Ranch *et al.*, 1985
Glycine max	IE, C	NAA, 2,4-D, BAP, ABA	SE, PD	Lazzeri *et al.*, 1987
	IE, C	NAA, KIN, BAP, Z, GA	SE, PD	Lazzeri *et al.*, 1987
	ZE	DIC, GA, KIN, IBA	SE, PD	Christou and Yang, 1989
	IC	2,4-D	SE, PD	Bailey *et al.*, 1993a, b
	C, H	NAA, 2,4-D	SE	Liu *et al.*, 1992
	IC	NAA, PIC, 2,4-D	SE	Griga, 1993
	IC	2,4-D	SE	Tian *et al.*, 1994
	C	2,4-D	SE, PD	Santos *et al.*, 1997
	IC	2,4-D	SE, PD	Li and Grabau, 1996
	IC	2,4-D	SE, PD	Samoylov *et al.*, 1998
	IC	2,4-D	CD, SE, PD	Simmonds and Donaldson, 2000

Table 1. Continued

Species name	Initial explants	Growth regulators used	Results	Reference
	IE	2,4-D, ABA	SE, PD	Distabanjong and Geneve, 1997
Hardwickia binata	C	NAA, KIN	CD, SE	Das *et al.*, 1995
Lens culinaris	IE	2,4-D	CD, SE, PD	Saxena and King, 1987
Lupinus angustifolius	IC	2,4-D, NAA,	SE, PD	Nadolska-Orczyk, 1992
Lupinus albus	IC	2,4-D, NAA	SE, PD	Nadolska-Orczyk, 1992
Lupinus mutabilis	IC	2,4-D, NAA	SE, PD	Nadolska-Orczyk, 1992
Medicago falcata	L	2,4-D, KIN	SE	Denchev *et al.*, 1990, 1991, 1993; Kuklin *et al.*, 1994
Medicago polymorpha	H	2iP, IAA	CD, SE, PD	Scarpa *et al.*, 1993
Medicago sativa	IE	BAP	SE	Maheswaran and Williams, 1984, 1985
	YP	NAA, KIN, 2,4-D	CD, SE, PD	Strickland *et al.*, 1987
	L, C	NAA, 2,4-D, KIN	CD, SE, PD	Chen *et al.*, 1987
	L, P, I	NAA, IAA, KIN	SE	Parrott and Bailey, 1993
	P	2,4-D, PAA	SE	Finstad *et al.*, 1993
	P	2,4-D, KIN	SE	Shetty and McKersie, 1993; Lecouteux *et al.*, 1993; Lai and McKersie, 1994 a, b, c; Horbowicz *et al.*, 1995; Senaratna *et al.*, 1995
	IE	BA	SE, PD	Ninkovic *et al.*, 1995
	L, P, I	NAA, IAA, KIN	SE	Parrott and Bailey, 1993
	L	2,4-D, BA	SE, PD	Trinh *et al.*, 1998
Medicago suffruticosa	L	2,4-D, BA	SE, PD	Li and Demarly, 1996
Medicago truncatula	L	NAA, BAP	CD, SE, PD	Nolan *et al.*, 1989
	F	2,4-D, Z	SE	das Neves *et al.*, 1999
	L	2,4-D, BA	SE, PD	Trinh *et al.*, 1998
Mimosa golois	C	2,4-D, NAA, BA, ABA	SE, PD	Chatterjee *et al.*, 1991
Phaseolus aureus	A	IAA, KIN	C, E	Bajaj *et al.*, 1980
	IS	BAP, NAA	SE, PD	Malik and Saxena, 1992
Phaseolus acutifolius, P. coccineus, P. wrightii	IS	BAP, NAA	SE, PD	Malik and Saxena, 1992

Table 1. Continued

Species name	Initial explants	Growth regulators used	Results	Reference
Phaseolus coccineus	IC	2iP, NOA	CD, SE, PD	Angelini and Allavena, 1989
Pisum sativum	IE, SA	PIC, 2,4-D, NAA	SE, PD	Kysely *et al.*, 1987
Pisum sativum	IE, SA	PIC, 2,4-D	SE, PD	Kysley and Jacobsen, 1990
	IC	NAA, 2,4-D	SE, PD	Ozcan *et al.*, 1993
	IE	2,4-D, NAA, PIC	SE, PD	Stejskal and Griga, 1993
	SA	PIC, BA, Z, 2,4-D KIN, NAA	SE, PD	Loiseau *et al.*, 1995
	SA, PC	2,4-D, PIC KIN, NAA	CD, SE, PD	Lehminger-Mertens and Jacobsen, 1989
	IE	2,4-D	SE, PD	Tetu *et al.*, 1990
Poinciana regia	A	IAA, KIN	SE	Bajaj and Dhanju, 1983
Psophocarpus tetragonolobus	L	NAA, BAP, 2,4-D,	CD, SE, PD	Ahmed *et al.*, 1996; Datta Gupta *et al.*, 1997
Robinia pseudoacacia	IS	2,4-D, BA	SE, PD	Arrillaga *et al.*, 1994
	IS	2,4-D, BA	SE, PD	Merkle and Wiecko, 1989
	C	2,4-D, BA	SE, PD	Chalupa, 1987
Sesbania rostrata	C	2,4-D, BA	SE	Hanower *et al.*, 1986
Trifolium repens,	IE	BAP	SE	Maheswaran and Williams, 1984, 1985
T. pratense	P	2,4-D, KIN	CD, SE, PD	Rybczynski *et al.*, 1994
Trifolium fragiferum	P, L, IE	2,4-D, NAA, Z	CD, SE, PD	Rybczynski, 1997
Trifolium subterraneum	SA	PIC	SE, PD	Heath *et al.*, 1993
Vicia narbonensis	ML	PIC, BAP	CD, SE, PD	Albrecht and Kohlenbach, 1989
	SA	2,4-D	CD, SE, PD	Pickardt *et al.*, 1989
Vicia narbonensis	E, SA	2,4-D, NAA, BA	CD, SE, PD	Tegeder *et al.*, 1996
Vigna aconitifolia	IC	NAA, PIC, 2,4-D	CD, SE, PD	Eapen and George, 1990
V. glabrescens × *V. radiata*	C	NAA, KIN, GA	SE, PD	Chen *et al.*, 1990
V. mungo	IC	NAA, PIC, 2,4-D	CD, SE, PD	Eapen and George, 1990
	H	2,4-D, NAA, BAP	CD, SE, PD	Geetha *et al.*, 1997

Table 1. Continued

Species name	Initial explants	Growth regulators used	Results	Reference
V. radiata	IC	NAA, PIC, 2,4-D	CD, SE, PD	Eapen and George, 1990
V. sinensis	IC	2,4-D, NAA, BAP, KIN, IAA	CD, SE, PD	Li *et al.*, 1995

Abbreviations: A – anthers; H – hypocotyl; E – epicotyl; IE – immature embryos; YP – young petiole; SA – shoot apex; L – leaflet; C – cotyledon; F – foliole; ZE – zygotic embryos; IC – immature cotyledon; IEM – immature endosperm; IS – intact seedlings; P – petiole; I – internode; ME – mature embryos; MS – mature seeds; ML – mature leaflet; PC – protoplast callus; C – callus; E – embryos; CD – callus development; Cx – centrophenoxine; SE – somatic embryos; PD – plantlet development; DIC – dicamba; Z – zeatin; GA – gibberellic acid; PIC – picloram; IAA – indole-3-acetic acid; NAA – 1-naphthaleneacetic acid; IBA – indole-3-butyric acid; KIN – kinetin; 2,4-D – 2,4-dichlorophenoxyacetic acid; ABA – abscisic acid; BAP – benzylaminopurine; BA – benzyladenine; 2-ip – isopentyladenine; TDZ – thidiazuron; 2,4,5-T – 2,4,5-trichlorophenoxyacetic acid.

legume somatic embryogenesis. Somatic embryogenesis has been defined by Williams and Maheswaran (1986) as a process by which haploid or diploid cells develop into differentiated plants through characteristic embryological stages without fusion of gametes. The stages are identical to zygotic embryogenesis stages. In dicots and conifers, somatic embryogenesis proceeds subsequently from (pre)globular, torpedo, cotyledonary to mature somatic embryos (Ammirato, 1987), and in monocots from (pre)globular, scutellar and coleoptilar stages to mature somatic embryos (Conger *et al.*, 1983). Plant embryogenesis and development is a particularly complex process. Plantlets can be obtained through the normal zygotic pathway or from somatic or androgenetic cells. Even at the zygotic level the mechanisms underlying the genesis of an embryo are as yet poorly understood (Dodeman *et al.*, 1997).

The family Leguminosae or Fabaceae is one of the largest and most diverse groups of plants represented by approximately 700 genera that encompass almost 18,000 species. Leguminosae family shows more promise for producing food for human consumption, pesticides and medicines, fodder for wildlife and domesticated livestock and wood for construction, paper products and fuels (Trigiano *et al.*, 1992, 1999; Lakshmi Sita, 1999). Attempts to obtain insect pest, fungal, viral resistant genotypes of legumes by conventional breeding methods have not been successful because of limited genetic variation and sexual incompatibility with wild relatives. Genetic transformation is a potential complementary technology that may be able to introduce traits, difficult to obtain via traditional plant breeding (e.g. pest, viral and fungal resistance) as genes related to these traits become obtainable. Biotechnological approaches have been useful to introduce genes coding for insecticidal proteins and viral resistance genes into different legume species in obtaining transgenic plants (Kar *et al.*, 1997; Singsit *et al.*, 1997; Yang *et al.*, 1998). So far, most gene transfer procedures require the availability of axenic material capable of plant regeneration such as embryogenic cells or protoplasts. For this reason, the study of "plant to plant" and "cell to plant" systems has been undertaken. Shoot regeneration from legume species has been reported to be difficult (Sen and Mukherjee, 1998;

Venkatachalam *et al.*, 1999a, b), compared to other crops or responses appeared genotype and species dependent (Geetha *et al.*, 1998; Dan and Reichert, 1998). For a long time grain legumes represented a very recalcitrant plant family concerning *de novo* regeneration *in vitro* and particularly the somatic embryogenesis pathway of regeneration (Griga, 1999). Regardless of the plant of interest, most plant transformation protocols need to be coupled with reliable *in vitro* regeneration protocols.

Somatic embryo cultures are one of the most important and convenient systems for plant genetic engineering and *in vitro* propagation. Unlike shoots derived via organogenesis, somatic embryos consist of complete propagules that do not require a separate rooting step. There are many benefits anticipated from somatic embryos based on their high potential for proliferacy, their bipolar structure with both root and shoot meristems and their ability to be singulated for synthetic seed (Janick, 1993). The fact that somatic embryos can regenerate from single cells provides opportunities for crop improvement exploiting biotechnological techniques such as cell selection, somatic hybridization and genetic transformation. As technology develops further, it may become possible to grow large numbers of somatic embryos in bioreactors for use as artificial seed to propagate elite or hybrid genotypes (Parrott *et al.*, 1995). Although published research and development of artificial seed technology in legumes have been limited to few legume species, advances in somatic embryo technology, including the ability to desiccate somatic embryos, should eventually facilitate their storage and use as artificial seed. The greatest limitation to artificial seed technology appears to be the requirement for high quality somatic embryos with high conversion rates, but one limitation continues to be the lack of accumulation of storage reserves in somatic embryos to the extent seen in zygotic embryos (Parrott *et al.*, 1995). Therefore, somatic embryogenesis has been reported in many plant species including legumes and extensive studies have been performed to establish and evaluate factors that may affect the induction and proliferation of embryogenic cultures for a number of legume crops (Samoylov *et al.*, 1998a). The limiting factor for the exploitation of somatic embryos still remains to be the lack of reliable protocols for induction, development and conversion (germination) especially in recalcitrant species such as legumes. To date, however, somatic embryogenesis in legume species has met with only limited success. This chapter offers an overview about the effect of various factors on somatic embryogenesis and plant regeneration in different legume species.

2. Types of somatic embryogenesis

Based on the morphology, the proliferating embryogenic cells or tissues can be divided in two groups. Direct and indirect somatic embryogenesis can be considered as two extremes of a continuum (Williams and Maheswaran, 1986; Carman, 1990; Emons, 1994). Two routes of somatic embryogenesis were postulated (Sharp *et al.*, 1980), with the main features also being observed in grain legume species (Parrott *et al.*, 1995). Once a cell or group of cells is determined to follow the embryogenic pathway, they will start to divide in a co-ordinated manner. When this co-ordinated behaviour is lost, then individual cells or groups of cells re-enter the embryogenic pathway and form a new generation of embryogenic tissue. The first is indirect somatic embryogenesis where the breakage of co-ordination of embryogenic determined groups of cells and their re-entrance into the embryogenic pathway occurs at an early stage of embryogenic development. As a result a callus-like mass structure is formed. The formation of such cultures has been defined by Sharp *et al.* (1980) as indirect somatic embryogenesis. The cells which become committed to embryogenesis after their de-differentiation in culture are termed induced embryogenic determined cells (IEDC) (Griga, 1999).

In the second type of proliferating cultures, the embryogenic determined cells develop into somatic embryos (globular shaped or more advanced) before the loss of group control allows specific cells to enter a new cycle of somatic embryogenesis. The formation of such cultures has been defined by Sharp *et al.* (1980) as direct somatic embryogenesis. Direct somatic embryos initiate directly from the cells of cultured tissue in the absence of callus proliferation. This occurs through so-called pre-embryogenic determinated cells (PEDC), where the cells are already committed to embryogenic development and need only to be released from spatial and physiological relations of the tissue-decorrelated (Erdelska and Vidovencova, 1992), so as to fully express their totipotency. The embryogenic cells show a number of common features, which are characteristic for rapidly dividing cells – these include small size, dense cytoplasmic contents, large nuclei with prominent enlarged nucleoli, small vacuoles and a profusion of starch grains. They exhibit intense RNA synthesis and metabolic activity (Griga, 1999). Somatic or asexual embryogenesis is the production of embryo-like structures from somatic cells. This process can proceed directly from sporophytic or gametophytic cells of a cultured explant when the embryos originate *in vitro* without an intervening callus, i.e. from pre-embryogenic-determined cells (PEDC). In both cases, however, somatic embryos result from indirectly determined embryogenic cells (IEDC). In both cases, the overall pattern of development of somatic embryos is similar to that of their zygotic counterparts, including the ability to germinate into whole plants (Sharp *et al.*, 1982).

3. Initiation and development of somatic embryos

The most important factors controlling somatic embryo induction were the developmental stage of initial explants, plant growth regulators type and concentrations used at the time of culture initiation. Other factors as e.g. mineral composition of culture media, type and concentration of carbohydrates and amino acids as well as some physical factors seem to play less important role in the process (Girga, 1999). Somatic embryogenesis in legumes appear to be a collection of diverse mechanisms. Some legumes, such as clovers (*Trifolium* spp.) have been reported to form somatic embryos upon exposure to cytokinins (Maheswaran and Williams, 1984). Others such as soybean (*Glycine max*), form somatic embryos in response to auxin and regenerate directly from zygotic embryo explant tissues without an intervening callus phase (Lazzeri *et al.*, 1985). Requirement of plant growth regulators for the induction of somatic embryogenesis is largely determined by the developmental state of the explant tissue (Lakshmanan and Taji, 2000). In most of the legume species, induction of somatic embryogenesis was achieved in the presence of auxins alone or in combination with cytokinins. Response of legume species to plant growth regulators varies significantly depending on the explant type and hormonal combinations. From the earlier reports on somatic embryogenesis in legume species it is apparent that exogenously supplied plant growth regulators play an important role in somatic embryo induction and development.

The majority of reports of successful somatic embryogenesis used embryonic axes as the explant source. In addition, immature cotyledons (Baker and Wetzstein, 1995; Eapen and George, 1993b), immature leaflet, embryo-derived leaflet (Baker and Wetzstein, 1992, 1994; Chengalrayan *et al.*, 1994), immature embryonal axis (Hazra *et al.*, 1989; Ramdev Reddy and Reddy, 1993; Wetzstein and Baker, 1993) and mature embryo axis (Baker *et al.*, 1995; McKently, 1991; Sellars *et al.*, 1990) have hitherto been successfully used for somatic embryogenesis in peanut. The ability to recover somatic embryos from axes of harvested, stored seeds represents significant advantages for the establishment of peanut embryogenic cultures, including the use of simple sterilization procedures and a constant source of explant tissue (Baker *et al.*, 1994, 1995). Durham and Parrott (1992) described that liquid embryogenic cultures can produce virtually unlimited numbers of somatic embryos. During somatic embryogenesis, from immature leaflets of peanut, globular, heart, torpedo stages were not clearly delimited (Chengalrayan *et al.*, 1994). In contrast, Venkatachalam *et al.* (1997, 1998) have observed various stages of somatic embryos (globular, heart, torpedo and cotyledonary stages) in both solid and liquid cultures of peanut (Fig. 1). Venkatachalam and Jayabalan (1996c) have reported plant regeneration from immature leaflet derived protoplasts via callus cultures of cultivated peanut through somatic embryogenesis. Somatic embryos began in the proembryo stage and proceeded from globular to dicotyledonary stage. Induction of somatic embryogenesis from young leaflets of cultivated and wild species of peanut was reported by Rani and Reddy (1996). Plant regeneration via somatic embryogenesis has been achieved using different explants of peanut by Venkatachalam *et al.* (1997). Recently, somatic embryogenesis and plant regeneration from immature leaflet derived cell suspension cultures of peanut has been developed by Venkatachalam *et al.* (1998). Thidiazuron induced high frequency direct somatic embryogenesis has been reported in peanut (Saxena *et al.*, 1992; Murthy *et al.*, 1995). Rey *et al.* (2000) have obtained embryogenic callus from leaf pieces of *Arachis pintoi* on medium containing picloram, KIN, zeatin, BA or 2iP.

Higher concentrations of 2,4-D or NAA in combination with BAP decreased the embryo induction frequency as well as number of somatic embryos/culture (Venkatachalam *et al.*, 1997, 1999a). Similar effect of 2,4-D in combination with BAP has been reported earlier in peanut (McKently, 1991; Ramdev Reddy and Reddy, 1993). McKently (1991) found in embryo axis cultures, that as auxin concentrations increased, the probability of obtaining normal-shaped peanut somatic embryos decreased. Higher concentrations of auxin not only decreased the embryos/explant but also delayed the embryogenesis (Ramdev Reddy and Reddy, 1993). 2,4-D has been found best for the induction of indirect somatic embryogenesis in peanut (McKently, 1991; Eapen *et al.*, 1993; Chengalrayan *et al.*, 1994; Venkatachalam *et al.*, 1996, 1997, 1998, 1999a, b, 2000). Although auxin appears to be the principal growth regulator suppressing histodifferentiation in most species examined (Merkle *et al.*, 1995), complete removal of BA from the induction medium was more critical for somatic embryo differentiation in *Medicago suffruticosa* (Li and Demarly, 1996). Plant regeneration via somatic embryogenesis in blackgram (*Vigna mungo*) has proved to be difficult (George and Eapen, 1990). In *Vigna mungo* somatic embryogenesis and plantlet development was first reported from immature

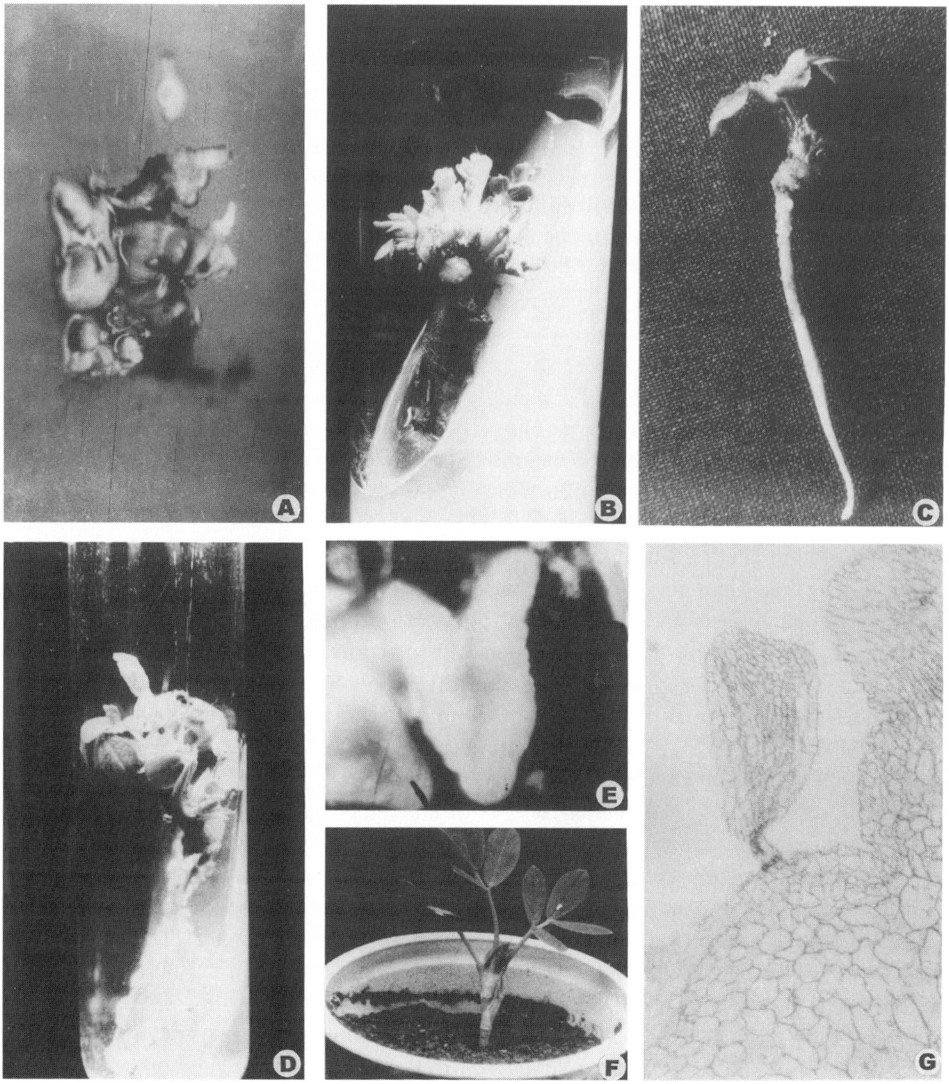

Figure 1. Induction of somatic embryos and plantlet regeneration in peanut. A. Embryo formation from hypocotyl cultures of peanut, B. Direct somatic embryo development on mature cotyledon explant of peanut, C. A single germinated somatic embryo with shoot and elongated radicle, D. In vitro *flowering on germinated somatic embryo derived plantlet, E. Cotyledonary stage embryo developed in leaflet derived cell suspension cultures of peanut, F. Somatic embryo derived plantlet established in plastic cup containing soil, G. Longitudinal section of globular and heart-shaped somatic embryos differentiated at 4 weeks.*

cotyledon explants by Eapen and George (1990). Subsequently, somatic embryogenesis and plantlet regeneration were also reported from hypocotyl explants of blackgram. However, the frequency of plant regeneration was low (Geetha, 1995; Geetha *et al.*, 1997). Eapen and George (1990) have reported the ontogeny of somatic embryos of *Vigna mungo* indicating that cells destined to develop somatic embryos divided to form filamentous proembryos. From the filamentous proembryo, the proper embryo developed either at single or multiple sites resulted in several embryos. They were connected by a common suspensor at the radicle. The continued divisions of the proembryos led to globular, heart shaped and cotyledonary stage of somatic embryos. Interestingly, in our study, somatic embryos of *Vigna mungo* were also developed in this way. We have traced the development of embryos through two distinct pathways, (1) division of round and oval cells to form spherical proembryos and (2) division of elongated cells to form filamentous structures which in turn produce single or multiple embryos (Geetha *et al.*, 1997). Sinha *et al.* (1983b) induced embryogenesis from mesophyll protoplasts of *Vigna mungo*. No reports are available on shoot developments from these embryos.

Bajaj *et al.* (1980) obtained callus showing a wide range of ploidy in the *in vitro* cultured anthers of pigeonpea. A suspension of pollen from such anthers incubated in drop cultures on agar medium developed further to form embryoids and colonies of callus. However, further plant regeneration was not reported. George and Eapen (1994) have reported organogenesis and somatic embryogenesis from seed and seedling explants of pigeonpea. Somatic embryogenesis has also been reported in pigeonpea by Patel *et al.* (1994) and Mallikarjuna *et al.* (1996). Patel *et al.* (1994) reported induction of somatic embryos in cotyledons of pigeonpea when cultured on MS and B_5 media supplemented with BAP (22.2 μM), KIN (2.3 μM) and AdS (271.0 μM). The distal halves of the isolated cotyledons expanded rapidly, turned green and induction of embryogenesis occurred. Venu *et al.* (1999) obtained 77% of embryogenesis with 30 embryos/explant, but germination of embryos into plants was not reported. Sreenivasu *et al.* (1998) reported effcient plant regeneration method via somatic embryogenesis in pigeonpea. Cotyledon and leaf explants from 10-day-old seedlings produced embryogenic callus and somatic embryos when cultured on MS medium supplemented with TDZ. Subsequent withdrawal of TDZ from the induction medium resulted in the maturation and growth of the embryos into plantlets on MS basal medium. Kulothungan *et al.* (1995) developed somatic embryos from cell suspension of cowpea on MS medium with 2,4-D and picloram. Recently, somatic embryos induced on MS medium supplemented with 2,4-D from cell suspension cultures of pigeonpea (Anbazhagan and Ganapathy, 1999).

Until recently, attempts to regenerate soybean had yielded only non-functional somatic embryos. Using immature embryos as explants, however, several instances of somatic embryogenesis and whole plant regeneration have now been reported. The earliest attempt to regenerate soybean plants via somatic embryogenesis was reported by Beversadorf and Bingham (1977) although somatic embryos were obtained only sporadically. Several groups (Phillips and Collins, 1981; Christianson *et al.*, 1983; Gamborg *et al.*, 1983; Lippmann and Lippmann, 1984; Li *et al.*, 1985) have attempted to regenerate soybean plants using different explants by modifying the components of the media, primarily the auxin source and concentration. However, little progress was made until 1985 when the first fertile regenerated plants were obtained by Lazzeri *et al.* (1985) and Ranch *et al.* (1985). Subsequently, numerous critical factors for each of the process in somatic embryogenesis have been investigated (Ranch *et al.*, 1985; Barwale *et al.*, 1986; Ghazi *et al.*, 1986; Lazzeri *et al.*, 1987a, b, 1988; Hammatt and Davey, 1987; Parrott *et al.*, 1988; Kerns *et al.*, 1986; Komatsuda *et al.*, 1992) to optimize the system. Although regenerated plants can be routinely obtained via auxin stimulated somatic embryogenesis, up to 76% of the explants respond producing 1 to 3.14 embryos per responding explant. In addition, all successful plant regeneration via somatic embryogenesis has been obtained exclusively from immature zygotic embryo cotyledons.

Hence, the availability of a suitable tissue culture technology for soybean embryo proliferation and regeneration may be a limiting step for efficient soybean genetic transformation (Samoylov *et al.*, 1998a). High efficiency of somatic embryogenesis has been obtained via somatic embryo cycling and the somatic embryo hypocotyl/radicle was used as the explant (Liu *et al.*, 1992). Similar somatic embryo cycling system was reported by Wright *et al.* (1991). The age of the immature zygotic embryo cotyledons also affected some somatic embryogenesis (Barwale *et al.*, 1986; Lippmann and Lippmann, 1984; Lazzeri *et al.*, 1985). The origin of somatic embryos induced from soybean immature zygotic cotyledons on NAA and 2,4-D has been reported by Hartweck *et al.* (1988), Hepher *et al.* (1988) and Liu *et al.* (1992). Although somatic embryogenesis in soybean has been reported by a number of laboratories (Christianson *et al.*, 1983; Lippman and Lippman, 1984; Lazzeri *et al.*, 1985; Ranch *et al.*, 1985; Parrott *et al.*, 1988; Buchheim *et al.*, 1989; Griga, 1993), most of the protocols described in these early studies have not been successfully used for genetic transformation work (Trick *et al.*, 1997). Transformation of soybean has been far from routine. The first two reports of soybean transformation used two very different methods to transform soybean. Hinchee *et al.* (1988) used *Agrobacterium*-mediated transformation of cotyledonary nodes while McCabe *et al.* (1988) used particle bombardment of shoot meristems. The putative transformed shoot apices were induced to form multiple shoots prior to whole plant regeneration. All of the transgenic plants described in this first report were chimeric. Somatic embryos offer an alternative target tissue for particle bombardment. Soybean transformation reports following these initial papers have been limited and the transformation efficiency for soybean has remained low (Trick *et al.*, 1997; Meurer *et al.*, 1998); only a handful of laboratories have had more than sporadic success. Recovery of transgenic soybean plants using this target tissue was first reported by Finer and McMullen (1991). This system has since been used by numerous laboratories to produce transgenic plants (Sato *et al.*, 1993; Parrott *et al.*, 1994; Hadi *et al.*, 1996; Stewart *et al.*, 1996; Liu *et al.*, 1996). Hadi *et al.* (1996) introduced 12 different plasmids into soybean embryogenic suspension cultures via particle bombardment method.

Finer and Nagasawa (1988) developed a medium for initiation of embryogenic cultures in soybean. Subsequently, liquid media are particularly adopted for large-scale proliferation of somatic embryos and their use has been incorporated into transformation systems for soybean (Finer and McMullen, 1991; Sato *et al.*, 1993; Parrott *et al.*, 1994; Stewart *et al.*, 1996; Samoylov *et al.*, 1998b). When embryogenic suspensions were directly compared to shoot apices as the target tissue for particle bombardment, embryogenic suspensions appeared to be more efficient of the two methods (Sato *et al.*, 1993). However, the use of the embryogenic suspensions does suffer from limitations. The major limitations of this method are the efforts required to establish liquid cultures and problems with sterility of plants regenerated from old embryogenic cultures (Hadi *et al.*, 1996). According to Simmonds and Donaldson (2000), 18 soybean genotypes screened for embryogenesis and only five genotypes produced embryogenic cultures which were used for transformation. Although fertile plants were recovered from embryogenic cultures, transgenic plants, which were derived from cultures 12–14 months of age, were sterile.

Although globular somatic embryos were observed in leaflet derived cultures of chickpea by Rao and Chopra (1989), further differentiation into plantlets had not been recorded. Until 1993, complete regeneration of mature plants and their successful transfer to soil via somatic embryogenesis has not been reported for chickpea. Regeneration of plants via somatic embryogenesis for the first time has been reported by Barna and Wakhlu (1993). Direct somatic embryogenesis and plant regeneration has also been reported by Suhasini *et al.* (1994) and Barna and Wakhlu (1995) in chickpea. Suhasini *et al.* (1994) reported that about 65% of the mature embryo axis explants produced globular stage somatic embryos directly as protuberances from the surface of the shoot tip region, of each explant bearing 12–15 embryos in chickpea. Vani and Reddy (1996) described somatic embryogenesis and plant regeneration using cotyledon derived cultures of chickpea on B_5 medium supplemented with 2.0 mg/L 2,4-D and 3% sucrose. Addition of glutamine (400 mg/L) along with 2,4-D increased the frequency of embryogenesis (54.2%). Anatomical studies showed the presence of globular, heart and torpedo shaped embryos.

Barna and Wakhlu (1993) obtained numerous globular embryos on the surface of callus on MS medium containing 25 μM 2,4-D in chickpea. Immature leaflets of chickpea were cultured on MS medium with 50 μM 2,4-D or picloram for somatic embryogenesis by Barna and Wakhlu (1995). Sagare *et al.* (1993) have reported embryogenesis from cotyledonary explants of three genotypes of chickpea. Subsequently, somatic embryogenesis has also been reported in chickpea by Eapen and George (1994) and Kumar *et al.* (1994). Eapen and George (1994) obtained plant regeneration through somatic embryogenesis in chickpea using immature cotyledons and immature embryonal axes as explants. Kumar *et al.* (1994) developed a method for plant regeneration via somatic embryogenesis from leaf explants of chickpea (cv. C235) on MS medium with 2,4-D, KIN and picloram. Kumar *et al.* (1995) cultured leaf explant of 12 genotypes of chickpea on MS medium containing 1.25 mg/L 2,4-D and 0.25 mg/L KIN under dark for 4 weeks. The calli were then transferred to maturation medium containing 0.125 mg/L IBA and 2.0 mg/L BA. Fully developed embryos were transferred to B_5 medium with 0.2 mg/L BA. Somatic embryogenesis was 28–97%. Ramana *et al.* (1996) described direct somatic embryogenesis from immature cotyledonary segments and genetic transformation via *Agrobacterium* in chickpea whereas Murthy *et al.* (1996) reported somatic embryogenesis in chickpea using mature seed as explant and TDZ and L-proline as induction factor. Recently, Suhasini *et al.* (1997) have attempted a comparative study of the development of zygotic and somatic embryos of chickpea. Immature cotyledonary segments and mature embryo axes were cultured on MS medium with 2,4,5-T for induction of somatic embryos. Somatic embryos were raised on a multiseriate suspensor like stalk. The size of a globular stage somatic embryo was much larger than zygotic embryo at a similar stage of development. The cells of apical dome of the zygotic embryo were flat in comparison to that of somatic embryos, which were curved. Germinated seedlings and converted somatic embryos showed similarity in organization, differing only in the size of cotyledons.

Somatic embryo development in pea was first obtained by Jacobsen and Kysely (1984) from leaf-derived callus with picloram (0.06 mg/L). Embryogenic suspension cultures were established from leaf and stem derived callus in a liquid medium containing picloram. Attempts to induce shoot morphogenesis from these embryos have been unsuccessful (Jacobsen and Kysely, 1984). First successful plant regeneration via somatic embryogenesis in pea was reported by Kysely *et al.* (1987) using shoot apices and immature zygotic embryos as explant source. In the same laboratory regeneration of pea plants through somatic embryogenesis was achieved from cultured protoplasts of zygotic embryonic axes (Lehminger-Mertens and Jacobsen, 1989). Whole plant regeneration via somatic embryogenesis was obtained in pea using explants from immature zygotic embryos or shoot apex segments. In pea somatic embryogenesis, the most effective auxins reported were picloram and 2,4-D added either in MS basal medium or in MS salts supplemented with B_5 vitamins. All somatic embryos failed to undergo further development when subcultured on induction medium. Transfer of embryogenic cultures to MS medium with 1.0 mg/L BA stimulated development of young embryo stages but generally repressed shoot growth. Shoot apices and immature zygotic embryos are now routinely used as initial explants for embryogenesis (Kysely and Jacobsen, 1990). Somatic embryogenesis was induced in immature zygotic embryos of pea, synthetic auxins NAA, 2,4-D, picloram being used. Later, somatic embryogenesis was reported by Ozcan *et al.* (1993) in pea. Somatic embryos were induced on MS medium containing high concentration of NAA (27–215 μM) and 2,4-D (23–181 μM). A high frequency of somatic embryos with a normal morphology was induced using NAA. Loiseau *et al.* (1995) obtained 95–100% embryogenesis from pea shoot apices. This frequency was much higher than those previously cited for pea (Kysely *et al.*, 1987; Kysely and Jacobsen, 1990; Tetu *et al.*, 1990; Stejskal and Griga, 1992).

The first report on somatic embryogenesis in tree legumes used *Albizzia lebbeck* (Gharyal and Maheswari, 1981). They reported direct somatic embryo formation from hypocotyl explants on MS medium with KIN and NAA. Typical globular, heart-shaped and dicotyledonary embryos were observed while other seedling explants namely cotyledons, leaflets and roots did not show any differentiation. In *Albizzia richardiana*, Tomar and Gupta (1988) reported somatic embryogenesis from hypocotyl explants of 12 day old seedlings, only two plantlets were obtained. Subsequently, Tomar and Gupta (1992) observed that ABA promoted embryogenic response in *Albizzia richardiana*. Das *et al.* (1995) have reported that the removal of both auxin and cytokinin from the induction medium was necessary to initiate embryo differentiation in *Hardwickia binata*. However, a reduction of auxin concentration favoured somatic embryo formation in *Psophocarpus tetragonologus* (Ahmed *et al.*, 1996). Rao and Lakshmi Sita (1996) obtained direct regeneration of embryos from immature zygotic embryos of rosewood on a medium with 2,4-D and KIN. Garg *et al.* (1996) obtained triploid plants from endosperm tissues of *Acacia nilotica* via somatic embryogenesis. Initiation of somatic embryogenic cultures for woody legumes has been restricted to immature zygotic embryos or seedlings part (Trigiano *et al.*, 1999), which consist of a few to many proembryogenic determined cells (PEDCs) (Merkle *et al.*, 1995). There have been a few studies of somatic embryo ontogeny in the woody legume species (Buckley and Trigiano, 1994; Das *et al.*, 1997; Geneve and Kester, 1990). Somatic embryos typically formed directly from cotyledons of immature zygotic embryos after 3–5 weeks of culture on medium containing 2,4-D (Trigiano *et al.*, 1988). Many stages of development intermediate between globular and cordate, including a transitional stage from radial to bilateral symmetry and the initiation of a bipolar embryo were found on *Cercis canadensis* explant (Trigiano *et al.*, 1999). In addition, somatic embryos also formed indirectly in callus cultures of *Cercis canadensis*. Not all the cells of a callus were embryogenic, in fact, in the case of *Cercis canadensis* most were not. Embryogenic cultures typically had densely stained cytoplasm with little or no apparent vacuolation, whereas the cytoplasm of non-embryogenic cultures were usually more lightly stained and more extensively vacuolated (Trigiano *et al.*, 1999).

4. Proliferation and maturation of somatic embryos

Once embryos induced, embryogenic cells continue to divide irregularly to form proembryonal structures or they divide in a highly organized manner to form somatic embryos, passing through globular, heart, torpedo and cotyledonary stages. Regular subculture of embryogenic calli or embryos is required for continuous proliferation of embryos. If not subcultured regularly, roots, will be developed and cultures will be turned brown in colour and will subsequently die. For the use of somatic embryogenesis in legume improvement, efficient methods must be developed for routine embryo maturation of high quality embryos. Repeated subculture of embryogenic cultures on a fresh medium is very essential to maintain them in an active stage. These embryogenic cultures are later used to produce mature cotyledonary embryos. When development of SE from globular to cotyledonary stages is completed, the growth of cell occurs as storage reserves accumulation (Parrott *et al.*, 1995). Embryo maturation seems to be an important factor for somatic embryo germination in plant species including legumes. Various physiological and biochemical changes associated with reserve accumulation and mobilization in somatic embryos showed remarkable similarities and differences with those events during zygotic embryogenesis (Lakshmanan and Taji, 2000). The efficiency of maturation of somatic embryos is dependent on the type of auxins and cytokinins. Addition of cytokinins (BAP or KIN) to the medium promoted the growth, development, maturation and conversion of SEs into plants. Positive effect of cytokinin on somatic embryogenesis in peanut was reported by Ramdev Reddy and Reddy (1993). In our laboratory, 2,4-D produced larger number of somatic embryos and NAA and BAP combinations favoured subsequent conversion into plants (Venkatachalam *et al.*, 1999a, b). Recently, Rey *et al.* (2000) reported that BA is necessary for embryo proliferation and further development in *Arachis pintoi*. Geetha *et al.* (1997) described a method for somatic embryogenesis and plant regeneration from hypocotyl derived cell suspension cultures of blackgram. Hypocotyl derived cell suspensions produced nodular structures within 2 weeks in liquid medium. The nodular structures appeared as green, light green, light cream or brown coloured globular embryos. When such cultures were transferred to embryo maturation medium different stages of somatic embryos were produced. Removal and replacement of spent medium with fresh medium at an interval of one week appeared to promote embryo development. On maturation medium, mature embryos differentiated into small plantlets with shoot-root meristems at opposite poles. But the frequency of such differentiation of plantlets with shoot-root meristems was found very low. Further growth of these plantlets was not observed in liquid medium. To test the further growth of these plantlets they were placed on semi-solid medium. In this medium, they were germinated further and developed into complete plantlets. Anbazhagan and Ganapathy (1999) observed different stages (globular, heart and torpedo shaped) of embryos in pigeonpea.

Initial attempts at soybean somatic embryogenesis using a liquid medium were made by Christianson *et al.* (1983). Later, Ranch *et al.* (1985) obtained proliferation of somatic embryos in MS liquid medium but could not maintain the proliferative state. Finer and Nagasawa (1988) reported the establishment and maintenance of an embryogenic suspension culture in soybean. Although maintenance of soybean embryogenic cultures in liquid medium was facilitated by Finer and Nagasawa (1988), the efficiency of soybean tissue culture manipulations *in vitro* still remains low relative to that of other crops. Parrott *et al.* (1988) described that a relatively short auxin treatment was sufficient to induce somatic embryos in soybean with increased proportion of morphological normality easily converting on hormone free medium. Griga (1998) reported that four weeks' exposure to picloram strongly blocked the normal conversion on hormone-free medium and application of TDZ was necessary in order to obtain complete plantlets. The detailed comparative analysis has shown that soybean somatic embryos had very similar characteristics of water content, water and osmotic potential, protein content, fatty acid composition and respiration of zygotic embryos (Slawinska and Obendorf, 1991). Shoemaker and Hammond (1988) have reported that variation in fatty acid composition of soybean somatic embryos was affected by the length of the maturation period and genotype as well. Accumulation of main seed storage proteins at the end of the maturation phase of somatic embryos was noticed by Komatsuda *et al.* (1992) and Stejskal and Griga (1992, 1995).

Exogenous ABA promotes embryo maturation (Denchev *et al.*, 1991), desiccation tolerance (Senaratna *et al.*, 1989) and subsequent conversion (Fujii *et al.*, 1990) of somatic embryos in alfalfa, permitting plant recovery from as many as 80% of somatic embryos. Horbowicz *et al.* (1995) observed soluble carbohydrate changes during embryo maturation in alfalfa which were similar to changes occurring in zygotic embryos, except for the lack of D-pinitol and its galactosyl derivatives. Lai and McKersie (1994) have reported that starch deposition was influenced mainly by the availability of sucrose in the maturation medium in alfalfa. Also these authors observed that the type of carbohydrate present in the maturation medium seems to influence somatic embryo development. The use of maltose, instead of sucrose, resulted in a remarkable enhancement in embryo vigour and subsequent conversion in soybean (Finer and McMullen, 1991) and alfalfa (Denchev *et al.*, 1991). Accumulation of proteins during embryo maturation is critical for high vigour and successful conversion of somatic embryos in legumes (Lai and McKersie, 1994). Lipid accumulation during somatic embryo maturation has been reported in developing embryos of peanut (Mhaske and Hazara, 1994) and soybean (Shoemaker and Hammond, 1988; Dahmer *et al.*, 1992). Mhaske *et al.* (1998) noticed the increased accumulation of triglycerides in peanut with increasing the medium osmolarity. Addition of mannitol enhanced normal development of embryos in pigeonpea (George and Eapen, 1994). Suhasini *et al.* (1994) used ABA for maturation of *Cicer* embryos.

Barna and Wakhlu (1993) have reported that the globular embryos differentiated into mature embryos upon removal of 2,4-D in chickpea. Maturation of embryos was achieved on a medium with BAP and IBA. Higher level of BAP (>3.0 mg/L) promoted recallusing of the embryos whereas, lower dose of BAP (1.0 mg/L) was not good for embryo maturation (Kumar *et al.*, 1994). Barna and Wakhlu (1995) transferred immature leaflet derived somatic embryos to plant growth regulators free MS medium for further development and proliferation. Exposure of explants to high 2,4-D level (200–2000 μM) for 3 days developed bottle-shaped somatic embryos, while exposure to low 2,4-D levels (below 50 μM) gave spherical-shaped somatic embryos. In pea, transfer of somatic embryos to media with cytokinins or GA_3 supported their development and maturation (Stejskal and Griga, 1992). Luo and Jia (1998) and Nolan *et al.* (1989) have reported that the presence of NAA and BA was necessary for maturation of somatic embryos. For somatic embryo maturation, either cytokinin alone or in combination with auxin was essential in *Cicer arietinum* (Kumar *et al.*, 1995), *Vigna aconitifolia* (Kumar *et al.*, 1988; Shekhawat and Galston, 1983), *Psophocarpus tetragonolobus* (Ahmed *et al.*, 1996) and *Dalbergia latifolia* (Rao and Lakshmi Sita, 1996). Gharyal and Maheswari (1981) described that the proliferation of embryos was observed on MS medium containing GA. 2,4-D alone or in combination with other plant growth regulators did not yield any embryo. Although well formed bipolar somatic embryos were obtained, conversion into plantlets was not achieved in this species. However, some authors have reported that maturation and further development of somatic embryos in many legumes occur only on growth hormone-free medium (Buchheim *et al.*, 1989; Durham and Parrott, 1992). Abscisic acid (ABA) also regulates somatic embryo maturation by increasing storage material accumulation and preventing precocious germination (Lecouteux *et al.*, 1993; Fujii *et al.*, 1990; Senaratna *et al.*, 1989; Suhasini *et al.*, 1994; Barna and Wakhlu, 1993; Nadolska-Orezyk, 1992; Gupta *et al.*, 1997). In contrast, Mhaske *et al.* (1998) reported that ABA treatment did not improve somatic embryo maturation or conversion in peanut, though its application markedly increased the accumulation of storage food reserves, including lipids.

5. Embryo germination and conversion into plantlets

Development of plantlets from germinated somatic embryos, a process referred to as conversion, is essential for the ultimate success of any system based on somatic embryogenesis. Another definition describes the germination as hypocotyl elongation, development of a root system and expansion of the primary leaves (Parrott *et al.*, 1988; Komatsuda *et al.*, 1992). The rate of embryo conversion differs widely across genotypes, species and culture systems. Although the germination of somatic embryos occurred frequently, the majority of germinated embryos did not have a viable apical meristem or well-developed epicotyl, forming mostly leaf-like, bud-like or cotyledonary structures, with or without a radicle on the opposite end. Subsequently, the conversion into normal plantlets is low. The regeneration of viable plantlets from somatic embryos is a frequent problem with many plant species, including legumes. The greatest successes and advances in this regard have been made in legume species, which has been discussed earlier (Parrott *et al.*, 1995; Griga, 1999; Lakshmanan and Taji, 2000). The accomplishment of full somatic embryo development and conversion has been an important aspect for utilizing somatic embryogenesis in plant propagation. In several legume species, embryo development and conversion are not always successful. Precocious germination and/or aberrant somatic embryo development is a common problem with legume cultures.

In peanut, several protocols for somatic embryogenesis have been developed, but plant recovery has been limited (Chengalrayan *et al.*, 1997). The higher plant recovery (50%) was achieved from immature zygotic embryos (Hazra *et al.*, 1989). To overcome these limitations, methods for the *in vitro* regeneration of plant from leaflet have been developed (Baker and Wetzstein, 1992; Chengalrayan *et al.*, 1994; Venkatachalam *et al.*, 1999b). To increase the frequency of plant recovery, attempts have been made to obtain normal somatic embryos by manipulating the culture medium prior to germination (Ozias-Akins *et al.*, 1992a). Plant formation remains a problem in that conversion of somatic embryos to plants, after transfer to basal medium, ranged from 0 to 18% overall experiments (Ozias-Akins, 1989). Eapen *et al.* (1993) reported that only 10% of the somatic embryos developed into complete plantlets in groundnut. Germination and conversion of somatic embryos occurred only in the presence of an auxin (Eapen and George, 1993a) or a combination of different cytokinin (Chengalrayan *et al.*, 1994, 1997; Venkatachalam, 1996). Similar process was noticed in winged bean somatic embryos by Ahmed *et al.* (1996). However, Chengalrayan *et al.* (1998) described that somatic embryo conversion appears to be genotype-dependent in peanut. Conversion of somatic embryos was enhanced by desiccation in soybean (Buchheim *et al.*, 1989), chickpea (Barna and Wakhlu, 1993), peanut (Durham and Parrott, 1992) and alfalfa (Anandarajah and McKersie, 1990; Senaratna *et al.*, 1995). ABA application, partial desiccation or exposure to cold, heat, water and osmotic stresses have been shown to increase the frequency of embryo germination and conversion in many legumes (Senaratna *et al.*, 1989, 1990; Nadolska-Orczyk, 1992; Durham and Parrott, 1992; Arrillaga *et al.*, 1994; Kuklin *et al.*, 1994). However, in chickpea somatic embryos required BAP for germination, but conversion occurred on hormone-free medium (Kumar *et al.*, 1995). Rao and Lakshmi Sita (1996) found that BAP was needed for both germination and conversion of somatic embryos in *Dalbergia latifolia*. Although embryo development was obtained in tree legumes (Skoleman and Mapes, 1976; Gharyal and Maheswari, 1981; Trigiano *et al.*, 1988), they failed to develop plantlets from the embryos. Recently, embryo germination and further plantlet development were achieved on hormone free medium in *Arachis pintoi* (Rey *et al.*, 2000) and in pigeonpea (Anbazhgan and Ganapathy, 1999). However, the plant conversion percentage was 5–10% in *Arachis* and 5–6% in pigeonpea.

Most of the literature on the conversion of peanut somatic embryos and survival of somatic embryo raised plantlets in soil described low frequency (Hazra *et al.*, 1989; Ozias-Akins, 1989; McKently, 1991, 1995; Baker and Wetzstein, 1992; Durham and Parrott, 1992; Ozias-Akins *et al.*, 1992a; Eapen *et al.*, 1993; Ramdev Reddy and Reddy, 1993; Chengalrayan *et al.*, 1994). However, failure in peanut embryos to undergo conversion to plants is primarily due to morphological abnormalities (Wetzstein and Baker, 1993). Wetzstein and Baker (1993) found that conversion of peanut somatic embryos derived from cotyledon cultures was related to embryo morphology. Baker *et al.* (1995) consistently have achieved conversion rates ranging from 15 to 40%. The best plant conversion frequency (81.7%) was obtained when 2,4-D was used for somatic embryo induction from groundnut hypocotyl explant (Venkatachalam *et al.*, 1997). Earlier workers could achieve only a very low plant conversion frequency in groundnut (Eapen and George,

1993a; Eapen *et al.*, 1993). Chengalrayan *et al.* (1994) reported that 100% of the embryos formed well differentiated roots; however, the frequency of conversion remained poor (25%).

Although somatic embryogenesis was obtained in blackgram, low efficiency of embryo conversion into plantlets remains a major problem (Geetha *et al.*, 1997). Development of somatic embryos in liquid medium was very much faster when compared to solid media of the same composition. From the visual observation of browning, it is suggested that in the solid medium, inhibitory phenolic substances produced by blackgram tissues are accumulated, in the explant vicinity. On the other hand, in the liquid medium these substances are diluted and distributed in the medium which may impose less constraints on the developing embryos. Somatic embryos induced on solid medium differentiated further into plantlets on maturation medium. There was no need of transferring mature embryos onto germination medium for plantlet development. But the embryos induced in liquid medium did not develop further into small plantlets, even though they have polarity (root-shoot meristems). So embryos induced in liquid medium were tested for further growth by plating on germination medium where they developed into tiny plantlets. Somatic embryogenesis and plant regeneration of blackgram species were possible although propagation number was still very low. Further work is necessary to develop consistent methods in increasing somatic embryogenesis of blackgram species for large scale plant propagation (Geetha *et al.*, 1997).

George and Eapen (1994) obtained altogether five normal mature embryos from pigeonpea cotyledon and one grew up to 1.5 cm to produce a small plant. However, plantlets capable of transfer to the field could not be obtained from the somatic embryos. Patel *et al.* (1994) described that reduction of cytokinins favoured maturation of somatic embryos. Complete withdrawal of cytokinins from the medium and addition of IBA and GA proved effective in germination of the embryos. Even though somatic embryogenesis was achieved in pigeonpea, only 3% of the embryos showed conversion into plants (Mallikarjuna *et al.*, 1996). Soybean somatic embryo development is improved by reducing explant exposure to auxins. Germination and conversion of soybean somatic embryos usually do not require exogenous growth regulators (Parrott *et al.*, 1988), but somatic embryos must undergo physiological maturation prior to germination (Samoylov *et al.*, 1998a). Komatsuda *et al.* (1992) and Buchheim *et al.* (1989) demonstrated that hypocotyl elongation stimulated by gibberellin increased the germination frequency, the removal of cotyledons could also have a stimulatory effect in soybean. The water content of somatic embryos may be reduced to 6% by various desiccation treatments without loss of viability, the transfer into fresh medium resulted in rapid and uniform germination of soybean (Buchheim *et al.*, 1989; Parrott *et al.*, 1988; Slawinska and Obendorf, 1991). The use of liquid medium based protocol relative to the standard procedure led to a four-fold increase in the number of cotyledonary stage embryos in soybean; however, the germination percentage was low (Samoylov *et al.*, 1998b). Although there has been one previous report on the use of liquid MS basal medium for soybean somatic embryogenesis, histodifferentiation and maturation (Rajasekaran and Pellow, 1997), a direct comparison had never been made between the use of solid and liquid media.

Barna and Wakhlu (1993) reported that only 2.8% of the embryos formed plants in chickpea. Desiccation of embryos for a period of 3 days increased their rate of conversion into plants. In mature somatic embryos, shoot apical meristem develops primary leaves only when transferred to a zeatin containing conversion medium (Suhasini *et al.*, 1994). Two percent of somatic embryos developed into plants when transferred to MS medium with 15 μM GA_3 and 1 μM IBA (Barna and Wakhlu, 1995). For germination of embryos basal B_5 medium with 0.25 mg/L BAP was found suitable. The germinated embryos developed into complete plants on basal B_5 medium. The malformed embryos produced on media containing higher BAP concentration did not germinate and showed poor conversion into plantlets. The conversion of embryos into plants is as high as 50% and more than 25 complete plants could be obtained from 100 explants. Sagare *et al.* (1995) studied the histology of somatic embryo initiation and development in chickpea from different explants. Immature cotyledon, immature embryo axes and mature embryo axes as explants were grown on MS medium fortified with 3.0 mg/L 2,4,5-T. Explants with different developmental stages were fixed for histological examination. Histological studies demonstrated that somatic embryos arose indirectly with an intervening callus phase from immature cotyledon and immature embryo axes and directly from mature embryo axes. Young leaflets formed somatic embryos by direct and indirect methods. Subsequently, the correlation between the aberrant morphologies and the lack of conversion of somatic embryos has also been studied by Suhasini *et al.* (1996). They observed different aberrant morphologies of somatic embryos along with normal somatic embryos. Kumar *et al.* (1995) reported plant regeneration via somatic embryogenesis from leaf explants of chickpea seedlings when cultured on B_5 medium augmented with 2,4-D and picloram (0.25 mg/L each) and 0.1 mg/L BA. A modified B_5 medium with 2.0 mg/L BA and 0.1 mg/L IAA was suitable for embryo maturation. The well formed somatic embryos germinated on B_5 medium with 0.25 mg/L BA and developed into healthy plantlets on B_5 basal medium. Normal, fertile, pod-bearing plants formed on transfer to the soil. In pea, induction of somatic embryos required picloram or 2,4-D. Germination of fully-developed embryos was accomplished by subculture on medium with only cytokinin and then on medium supplemented with cyotokinins in combination with a reduced auxin concentration (Kysely *et al.*, 1987). Kysely *et al.* (1987) regenerated only 15 plants, 5 of them being obtained from shoot apices.

In woody leguminous species, somatic embryos either germinate or undergo conversion to plants at low frequencies (Trigiano *et al.*, 1995; Weaver and Trigiano, 1991) or do not germinate at all regardless of the treatment applied (Curir *et al.*, 1990; Das *et al.*, 1997). Germination media typically were not supplemented with plant growth regulators (Geneve and Kester, 1990; Weaver and Trigiano, 1991), but sometimes low levels of cytokinins were added (Rao and Lakshmi Sita, 1996; Trigiano *et al.*, 1988). Desiccation and cold treatment proved to be detrimental to *Cercis canadensis* embryo germination (Trigiano *et al.*, 1995).

6. Effect of explants and genotypes on somatic embryogenesis

The choice of appropriate explants, excised at defined stages of development, is often very important for the initiation of embryogenic cultures. However, factors limiting the competence of explant cells are still unknown. Legume embryogenic cultures are mostly initiated from young or immature tissue, most frequently from embryonic tissue. The most effective explants used for induction of embryogenic cultures were immature zygotic embryos, cotyledons, axes, immature leaflets, etc. The ability to produce embryonic cultures differed among genotypes. A genotypic effect on embryogenic explant production has been seen in many plant species and legumes are no exception. Although a genotypic effect is observed in legume species, it is possible to obtain embryogenic calli or embryos irrespective of the genotype, more importantly with fairly high response levels. This characteristic causes legume plants to respond well to somatic embryogenesis. Either no or poor response of certain genotypes has been reported for most species, and this is one of the main obstacles to the routine use of this technique in mass production of plants. So far several genotypes and explants of legume species have been tested for their embryogenic competence. The most successful of these is immature zygotic embryos/ tissues. Nevertheless, stem, leaflet, cotyledon, hypocotyl, epicotyl, seed explants of various legume species could also produce somatic embryos and subsequent plant regeneration. A number of recent reports describe somatic embryogenesis in groundnut using a variety of different explants, including leaves (Baker and Wetzstein, 1992), immature cotyledons (Ozias-Akins, 1989; Ozias-Akins *et al.*, 1992a; Durham and Parrott, 1992; Wetzstein and Baker, 1993; Baker and Wetzstein, 1994; Eapen *et al.*, 1993; Baker *et al.*, 1994; Baker and Wetzstein, 1995), immature embryo axes (Hazra *et al.*, 1989; Ozias-Akins *et al.*, 1992b; Ramdev Reddy and Reddy, 1993; Eapen *et al.*, 1993), embryo axes collected at harvest (McKently, 1991) and whole embryos (Sellars *et al.*, 1990). Although the use of explants from immature embryo explants for somatic embryogenesis and plantlet regeneration has become routine, several inherent difficulties are encountered in using such materials in peanut (Baker *et al.*, 1995; Baker and Wetzstein, 1998). These include the need to maintain appropriately aged parent material in the greenhouse or field and the need to categorize embryos at specific developmental stages. In addition, high contamination rates associated with the use of explants from a below-ground organ, the peanut peg. Since immature embryo explants are not always readily available, attempts have been made to regenerate plants from other explants including hypocotyls, leaves and mature cotyledons (Venkatachalam *et al.*, 1997, 1998, 1999a, b).

In legumes, as with other groups of plants, immature, meristematic tissues proved to be the most suitable explant for somatic embryogenesis. Tissues associated with reproduction, such as ovary or anther tissue, or tissue from young seedlings, may be more easily returned to an embryogenic state than more mature tissue (Williams *et al.*, 1990). Occasionally, some legumes, such as *Trifolium rubens*, can be identified which retain the capacity for direct embryogenesis from mature tissues (Cui *et al.*, 1988). Immature zygotic embryos and cotyledons were used as initial explants for embryogenesis in various legume species (Parrott, 1991; Ahmed *et al.*, 1996; Sagare *et al.*, 1995). Somatic embryos were obtained from different explants namely mature seeds (Baker *et al.*, 1995), shoot apices (Kysley and Jacobsen, 1990), intact seedlings (Saxena *et al.*, 1992), leaves (Kumar *et al.*, 1994; Baker and Wetzstein, 1992; Venkatachalam *et al.*, 1999b), petioles (Senaratna *et al.*, 1995), hypocotyls (Dhanalakshmi and Lakshmanan, 1992; Venkatachalam *et al.*, 1997), internodes (Parrott and Bailey, 1993), cell suspensions (Rybczynski, 1997), protoplasts (Tegeder *et al.*, 1996; Venkatachalam and Jayabalan, 1996). In many legumes cells of immature zygotic embryos develop directly into somatic embryos (Rao and Lakshmi Sita, 1996; Hazra *et al.*, 1989; Finer, 1988), whereas highly differentiated explants usually undergo a callus phase to acquire embryogenic potential (Barna and Wakhlu, 1993;

Chengalrayan *et al.*, 1994; Ahmed *et al.*, 1996; Trinh *et al.*, 1998; Sreenivasu *et al.*, 1998). Direct somatic embryogenesis has been achieved on growth regulator-free medium from hypocotyl segments of *Albizia lebbeck* (Gharyal and Maheshwari, 1983) and *Clitoria ternata* (Dhanalakshmi and Lakshmanan, 1992).

Tetu *et al.* (1990) demonstrated that the morphogenesis in pea depends on three major factors: explant size, cultivars/genotypes and nutrient media. High frequency regeneration was achieved via direct somatic embryogenesis and organogenesis using zygotic embryos. Somatic embryos and buds described here developed directly from the cotyledon of whole immature zygotic embryos. The developmental stage of the zygotic embryos was very critical for the initiation of somatic embryos. Moreover, for efficient development of somatic embryos in pea, it was very important that the somatic embryos were transferred to germinating media; otherwise, the somatic embryos either did not develop any further or produced calli. Stejskal and Griga (1992) found that the whole zygotic embryos were the best explant for pea somatic embryo induction. If only cotyledons without organized meristematic zones were placed on induction media, no embryogenesis, but browning of the explants, was observed. This fact suggests that meristematic tissues are involved in somatic embryogenesis. Rao and Lakshmi Sita (1996) observed that the maximum number of embryos produced by cotyledon explants of rosewood and mature cotyledons from dry seeds did not respond to treatment, turning brown and callused after 4 weeks. Trigiano *et al.* (1988) found that in red bud (*Cercis cannadesis*) zygotic embryos explanted 117 days after anthesis produced only callus or roots and no embryos, while somatic embryos developed directly from 96 and 110 days of post-anthesis explants. In black locust, however, somatic embryos developed within 14–21 days from post-anthesis explants (Arrillaga *et al.*, 1994).

Genotype remains the factor with the greatest influence on frequency of somatic embryos from mature axes (Baker *et al.*, 1995). Similar genotype effects had also been reported for embryogenesis from other explants of peanut (Sellars *et al.*, 1990; Ozias-Akins *et al.*, 1992a; George and Eapen, 1993; Venkatachalam *et al.*, 1997). The failure of somatic embryos to undergo conversion to plantlets could be a genotype-dependent characteristic (Chengalrayan *et al.*, 1998). Genotype-dependent variation during somatic embryogenesis has also been observed in peanut (Ozias-Akins *et al.*, 1992b; George and Eapen, 1993; Ramdev Reddy and Reddy, 1993; McKently, 1995; Rani and Reddy, 1996). In contrast, Ozias-Akins *et al.* (1992a) suggested that all of the peanut genotypes tested were found competent for somatic embryogenesis. Using 14 genotypes of peanut, McKently (1995) analysed the embryogenic response and suggested that significant differences exist among genotypes and varieties for somatic embryo formation and plantlet regeneration. The failure of somatic embryos to convert into plantlets which is often attributed to malformation of the plumule or to a lack of maturation, may as well be a genotypic characteristic of cultivar. Geetha *et al.* (1997) found that the number of globular embryos formed per callus piece varied with three blackgram cultivars. Mallikarjuna *et al.* (1996) obtained somatic embryogenesis in pigeonpea from different explants, viz. immature leaflet segment, root discs, epicotyl, hypocotyl segments and mature cotyledons on a medium containing NAA (1 to 50 mg/L) and BA (0.1 to 10 mg/L). However, plant regeneration was obtained only from cotyledon cultures (Mallikarjuna *et al.*, 1996). Sreenivasu *et al.* (1998) have reported that somatic embryogenesis was genotype independent in pigeonpea. Venu *et al.* (1999) evaluated two genotypes for direct somatic embryogenesis from cotyledon explants of pigeonpea. Within 2–3 weeks, embryo formation occurred on cotyledons without an intervening callus phase. However, embryos de-differentiated into callus, if they were maintained on the same medium.

The efficiency of somatic embryogenesis depends on the genotype used (Parrott *et al.*, 1989). Considerable variation in embryogenesis was also found to exist between individual genotypes of *Glycine max* (Komatsuda and Ohyama, 1988; Parrott *et al.*, 1989; Santos *et al.*, 1997; Komatsuda *et al.*, 1991; Bailey *et al.*, 1993; Li and Grabau, 1996), *Medicago sativa* (Fuentes *et al.*, 1993; Mitten *et al.*, 1984; Meijer and Brown, 1987; Brown and Atanassov, 1985; Chen and Marowitch, 1987; Chen *et al.*, 1987; Binachi *et al.*, 1988; Kielly and Bowley, 1992) and *Trifolium pratense* (Keyes *et al.*, 1980; Quesenberry and Smith, 1993; McLean and Nowak, 1998). Genotype differences in somatic embryogenesis from immature cotyledons were reported by Parrott *et al.* (1989), Komatsuda and Ohyama (1988), Komatsuda and Ko (1990), Komatsuda *et al.* (1991), Shoemaker *et al.* (1991), Griga (1993) and Bailey *et al.* (1993). For some genotypes, this system can provide a large number of somatic embryos each of which originates from two to four cells in the surface of the embryogenic tissue (Finer and McMullen, 1991). Using different genotypes and several concentrations of 2,4-D, Shoemaker *et al.* (1991) obtained as high as 16 to 92% embryogenic responsiveness with 3.4 to 17 somatic embryos per responding cotyledon from field grown zygotic embryos. It is nearly impossible to obtain a consistent response for somatic embryogenesis even using the same kind of explant from the same genotype without a carefully controlled environment in which the donor plants are grown. Li and Grabau (1996) compared six commercially important soybean cultivars and results indicated that the efficiency of embryo induction and yield of primary somatic embryos were genotype-dependent. In contrast, no dependence on genotype was observed for the conversion of embryo to form roots and shoots. Conversion frequency, expressed as the percentage of mature embryos that gave rise to roots and shoots, were genotype independent, which differed from previous reports (Ranch *et al.*, 1985; Komatsuda and Ohyama, 1988; Bailey *et al.*, 1993). A soybean genotype which require very specific conditions for somatic embryogenesis and plant regeneration, as it does not respond to those conditions to which most other soybean genotypes respond, was previously identified (Santos *et al.*, 1997).

Sagare *et al.* (1993) induced somatic embryos from both immature cotyledons and immature embryo axes. The maximum frequency of globular embryo formation occurred in cotyledonary segments on MS medium with 2,4,5-T. Further development was achieved only in somatic embryo derived from cotyledonous segments. Globular embryos derived from immature embryo axes de-differentiated and formed callus (Sagare *et al.*, 1993). They found that zeatin is necessary for germination of somatic embryos. The embryo maturation was significantly affected by pH, photoperiod, ABA and genotypes. However, the rate of conversion of somatic embryos into plantlets was not influenced by the genotype (Barna and Wakhlu, 1993). Sagare *et al.* (1993) evaluated five genotypes of chickpea for induction of somatic embryos. However, globular somatic embryo formation was obtained with only one genotype which was developed into cotyledonary stage with a low frequency (3%), of which only half developed further. Also this protocol is not suitable for another cultivar because the globular embryos did not develop into plantlets and immature cotyledons used as the source of explant are available for only a limited period in the season. Of the

12 genotypes used, 3 genotypes were non-responsive (Kumar *et al.*, 1995). Genotypic difference was also noticed between cultivars (Vani and Reddy, 1996). Although there were successful reports in obtaining plantlets through somatic embryogenesis in chickpea, the frequency of plantlet development was low (Eapen and George, 1994).

Genotypic effects for somatic embryogenesis and plant regeneration from immature embryos among the 3 genotypes have been reported in pea (Kysely *et al.*, 1987; Van Doorne *et al.*, 1995; Griga, 1998). Among 46 pea genotypes tested, only one HM-6, was potential, but on MS medium with 2.26 μM 2,4-D, they obtained only 14.6% embryogenic calli, with a yield of 0.30 embryos per callus (Kysely and Jacobsen, 1990). Stejskal and Griga (1992) demonstrated that of the 46 genotypes tested, only one genotype gave rise to several somatic embryos. Six lines of pea were tested for their *in vitro* ability to produce somatic embryos from apices for application in improved regeneration for the production of artificial seeds (Bencheikh and Gallais, 1996a). In another report, Bencheikh and Gallais (1996b) used 30 lines of pea for induction of somatic embryogenesis. Callus and embryos were induced on MS medium containing 0.5 mg/L picloram. Abnormal somatic embryos and secondary embryogenesis constituted the principal obstacle to the development of the embryos. However, in most of the reports concerning somatic embryogenesis in pea, the rate of somatic embryos forming plants was low and limited to few cultivars. As in most of the other genera, success in plant regeneration and especially in somatic embryogenesis is highly dependent on genotype.

7. Effect of plant growth regulators on somatic embryogenesis

Legume species differ in embryogenic response to various plant growth regulators. The study of embryo formation was markedly dependent on the type and concentration of plant growth regulators in the nutrient medium. Limited comparisons have been made of growth regulator effects on embryogenesis from leaves (Baker and Wetzstein, 1992; Gill and Saxena, 1992; Chengalrayan *et al.*, 1994, 1995, 1997). George and Eapen (1993) have reported that seedling explants such as primary leaves, hypocotyl, epicotyl and petiole did not produce somatic embryos. Gill and Saxena (1992), Saxena *et al.* (1992) and Murthy *et al.* (1995) used thidiazuron and BAP for induction of somatic embryos in groundnut. Somatic embryos formed from immature leaflets of peanut demonstrated a high degree of morphological abnormality (Chengalrayan *et al.*, 1994, 1995, 1997, 1998). Meristematic activity in the shoot apices was stimulated by exposing the embryos to 2.2 μM BA and 2.3 μM KIN and shoot formation occurred in 25% of the rooted embryos. Plantlets often had multiple shoot/roots. This may be primarily due to the multiple meristem existing in the stimulated embryos (Chengalrayan *et al.*, 1994, 1997). While a high frequency of abnormalities and a low frequency of conversion were noted in the embryos obtained, plantlet formation could be induced in 86–92% of the abnormal embryos in media containing cytokinin or thidiazuron (Chengalrayan *et al.*, 1997). Abnormalities in peanut somatic embryogenesis were frequently observed by several researchers (Baker and Wetzstein, 1992; Wetzstein and Baker, 1993; Chengalrayan *et al.*, 1994, 1995, 1997, 1998). In peanut, somatic embryogenesis was promoted by addition of kinetin (0.5 mg/L) along with 2,4-D (Venkatachalam *et al.*, 1997).

Geetha *et al.* (1997) have reported that hypocotyl explants formed numerous globular embryos on medium containing 2,4-D, NAA and BAP within 4 weeks of culture. The frequency of calli bearing globular embryos was greatest on a medium supplemented with 2,4-D (4.0 mg/L), NAA (2.0 mg/L) and BAP (0.5 mg/L). George and Eapen (1994) obtained somatic embryos in the presence of NAA, some of the cotyledons produced thick roots in addition to embryos. Picloram induced highest frequency of embryos with cotyledon explants whereas 2,4-D was more favourable for embryogenesis than NAA or picloram. Embryos were transferred to fresh medium containing $AgNO_3$, zeatin, GA and ABA. In chickpea cultivars, 2,4-D was found to be the auxin of choice for the induction of somatic embryos although picloram and dicamba were also effective. NAA was not effective in the induction of somatic embryos at the concentrations tested. Addition of $AgNO_3$ and mannitol to the medium promoted the growth and development of somatic embryos in culture. The well developed somatic embryos of chickpea were transferred to medium supplemented with GA and ABA for further development into plants. The frequency of somatic embryogenesis was found to be the result of interaction between the genotype, source material and the type and concentration of auxin used.

In legumes, with a few exceptions (Gharyal and Maheshwari, 1983; Dhanalakshmi and Lakshmanan, 1992; Malik and Saxena, 1992; Sreenivasu *et al.*, 1998; Murthy *et al.*, 1995), induction of somatic embryogenesis was achieved in the presence of auxins in the medium. Response of legumes to auxins varies significantly depending on the species and the type of auxins used. 2,4-D and picloram are known to induce somatic embryos from immature zygotic embryos (Kysely *et al.*, 1987; Hartweck *et al.*, 1988; Kysely and Jacobsen, 1990; Stejskal and Griga, 1992) and from protoplasts (Lehminger-Mertens and Jacobsen, 1989). However, 2,4-D was totally ineffective in *Chamaecytisus purpureus* and *C. austriacus* (Greinwald and Czygan, 1991) and it completely inhibited the production of embryogenic callus in *Hardwickia binata* (Das *et al.*, 1995). Rao and Lakshmi Sita (1996) reported that varying concentrations of 2,4-D was necessary to induce somatic embryogenesis in rosewood while different auxin combinations were needed to develop embryos in some genotypes of *Medicago sativa* (Parrott and Bailey, 1993). NAA was not able to induce somatic embryos from shoot apices, as has been shown with immature zygotic embryo explants (Kysely and Jacobsen, 1990; Stejskal and Griga, 1992), though NAA seems to be efficient on immature zygotic embryos of pea (Tetu *et al.*, 1990; Ozcan *et al.*, 1993) and soybean (Barwale *et al.*, 1986; Lazzeri *et al.*, 1987a). Cytokinins seem to promote pea embryo quality when they are introduced into development medium (Kysely *et al.*, 1987; Kysely and Jacobsen, 1990). In legumes, limited reports are available on successful induction of somatic embryogenesis by cytokinin alone (Maheswaran and Williams, 1985; Malik and Saxena, 1992; Ninkovic *et al.*, 1995). Further, the effect of cytokinin appears to be mostly enhanced in indirect somatic embryogenic systems requiring extensive cell proliferation prior to embryo formation (Trinh *et al.*, 1998; Nagarajan *et al.*, 1986; Li and Demarly, 1996). Species specific cytokinin activity on somatic embryogenesis was reported for *Medicago* (Denchev *et al.*, 1993; Senaratna *et al.*, 1995; Trinh *et al.*, 1998; Nolan *et al.*, 1989; Li and Damarly, 1996; Scarpa *et al.*, 1993) and *Trifolium* (Rybcsynski, 1997; Radionenko *et al.*, 1994; Heath *et al.*, 1993). In contrast, cytokinin supply in induction medium strongly inhibited somatic embryogenesis, as previously reported in pea (Kysely and Jacobsen, 1990), in soybean (Lippmann and

Lippmann, 1984; Lazzeri *et al.*, 1987b) and *Coronilla varia* (Arcioni and Marriotti, 1982). Although somatic embryos were induced and developed on the same auxin-containing medium, approximately 40% of them could be converted into plants when they were transferred on to a medium devoid of growth regulators (Loiseau *et al.*, 1995). In most of the previous reports concerning somatic embryogenesis in pea, the rate of somatic embryos forming plants was not stated precisely (Lehminger-Mertens and Jacobsen, 1989; Kysely and Jacobsen, 1990; Tetu *et al.*, 1990; Stejskal and Griga, 1992).

In some legumes as soybean, pea, peanut, the auxin (2,4-D, 2,4,5-T, NAA, MCPA, Picloram) was necessary for direct somatic embryogenesis from ontogenetically very young stages (Lippmann and Lippmann, 1984; Lazzeri *et al.*, 1985; Finer, 1988; Griga, 1990; Tetu *et al.*, 1990; Hazra *et al.*, 1989; Ozias-Akins, 1989; Sellars *et al.*, 1990; Loiseau *et al.*, 1995; Griga, 1998) and cytokinins (BAP, Zeatin, KIN) reduced or strongly inhibited somatic embryogenesis (Lippmann and Lippman, 1984; Komatsuda and Ohyama, 1988; Kysely and Jacobsen, 1990; Tetu *et al.*, 1990; Loiseau *et al.*, 1995). Use of 2,4-D for somatic embryogenesis resulted in high frequency embryo formation but a large number of these embryos were morphologically abnormal and failed to convert into complete plantlets (Chengalrayan *et al.*, 1994; Garg *et al.*, 1996; Weaver and Trigiano, 1991). On the other hand, application of cytokinin alone was able to induce somatic embryos even in ontogenetically more developed and differentiated tissues of intact seedlings of bean, peanut or chickpea and mature cotyledons of groundnut (Malik and Saxena, 1992; Murthy *et al.*, 1995, 1996; Venkatachalam *et al.*, 1999a). 2,4-D was effective for inducing high frequency somatic embryogenesis and other auxins (IAA, NAA and IBA) did not favour embryogenesis in pigeonpea (Anbazhagan and Ganapathi, 1999). Substituted phenylurea compounds, such as CPPU (Murthy and Saxena, 1994) and thidiazuron (Gill and Saxena, 1992; Murthy *et al.*, 1995), have cytokinin-like activity and have been used to induce somatic embryos from peanut.

8. Effect of carbohydrates on somatic embryogenesis

The carbon source plays an important role on somatic embryogenesis. It is common to use 3% sucrose in the culture media for the induction of somatic embryogenesis. Other sugars such as maltose, glucose, fructose, can also support somatic embryo induction. However, in some cases, high sucrose levels have increased the frequency of somatic embryogenesis. The precise role of sucrose on the improvement of somatic embryo differentiation is poorly understood. Several authors (Ammirato, 1983; Levi and Sink, 1990) have pointed out that carbohydrates have a dual role, acting both as a carbon source and as an osmotica. Among the different carbon sources tested for somatic embryogenesis in groundnut, sucrose evoked the best response followed by fructose, dextrose and maltose. Sucrose at 6% produced the best results in groundnut (Venkatachalam, 1996). Eapen and George (1993a) described the influence of sucrose on somatic embryogenesis and morphology. Mannitol favoured normal embryo development in pigeonpea (Patel *et al.*, 1994). No other carbohydrate other than sucrose (2–3%) has been used for somatic embryogenesis in pea, and media were sometimes supplemented with casein hydrolysate (Lehminger-Mertens and Jacobsen, 1989). In all reports concerning pea somatic embryogenesis, only 88 mM sucrose has been used and Loiseau *et al.* (1995) obtained 3 to 4 fold higher number of somatic embryos in pea, if 84 mM sucrose is substituted by 252–504 mM fructose.

The role of sugars as a source of reduced carbon seems to be more important in the phase of somatic embryogenesis. The most frequently used saccharide is sucrose, but some other mono- or disaccharides can have special effects in particular legume species (Griga, 1999). The lower concentrations (0.5–1%) were more suitable for embryo induction in soybean and considerably improved the ratio of normal two-cotyledon types; a significant interaction between sucrose concentration and soybean genotype was also reported (Komatsuda, 1990; Komatsuda *et al.*, 1991). Novak *et al.* (1987) observed an interaction between sucrose concentration and age of the soybean explants. Low sucrose concentration (1%) resulted in a significantly higher frequency of embryogenesis as well as better embryo development as compared to higher concentration of sucrose (4%). In contrast, fructose in high concentrations (4.5, 6 and 9%) significantly increased the frequency of pea embryogenic explants as well as the number of somatic embryos per explant (Loiseau *et al.*, 1995). Finer and Nagasawa (1988) found that high sucrose concentration (6%) was necessary for initiation of embryos from soybean suspension cultures, whereas similar results were observed in soybean with 2% sucrose by Christianson *et al.* (1983). Maltose was the most favourable carbohydrate, as compared to glucose, fructose or sucrose, for somatic embryogenesis in alfalfa (Parrott and Bailey, 1993). Genga and Allavena (1991) reported that no single carbohydrate was best for all alfalfa genotypes but, overall, the highest embryo yield was obtained with maltose. Glucose is superior to sucrose for the induction of somatic embryos of scarlet runner bean, *Phaseolus coccineus* (Genga and Allavena, 1991). The use of maltose instead of sucrose has been reported to enhance somatic embryo production and conversion in alfalfa (Stickland *et al.*, 1987; Denchov *et al.*, 1991) and soybean (Finer and McMullen, 1991).

In *Albizzia lebbeck*, sucrose was more effective compared to mannitol for embryogenesis (Gharyal and Maheswari, 1981). Maltose rather than sucrose enhanced alfalfa somatic embryo production and conversion (Stickland *et al.*, 1987) and maturation protocols for soybean (Finer and McMullen, 1991) and alfalfa somatic embryos used maltose (Denchov *et al.*, 1991). Rao and Lakshmi Sita (1996) found that high sucrose (10%) seems to play an important role in the induction of somatic embryogenesis in rosewood. High sucrose has been used in forage legumes such as *Medicago sativa* and *Trifolium* spp. (Maheswaran and Williams, 1984), grain legumes like groundnut (Hazra *et al.*, 1989) and pea (Jacques *et al.*, 1995) to induce somatic embryogenesis. Weissinger and Parrott (1993) reported that conversion of white clover somatic embryos is significantly enhanced when 6% maltose is used instead of the more common 3% sucrose. Jaques *et al.* (1995) observed that the frequency of embryogenesis was increased with an increase in the glucose and fructose concentrations in pea. Recently, Anbazhagan and Ganapathy (1999) reported that the frequency of embryos improved by increasing concentrations of carbohydrates and sucrose at 87.64 mM was most effective carbohydrate supplement for induction of embryos in pigeonpea.

9. Repetitive or secondary somatic embryogenesis

Other commonly used descriptions for secondary somatic embryogenesis are repetitive, cyclic, recurrent, accessory or adventitious somatic embryogenesis. Secondary somatic embryo formation can proceed in two ways: (1) through callus proliferation in former somatic embryos followed by somatic embryo differentiation and (2) directly on the cotyledon surface of previously formed somatic embryos. The first pathway is more common when an auxin is present in the culture media but both types could be found in the same culture conditions. Secondary embryogenesis offers a great potential for large-scale propagation and more experiments are now being carried out for better understanding of the conditions in which this kind of morphogenesis is expressed. The capacity for secondary embryogenesis in plants is inversely proportional to the maturation state of the somatic embryo (Williams and Maheswaran, 1986). Proliferation of embryogenic tissue starts with subculture of primary somatic embryos for producing secondary somatic embryogenesis on a medium supplemented with auxins. Secondary somatic embryogenic cultures have been used for plant transformation. Besides the synthetic auxins, 2,4-D, picloram and dicamba, NAA has also the capacity to induce secondary somatic embryogenesis. Most probably, the origin of secondary somatic embryos was multicellular and from subepidermal cell layers.

Repetitive somatic embryogenesis has only been reported in a few legume species, such as alfalfa (Lupotto, 1993), *Trifolium* (Maheswaran and Williams, 1986), soybean (Finer and Nagasawa, 1988), barrel medic (Das Neves *et al.*, 1999) and peanut (Durham and Parrott, 1992; Ozias-Akins *et al.*, 1993; Baker and Wetzstein, 1994, 1995; Little *et al.*, 2000). However, this characteristic has been analysed extensively in alfalfa (Lupotto, 1983; Parrott and Bailey, 1993; Lai and McKersie, 1994; Ninkovic *et al.*, 1995). Maintenance of an embryogenic state may require exogenous auxin or it may occur in the absence of exogenous growth regulators (Parrott *et al.*, 1995). The amount of auxin required for somatic embryogenesis may vary with different protocols. For example, on solid medium, soybean and white clover require 20–40 mg/L, 2,4-D (Parrott *et al.*, 1995). In liquid medium, soybean (Finer and Nagasawa, 1988) and peanut (Durham and Parrott, 1992) undergo repetitive embryogenesis at 5 mg/L 2,4-D. Alfalfa produced repetitive embryogenesis on solid medium devoid of all growth regulators (Dos Santos *et al.*, 1983; Lupotto, 1983, 1986). These particular species also differ in the developmental stage at which they undergo repetitive embryogenesis. White clover and soybean somatic embryos reach a globular stage of development before secondary embryos appear, while peanut somatic embryos reach a heart to torpedo stage before the appearance of new embryos. For white clover, peanut and soybean, secondary embryogenesis is inhibited and embryo differentiation and maturation are promoted when auxin is removed from the medium (Parrott *et al.*, 1995). Alfalfa somatic embryos reach the secondary stage and may even start to germinate before becoming repetitive. Genotypic differences for repetitive embryogenesis have been reported in peanut (Ozias-Akins *et al.*, 1992; Little *et al.*, 2000) and in soybean (Bailey *et al.*, 1993).

Previously, 2,4-D was used to achieve a repetitive somatic embryogenesis system in either liquid (Durham and Parrott, 1992) or solid (Baker and Wetzstein, 1995) medium, but very few of the embryos induced were in the repetitive globular stage useful for transformation. Picloram at 3 mg/L (12.4 mM) was used to develop repetitive embryogenic cultures on solid medium for use in a transformation system (Ozias-Akins *et al.*, 1993). Recently, the effects of 11 auxins and one cytokinin-like compound were tested at four concentrations for their ability to induce primary and repetitive somatic embryogenesis from mature, dry peanut epicotyls (Little *et al.*, 2000). High levels (83.0 or 124.4 mM) of picloram or centrophenoxine in solid medium reliably induces highly repetitive globular-stage somatic embryos from mature peanut axes and maintains the long term (five or more months) continual production of embryos in the globular stage to serve as a convenient source of tissue for a gene transfer system (Little *et al.*, 2000). CPPU was not useful for induction of repetetive somatic embryogenesis in peanut, while dicamba was less effective.

Somatic embryos isolated from primary explants or suspension cultures and maintained on auxin medium can produce secondary embryos. The ability of somatic embryos to become repetitive makes secondary somatic embryogenesis a powerful tool. Long-term maintenance of somatic embryos and their clumps on auxin medium results in continual production of somatic embryos (Finer and Nagasawa, 1988; Griga and Stejskal, 1994). Recently, Griga (1999) reported that more than 10-year-old soybean embryogenic cultures are maintained on 10 mM 2,4-D medium without any loss of embryogenic potential. Apart from its obvious potential for mass propagation, repetitive embryogenesis can be used for genetic transformation, especially in those species that do not regenerate readily from callus (Finer and McMullen, 1991). Somatic embryogenic systems may be additionally modified for mass propagation by using micropropagation techniques and exposing somatic embryos to high cytokinin : low auxin ratios to induce multiple shoot formation from the apical meristem, thus increasing the number of plants obtainable from one embryo (Collins and Phillips, 1982).

The use of repetitive embryo system has been effective for the transformation of soybean using microprojectile bombardment (Finer and McMullen, 1991; Sato *et al.*, 1993; Parrott *et al.*, 1994). However, this method has not come into general use as the number of stable transformation events recovered can be highly variable and is generally low when cell lines under 6 months old are used (Stewart *et al.*, 1996). A second limitation to this system is the recovery of cell lines which will not convert or will convert into plants wich are either sterile or which have limited fertility (Cho *et al.*, 1995; Simmonds and Donaldon, 2000), a problem that traditionally has been greater when older cell lines are used. Stejskal and Griga (1992) reported that secondary somatic embryos appeared on media used for conversion in pea. Recently, Simmonds and Donaldson (2000) reported that five genotypes produced embryogenic cultures which proliferated embryogenic capacity and formed secondary globular embryos in soybean. The influence of genotype on capacity for proliferative or secondary embryogenesis remains unclear (Simmonds and Donaldson, 2000).

10. Artificial or synthetic seed technology

In the last 15 years, there has been a great interest in the production of encapsulated somatic embryos usually known as artificial, synthetic or manufactured seeds. Synthetic seed (synseed) refers, in a narrow sense, to encapsulated somatic embryos which functionally mimic seeds and can develop into seedlings under suitable conditions. They would have a great potential for storage and handling of somatic embryos. However, the development of this technology has been limited by poor embryo quality (Merkle *et al.*, 1995). Another constraint is the absence of a seed coat-like structure that simultaneously protects the embryo from mechanical damage and water loss. The difficulty to find a reserve compound that could be maintained in the manufactured seeds and feed the embryo during early stages of germination is another limitation. Somatic embryos of legume species have been encapsulated but no commercial application of this technology is available yet. For synthetic seed system to be successful, the quality of the somatic embryo is very important for conversion and development of plantlets. To achieve uniform quality embryos, it is necessary to have a synchronously developing culture from which embryos at a specific developmental stage can be harvested. Research on synthetic seed technology in legume crops, using somatic embryos, is being carried out with considerable success and the coming years may find application in legume improvement.

Although published research and development of artificial seed technology in legumes have been limited to alfalfa, advances in somatic embryo technology, including the ability to desiccate somatic embryos, should eventually facilitate their storage and use as artificial seed. Padmaja *et al.* (1995) produced artificial seeds of peanut by encapsulating 5–30-day-old somatic embryos in calcium alginate. The highest germination percentage (33.3%) and plant regeneration percentage (25.4%) were recorded. Production of artificial seeds using somatic embryos remains another area requiring considerable research input, as successful commercial production and release of synseeds is currently limited to very few legumes (Senaratna *et al.*, 1990; Parrott *et al.*, 1995). The greatest limitation to artificial seed technology appears to be the requirement for high quality somatic embryos with high conversion rates (Parrott *et al.*, 1995). Other obstacles to overcome include the efficient mass production of synchronized somatic embryos, the elimination of somaclonal variation and the price of aritificial seed technology.

11. Somaclonal variation and evaluation of genetic stability of somatic embryo-derived plants

Somaclonal variation manifested as phenotypic changes and biochemical alterations arising from tissue culture has been documented (Armstrong and Phillips, 1988). Somaclonal variation is a major obstacle in the production of genetically uniform plants from somatic embryos. Many factors can influence the frequency and nature of this variation but, to date, no comprehensive studies have been done in any somatic embryogenic systems to analyse the effects of manipulating all the possible parameters to minimize somaclonal variation. *In vitro* culture conditions appear to affect the stability of plant genomes, with different plant species and genotypes responding in various ways. Somaclonal variation may, therefore, provide a valuable source of genetic variation for plant breeding, but it is undesirable when genetic stability is required for production of cloned plantlets. Somaclonal variation may be the result of pre-existing genetic variation in somatic cells (Walbot, 1985) or may be due to alterations of nuclear or organellar genomes caused by the *in vitro* culture system. Several factors influence the level of somaclonal variation that occurs as a result of *in vitro* culture. The genotype of the explant is an important factor influencing the frequency of genetic variation, as is the presence of any pre-existing cytological variation. Genetic instability is more frequently associated with cell suspension or callus growth than with organized cultures from meristems. Somatic embryogenesis may reduce the frequency of genetic variation in regenerated plants due to developmental constraints during embryogenesis that select against genetically altered cells. A number of theories have been proposed to explain the increase of genetic instability during *in vitro* culture of explant tissue.

Very little effort has been expended to analyze somaclonal variation in legumes produced through somatic embryogenesis. The tissue culture medium is composed of substances that are almost certainly mutagenic. Both 2,4-dichlorophenoxyacetic acid (2,4-D) and 1-naphthaleneacetic acid (NAA) cause chromosomal aberrations when applied to cell cultures at elevated concentrations. The auxin 2,4-D has also been implicated as the cause of heritable variation in soybean plantlets regenerated from somatic embryogenic culture with lower concentrations of 2,4-D producing more variation than higher concentrations. Teratogenic effects on embryo morphology and development were also greater at lower 2,4-D concentrations (Shoemaker *et al.*, 1991). Apparently, in soybean somaclonal variation rate decreased with increasing the concentration of 2,4-D in the induction medium (Shoemaker *et al.*, 1991). However, such a correlation between auxin level and somaclonal variation was not evident in alfalfa (Nagarajan and Walton, 1987) and red clover (Wang and Holl, 1988). Bingham *et al.* (1988) reported an increased incidence of genetic variation with increase in culture duration in alfalfa plants regenerated from somatic embryos.

Kysely *et al.* (1987) obtained diploid and tetraploid pea plants via somatic embryogenesis from calli derived from immature embryos and shoot apices. Stejskal and Griga (1992) and Griga *et al.* (1995) observed variations with altered habit (leaflet shape, one pair of leaflets, abnormal flower morphology, reduced flower stalk, shortened internodes) within the regenerants of pea obtained by somatic embryogenesis from immature zygotic embryos. Griga *et al.* (1995) and Griga and Letal (1995, 1996) compared somaclones obtained by somatic embryogenesis and organogenesis and mainly morphological changes were recorded when evaluating qualitative traits in plants of both origin. All plants exhibiting such phenomena were chimaeric and the altered trait occurred randomly or was usually lost in the next generations. Barwale and Widholm (1987) analysed plants regenerated from embryogenic and organogenic cultures of nine soybean genotypes and found extensive variation in qualitative traits. Variations observed in later selfed generations included twin seeds, multiple shoots, dwarfs, abnormal leaf morphology and leaf number, wrinkled leaves, sterility. Similar type of variations was observed by Amberger *et al.* (1992). Hildebrand *et al.* (1989) estimated variation in fatty acid composition of the seeds and morphological traits in soybean regenerants obtained via somatic embryogenesis. Jin *et al.* (1996) developed somaclonal variants in soybean with respect to the pathogen resistance from embryogenic suspension cultures. However, additional studies are required to confirm the stability and inheritance of the regenerated somaclones.

Somaclonal variation appears to increase in frequency with duration of *in vitro* culture, especially when the cells are maintained as callus or cell suspension. Reducing the length of time required to produce regenerated plantlets from the initial induction of somatic embryos, as well as attempting to induce the production of somatic embryos directly on the explant tissue without the intervening callus stage, could minimize the occurrence of somaclonal variation. The cryopreservation would reduce the duration of the *in vitro* maintenance of cells in an unorganized state and perhaps periods in culture.

The complex compositions of the various culturing media may play an important role in the control of somaclonal variation. In addition, the type of auxin, its concentration and duration of application can be altered to reduce the effects of these toxic compounds on the genetic stability of the cells. Somaclonal variation has been detected in somatic embryogenesis of legumes. However, somaclonal variation may not always be detected during somatic embryogenic culture or during regeneration of plants. There is no single technique that will unequivocally demonstrate the presence or absence of somaclonal variation in cell cultures or regenerated plants. Somaclonal variation can be assessed by analysis of phenotype, chromosome number and structure, proteins or direct DNA evaluation of plants regenerated from *in vitro* culture and their progeny. Changes in the structure and organization of nuclear and organellar DNA can be directly evaluated by restriction fragment length polymorphism (RFLPs) and random amplified polymorphic DNA (RAPDs) analysis. The effect of *in vitro* culture on proteins can be evaluated by isozyme electrophoresis or analysis of seed storage protein patterns. Genetic changes arising during *in vitro* somatic embryogenic culture that remain undetected in regenerated plants could lead to large investments in propagated plants that are genetically defective. To prevent potential losses, these somatic embryo derived plants need to be screened for genetic integrity during early stages of regeneration. Evaluations using morphological markers, chromosome counts and protein electrophoresis indicate that somatic embryo derived *Trifolium repens* plants maintain genetic integrity and clonal fidelity (Maheswaran and Williams, 1987), while studies of soybean plants have detected somaclonal variation (Amberger *et al.*, 1992). Studies on mutagenesis and induction of somaclonal variation are lacking in woody legume trees (Lakshmi Sita, 1999). Morphological abnormality is observed in several regenerated germinants of somatic embryos of legumes. It is important to measure genetic stability of somatic seedlings for propagation of true-to-type selected mother plants. Random amplified polymorphic DNA (RAPD) markers are used to determine the genetic stability of the somatic embryogenic materials. However, these molecular markers are not good enough for identifying any point mutation between the primers. Therefore, still a better molecular system is needed to find difference at molecular level. It is important to control the genetic stability of the culture during somatic embryogenesis phase if the selected materials need to be propagated.

12. Biochemical and molecular studies on somatic embryogenesis

Superficially, somatic embryos mimic the developmental stages of zygotic seeds (globular, heart, torpedo, cotyledonary stages); however, despite these gross morphological similarities, somatic embryos sometimes exhibit other features suggesting an aberrant development (truncated cotyledonary development, precocious germination, recallusing, multiple or fused cotyledons, inability to germinate, etc.). The extent to which somatic seed development represents as faithful recapitulation of zygotic seed development is still largely unknown. Some evidence from pea (*Pisum sativum*) suggests that soluble auxin-binding proteins play a role in somatic embryogenesis. Picloram, which does not affect cell wall elongation in pea, but does induce somatic embryogenesis, only binds to cytoplasmic receptors. Furthermore, soluble auxin-binding proteins are found in zygotic embryos only at the developmental stage which will give rise to somatic embryos and are limited to the axis, the region from which somatic embryos originate (Jacobsen, 1991). Krochko *et al.* (1992) investigated the expression of the three major storage proteins in alfalfa somatic embryogenesis. Despite these similarities, they differ from zygotic embryos in the extent of cotyledonary development and the pattern of synthesis and quantitative expression of seed-specific storage proteins (7S, 11S and 2S proteins). Alfin (7S) is the first storage protein synthesized in developing zygotic embryos. The 11S (medicagin) and 2S (low molecular weight, LMW) storage proteins are not detectable until the following stage of development, although all three are present before the completion of embryo enlargement. Likewise, the 7S storage protein is the first to be synthesized in developing somatic embryos (day 5). Medicagin is evident by day 7 and the LMW protein by day 10. In contrast to zygotic embryos, alfin remains the predominant storage protein in somatic embryos throughout development. Not only the relative amounts of medicagin and the LMW protein reduced in somatic embryos but the LMW protein is accumulated much later than the other proteins. Further, quantification of the storage protein mRNAs (7S, 11S and 2S) by northern blot analysis confirms that there are substantial differences in the patterns of message accumulation in zygotic and somatic embryos of alfalfa (*Medicago sativa*). In somatic embryos, the predominance of the 7S storage protein is correlated with increased accumulation of its mRNA, whereas the limited synthesis of the 11S storage protein is associated with much lower steady-state levels of its message. The mRNA for the LMW protein is present already by 3 days after transfer to hormone-free media, yet that protein is not evident on stained gels until day 10. Thus, both transcriptional and posttranscriptional events appear to be important in determining the protein complement of these seed tissues. On the basis of storage protein and mRNA accumulation, mature (14 days) somatic embryos most closely resemble stage VI zygotic embryos. The results of the developmental comparison also suggest that the patterns of synthesis of the individual storage proteins (7S, 11S or 2S) are regulated independently of each other during embryogenesis in alfalfa (Krochko *et al.*, 1992).

In legume species, accumulation of specific storage proteins is a very reliable distinctive marker between somatic embryogenesis and organogenesis (Stirn and Jacobsen, 1987; Christou and Yong, 1989; Stejskal and Griga, 1992). Tissue specific (cotyledon versus hypocotyl) synthesis of storage proteins was observed in maturing soybean somatic as well as zygotic embryos and thus storage proteins may be useful in monitoring the degree of somatic embryo maturation in soybean (Komatsuda *et al.*, 1992; Stejskal and Grigam, 1992). Griga (1999) reported that b-conglycinin and glycinin – the main components of soybean seed storage protein – are synthesized at the end of the maturation phase of somatic embryos. Poulsen *et al.* (1996) described that the synthesis of at least three different 2,4-D-induced proteins during the induction of embryos, which displayed differential expression in the histodifferentiation stage of embryogenesis. According to Jacobsen (1991), a specific auxin binding protein was involved in auxin-dependent somatic embryogenesis in pea. Reisch and Bingham (1980), Wan *et al.* (1988), Hernandez-Fernandez and Christie (1989) and McLean and Nowak (1998) found that different dominant gene pairs are thought to be involved in the control of somatic embryogenesis in both *Medicago sativa* and *Trifolium pratense*.

In order to understand the process of direct somatic embryogenesis it is important to determine the changes in gene expression which occur very soon after induction, with the aim of identifying potential regulatory sequences. Several different genes which are induced during indirect somatic embryogenesis in the carrot system have been isolated, typically by differential screening of cDNA libraries. These include genes encoding Late Embryogenesis Abundant (LEA) proteins, proline- and glycine-rich cell wall-type proteins such as the EP2 lipid transfer protein, and the EP3 endochitinase. None of these genes is specific to the very early stages of somatic embryo formation and, apart from the chitinase (Schmidt *et al.*, 1994), appear to have structural rather than regulatory roles (Fowler *et al.*, 1998). Recently, RNA arbitrarily primed PCR (RAP-PCR) method was used for the identification of somatic embryogenesis regulatory genes in alfalfa (Fowler *et al.*, 1998). Changes in gene expression following induction of direct somatic embryogenesis were investigated by RAP-PCR with a view to identifying genes involved in controlling the onset of somatic embryo development in alfalfa. Amongst the induced sequences identified were a calnexin-like sequence (which was induced within 2 d), and a novel plant homologue of the yeast SNQ2 multi-drug resistance protein (which did not appear until 10 d after induction). The identification of an SNQ2-like gene which is specifically expressed at the start of the globular stage of embryo formation is intriguing since no similar homologue has yet been reported in plants. RT-PCR amplification of these sequences confirmed the expression patterns revealed by RAP-PCR and was used to show that the calnexin-like gene is induced by 2,4-D, but that the level of expression is enhanced by wounding during direct somatic embryo induction in alfalfa (Fowler *et al.*, 1998).

13. Conclusions and future prospects

Somatic embryogenesis is considered to be an appropriate method of producing large numbers of plants in a short time. The embryogenic regeneration system is important in legume species improvement. Significant gains in legume breeding can be obtained using efficient tissue culture system and molecular biology techniques. The development of *in vitro* propagation by somatic embryogenesis improved the possibility of applications of gene transformation techniques in legume breeding. The results on the induction of somatic embryos in legumes indicate the ability for embryogenic development within this legume genotype. However, the fact remains that the plant regeneration via somatic embryogenesis is still restricted to few genotypes and explants. This shows that further research studies are needed. The application of stress or chemical treatments may be the key to trigger the somatic embryogenic process in recalcitrant genotypes. Special attention should be given to economically important legumes to develop an efficient and reproducible somatic embryogenesis system and furthermore high maturation and germination rates of somatic embryos into fully developed plants, only when this goal is achieved will somatic embryogenesis become an excellent system of clonal propagation and genetic manipulation of legumes.

In most cases, these embryogenic systems need to be optimized before an effective protocol for regeneration by somatic embryogenesis would be envisaged. Particular attention should be paid to various parameters that could improve the efficiency of the process such as reduction of embryo abnormalities, increase of induction frequencies, better rates of somatic embryo conversion and increased rates of secondary embryogenesis induction. Another problem is the systematic use of immature zygotic embryos as explant source for somatic embryogenesis induction which limits the propagation of selected genotypes. When plants are regenerated via somatic embryogenesis, the level of variability among the regenerants may be evaluated by molecular markers such as random amplified polymorphic DNA and restriction fragment length polymorphisms. Although somatic embryos at all developmental stages and plantlet regeneration could be achieved, the plant conversion rate is still very low. More research is needed in this direction to develop an effective method for high conversion rate of somatic embryos into plantlets. The identification of genes responsible for embryogenesis is of great importance as well as more exact knowledge of the interaction of hormones in signal transduction during this pathway.

Somatic embryogenesis has been induced on immature zygotic embryo explants of various legume species. The embryogenic cultures can be multiplied via secondary embryogenesis by regular subculture on the culture medium having low level of cytokinin. However, embryo development is asynchronous and germination rate is relatively low. Conversion of somatic embryos into plantlets is limited due to poor shoot development. Further, limited information is available on genetic control of embryogenesis and gene expression studies in legumes species. Therefore much attention should be given for synchronous somatic embryo development with high frequency of plant conversion and regulation of gene expression during embryogenesis. Given the importance of legume species for food, fodder and fuel, there is an urgent need for these barriers to be overcome. Furthermore, the potential for application of woody legume embryogenic cultures as target material for genetic transformation has yet to be realized. In the future, we believe that the combination of techniques of genetic engineering with efficient plant regeneration system via somatic embryogenesis may result in the obtaining of genetically modified leguminous plants showing new favourable traits.

References

Ahmed R, Gupta S D and De D N (1996) Somatic embryogenesis and plant regeneration from leaf derived callus of winged bean (*Psophocarpus tetragonolobus* L.). *Plant Cell Rep.*, **15**: 531–535.

Amberger A L, Palmer R G and Shoemaker R C (1992) Analysis of culture-induced variation in soybean. *Crop Sci.*, **32**: 1103–1108.

Ammirato P V (1983) Embryogenesis. In: *Handbook of Plant Cell Culture, Vol. I* (Eds Evans D A, Sharp W R, Ammirato P V and Yamada Y), MacMillan, New York, pp. 82–123.

Ammirato P V (1987) Organisational events during somatic embryogenesis. In: *Plant Biology, Vol. 3, Plant Tissue and Cell Culture* (Eds Green C E, Somers D A, Hackett W P and Biesboer D D), Alan R. Liss, New York, pp. 57–81.

Anandarajah K and McKersie B D (1990) Manipulating the desiccation tolerance and vigor of dry somatic embryos of *Medicago sativa* L. with sucrose, heat shock, and abscisic acid. *Plant Cell Rep.*, **9**: 451–455.

Anbazhagan V R and Ganapathi A (1999) Somatic embryogenesis in cell suspension cultures of pigonpea (*Cajanus cajan*). *Plant Cell Tiss. Org. Cult.*, **56**: 179–184.

Arcioni S and Mariotti D (1982) Tissue culture and plant regeneration in the forage legumes *Onobrychis viciaefolia* Scop., *Coronilla varia* and *Lotus corniculatus* L. In: *Plant Tissue Culture, Vol. 11* (Ed Fujiwara A), Maruzen, Tokyo, pp. 707–708.

Arrillaga I, Tobolski S A and Merkle S A (1994) Advances in somatic embryogenesis and plant production of black locust. *Plant Cell Rep.*, **13**: 171–175.

Bailey M A, Boerma H R and Parrott W A (1993) Genotype effects on proliferative embryogenesis and plant regeneration of soybean. In Vitro *Cell. Dev. Biol. Plant.*, **29**: 102–108.

Bajaj Y P S (1995) Somatic embryogenesis and its applications in crop improvement. In: *Biotechnology in Agriculture and Forestry, Vol. 30, Somatic Embryogenesis and Synthetic Seed* (Ed Bajaj Y P S), Springer-Verlag, Berlin, pp. 105–125.

Bajaj Y P S, Singh H and Gosal S S (1980) Haploid embryogenesis in another culture of pigeonpea (*Cajanus cajan* L.). *Theort. Appl. Genet.*, **58**: 157–159.

Baker C M and Wetzstein H Y (1992) Somatic embryogenesis and plant regeneration from leaflets of peanut (*Arachis hypogaea* L.). *Plant Cell Rep.*, **11**: 71–75.

Baker C M and Wetzstein H Y (1994) Influence of auxin type and concentration on peanut somatic embryogenesis. *Plant Cell Tiss. Org. Cult.*, **36**: 361–368.

Baker C M and Wetzstein H Y (1998) Leaflet development, induction time and medium influence somatic embryogenesis in peanut (*Arachis hypogaea* L.). *Plant Cell Rep.*, **17**: 925–929.

Baker C M and Wetzstein H Y (1995) Repetitive somatic embryogenesis in peanut cotyledon cultures by continual exposure to 2,4-D. *Plant Cell Tiss. Org. Cult.*, **40**: 249–254.

Baker C M, Burns J A and Wetzstein H Y (1994) Influence of photoperiod and medium formulation on peanut somatic embryogenesis. *Plant Cell Rep.*, **13**: 159–163.

Baker C M, Durham R E, Burns J A, Parrott W A and Wetzstein H Y (1995) High frequency somatic embryogenesis in peanut (*Arachis hypogaea* L.) using mature, dry seed. *Plant Cell Rep.*, **15**: 38–42.

Barna K S and Wakhlu A K (1993) Somatic embryogenesis and plant regeneration from callus cultures of chickpea (*Cicer arietinum* L.). *Plant Cell Rep.*, **12**: 521–524.

Barna K S and Wakhlu A K (1995) Direct somatic embryogenesis and plantlet regeneration from immature leaflets in chickpea. In Vitro *Cell. Dev. Biol. Plant.*, **31**: 137–139.

Barwale U B, Kerns H R and Widholm J M (1986) Plant regeneration from callus cultures of several soybean genotypes via embryogenesis and organogenesis. *Planta*, **167**: 473–481.

Barwale U B and Widholm J M (1987) Somaclonal variation in plants regenerated from cultures of soybean. *Plant Cell Rep.*, **6**: 365–368.

Bencheikh M and Gallais A (1996a) Somatic embryogenesis in pea (*Pisum sativum* L. and *Pisum arvense* L.) diallele analysis and genetic control. *Euphytica*, **90**: 257–264.

Bencheikh M and Gallais A (1996b) Study of the variation in the somatic embryogenesis ability of some pea lines (*Pisum sativum* L. and *Pisum arvense* L.). *Euphytica*, **90**: 251–256.

Beversadorf W D and Bingham E T (1977) Degrees of differentiation obtained in tissue cultures of *Glycine* species. *Crop Sci.*, **17**: 307–311.

Binachi S, Flament P and Datee Y (1988) Somatic embryogenesis and organogenesis in alfalfa: Genotype variations in regeneration ability. *Agronomy*, **8**: 121–126.

Bingham E T, McCoy T J and Walker K A (1988) Alfalfa tissue culture. In: *Alfalfa and Alfalfa Improvement* (Eds Hanson A A, Barnes D K and Hill Jr. R R), Madison, WI:ASA-CSSA-SSA, pp. 903–929.

Brown D C W and Atanassov A (1985) Role of genetic background in somatic embryogenesis in *Medicago. Plant Cell Tiss. Org. Cult.*, **4**: 111–222.

Buchheim J A, Colburn S M and Ranch J P (1989) Maturation of soybean somatic embryos and the transition to plantlet growth. *Plant Physiol.*, **89**: 768–775.

Buckley L G and Trigiano R N (1994) Changes in ovule protein profiles associated with somatic embryogenesis in *Cercis canadensis* (Redbud). *Plant Cell Rep.*, **14**: 27–30.

Carman J (1990). Embryogenic cells in plant tissue cultures: Occurrence and behaviour. In Vitro *Cell. Dev. Biol. Plant.*, **26**: 746–753.

Chen T H H and Marowitch J (1987) Screening of *Medicago falcata* germplasm for *in vitro* regeneration. *J. Plant Physiol.*, **128**: 271–277.

Chen T H H, Marowitch J and Thompson B G (1987) Genotype effects on somatic embryogenesis and plant regeneration from callus cultures of alfalfa. *Plant Cell Tiss. Org. Cult.*, **8**: 73–81.

Chengalrayan K, Mhaske V B and Hazra S (1995) *In vitro* regulation of morphogenesis in peanut (*Arachis hypogaea* L.). *Plant Sci.*, **110**: 259–268.

Chengalrayan K, Mhaske V B and Hazra S (1997) High frequency conversion of abnormal peanut somatic embryos. *Plant Cell Rep.*, **16**: 783–786.

Chengalrayan K, Mhaske V B and Hazra S (1998) Genotypic control of peanut somatic embryogenesis. *Plant Cell Rep.*, **17**: 522–525.

Chengalrayan K, Sathaye S S and Hazra S (1994) Somatic embryogenesis from mature zygotic embryo-derived leaflets of peanut (*Arachis hypogaea* L.). *Plant Cell Rep.*, **13**: 578–581.

Cho M J, Widholm J M and Vodkin L O (1995) Cassettes for seed specific expression tested in transformed embryogenic cultures of soybean. *Plant Mol. Biol. Rep.*, **13**: 255–269.

Christianson M L, Warnick D A and Carlson P S (1983) A morphogenetically competent soybean suspension culture. *Science*, **222**: 632–634.

Christou P and Yang N S (1989) Developmental aspects of soybean (*Glycine max*) somatic embryogenesis. *Ann. Bot.*, **64**: 225–234.

Collins G B and Phillips G C (1982) *In vitro* tissue culture and plant regeneration in *Trifolium pratense* L. In: *Variability in Plants Regenerated from Tissue Culture* (Eds Earle E D and Demarly Y), Praeger, New York, pp. 22–34.

Conger B V, Hanning G E, Gray D J and McDaniel J K (1983) Direct somatic embryogenesis from mesophyll cells of orchard grass. *Science*, **221**: 850–851.

Curir P, Ruffoni B, Massabo F and Damiano (1990) Induction of somatic embryogenesis in *Genista monosperma* Lam. *Acta Hort.*, **280**: 113–116.

Dahmer M L, Hildebrand D F and Collins G B (1992) Comparative protein accumulation patterns in soybean somatic and zygotic embryos. In Vitro *Cell Dev. Biol. Plant.*, **28**: 106–111.

Dan Y and Reichert N A (1998) Organogenic regeneration of soybean from hypocotyl explants. In Vitro *Cell. Dev. Biol. Plant.*, **34**: 14–21.

Das A B, Rout G R and Das P (1995) *In vitro* somatic embryogenesis from callus culture of the timber yielding tree *Hardwickia binata* Roxb. *Plant Cell Rep.*, **15**: 147–149.

Das P, Samantaray S, Roberts A V and Rout G R (1997) *In vitro* somatic embryogenesis of *Dalbergia sissoo* Roxb. – a multipurpose timber yielding tree. *Plant Cell Rep.*, **16**: 578–582.

Das Neves L O, Duque S R L, Almeida J S and Fevereiro S (1999) Repetitive somatic embryogenesis in *Medicago truncatula* sp. Narbonensis and *M. truncatula* Gaertn cv. Jemalong. *Plant Cell Rep.*, **18**: 398–405.

Denchev P, Velcheva M and Atenssov A (1991) A new approach to direct somatic embryogenesis in *Medicago. Plant Cell Rep.*, **10**: 338–341.

Dhanalakshmi S and Lakshmanan K K (1992) *In vitro* somatic embryogenesis and plant regeneration in *Clitoria ternatea. J. Exp. Bot.*, **43**: 213–219.

Dodeman V L, Ducreux G and Kreis M (1997) Zygotic embryogenesis versus somatic embryogenesis. *J. Exp. Bot.*, **48**: 1493–1509.

Dos Santos A V P, Cutter E G and Davey M R (1983) Origin and development of somatic embryos in *Medicago sativa* (alfalfa). *Protoplasma*, 117: 107–115.

Durham R E and Parrott W A (1992) Repetitive somatic embryogenesis in liquid cultures of peanut. *Plant Cell Rep.*, **11**: 122–125.

Eapen S and George L (1990) Ontogeny of somatic embryos of *Vigna aconitifolia*, *Vigna mungo* and *Vigna radiata. Ann. Bot.*, **66**: 219–226.

Eapen S and George L (1993a) Somatic embryogenesis in peanut: Influence of growth regulators and sugars. *Plant Cell Tiss. Org. Cult.*, **35**: 151–156.

Eapen S and George L (1993b) Plant regeneration from leaf discs of peanut and pigeonpea: Influence of benzyladenine, indoleacetic acid and indole acetic acid-amino conjugates. *Plant Cell Tiss. Org. Cult.*, **35**: 223–227.

Eapen S and George L (1994) Somatic embryogenesis in *Cicer arietinum* L.: Influence of genotype and auxins. *Biol. Plant.*, **36**: 343–349.

Eapen S, George L and Rao P S (1993) Plant regeneration through somatic embryogenesis in peanut (*Arachis hypogaea* L.). *Biol. Plant.*, **35**: 499–504.

Emons A M C (1994) Somatic embryogenesis: Cell biological aspects. *Act. Bot. Neerl.*, **43**: 1–14.

Erdelska O and Vidovencova Z (1992) The polyembryonic character of somatic embryogenesis. In: *Regulation of Plant Somatic Embryogenesis* (Eds Griga M and Tejklova E), VUTPL, Sumperk, CSFR, pp. 78–79.

Finer J J (1988) Apical proliferation of embryogenic tissue of soybean (*Glycine max* (L.) Merril). *Plant Cell Rep.*, **7**: 238–241.

Finer J J and McMullen M D (1991) Transformation of soybean via particle bombardment of embryogenic suspension culture tissue. In Vitro *Cell Dev. Biol. Plant*, **27**: 175–182.

Finer J J and Nagasawa A (1988) Development of an embryogenic suspension culture of soybean [*Glycine max* (L.) Merrill]. *Plant Cell Tiss. Org. Cult.*, **15**: 125–136.

Fowler M R, Ong L M, Russinova E, Atanassov A I, Scott N W, Slater A and Elliott M C (1998) Early changes in gene expression during direct somatic embryogenesis in alfalfa revealed by RAP-PCR. *J. Exp. Bot.*, **49**: 249–253.

Fuentes S I, Suarez R, Villegas T, Acero L C and Hernendez G (1993) Embryogenic response of Mexican alfalfa (*Medicago sativa*) varieties. *Plant Cell Tiss. Org. Cult.*, **34**: 299–302.

Fujii J A, Slade D, Olsen R, Ruzin S E and Redenbaugh K (1990) Alfalfa somati embryo maturation and conversion into plants. *Plant Sci.*, **72**: 93–100.

Gamborg O L, Davis B P and Stahlhut R W (1983) Somatic embryogenesis in cell cultures of *Glycine* species. *Plant Cell Rep.*, **2**: 209–212.

Garg L, Bhandari N N, Vijaya R and Bhojwani S S (1996) Somatic embryogenesis and regeneration of triploid plants in endosperm cultures of *Acacia nilotica. Plant Cell Rep.*, **15**: 855–858.

Geetha N (1995) *In vitro* plant regeneration and selection for salt and drought tolerance in *Vigna mungo* (L.) Hepper. Ph.D. thesis, Bharathidasan University, Tiruchirappalli, India.

Geetha N, Venkatachalam P, Prakash V and Lakshmi Sita G (1998) High frequency induction of multiple shoots and plant regeneration from seedling explants of pigeonpea (*Cajanus cajan* L.). *Curr. Sci.*, **75**: 1036–1041.

Geetha N, Venkatachalam P and Rao G R (1997) Somatic embryogenesis and plant regeneration from cell suspension cultures of blackgram [*Vigna mungo* (L.) Hepper]. *Physiol. Mol. Biol. Plants*, **3**: 25–30.

Geneve R L and Kester S T (1990) The initiation of somatic embryos and adventitious roots from developing zygotic embryos explants of *Cercis canadensis* L. cultured *in vitro. Plant Cell Tiss. Org. Cult.*, **22**: 71–76.

Genga A and Allavena A (1991) Factors affecting morphogenesis from immature cotyledons of *Phaseolus coccineus* L. *Plant Cell Tiss. Org. Cult.*, **27**: 189–196.

George L and Eapen S (1993) Influence of genotype and explant source on somatic embryogenesis in peanut. *Oleagineux*, **48**: 361–363.

George L and Eapen S (1994) Organogenesis and embryogenesis from diverse explants in pigeonpea (*Cajanus cajan* L.). *Plant Cell Rep.*, **13**: 417–420.

Gharyal P K and Maheswari S C (1983) *In vitro* differentiation of somatic embryos in a leguminous tree, *Albizia lebbeck* L. *Naturwiss.*, **68**: 379–380.

Ghazi T D, Cheema H K and Nabors M W (1986) Somatic embryogenesis and plant regeneration from embryogenic callus of soybean, *Glycine max* L. *Plant Cell Rep.*, **5**: 452–456.

Gill R and Saxena P K (1992) Direct somatic embryogenesis and regeneration of plants from seedling explants of peanut (*Arachis hypogaea* L.): Promotive role of thidiazuron. *Can. J. Bot.*, **70**: 1186–1192.

Greinwald R and Czygan F C (1991) Regeneration of plants from callus cultures of *Chamaecytisus purpureus* and *C. austriacus* (Leguminosae). *Bot. Acta*, **104**: 64–67.

Griga M (1990) The study of *in vitro* regeneration systems in pea, horse bean and soybean, Ph.D. thesis, Masaryk University, Brno, Czech.

Griga M (1993) Some factors affecting somatic embryogenesis efficiency in soybean [*Glycine max* (L.) Merr]. *Biol. Plant.*, **35**: 179–189.

Griga M (1998) Direct somatic embryogenesis from shoot apical meristems of pea and thidiazuron induced high conversion rate of somatic embryos. *Biol. Plant.*, **41**: 481–495.

Griga M (1999) Somatic embryogenesis in grain legumes. *Adv. Regulat. Plant Growth Dev.*, **1**: 233–250.

Griga M and Letal J (1995) Somaclonal variation study of pea (*Pisum sativum* L.). In: *Recent Advances in Plant Biotechnology*, Inst. Plant Genet., Nitra, pp. 130–135.

Griga M and Letal J (1996) Improved protocol for pea (*Pisum sativum* L.) somatic embryogenesis and field performance of pea soma clones. In: *Second Asia–Pacific Conf. Plant Cell Tiss. Cult.*, Beijing, China, p. 4.

Griga M and Stejskal J (1994) Auxin and sucrose effects on maturation of soybean somatic embryos from long-term repetitive culture. *Biol. Plant.*, **36**: 81.

Griga M, Stejskal J and Beber K (1995) Analysis of tissue culture-derived variation in pea (*Pisum sativum* L.) – preliminary results. *Euphytica*, **85**: 335–339.

Gupta S D, Ahmed R and De D N (1997) Direct somatic embryogenesis and plantlet regeneration from seedling leaves of winged bean, *Psophocarpus tetragonolobus* (L.) DC. *Plant Cell Rep.*, **16**: 628–631.

Hadi M Z, McMullen M A and Finer J J (1996) Transformation of 12 different plasmids into soybean via particle bombardment. *Plant Cell Rep.*, **15**: 500–505.

Hammatt N and Davey M R (1987) Somatic embryogenesis and plant regeneration from cultured zygotic embryos of soybean (*Glycine max* (L.) Merr). *J. Plant Physiol.*, **128**: 219–226.

Hartweck L M, Lazzeri P A, Cui D, Collins G B and Williams E G (1988) Auxin-orientation effects on somatic embryogenesis from immature soybean cotyledons. *In Vitro Cell. Dev. Biol. Plant*, **24**: 821–828.

Hazra S, Sathaye S S and Mascarenhas A F (1989) Direct somatic embryogenesis in peanut (*Arachis hypogaea* L.). *Bio/Technology*, **7**: 949–951.

Heath L C, Chin S, Spencer D and Higgins T J V (1993) *In vitro* regeneration of commercial cultivars of subterranean clover. *Plant Cell Tiss. Org. Cult.*, **35**: 43–48.

Hepher A, Boulter M E, Harris N and Nelson R S (1988) Development of a superficial meristem during somatic embryogenesis from immature cotyledons of soybean (*Glycine max* L.). *Ann. Bot.*, **62**: 513–519.

Hernandez-Fernandez M M and Christie B R (1989) Direct somatic embryogenesis in alfalfa (*Medicago sativa* L.) somatic embryos, leaflets and mature seeds. *Plant Sci.*, **109**: 191–198.

Hildebrand D F, Adams T R, Dahmer M L, Williams E G and Collins G B (1989) Analysis of lipid composition and morphological characteristics in soybean regenerants. *Plant Cell Rep.*, **7**: 701–703.

Hinchee M A W, Connor-Ward D V, Newell C A, McDonnell R E, Sato S J, Gasser C S, Fishhoff D A, Re D B, Fraley R T and Horsch R B (1988) Production of transgenic soybean plants using *Agrobacterium*-mediated gene transfer. *Bio/Technology*, **6**: 915–922.

Horbowicz M, Obendorf R L, McKersie B D and Viands D R (1995) Soluble saccharides and cyclitols in alfalfa (*Medicago sativa* L.) somatic embryos, leaflets and mature seeds. *Plant Sci.*, **109**: 191–198.

Jacobsen H J (1991) Somatic embryogenesis in seed legumes: The possible role of soluble auxin receptors. *Israel J. Bot.*, **40**: 139–143.

Jacobsen H J and Kysely W (1984) Induction of somatic embryos in pea, *Pisum sativum* L. *Plant Cell Tiss. Org. Cult.*, **3**: 319–324.

Jacques L C, Marche Y L and Deunff Y L (1995) Effects of auxin, cytokinin, carbohydrates and aminoacids on somatic embryogenesis induction from shoot apices of pea. *Plant Cell Tiss. Org. Cult.*, **4**: 267–275.

Janick J (1993) Agricultural uses of somatic embryos. *Act. Hort.*, **336**: 207–215.

Jin H, Hartman G L, Huang Y H, Nickell C D and Widholm J M (1996) Regeneration of soybean plants from embryogenic suspension cultures treated with toxic culture filtrate of *Fusarium solani* and screening of regenerants for resistance. *Phytopathology*, **86**: 714–718.

Kar S, Basu D, Das S, Ramakrishnan N A, Mukherjee R, Nayak P and Sen S K (1997) Expression of CryIA(C) gene of *Bacillus thuringiensis* in transgenic chickpea plants inhibits development of pod-borer (*Heliothis armigera*) larvae. *Trans. Res.*, **6**: 177–185.

Kerns H R, Barwale U B, Meyer M M Jr. and Widholm J M (1986) Correlation of cotyledonary node shoot proliferation and somatic embryoid development in suspension cultures of soybean [*Glycine max* (L.) Merr.]. *Plant Cell Rep.*, **5**: 140–143.

Keyes G J, Collins G B and Taylor N L (1980) Genetic variations in tissue cultures of red clover. *Theort. Appl. Genet.*, **58**: 265–271.

Kielly G A and Bowley S R (1992) Genetic control of somatic embryogenesis in alfalfa. *Genome*, **35**: 474–477.

Komatsuda T (1990) Ability of soybean (*Glycine max* (L.) Merr.) genotypes to produce somatic embryos on a medium containing a low concentration of sucrose. *Jpn. J. Breed.*, **40**: 371–375.

Komatsuda T, Kanebo K and Oka S (1991) Genotype × sucrose interactions for somatic embryogenesis in soybean. *Crop Sci.*, **31**: 333–337.

Komatsuda T and Ko S W (1990) Screening of soybean [*Glycine max* (L.) Merrill] genotype for somatic embryo production from immature embryos. *Jpn. J. Breed.*, **40**: 249–251.

Komatsuda T, Lee W and Oka S (1992) Maturation and germination of somatic embryos as affected by sucrose and plant growth regulators in soybean, *Glycine gracilis* Skvortz and *Glycine max* (L.) Merr. *Plant Cell Tiss. Org. Cult.*, **28**: 103–113.

Komatsuda T and Ohyama K (1988) Genotypes of high competence for somatic embryogenesis and plant regeneration in soybean, *Glycine max* L. *Theort. Appl. Genet.*, **75**: 695–700.

Krochko J E, Pramanik S K and Bewley J D (1992) Contrasting storage protein synthesis and mRNA accumulation during development of zygotic and somatic embryos of alfalfa (*Medicago sativa* L.). *Plant Physiol.*, **99**: 46–53.

Kuklin A I, Denchev P D, Atanassov A I and Scragg A H (1994) Alfalfa embryo production in airlift vessels via direct somatic embryogenesis. *Plant Cell Tiss. Org. Cult.*, **38**: 19–23.

Kulothungan S, Ganapathi A, Shajahan A and Kathiravan K (1995) Somatic embryogenesis in cell suspension culture of cowpea (*Vigna unguiculata* (L.) Walp). *Israel J. Plant Sci.*, **43**: 385–390.

Kumar A S, Gamborg O L and Nabors M W (1988) Plant regeneration from cell suspension cultures of *Vigna aconitifolia. Plant Cell Rep.*, **7**: 138–144.

Kumar V D, Kirti P B, Sachan J K S and Chopra V L (1994) Plant regeneration via somatic embryogenesis in chickpea (*Cicer arietinum* L.). *Plant Cell Rep.*, **13**: 468–472.

Kumar V, Kirti P B, Sachan J K S and Chopra V L (1995) Picloram induced somatic embryogenesis in chickpea (*Cicer arietinum* L.). *Plant Sci.*, **109**: 207–213.

Kysely W and Jacobsen H J (1990) Somatic embryogenesis from pea embryos and shoot apices. *Plant Cell Tiss. Org. Cult.*, **20**: 7–14.

Kysely W, Myers J R, Lazzeri P A, Collins G B and Jacobsen H J (1987) Plant regeneration via somatic embryogenesis in pea (*Pisum sativum* L.). *Plant Cell Rep.*, **6**: 305–308.

Lai F and McKersie B D (1994) Regulation of starch and protein accumulation in alfalfa (*Medicago sativa* L.) somatic embryos. *Plant Sci.*, **100**: 211–219.

Lakshmanan P and Taji A (2000) Somatic embryogenesis in leguminous plants. *Plant Biol.*, **2**: 136–148.

Lakshmi Sita G (1999) Somatic embryogenesis in rosewood and other Indian tree legumes. In: *Somatic Embryogenesis in Woody Plants, Vol. 5* (Eds Jain S M, Gupta P K and Newton R J), Kluwer Academic Publishers, The Netherlands, pp. 95–112.

Lazzeri P A, Hildebrand D F and Collins G B (1985) A procedure for plant regeneration from immature cotyledon tissue of soybean. *Plant Mol. Biol. Rep.*, **3**: 160–167.

Lazzeri P A, Hildebrand D F and Collins G B (1987a) Soybean somatic embryogenesis: Effects of hormones and culture manipulations. *Plant Mol. Biol. Rep.*, **10**: 197–208.

Lazzeri P A, Hildebrand D F and Collins G B (1987b) Soybean somatic embryogenesis: Effects of nutritional, physical and chemical factors. *Plant Mol. Biol. Rep.*, **10**: 209–220.

Lazzeri P A, Hildebrand D F, Sunega J, Williams E G and Collins G B (1988) Soybean somatic embryogenesis: Interactions between sucrose and auxin. *Plant Cell Rep.*, **7**: 517–520.

Lecouteux C G, Lai F and McKersie B D (1993) Maturation of alfalfa (*Medicago sativa* L.) somatic embryos by abscisic acid, sucrose and chilling stress. *Plant Sci.*, **94**: 207–213.

Lehminger-Mertens R and Jacobsen H J (1989) Plant regeneration from pea protoplasts via somatic embryogenesis. *Plant Cell Rep.*, **8**: 379–382.

Levi A and Sink K C (1990) Differential aspects of sucrose, glucose and fructose during somatic embryogenesis in Asparagus. *J. Plant Physiol.*, **137**: 184–189.

Li X and Demarly Y (1996) Somatic embryogenesis and plant regeneration in *Medicago suffruticosa. Plant Cell Tiss. Org. Cult.*, **44**: 79–81.

Li B J, Langridge W H R and Szalay A A (1985) Somatic embryogenesis and plant regeneration in the soybean, *Glycine max* L. *Plant Cell Rep.*, **4**: 344–347.

Li J and Grabau E A (1996) Comparison of somatic embryogenesis and embryo conversion in commercial soybean cultivars. *Plant Cell Tiss. Org. Cult.*, **44**: 87–89.

Lippmann B and Lippmann G (1984) Induction of somatic embryos in cotyledonary tissue of soybean, *Glycine max* L. Merr. *Plant Cell Rep.*, **3**: 215–218.

Little E L, Magbanua Z V and Parrott W A (2000) A protocol for repetitive somatic embryogenesis from mature peanut epicotyls. *Plant Cell Rep.*, **19**: 351–357.

Liu W, Morre P J and Collins G B (1992) Somatic embryogenesis in soybean via somatic embryo cycling. In Vitro *Cell. Dev. Biol. Plant*, **28**: 153–160.

Liu W, Torisky R S, McAllister K P, Avdiushko S, Hildebrand D and Collins G B (1996) Somatic embryo cycling: evaluation of a novel transformation and assay system for seed specific gene expression in soybean. *Plant Cell Tiss. Org. Cult.*, **47**: 33–42.

Loiseau J, Marche C and Deunff Y L (1995) Effect of auxins, cytokinins, carbohydrates and amino acids on somatic embryogenesis from shoot apices of pea. *Plant Cell Tiss. Org. Cult.*, **41**: 267–275.

Luo J P and Jia J F (1998) Plant regeneration from callus protoplasts of the forage legume *Astrafalus adsurgens* Pall. *Plant Cell Rep.*, **17**: 313–317.
Lupotto E (1983) Propagation of an embryogenic culture of *Medicago sativa* L. *Pflanzenphysiol.*, **111**: 95–104.
Lupotto E (1986) The use of single somatic embryo culture in propagating and regenerating lucerne (*Medicago sativa* L.). *Ann. Bot.*, **57**: 19–24.
Maheswaran G and Williams E G (1984) Direct somatic embryoid formation on immature embryos of *Trifolium repens, T. pratense* and *Medicago sativa* and rapid clonal propagation of *T. repens. Ann. Bot.*, **54**: 201–211.
Maheswaran G and Williams E G (1985) Origin and development of somatic embryoids formed directly on immature embryos of *Trifolium repens in vitro. Ann. Bot.*, **56**: 618–630.
Maheswaran G and Williams E G (1986) Direct secondary somatic embryogenesis from immature sexual embryos of *Trifolium repens* cultured *in vitro. Ann. Bot.*, **57**: 109–117.
Maheswaran G and Williams E G (1987) Uniformity of plants regenerated by direct somatic embryogenesis from zygotic embryos of *Trifolium repens. Ann. Bot.*, **59**: 93–97.
Malik K A and Saxena P K (1992) Somatic embryogenesis and shoot regeneration from intact seedlings of *Phaseolus acutifolius, P. aureus* (L.) Wilczek, *P. coccineus* and *P. wrightii* L. *Plant Cell Rep.*, **11**: 163–168.
Mallikarjuna N, Reena M J T, Sastri D C and Moss J P (1996) Somatic embryogenesis in pigeonpea, *Cajanus cajan* L. *Indian J. Exp. Biol.*, **34**: 282–284.
McCabe D E, Swain W F, Martinell B J and Christou P (1988) Stable transformation of soybean (*Glycine max* L.) by particle acceleration. *Bio/Technology*, **6**: 923–926.
McKently A H (1991) Direct somatic embryogenesis from axes of mature peanut embryos. In Vitro *Cell. Dev. Biol. Plant*, **27**: 197–200.
McKently A H (1995) Effect of genotype on somatic embryogenesis from axes of mature peanut embryos. *Plant Cell Tiss. Org. Cult.*, **42**: 251–254.
McLean N L and Nowak J (1998) Inheritance of somatic embryogenesis in red clover (*Trifolium pratense* L.). *Theort. Appl. Genet.*, **97**: 557–562.
Meijer E G M and Brown D C W (1987) Role of exogenous reduced nitrogen and sucrose in rapid high frequency somatic embryogenesis in *Medicago sativa. Plant Cell Tiss. Org. Cult.*, **10**: 11–20.
Merkle S A, Parrott W A and Flinn B S (1995) Morphogenic aspects of somatic embryogenesis. In: *In Vitro Embryogenesis in Plants* (Ed Thorpe T A), Kluwer Academic Publishers, Dordrecht, The Netherlands, pp. 155–203.
Meurer C A, Dinkins R D and Collins G B (1998) Factors affecting soybean cotyledonary node transformation. *Plant Cell Rep.*, **18**: 180–186.
Mhaske V B, Chengalrayan K and Hazra S (1998) Influence of osmotica and abscisic acid on triglyceride accumulation in peanut embryos. *Plant Cell Rep.*, **17**: 742–746.
Mhaske V B and Hazra S (1994) Appearance of storage lipids (triglycerides) in somatic embryos of peanut (*Arachis hypogaea* L.). *In Vitro Cell Dev. Biol. Plant*, **30**: 113–116.
Mitten D H, Sato S J and Skokut T A (1984) *In vitro* regenerative potential of alfalfa germplasm sources. *Crop Sci.*, **24**: 943–945.
Murthy B N S, Murch S J and Saxena P K (1995) Thidiazuron induced somatic embryogenesis in intact seedlings of peanut (*Arachis hypogaea* L.): Endogenous growth regulator levels and significance of cotyledons. *Physiol. Plant.*, **94**: 268–276.
Murthy B N S and Saxena P K (1994) Somatic embryogenesis in peanut (*Arachis hypogaea* L.): Stimulation of direct differentiation of somatic embryos by forchlorfenuron (CPPU). *Plant Cell Rep.*, **14**: 145–150.
Murthy B N S, Victor J, Singh R P, Fletcher R A and Saxena P K (1996) *In vitro* regeneration of chickpea (*Cicer arietinum* L.): Stimulation of direct organogenesis and somatic embryogenesis by thidiazuron. *Plant Growth Reg.*, **19**: 233–240.
Nadolska-Orezyk A (1992) Somatic embryogenesis of agriculturally important lupin species (*Lupinus angustifolius, L. albus, L. mutabilis*), *Plant Cell Tiss. Org. Cult.*, **28**: 19–25.
Nagarajan P, McKenzie J S and Walton P D (1986) Embryogenesis and plant regeneration of *Medicago* species in tissue culture. *Plant Cell Rep.*, **5**: 77–80.
Nagarajan P and Walton P D (1987) A comparison of somatic chromosomal instability in tissue culture regenerants from *Medicago media* pers. *Plant Cell Rep.*, **6**: 109–113.
Ninkovic S, Miljus-Djukic J and Mirjana N (1995) Genetic transformation of alfalfa somatic embryos and their propagation through repetitive somatic embryogenesis. *Plant Cell Tiss. Org. Cult.*, **42**: 255–260.

Nolan K E, Rose R J and Gorst J R (1989) Regeneration of *Medicago truncatula* from tissue culture: Increased somatic embryogenesis using explants from regenerated plants. *Plant Cell Rep.*, **8**: 278–281.

Novak M, Griga M and Tejklova E (1987) Induction of somatic embryos from immature zygotic embryos of *Glycine max* (L.) Merr. In Vitro *Sci. Agr. Bohem.*, **19**: 233–241.

Ozcan S, Barghchi M, Firek S and Draper J (1993) Efficient adventitious shoot regeneration and somatic embryogenesis in pea. *Plant Cell Tiss. Org. Cult.*, **34**: 271–277.

Ozias-Akins P (1989) Plant regeneration from immature embryos of peanut (*Archis hypogaea* L.). *Plant Cell Rep.*, **8**: 217–218.

Ozias-Akins P, Anderson W F and Holbrook C C (1992a) Somatic embryogenesis in *Arachis hypogaea* L.: genotype comparison. *Plant Sci.*, **83**: 103–111.

Ozias-Akins P, Singsit C and Branch W B (1992b) Interspecific hybrid inviability in crosses of *Arachis hypogaea* × *A. stenosperma* can be overcome by *in vitro* embryo maturation or somatic embryogenesis. *J. Plant Physiol.*, **140**: 207–217.

Padmaja G, Reddy L R and Reddy G M (1995) Plant regeneration from synthetic seeds of groundnut, *Arachis hypogaea* L. *Indian J. Exp. Biol.*, **33**: 967–971.

Parrott W A (1991) Auxin-stimulated somatic embryogenesis from immature cotyledons of white clover. *Plant Cell Rep.*, **10**: 17–21.

Parrott W A, All J N, Adang M J, Bailey M A, Boerma H R and Stewart Jr C N (1994) Recovery and evaluation of soybean (*Glycine max* [L.] Merr.) plants transgenic for a *Bacillus thuringiensis* var. Kurstaki insecticidal gene. In Vitro *Cell. Dev. Biol. Plant*, **30**: 144–149.

Parrott W A and Bailey M A (1993) Characterization of recurrent somatic embryogenesis of alfalfa on auxin-free medium. *Plant Cell Tiss. Org. Cult.*, **32**: 69–76.

Parrott W A, Dryden G, Vogt S, Hilderbrand D F, Collins G B and Williams E G (1988) Optimisation of somatic embryogenesis and embryo germination in soybean. In Vitro *Cell. Dev. Biol. Plant*, **24**: 817–820.

Parrott W A, Williams E G, Hildbrand D F and Collins G B (1989) Effect of genotype on somatic embryogenesis from immature cotyledons of soybean. *Plant Cell Tiss. Org. Cult.*, **16**: 15–21.

Parrott W A, Durham R E and Bailey M A (1995) Somatic embryogenesis in legumes. In: *Biotechnology in Agriculture and Forestry, Vol. 31, Somatic embryogenesis and synthetic seed II* (Ed Bajaj Y P S), Springer-Verlag, Berlin, pp. 199–227.

Patel D B, Barve D M, Nagar N and Mehta A R (1994) Regeneration of pigeonpea (*Cajanus cajan* L.) through somatic embryogenesis. *Indian J. Exp. Biol.*, **32**: 740–744.

Phillips G C and Collins G B (1981) Induction and development of somatic embryos from cell suspension cultures of soybean. *Plant Cell Tiss. Org. Cult.*, **1**: 123–129.

Poulsen G B, Frugis G, Albrechtsen M and Mariotti D (1996) Synthesis of extracellular proteins in embryogenic and non-embryogenic cell cultures of alfalfa. *Plant Cell Tiss. Org. Cult.*, **44**: 257–260.

Quesenberry K H and Smith R R (1993) Recurrent selection for plant regeneration from red clover tissue culture. *Crop Sci.*, **33**: 585–589.

Radionenko M A, Kuchuk A N, Khvedynich O A and Gleba Y Y (1994) Direct somatic embryogenesis and plant regeneration from protoplast of red clover (*Trifolium pratense* L.). *Plant Sci.*, **97**: 75–81.

Rajasekaran K and Pellow J W (1997) Somatic embryogenesis from cultured epicotyls and primary leaves of soybean (*Glycine max* L.). In Vitro *Cell. Dev. Biol. Plant*, **33**: 88–91.

Ramana R V, Venu C H, Jayasree T and Sadanandam A (1996) Direct somatic embryogenesis and transformation in *Cicer arietinum* L. *Indian J. Exp. Biol.*, **34**: 716–718.

Ramdev Reddy L and Reddy G M (1993) Factors affecting somatic embryogenesis and plant regeneration in groundnut, *Arachis hypogaea* L. *Indian J. Exp. Biol.*, **31**: 57–60.

Ranch J P, Oglesby L and Zielinski A C (1985) Plant regeneration from embryo-derived tissue cultures of soybean by somatic embryogenesis. In Vitro *Cell. Dev. Biol. Plant*, **21**: 653–657.

Rani A S and Reddy G M (1996) Induction of somatic embryos from young leaflets of cultivated and wild species of groundnut. *Indian J. Exp. Biol.*, **34**: 569–571.

Rao B G and Chopra V L (1989) Regeneration in chickpea (*Cicer arietinum* L.) through somatic embryogenesis. *J. Plant Physiol.*, **134**: 647–638.

Rao M M and Lakshmi Sita G (1996) Direct somatic embryogenesis from immature embryos of rosewood (*Dalbergia latifolia* Roxb.). *Plant Cell Rep.*, **15**: 355–359.

Reinert J (1958) Untersuchungen uber die morphogenese an gewbakulturen. *Ber. Dtsch. Bot. Ges.*, **71**: 15.

Reisch B and Bingham E T (1980) The genetic control of bud formation from callus culture of diploid alfalfa. *Plant Sci. Lett.*, **20**: 71–77.

Rey H Y, Scocchi A M, Gonzalez A M and Mroginski L A (2000) Plant regeneration in *Arachis pintoi* (Leguminosae) through leaf culture. *Plant Cell Rep.*, **19**: 856–862.

Rout G R, Samantaray S and Das P (1995a) Effect of growth regulators and culture environment on somatic embryogenesis of *Acacia catechu* Willd. *Israel J. Plant Sci.*, **43**: 263–269.

Rout G R, Samantaray S and Das P (1995b) Somatic embryogenesis and plant regeneration from callus cultures of *Acacia catechu* – a multipurpose leguminous tree. *Plant Cell Tiss. Org. Cult.*, **42**: 283–285.

Rybczynski J J (1997) Plant regeneration from highly embryogenic callus, cell suspension and protoplast cultures of *Trifolium fragiferum. Plant Cell Tiss. Org. Cult.*, **51**: 159–170.

Sagare A P, Suhasini K and Krishnamurthy K V (1993) Plant regeneration via somatic embryogenesis in chickpea (*Cicer arietinum* L.). *Plant Cell Rep.*, **12**: 521–524.

Sagare A P, Suhasini K and Krishnamurthy K V (1995) Histology of somatic embryo initiation and development in chickpea (*Cicer arietinum* L.). *Plant Sci.*, **109**: 87–93.

Samoylov V M, Tucker D M and Parrott W A (1998a) Soybean [*Glycine max* (L.) Merrill] embryogenic cultures: The role of sucrose and total nitrogen content on proliferation. In Vitro *Cell Dev. Biol. Plant*, **34**: 8–13.

Samoylov V M, Tucker D M, Thibaud-Nissen F and Parrott W A (1998b) A liquid medium based protocol for rapid regeneration from embryogenic soybean cultivars. *Plant Cell Rep.*, **18**: 49–54.

Santos K G B, Mundstock E and Bodansese-Zanettini M H (1997) Genotype-specific normalization of soybean somatic embryogenesis through the use of an ethylene inhibitor. *Plant Cell Rep.*, **16**: 859–864.

Sato S, Newell C, Kolacz K, Tredo L, Finer J J and Hinchee M (1993) Stable transformation via particle bombardment in two different soybean regeneration systems. *Plant Cell Rep.*, **12**: 408–413.

Saxena P K, Malik K A and Gill R (1992) Induction by thidiazuron of somatic embryogenesis in intact seedlings of peanut. *Planta*, **187**: 421–424.

Scarpa G M, Pupilli F, Damiani F and Arcioni S (1993) Plant regeneration from callus and protoplasts in *Medicago polymorpha. Plant Cell Tiss. Org. Cult.*, **35**: 49–57.

Schmidt E D L, De Jong A J and DeVries S C (1994) Signal molecules involved in plant embryogenesis. *Plant Mol. Biol.*, **26**: 1305–1313.

Sellars R M, Southward G M and Phillips G C (1990) Adventitious somatic embryogenesis from cultured immature zygotic embryos of peanut using soybean as a model system. *Crop Sci.*, **30**: 408–414.

Sen J and Mukherjee S G (1998) *In vitro* induction of multiple shoots and plant regeneration in *Vigna*. In Vitro *Cell. Dev. Biol. Plant*, **34**: 276–280.

Senaratna T, Mckersie B D and Bowley S R (1989) Desiccation tolerance of alfalfa (*Medicago sativa* L.) somatic embryos: Influence of abscisic acid, stress pretreatments and drying rates. *Plant Sci.*, **65**: 253–259.

Senaratna T, Mckersie B D and Bowley S R (1990) Artificial seeds of alfalfa (*Medicago sativa* L.): Induction of desiccation tolerance in somatic embryos. In Vitro *Cell Dev. Biol. Plant*, **26**: 85–90.

Senaratna T, Saxena P K, Rao M V and Afela J (1995) Significance of the zygotic seed coat on quiescence and desiccation tolerance of *Medicago sativa* L. somatic embryos. *Plant Cell Rep.*, **14**: 375–379.

Sharp W R, Evans D A and Soundahl M R (1982) Application of somatic embryogenesis to crop improvement. In: *Plant Tissue Cult. Proc. 5th Inter. Cong. Plant Tiss. & Cell Cult.* (Ed Fujiwara A), *Jap. Assoc. Plant Tiss. Cult.*, Japan, pp. 759–762.

Sharp W R, Sondahl M R, Caldas I S and Maraffa S B (1980) The physiology of *in vitro* asexual embryogenesis. *Hort. Rev.*, **2**: 268–310.

Shekhawat N S and Galston A W (1983) Isolation, culture and regeneration of moth bean (*Vigna aconitifolia*) leaf protoplasts. *Plant Sci. Lett.*, **32**: 43–51.

Shoemaker R C, Amberger L A, Palmer R G, Oglesby L and Ranch J P (1991) Effect of 2,4-dichlorophenoxyacetic acid concentration on somatic embryogenesis and heritable variation in soybean [*Glycine max* (L.) Merr]. In Vitro *Cell. Dev. Biol. Plant*, **27**: 84–88.

Shoemaker R C and Hammond E G (1988) Fatty acid composition of soybean (*Glycine max* (L.) Merr) somatic embryos. In Vitro *Cell Dev. Biol. Plant*, **24**: 829–832.

Simmonds D H and Donaldson P A (2000) Genotype screening for proliferative embryogenesis and biolistic transformation of short-season soybean genotypes. *Plant Cell Rep.*, **19**: 485–490.

Singsit C, Adang M J, Lynch R E, Anderson W F, Wang A, Cardineau G and Ozias-Akins P (1997) Expression of a *Bacillus thuringiensis* cry IA(C) gene in transgenic peanut plants and its efficacy against lesser cornstalk borer. *Transgenic Res.*, **6**: 169–176.

Sinha R R, Das K and Sen S K (1983b) Embryoids from mesophyll protoplasts of *Vigna mungo* (L.) Hepper: A seed legume crop plant. *Basic Life Sci.*, **22**: 209–214.

Skoleman R G and Mapes M O (1976) Acacia koa gray plantlets from somatic callus tissue. *J. Heredity*, **67**: 114–115.

Slawinska J and Obendorf R L (1991) Soybean somatic embryo maturation: composition, respiration, and water relations. *Seed. Sci. Res.*, **1**: 251–262.

Sreenivasu K, Malik S K, Kumar P A and Sharma R P (1998) Plant regeneration via somatic embryogenesis in pigeonpea [*Cajanus cajan* (L.) Millsp.]. *Plant Cell Rep.*, **17**: 294–297.

Stejskal J and Griga M (1992) Somatic embryogenesis and plant regeneration in *Pisum sativum* L. *Biol. Plant.*, **34**: 15–22.

Stejskal J and Griga M (1995) Comparative analysis of some isozymes and proteins in somatic and zygotic embryos of soybean (*Glycine max* (L.) Merr). *J. Plant Physiol.*, **146**: 497–502.

Stewart C N Jr, Adang M J, All J N, Boerma H R, Cardineau G, Tucker D and Parrott W A (1996) Genetic transformation, recovery and characterization of soybean [*Glycine max* (L.) Merrill] transgenic for a synthetic *Bacillus thuringiensis* CRY IA (C) gene. *Plant Physiol.*, **112**: 121–129.

Stewart F C, Mapes M O and Mears K (1958) Growth and organized development of cultured cells. II. Organization in cultures grown from freshly suspended cells. *Am. J. Bot.*, **45**: 705–708.

Stickland S G, Nichol J W, McCall C M and Stuert D A (1987) Effects of carbohydrate source on alfalfa somatic embryogenesis. *Plant Sci.*, **48**: 113–121.

Stirn S and Jacobsen H J (1987) Marker proteins for embryogenesis differentiation patterns in pea callus. *Plant Cell Rep.*, **6**: 50–54.

Suhasini K, Sagare A P and Krishnamurthy K V (1994) Direct somatic embryogenesis from mature embryo axes in chickpea (*Cicer arietinum* L.). *Plant Sci.*, **102**: 189–194.

Suhasini K, Sagare A P and Krishnamurthy K V (1996) Study of aberrant morphologies and lack of conversion of somatic embryos of chickpea (*Cicer arietinum* L.). In Vitro *Cell. Dev Biol. Plant*, **32**: 6–10.

Suhasini K, Sagare A P, Sainkar S R and Krishnamurthy K V (1997) Comparative study of the development of zygotic and somatic embryos of chickpea (*Cicer arietinum* L.). *Plant Sci.*, **128**: 207–216.

Tegeder M, Kohn H, Nibbe M, Schieder O and Pickardt T (1996) Plant regeneration from protoplasts of *Vicia narbornensis* via somatic embryogenesis and shoot organogenesis. *Plant Cell Rep.*, **16**: 22–25.

Tetu T, Sangwan R S and Sangwan-Norreel B S (1990) Direct somatic embryogenesis and organogenesis in cultured immature zygotic embryos of *Pisum sativum* L. *J. Plant Physiol.*, **137**: 102–109.

Tomar U K and Gupta S C (1988) Somatic embryogenesis and organogenesis in callus cultures of a tree legume – *Albizzia richardiana* King. *Plant Cell Rep.*, **7**: 70–73.

Tomar U K and Gupta S C (1992) Factors affecting somatic embryogenesis in four-year old callus of a Fabaceous tree – *Albizzia richardiana* Kin. In: *Tissue Culture of Forest Tree Species: Recent researches in India* (Eds Dhawan V, Ganapathy P M and Khurana D K), IDRC-TIFNET, pp. 38–50.

Trick H N, Dinkins R B, Santarem E R, Di R, Samoylov V, Meurer C A, Walker D R, Parrott W A, Finer J J and Collins G B (1997) Recent advances in soybean transformation. *Plant Tiss. Cult. Biotech.*, **3**: 9–26.

Trigiano R N, Beaty R M and Graham E T (1988) Somatic embryogenesis from immature embryos of redbud (*Cercis canadensis*). *Plant Cell Rep.*, **7**: 148–150.

Trigiano R N, Buckley L G and Merkle S A (1999) Somatic embryogenesis in woody legumes. In: *Somatic Embryogenesis in Woody Plants, Vol. 4* (Eds Jain S M, Gupta P K and Newton R J), Kluwer Academic Publishers, Dordrecht, The Netherlands, pp. 189–208.

Trigiano R N, Geneve R L, Merkle S A and Preece J E (1992) Tissue and cell cultures of woody legumes. *Hort. Rev.*, **14**: 265–332.

Trigiano R N, Geneve R L and Buckley L G (1995) Somatic embryogenesis in eastern redbud *Cercis canadensis*. In: *Somatic Embryogenesis in Woody Plants, Vol. 2* (Eds Jain S M, Gupta P K and Newton R J), Kluwer Academic Publishers, Dordrecht, The Netherlands, pp. 471–482.

Trinh T H, Patel P, Kondorosi K, Durand P, Kamate K, Baner P and Kondorosi A (1998) Rapid and efficient transformation of diploid *Medicago truncatula* and *Medicago sativa* ssp. Falcata lines improved in somatic embryogenesis. *Plant Cell Rep.*, **17**: 345–355.

VanDoorne L E, Marshall G and Kirkwood R C (1995) Somatic embryogenesis in pea (*Pisum sativum* L.): Effect of explant, genotype and culture conditions. *Ann. Appl. Biol.*, **126**: 169–179.

Vani A K S and Reddy V D (1996) Morphogenesis from callus cultures of chickpea (*Cicer arietinum* L.). *Indian J. Exp. Biol.*, **34**: 285–287.

Venkatachalam P (1996) Tissue culture, mutational and genetic studies in groundnut (*Arachis hypogaea* L.) to develop cultivars for high yield, oil content and disease resistance. Ph.D. thesis, Bharathidasan University, Tiruchirappalli, India.

Venkatachalam P, Geetha N and Jayabalan N (1998) Induction of somatic embryos and plantlet development in cell suspension cultures of *Arachis hypogaea* L. *Breed. Sci.*, **48**: 231–236.

Venkatachalam P, Geetha N, Khandelwal A, Shaila M S and Lakshmi Sita G (1999a) Induction of direct somatic embryogenesis and plant regeneration from mature cotyledon explants of *Arachis hypogaea* L. *Curr. Sci.*, **77**: 269–273.

Venkatachalam P and Jayabalan N (1996) Efficient callus induction and plant regeneration via somatic embryogenesis from immature leaf-derived protoplasts of groundnut (*Arachis hypogaea* L.). *Israel J. Plant Sci.*, **44**: 387–396.

Venkatachalam P, Kavi Kishor P B, Geetha N, Thangavelu M and Jayabalan N (1999b) A rapid protocol for somatic embryogenesis from immature leaflets of groundnut (*Arachis hypogaea* L.). In Vitro *Cell. Dev. Biol. Plant*, **35**: 409–412.

Venkatachalam P, Kavi Kishor P B and Jayabalan N (1997) High frequency somatic embryogenesis and efficient plant regeneration from hypocotyl explants of groundnut (*Arachis hypogaea* L.). *Curr. Sci.*, **72**: 271–275.

Venkatachalam P, Geetha N, Khandelwal A, Shaila M S and Lakshmi Sita G (2000) *Agrobacterium*-mediated genetic transformation and regeneration of transgenic plant cotyledon explants of groundnut (*Arachis hypogaea* L.) via somatic embryogenesis. *Curr. Sci.*, **78**: 1130–1136.

Venu C H, Pavan U, Jayasree T, Ramana R V, Cheralu C and Sadanandam A (1999) Genotype-dependent embryogenesis, organogenesis and *Agrobacterium*-mediated transformation in pigeonpea (*Cajanus cajan* L.). *Plant Tiss. Cult.*, **9**: 89–95.

Wan Y, Sorensen E L and Liang G H (1988) Genetic control of *in vitro* regeneration in alfalfa (*Medicago sativa* L.). *Euphytica*, **39**: 3–10.

Wang H and Holl F B (1988) *In vitro* culture and incidence of somaclonal variation in regenerated plants of *Trifolium pratense* L. *Plant Sci.*, **55**: 159–167.

Weaver L A and Trigiano R N (1991) Regeneration of *Cladastris lutea* (Fabaceae) via somatic embryogenesis. *Plant Cell Rep.*, **10**: 183–186.

Weissinger A K H and Parrott W A (1993) Repetitive somatic embryogenesis and conversion to plants in white clover. *Plant Cell Rep.*, **12**: 125–128.

Wetzstein H Y and Baker C M (1993) The relationship between somatic embryo morphology and conversion in peanut (*Arachis hypogaea* L.). *Plant Sci.*, **92**: 81–89.

Williams E G, Collins G B and Myers J R (1990) Clovers (*Trifolium* spp.). In: *Biotechnology in Agriculture and Forestry, Vol. 10, Legumes and Oilseed Crops I* (Ed Bajaj Y P S), Springer; Berlin, Heidelberg, New York, pp. 242–287.

Williams E G and Maheswaran G (1986) Somatic embryogenesis: Factors influencing coordinated behaviour of cells as an embryogenic group. *Ann. Bot.*, **57**: 443–462.

Wright M S, Launis K L and Novitzky R (1991) A simple method for the recovery of multiple fertile plants from individual somatic embryos of soybean [*Glycine max* (L.) Merrill]. In Vitro *Cell. Dev. Biol. Plant*, **27**: 153–157.

Yang H, Singsit C, Wang A, Gonsalves D and Ozias-Akins P (1998) Transgenic peanut plants containing a nucleocapsid protein gene of tomato spotted wilt virus show divergent levels of gene expression. *Plant Cell Rep.*, **17**: 693–699.

PROTOPLASTS OF GRAIN AND FORAGE LEGUMES: THEIR EXPLOITATION IN GENETIC MANIPULATION, PHYSIOLOGICAL INVESTIGATIONS AND PLANT-PATHOGEN INTERACTIONS

M. R. DAVEY, R. MARCHANT AND J.B. POWER
Plant Science Division, School of Biosciences,
University of Nottingham, University Park,
Nottingham NG7 2RD, UK
e-mail: mike.davey@nottingham.ac.uk

Abstract

During the last decade, the isolation and culture of protoplasts has been described for grain legumes of the genera *Cajanus*, *Pisum* and *Vigna*, while in forage legumes, protoplasts of *Astragalus*, *Medicago*, *Onobrychis* and *Trifolium* have received most attention. Reports on the transformation of legume protoplasts are limited. Somatic hybridisation has been directed towards the grain legumes *Lathryus*, *Pisum* and *Vigna*, with *Lotus*, *Medicago*, *Onobrychis* and *Trifolium* being targeted in forage crops. Several reports have described the use of legume protoplasts as experimental material in physiological investigations and in studies of plant-pathogen interactions.

1. Introduction

Since grain and forage legumes contribute significantly to the nutrition of humankind and livestock, it is not unexpected that much effort has been directed towards the progressive genetic improvement of these crops through conventional breeding approaches. In recent years, an important adjunct to established breeding practices has been provided by somatic cell techniques, themselves underpinned by molecular procedures. Earlier reviews have summarised the literature relating to such tissue culture-based approaches for the genetic improvement of legumes (Kumar and Davey, 1991). Broader perspectives have also focused on the exploitation of *in vitro* techniques coupled with recombinant DNA technology to engineer crops, including legumes, for specific traits such as salt tolerance (Chowdhury *et al.*, 1993). However, during the last decade there have been relatively few innovative reports describing the development *per se* of legume protoplast-to-plant systems. Similarly, there has been limited use of isolated legume protoplasts to generate transgenic plants through direct uptake of DNA, especially when compared

P.K. Jaiwal and R.P. Singh (eds.), Improvement Strategies for Leguminosae Biotechnology, 133–153.

to other mainstream crops. Some effort, though, has been invested in exploiting isolated protoplasts to overcome sexual incompatibility through the generation of somatic hybrid plants (Arcioni *et al.*, 1997). Interestingly, the main focus, as evidenced by the recent literature, has been in the use of legume protoplasts for cell physiological studies, ranging from the effects of salinity to aspects of photosynthesis and respiration, and the interaction of plant cells with viral and bacterial pathogens.

2. Culture of protoplasts of grain and forage legumes

2.1. GRAIN LEGUMES

Species of the genera *Cajanus*, *Vigna* and *Pisum* have featured predominantly in investigations involving grain legumes. In several cultivars of *Cajanus cajan* (pigeon pea), high yields of mesophyll protoplasts were obtained from leaves of axenically-grown shoots, with protoplast-derived cells producing colonies, usually at low efficiency, followed by limited differentiation of chlorophyll-containing calli after culture in nutrient-rich media, such as KM8P medium, the latter based on the formulation of Kao and Michyaluk (1975) as modified by Gilmour *et al.* (1989). However, protoplast-derived tissues failed to complete the differentiation process leading to regenerated shoots (Sarangi *et al.*, 1992).

Lehminger-Mertens and Jacoben (1993) and Böhmer *et al.* (1995) described the successful regeneration of plants from leaf and lateral shoot bud protoplasts of pea (*Pisum sativum*). In the later study, protoplasts were isolated from the lateral shoot buds of cotyledon-free embryo axes, with protoplast-derived cells ultimately giving totipotent calli. Twelve percent of such calli regenerated shoots after 16 weeks on medium suplemented with thidiazuron (TDZ). Although N-isopentenyl adenine and indole-3-butyric acid (IBA) as supplements in the culture medium stimulated shoot elongation, such shoots were almost impossible to root and could only be induced to develop into fertile plants following grafting onto recipient pea seedlings as rootstocks. In attempts to predict expressable totipotency of cultured pea protoplasts, Ochatt *et al.* (2000) isolated viable protoplasts from five genotypes. Protoplast-derived tissues exhibited considerable differences in proliferation rates and in their competence to regenerate shoots, both within and between genotypes. Flow cytometric analyses of genomic DNA demonstrated a correlation between endoreduplication and the inability to regenerate shoots, since fertile plants were regenerated only from tissues with a normal DNA content. This technique may see more general application in predicting the totipotency of tissues derived from protoplasts and other sources, not only in legumes but also in other major crops.

The nutrient-rich medium of Kao and Michayluk (1975) supplemented with 0.5 mg/L 2,4-dichlorophenoxyacetic acid (2,4-D), naphthaleneacetic acid (NAA) and 6-benzylaminopurine (BAP), was also used to promote division of cells derived from protoplasts isolated from etiolated shoot tips of 10 cultivars of *Vicia faba* (Tegeder *et al.*, 1995). Protoplast division frequencies of up to 40% were recorded, although these were again cultivar dependent. Only protoplast-derived tissues of the cv. Mythos regenerated shoots, following transfer of tissues to medium semi-solidified with Gelrite. Interestingly, as in the case of pea, thidiazuron was employed in these experiments, in combination with NAA, to stimulate shoot formation from protoplast-derived tissues. As in the studies reported for pea protoplasts by Böhmer *et al.* (1995), there was a necessity to graft regenerated shoots of *Vicia* onto young seedlings in order to convert shoots to fertile plants.

Most attention has been directed towards protoplasts of species in the genus *Vigna*. For example, Sato *et al.* (1993) released protoplasts from cell suspensions of five cultivars of adzuki bean (*V. angularis*), and then cultured the protoplasts on Murashige and Skoog (MS; 1962)-based medium supplemented with 2,4-D and BAP, both at 1.0 mg/L. The first cell divisions were observed within 3 days of culture, with about 30% of the protoplasts undergoing mitotic division by day 10 of culture. The protoplast plating efficiency was higher on agarose-solidified medium than in liquid culture. Again, differences in the frequency of protoplast division were observed. Reduction of the 2,4-D concentration in the culture medium to 0.5 mg/L promoted growth of protoplast-derived tissues; the latter regenerated shoots following transfer to medium with BAP, kinetin, trans-zeatin and 3-indoleacetic acid (IAA). Shoots developed roots when transferred to MS-based medium (Murashige and Skoog, 1962) lacking growth regulators and, following *ex vitro* acclimation and potting, produced flowers leading to seed set.

In their studies with leaf protoplasts of mung bean (*V. radiata*), Iriawati *et al.* (1994) obtained microcolonies within 15 days of culture; protoplast-derived tissues were large enough to transfer to a callus induction medium 1.5 months post-isolation. Globular calli were produced on MS-based medium with NAA and benzyladenine (BA) or on MS-based medium supplemented with B_5 vitamins and 2,4-D. Unfortunately, shoot regeneration was not reported. Similarly, protoplast-derived tissues were obtained for rice bean (*V. umbellata*) by Bhadra *et al.* (1991), but these also failed to regenerate shoots. In the latter experiments, viable protoplasts were released from epicotyls of axenic seedlings using an enzyme mixture consisting of Rhozyme HP 150 (0.4%, w/v), 0.2% Cellulase R10 and Pectolyase Y23 (0.1%) dissolved in a solution containing CPW salts (Frearson *et al.*, 1973) and 9.0% w/v mannitol. Protoplast yields were dependent upon the age of the seedlings, with maximum yields being obtained from seedlings harvested after 5 days of growth. As in studies with pea and pigeon pea, the use of KMP8 medium semi-solidified with agarose was important in stimulating sustained division of isolated protoplasts, with division and plating efficiencies being optimal at a plating density of 2.0×10^4 protoplasts/mL. Although Bhadra *et al.* (1991) failed to develop a protoplast-to-plant system for rice bean, they discussed the relevance of their results in relation to the genetic improvement of this important pulse crop. Subsequently, the same group achieved success in regenerating shoots from protoplasts of *V. sublobata* (Bhadra *et al.*, 1994). In this case, protoplasts were isolated enzymatically from hypocotyls of axenic seedlings. Again, yields of viable protoplasts were dependent on seedling age with maximum yields, in this case, from 6-day-old seedlings; the density at which the protoplasts were plated in K8P medium influenced the division and plating efficiencies. Dividing protoplasts produced glossy green compact nodular calli on transfer to agar-solidified medium containing MS salts, B_5 organic components, NAA (0.2–0.5 mg/L), zeatin riboside (0.5–2.0 mg/L) and gibberellic acid (GA_3; 0.5–1.0 mg/L). Shoot buds elongated after the tissues were transferred to B_5-based medium supplemented with indole-butyric acid (IBA) at 0.01 mg/L and BAP (1.0 mg/L). Elongated shoots developed roots on hormone-free MS-based medium.

Other workers have regenerated plants, through somatic embryogenesis rather than organogenesis, from protoplasts isolated from immature cotyledons of seedlings of *V. sinensis* (Li *et al.*, 1995). Using an MS-based liquid medium containing 0.2 mg/L 2,4-D, 1.0 mg/L NAA and 0.5 mg/L BAP, protoplasts underwent division within 3–5 days when cultured in the dark at a density of 1.0×10^5 protoplasts/mL. Initial cell division and plating efficiences of 27.7 and 1.7% respectively were recorded. About 5% of protoplast-derived tissues became embryogenic following transfer to MS-based medium with 2 mg/L 2,4-D and 0.5 mg/L BAP. Tissues remained embryogenic when maintained on this medium, but could be induced to form somatic embryos following transfer to medium with 0.1 mg/L IAA and 0.5 mg/L kinetin. Some cotyledonary-stage somatic embryos germinated and developed into plants which were rooted on half-strength MS-based medium with 0.1–0.3 mg/L IBA. Plants transferred to soil flowered and were fertile.

2.2. FORAGE LEGUMES

Studies in forage legumes have focused mainly upon four genera, namely *Astragalus, Onobrychis, Trifolium* and, more specifically, *Medicago*. Friable calli were used as source materials in the case of *A. adsurgens* (Luo and Jia, 1998), with isolated protoplasts undergoing sustained division to form cell colonies when cultured in either liquid or agarose-solidified KM8P medium supplemented with 1.5 mg/L 2,4-D and 0.5 mg/L BA. Protoplast plating efficiencies were higher in agarose-bead culture, with a value of 6.5 ± 0.7% compared to 0.8 ± 0.5% for liquid medium. More than 50% of protoplast-derived tissues produced plants via somatic embryogenesis on MS-based medium with 0.1 mg/L NAA and 2.0 mg/L BA.

Suspension cultures, which had been both newly initiated or those recovered from cryopreservation, readily released viable protoplasts of *O. viciaefolia* following digestion with 0.5% (w/v) Cellulase Onozuka RS and 0.1% Pectolyase Y-23. A high plating efficiency (60–70%) was recorded after 10 days of culture in medium with 0.2 mg/L 2,4-D, 0.4 mg/L NAA and 1.0 mg/L BAP or zeatin with shoot regeneration from protoplast-derived callus on agar-solidified medium supplemented with 0.5 mg/L BAP and 0.5 mg/L zeatin (Lu *et al.*, 1992). In detailed studies with the wild *Trifolium* species, *T. fragiferum*, Rybczynski (1997) compared the responses of protoplasts isolated from different source tissues to a range of media. Somatic embryogenesis was observed in tissues derived from protoplasts isolated from leaf mesophyll, cell suspensions and, interestingly, somatic embryos. An important observation was that tissues retained their embryogenic potential for more than two years on MS-based medium supplemented with 0.5 mg/L BA and 0.05 mg/L NAA, these long-term cultures being capable of regenerating fertile plants. In his report, Rybczynski (1997) compared the response, in culture, of protoplasts of *T. fragiferum* with that of protoplasts of other *Trifolium* species decribed previously in the literature. He concluded that protoplasts of *T. fragiferum* were significantly more amenable to culture and plant regeneration than those of other *Trifolium* species.

Three reports have summarised studies with species of *Medicago*. For example, Zafar *et al.* (1995) described plant regeneration from callus derived from protoplasts isolated from cotyledons and expanded leaflets of axenic shoots and also from callus of *M. littoralis* cv. Harbinger 1886, an annual legume resistant to the fungal pathogen *Pseudopeziza medicaginis*. Protoplasts from cotyledons and leaves exhibited similar plating efficiencies (5%) when embedded in agarose-solidified medium. Shoot regeneration was stimulated by 2-isopentenyl adenine with either IAA or 1,2-benzisoxazole-3-acetic acid, or by BA in combination with NAA. Interestingly, the pathway of plant regeneration was influenced by the cytokinin:auxin ratio in the culture medium. A ratio greater than 1 induced somatic embryogenesis, while a ratio of 1 stimulated organogenesis. Transfer of tissues to medium lacking growth regulators resulted in the development of somatic embryos and root formation by the regenerated shoots. In the case of *M. polymorpha*, the behaviour was evaluated during the culture of protoplasts from 17 ecotypes adapted to Sardinian (Italian) field conditions. Although leaf mesophyll cells gave the highest yields of protoplasts, only protoplasts from roots underwent sustained division to regenerant plants (Scarpa *et al.*, 1993). Such experiments provided a foundation for subsequent somatic hybridisation investigations to combine agronomically important genetic traits, such as adaptation to specific field conditions and climate, of *M. polymorpha* with those of *M. sativa*.

The annual wild legume *M. truncatula* (barrel medic) has also received attention, with Rose and Nolan (1995) focusing their studies on the nature of the source material, the use of media semi-solidified with agarose and protoplast plating density. Plants used as source of protoplasts were grown in a controlled environment. Enzymatically isolated protoplasts were purified by a Percoll-based flotation procedure, followed by embedding in 100 μl agarose droplets containing 25 μM NAA and 4 μM BAP. Maximum cell colony formation occurred at a plating density of $6–8 \times 10^5$ protoplasts/mL. In the same publication, the authors also isolated protoplasts from plants of *M. truncatula* transformed previously for resistance to kanamycin sulphate. The protoplast-to-plant regeneration protocol developed for *M. truncatula* was applied successfully to protoplasts of other *Medicago* species, including *M. intertexta* and *M. scutella*.

3. Isolated protoplasts for the genetic manipulation of legumes

Isolated protoplasts provide source material for the generation of transformed tissues and transgenic plants by DNA uptake, and also for the production of somatic hybrid and cybrid plants by protoplast fusion.

3.1. TRANSFORMATION

The reports are limited of the exploitation of isolated protoplasts in transformation studies during the last decade. Puonti-Kaerlas *et al.* (1992) demonstrated that protoplasts of the pea cultivars Belman and Filby could be stably transformed by direct gene transfer using electroporation, these workers studying the effects of field strength during electroporation on survival and division rates of the isolated protoplasts. Transgenic calli were recovered when hygromycin was utilised as the selective agent; transgenic tissues were not recovered when kanamycin was employed. The efficiency of transformation was assessed using a histochemical assay to monitor expression of the ß-glucuronidase (*gus*) gene, with DNA analysis confirming the transgenic nature of hygromycin-resistant tissues. The transformed calli could not be converted to plants. Electroporation was also used by Hashimoto *et al.* (1992) to introduce the chloramphenicol acetyltransferase (*cat*) gene, fused to a CaMV35S promoter, into protoplasts of pea. In the same study, the conditions for elicitor induction of a member of the phenylalanine ammonia-lyase (PAL) gene family was investigated using a chimeric gene carrying a 480 bp region of the putative promoter region of the PAL1 gene driving the *cat* gene. CAT activity could be induced by treatment with a fungal elicitor isolated from the pea pathogen *Mycosphaerella pinodes*.

The *gus* gene has also been used as a readily assayable reporter to study promoter function using pea protoplasts as the experimental system. For example, Kalbin *et al.* (1999) introduced, by electroporation into pea protoplasts, the *gus* gene driven by different parsley (*Petroselinum crispum*) chalcone synthase promoter constructs, with high constitutive *gus* gene expression and loss of the light-inducibile gene expression observed in an homologous parsley protoplast system. The results of these experiments indicated that unit 1 of the parsley chalcone synthase promoter is only partly responsible for the GUS activity detected, with additional *cis*-elements located downstrean within 100 bp of the transcription start site mediating de-repression in pea protoplasts. Kalbin *et al.* (1999) concluded that their results with pea protoplasts implied that the isolation of protoplasts from pea leaf cells itself induces de-repression as a result of stress, this proposal being strengthened by the finding that mRNA levels of endogenous chalcone synthase were increased as a result of protoplast isolation. The limited number of investigations which have exploited DNA uptake to transform isolated protoplasts probably reflect the fact that other procedures, such as *Agrobacterium*- and Biolistic-mediated transfer of genes into totipotent cells, provide more reliable and less labour-intensive procedures for generating transgenic plants.

3.2. SOMATIC HYBRIDISATION

3.2.1. Grain legumes

The study by Dimitrova and Christov (1992), involving the self-fusion of mesophyll protoplasts of *Pisum sativum* cv. Ran 1 (Table 1), was designed to assess the effects of basic parameters of the fusion process, in this case, the effects of sine waves (2V peak to peak; 500–0.1 kHz) and an alternating pulse (1V amplitude; 0.1–0.4 kHz) field on protoplasts during electrofusion. An increase in the number of dielectrophoretically paired protoplasts was found in pulse fields. By observing protoplast fusion in the presence of different concentrations of calcium chloride, the authors concluded that it was possible to fuse protoplasts electrically with alternating pulse fields at cation concentrations that normally prevent electrofusion with sine wave fields.

Since adzuki bean (*V. angularis*) is a major crop in Japan, with the main area of production in Hokkaido, it is not surprising that there has been discussion of the potential of somatic cell approaches for the genetic manipulation of this legume (Sato, 1995). A subsequent paper by Iriawati *et al.* (1996a) reported the fusion of cell suspension-derived protoplasts of adzuki bean with mesophyll protoplasts of mung bean (*V. radiata*), the aim of such experiments being to characterise the fate of organelles, particularly plastids,

Table 1. Examples of protoplast fusion/somatic hybridisation studies in legumes

Source of protoplasts							
Parent one		Parent two		Protoplast fusion technique	Plant regeneration?	Notes	Reference
Species	Donor tissue	Species	Donor tissue				
Medicago arborea	Leaf mesophyll	*Medicago sativa*	Cell suspension/ callus	Electrical	No	Heterokaryons by their dual fluorescein isothiocyanate-chlorophyll fluorescence	Pupilli *et al.*, 1991
Pisum sativum	Leaf mesophyll	n.a.	n.a.	Electrical	No	Self-fusion; comparison of sine wave and alternating pulse fields on protoplast fusion	Dimitrova and Christov, 1992
Medicago sativa	Callus	*Oryza sativa*	Callus	Chemical – PEG	No	Wide hybridisation involving protoplasts of a dicotyledon with those of a monocotyledon	Niizeki *et al.*, 1992
Medicago sativa	Leaf mesophyll	*Medicago coerulea*	Callus	Electrical	Yes	Somatic hybrids confirmed by cytological and isozyme analyses	Pupilli *et al.*, 1992
Medicago sativa	Leaf mesophyll; -iodoacetamide treated	*Onobrychis viccifolia*	Cell suspensions; gamma irradiated	Electrical	Yes	Southern blotting confirmed asymmetric somatic hybrids	Li *et al.*, 1993
Medicago sativa	Leaf mesophyll	*Medicago coerulea*	Callus	Electrical	Yes	Somatic hybrids displayed formation of univalents and lagging chromosomes	Pupilli *et al.*, 1995
Medicago sativa	Leaf mesophyll	*Medicago arborea*	Callus	Electrical	Yes	Somatic hybrid calli selected manually by their dual fluorescence; somatic hybrid plants exhibited extensive genome rearrangements	Nenz *et al.*, 1996
Onobrychis viciifolia	Cultured cells	*Medicago sativa*	Cultured cells	Chemical – PEG	Buds but not intact plants	Somatic hybridity of protoplast-derived tissues confirmed by isozyme analysis	Xu and Jia, 1997

Table 1. Continued

Source of protoplasts							
Parent one		Parent two		Protoplast fusion technique	Plant regeneration?	Notes	Reference
Species	Donor tissue	Species	Donor tissue				
Vigna radiata	Leaf mesophyll	*Vigna angularis*	Cell supension	Electrical	No	Comparison of the ultra-structure of heterokaryons and parental protoplasts	Mikaye *et al.*, 1996
Medicago sativa	-iodoacetamide treated	*Medicago sativa* – CMS line L-2905		Electrical	Somatic hybrid tissues produced	Asymmetric somatic hybridisation assessed for transfer of organelle-specific genetic material	Ando *et al.*, 1997
Medicago sativa	Leaf mesophyll	*Medicago falcata*	Leaf-derived calli	Electrical	Yes	Somatic hybrids confirmed by RAPD analysis; somaclonal variation observed by RFLP of NOR	Crea *et al.*, 1997
Medicago sativa	-X-irradiated	*Lotus corniculatus*	-iodoacetamide treated	Chemical – PEG	Yes	Isozyme analysis confirmed tissues were somatic hybrid; isozymes were characteristic of *L. corniculatus* in regenerated plants	Kaimori *et al.*, 1998
Medicago trunculata	Leaf mesophyll; -iodoacetamide treated	*Medicago scutellatta*	Cotyledons; gamma irradiated	Chemical – PEG	Yes	Two asymmetric hybrids recovered; confirmed by AFLP	Tian and Rose, 1999
Pisum sativum	Young shoots from embryo axes	*Lathyrus sativus*	Young shoots from embryo axes	Electrical and Chemical – PEG	No	Assessment of fusion procedures; PEG most efficient and reproducible	Durieu and Ochatt, 2000

n.a., not applicable; PEG, polyethylene glycol; CMS, cytoplasmic male sterility; RAPD, random amplified polymorphic DNA; RFLP, restriction fragment length polymorphism; AFLP, amplified fragment length polymorphism; NOR, nucleolar organiser region.

from the parental protoplasts in heterokaryons and in protoplast-derived cells using light and electron microscopy. An interesting observation in this study was that, apparently, electrical treatment stimulated the fusion of organelles within the heterokaryons. Some of the fused protoplasts underwent division during the first 5 days of culture, but plant regeneration was not reported in this system. The same workers also investigated the effects of iodoacetamide on the fine structure of protoplasts of adzuki bean, both before and following fusion with protoplasts of mung bean (Iriawati *et al.*, 1996b). Excessive formation of rough endoplasmic reticulum and circular Golgi complexes were present in iodoacetamide-treated protoplasts, followed by lipid accumulation in the cytoplasm. These ultrastructural perturbations were not observed in heterokaryons generated by fusing iodoacetamide-treated protoplasts of adzuki bean with untreated protoplasts of mung bean.

Peas, which are rich in proteins, are being used increasingly as animal feed. However, there is a requirement to develop new varieties with increased tolerance to various biotic and abiotic stresses. Grass pea (*Lathyrus sativus*) is a wild relative of the cultivated pea that could contribute to the gene pool of the cultivated crop, especially in terms of disease resistance. In this context, Durieu and Ochatt (2000) reported a detailed study in which different conditions were evaluated for the fusion of protoplasts of the cv. Frisson with those of grass pea. The fusion procedures that were compared were based on electrofusion, the use of chemical fusion employing micro- and macro-methods involving polyethylene glycol (PEG) of different molecular weights (1,540 or 6,000), or treatment of protoplast mixtures with a glycine-high pH solution. Seedling epicotyls provided the source material for protoplast isolation, with isolated protoplasts being stained with fluorescein diacetate (FDA) (*Lathyrus*) or rhodamine B (*Pisum*) to give yellow-green and red fluorecences respectively to aid the identification of heterokaryons following fusion treatment. Treatment of parental protoplast mixtures with PEG was the most efficient and reproducible, with more than 20% of viable heterokaryons being generated using this procedure. The authors could not detect any statistical differences between macro- and micro-procedures, with PEG 6000 being as efficient as PEG of the lower molecular weight of 1,540. The heterokaryons underwent mitotic division with the formation of protoplast-derived tissues. In electrofusion, the production of heterokaryons was optimal at 1,500 V cm^{-1}, although large variation was reported between experiments. When a glycine high-pH solution was employed as the fusogen, crystals and agglomerations of debris were observed, which entrapped the heterokaryons and curtailed their subsequent development. Somatic hybrid plants were not regenerated in these experiments, since the authors predicted that shoot formation from heterokaryon-derived tissues would require at least 12–15 months of further culture which was outside the remit of the investigation. Nevertheless, such experiments provide a baseline for the creation of genetically novel plants in which it should be possible to combine the grain quality of *Pisum* with traits, such as stress tolerance, from *Lathyrus*.

3.2.2. *Forage legumes*

The most recent studies, in chronological order, of protoplast fusion to generate somatic hybrid plants in forage legumes are summarised in Table 1. *Medicago* species, particularly *M. sativa*, has featured as one of the parental species in most of the reports, probably because of the economic importance of this species (alfalfa, lucerne) as a cultivated crop, especially in North America. Likewise, in most of the experiments reported, electrical pulses have been used to fuse the parental protoplasts rather than chemical procedures involving the use of polyethylene glycol (PEG) as fusogen.

Arcioni *et al.* (1997) discussed the benefits of somatic hybridisation and embryo rescue for the introduction of genetic material from wild germplasm into cultivated species of *Lotus, Medicago* and *Trifolium*. The widest species combinations attempted have involved the fusion of protoplasts of *M. sativa* with those of the major cereal, rice (*Oryza sativa*). However, it is not surprising that somatic hybrid plants were not regenerated in this wide combination involving protoplasts of a dicotyledon with those of a monocotyledon. Nevertheless, asymmetric hybrid callus was produced, with the nuclear genome of alfalfa. Southern blot analysis of chloroplast DNA using three fragments of rice chloroplast DNA as probes, and of mitochondrial DNA using three mitochondrial genes, indicated intergenomic and/or intragenomic recombination of mitochondrial DNA in the hybrid cells (Niizeki *et al.*, 1992). The chloroplast DNA fingerprints of all hybrid cell lines were identical to those of alfalfa; reduction and/or rearrangement of chloroplast DNA did not occur in these hybrid cells, confirming a not uncommon unidirectional sorting out of plastids during culture of the heterokaryons with elimination of those of the rice parent.

Other wide fusion combinations have been attempted within the Leguminosae. For example, Li *et al.* (1993) generated intergeneric somatic hybrid plants between species in the sexually incompatible genera *Medicago* and *Onobrychis*. This study also involved a selection procedure based on biochemical complementation to recover hybrid tissues. Thus, leaf mesophyll protoplasts of *M. sativa* were inactivated by treatment with iodoacetamide prior to fusion with gamma-irradiated suspension cell protoplasts of *O. viciifolia* (sainfoin). Complementation permitted only heterokaryons to survive, their growth being stimulated by culture in the presence of alfalfa nurse cells. Although the aim of these experiments was to introduce condensed tannins from sainfoin, a bloat-safe legume, into alfalfa, tannin-positive plants were not regenerated from fusion-treated protoplasts. All regenerated plants resembled the alfalfa parent, although Southern analysis showed low levels of sainfoin-specific DNA in 43 of 158 regenerated plants analysed. Seventeen percent of the plants were euploid ($2n = 32$), while 83% were aneuploid ($2n = 30$, 33–78). An important observation was that the majority of the somatic hybrids were fertile.

A subsequent study involving the same parental protoplasts with selection based on complementation of hydroxyproline-resistance of sainfoin and growth regulator independence of a transformed line of alfalfa resulted in the generation of seven somatic hybrid plants with chromosome complements of 30–60 (Xu and Jia, 1997). Cytologically, the hybrid plants were also identified by the presence within their nuclear genomes of a characteristic marker chromosome from sainfoin and two chromosomes, with easily observed constrictions, from alfalfa.

Other somatic hybrid combinations in forage legumes have involved the fusion of protoplasts of *M. sativa* with those of other *Medicago* species. In the first of several studies, Pupilli *et al.* (1991) electrofused protoplasts from cell suspensions and/or callus of *M. sativa* with leaf mesophyll protoplasts of *M. arborea* ($2n = 2x = 32$). Heterokaryons, identified by their dual fluorescein isothiocyanate-chlorophyll fluorescence, were selected manually prior to subsequent culture. Hybrid tissues were identified by their isozyme patterns, with plants being regenerated from such tissues. In subsequent studies by the same research group, isolated leaf protoplasts of *M. sativa* were electrofused with protoplasts from callus of

M. coerulea (Pupilli *et al.*, 1992) or *M. arborea* (Nenz *et al.*, 1996). Again, heterokaryons and hybrid cells were selected manually based on their dual fluorescence. Cytological and isoenzyme analyses (Pupilli *et al.*, 1992) confirmed the hybridity of regenerated plants which, although morphologically different in stature to the parental plants, were fertile and set seed. In more detailed analyses by Pupilli *et al.* (1995) of the same parental combinations, but involving protoplasts of tetraploid *M. sativa* ($2n = 4x = 32$) with those of the diploid species *M. coerulea* ($2n = 2x = 16$), RFLP fingerprints were used to establish the nuclear composition of the somatic hybrid plants. Although the somatic hybrids retained all parental chromosomes, molecular analysis revealed incomplete incorporation of alleles of the *M. coerulea* parent into the fusion products. Cytologically, the somatic hybrids exhibited minor abnormalities, with the formation of univalents and lagging chromosomes. Somatic hybrid plants between *M. sativa* and *M. arborea* also showed extensive genome rearrangements as revealed by isozyme and RFLP analyses (Nenz *et al.*, 1996). Of significance was the fact that in field trials, the somatic hybrids between tetraploid *M. sativa* and diploid *M. coerulea* were vigorous in their growth, with forage yields equivalent to those of the most productive parental plants. Pupilli *et al.* (1995) emphasised the importance of these somatic hybrid plants as starting material for breeding alfalfa at the hexaploid level.

The fusion of protoplasts from tetraploid and diploid species has also been extended to *M. sativa* (tetraploid) in combination with *M. falcata* (diploid; $2n = 2x = 16$) (Crea *et al.*, 1997). In order to facilitate the recovery of hybrid plants, protoplasts were isolated from *M. sativa* cv. Rangelander and from *M. falcata* cv. Wisfal-1, both selected previously for their shoot regeneration ability. Detailed cytological analyses confirmed that the somatic hybrid plants were hyperaneuploid ($2n = 33$) and contained one extra long chromosome, indicating that translocation had occurred. Molecular analyses (RAPD) confirmed the hybridity of the plants, while rearrangements within the parental genomes and the presence of somaclonal variation in the hybrids were observed through an RFLP analysis of the nucleolar organizing regions (NORs). Crea *et al.* (1997) discussed the possible causes of the gross genomic alterations in the hybrid plants and of the relevance of somatic hybridisation to transfer agronomically useful traits from wild *Medicago* species into alfalfa.

More recently, protoplast fusion has been directed to the generation of asymmetric hybrids between the annual *Medicago* species, *M. truncatula* and *M. scutellata* (Tian and Rose, 1999). Leaf protoplasts of the former species, which are normally totipotent, were inactivated by iodoacetamide prior to fusion with protoplasts, incapable of regenerating shoots, isolated from gamma-irradiated cotyledons excised from 9– to 13-day-old seedlings. AFLP and cytological analyses identified two asymmetric hybrid plants within a population of 20 regenerants examined. As the authors emphasised, this procedure enables genetic traits to be transferred into *M. truncatula* from sexually incompatible wild species. Presumably, more detailed characterisation of these asymmetric somatic hybrids at the molecular level, and in terms of their field performance, will be available as these studies progress.

4. Physiological investigations with legume protoplasts

Isolated legume protoplasts have been used as source material in several investigations of cell physiology and plant-pathogen interactions.

4.1. STUDIES OF CELL MEMBRANES

In order to circumvent any effects of extracellular enzyme activity on the plasma membrane during protoplast isolation, Kurkdjian *et al.* (1993) employed UV laser microsurgery to perforate the walls of roots hairs of *M. sativa* under plasmolysing conditions. The protoplasts of treated cells could be extruded or maintained *in situ* by adjusting the calcium ion concentration in the bathing medium. This procedure provided material suitable for electrophysiological studies using patch-clamp techniques, since it is preferable that the plasma membrane has not been disturbed by enzymatic treatment by cell wall degrading enzymes normally employed for protoplast isolation.

Whilst exposure of the protoplast plasma membrane by UV microsurgery, as demonstrated by Kurkdjian *et al.* (1993) may provide, as already emphasised, the most suitable material for patch clamp studies, enzymatically isolated mesophyll protoplasts of pea have been used, nevertheless, in studies of light-controlled anion channels in the plasma membrane (Elzenga and Van Volkenburgh, 1997). Presumably, protoplasts were isolated enzymatically because of the absence of laser technology in the authors' laboratory. In isolated pea mesophyll protoplasts, light induces a transient depolymerisation that is partly due to increased plasma membrane conductance for anions. Using the patch clamp technique, these authors identified channels in the plasma membrane, one of which was an anion channel which opened during illumination of the protoplast preparation. Indeed, this light activation of the 32-picasiemen anion channel was a strong indication that this channel does, in fact, conduct the light-induced depolarizing current. Activity of the membrane channel was calcium ion dependent, indicating a role of cytoplasmic calcium ions in its behaviour.

4.2. USE OF PROTOPLASTS TO MONITOR OSMOTIC CHANGES

Protoplasts have also been used as unique source material to develop new approaches to study cell physiology. For example, Willmer *et al.* (1999) devised a rapid and accurate method to monitor changes in the volume of pea mesophyll protoplasts by differences in their absorbance at 440 and 750 nm in response to changes in osmotic pressure of the external medium. Absorbance measurements also permitted the monitoring of the kinetics of protoplast expansion and contraction, together with the effects of the water channel blocking agent, mercuric chloride.

The question as to whether or not apoplastic sucrose enters protoplasts has been studied by Opaskornkul *et al.* (1994), also using mesophyll protoplasts of pea, in this case the cv. Fenomen. Photosynthesis, dark respiration, sucrose and starch contents did not change with increased apoplastic sucrose, although the concentrations of glucose and fructose increased with the external sucrose concentration. Sucrose efflux from the protoplasts followed the increase in external sucrose concentrations between 1 and 5 mM, but decreased above this range. The results of their experiments led the authors to conclude that an active sucrose efflux from the protoplasts occurs at physiological concentrations of apoplastic sucrose, with a simple diffusion mechanism at concentrations above 5 mM.

Guard cell protoplasts prepared from young leaves of pea have provided a convenient system with which to study the behaviour of vacuoles during osmotic stress (Diekmann *et al.*, 1993). In hypertonic solution, these protoplasts shrink with endocytotic ("osmocytotic") vacuoles being formed by invagination of the plasma membrane. Such vacuoles resemble the smaller vacuoles which arise by fragmentation of the larger central vacuole. However, the two types can be distinguished by labelling the central vacuole with Neutral Red and by labelling the external medium during osmotic shrinkage with Lucifer Yellow or Cascade Blue. Internally, osmocytotic vacuoles do not fuse with the plasma membrane or with vacuoles derived from the central vacuole when the protoplasts are allowed to swell in a solution of lower osmotic pressure. Osmocytotic vacuoles appear to develop only during the isolation of guard cell protoplasts, as they were not observed in intact guard cells within strips of leaf epidermal tissue.

Whilst not related strictly to osmotic changes *per se*, it is known that starch accumulates in cultured cells and protoplasts as a result of the metabolism of carbohydrates from the culture medium. Starch accumulation and protoplast size have been monitored, by image analysis, in relation to their effects on division of pea protoplasts of the var. Belman isolated from embryogenic shoot tips (Gram *et al.*, 1996). Starch accumulated rapidly in protoplasts during the first 3 days of culture prior to the onset of mitotic division on day 4. Interestingly, subsequent periods of rapid division of protoplasts were always preceded by further starch accumulation.

4.3. STUDIES OF GENE EXPRESSION AND ENZYME ACTIVITY

The differential expression of histone genes has been demonstrated in protoplast-derived cells of *M. varia*, with Northern analysis revealing differences in mRNA accumulation (Kapros *et al.*, 1992). The maximum level of histone H3-1 gene transcription coincided with the peak of H^3-thymidine incorporation into DNA, with stress factors such as replacement of the culture medium, enzymatic digestion of cell walls, osmotic shock and auxin treatment increasing gene transcription. The observations of Kapros *et al.* (1992) suggested that the histone H3-1 gene was a constitutively expressed replacement-type gene, which belongs to a class of replication-dependent histone genes. Subsequent studies from the same group (Kapros *et al.*, 1993) extended their earlier report and confirmed that cell division was dependent on the activity of a histone H^3 promoter from *M. sativa*, following introduction into a cell line of *M. varia* of the promoter fused to the coding region of the *gus* gene. GUS activity was detected histochemically in shoot and root meristems of transgenic plants and in cells of young leaves and callus. Increased histone H3 promoter activity was detected in 4–5-day-old protoplast-derived cells, which correlated with elevated incorporation of H^3-thymidine into DNA.

Protoplasts have provided a unique experimental system with which to monitor enzyme activity in cells containing *Rhizobium* bacteroids and in uninfected cells from the central nitrogen fixing tissue of root nodules of French bean (*Phaseolus vulgaris* cv. Contender). The highest yields of protoplasts were obtained from nodules 4 weeks after their initiation (Pladys *et al.*, 1991). Protease activity, with leghaemoglobin or azocasein as substrate, was detected only in infected cells containing the *Rhizobium* bacteroids.

Western blotting of protease probed with purified anti-protease immunoglobulin G showed the molecular mass of the protease to be 58 kilodaltons.

4.4. INVESTIGATIONS OF THE EFFECTS OF GROWTH REGULATORS ON CELL METABOLISM

Estruch and Beltran (1991) demonstrated that gibberellic acid stimulated acid invertase secretion from protoplasts of the mesocarp of unpollinated pea ovaries into the surrounding medium. The acid invertase was associated with the microsomal fractions, with secretion being an active process sensitive to metabolic inhibitors and to tunicamycin. GA_3 stimulated invertase secretion was a result of increased enzyme activity from the endoplasmic reticulum to the Golgi apparatus within the cell cytoplasm.

The effects of abscisic acid have also been studied on light-induced carbon dioxide dependent oxygen evolution, sucrose synthesis, efflux of sucrose and changes in cytosolic proton concentration in mesophyll protoplasts of the pea cv. Fenomen (Opaskornkul *et al.*, 1999). The results of the studies by these workers indicated that a relationship exists between the external environment, the sucrose efflux and cytosolic proton concentration in pea mesophyll protoplasts, with a sucrose/proton symport at the protoplast plasma membrane. Protoplasts isolated from pea have also been used to develop a capillary electrophoretic and amperometric procedure to determine the concentration of ascorbic acid in single cells (Olsson *et al.*, 1998). Since abscisic acid is often elevated in vegetative tissues in response to environmental stress, such as drought, it is sometimes considered a stress hormone and a potential target for genetic manipulation (Taylor *et al.*, 2000).

4.5. INVESTIGATIONS OF SALINITY ON LEGUME CELL PHYSIOLOGY

In studies of the effects of salt on plant cell metabolism, Fedina *et al.* (1994) demonstrated that fixation of carbon dioxide was inhibited by about 60% in protoplasts isolated from 10-day-old salt-stressed plants of the pea cv. Ran 1, but this effect of salt stress could be negated by abscisic acid treatment of the source plants. The authors concluded that reduced fixation of carbon dioxide by protoplasts of salt-stressed plants indicated a direct effect of sodium chloride on photosynthesis which was independent of stomatal closure. Other investigations have targeted protoplasts of *Wollastonia biflora*, a plant that accumulates the osmoprotectant 3-dimethylsulfoniopropionate, known to be synthesised in chloroplasts from S-methylmethionine imported from the cytosol during salt stress. In order to determine the sizes of the pools of these compounds both within and outside the chloroplasts, mesophyll protoplasts were exposed to 30% artificial seawater. Trossat *et al.* (1998) reported that about one half of the 3-dimethylsulfoniopropionate was in the plastids, which could protect photosynthesis from stress injury. About 40% of S-methylmethionine was in the chloroplasts of untreated plants, this increasing to about 80% under conditions of salt stress. In contrast, at least 80% of S-methylmethionine was outside the chloroplasts in pea and spinach, both of which lack 3-dimethylsulfoniopropionate. Thus, in *Wollastonia*, synthesis of the latter compound may be associated with enhanced accumulation of S-methylmethionine in the chloroplasts.

4.6. STUDIES OF PHOTOSYNTHESIS AND CELL RESPIRATION

Several workers have employed legume protoplasts as experimental material to study aspects of photosynthesis. Saradadevi and Raghavendra (1994) investigated the effects of reduced osmotic potential on photosynthesis and respiration in pea mesophyll protoplasts. When osmotic stress was induced by increasing the sorbitol concentration in the medium, protoplasts lost up to 35% of their maximum capacity for photosynthetic carbon assimilation. A stimulation of respiration was induced with osmotic stress at 25°C, but the process was inhibited at 0°C. Photosynthesis was also more sensitive to osmotic stress at 0°C. These osmotic stress-induced changes were reversible, since both photosynthesis and respiration recovered when protoplasts were transferred from hyperosmotic to iso-osmotic medium. In a subsequent study from the same laboratory, Saradadevi *et al.* (1996) evaluated the quantitative relationship between the decrease in volume of pea mesophyll protoplasts and the reduction in photosynthesis under osmotic stress induced by sorbitol. A correlation was found between a reduction in volume and the decrease in photosynthesis, which led the authors to suggest that a decrease in absorbance of protoplasts at 440 nm can be used as a simple measure to predict the inhibition of photosynthesis resulting from osmotic stress.

Ignatova and Romanova (1992) employed an amperometric procedure to study photosynthetic oxygen evolution by suspensions of pea leaf protoplasts at saturating (0.1 mM) and superoptimal (0.6 mM) concentrations of carbon dioxide in the medium. The maximum rate of photosynthesis was at pH 6 to 8. Carbonic anhydrase was involved in the inhibition of photosynthesis by excess carbon dioxide, this enzyme activity being inhibited by acetazolamide. Washing of untreated protoplasts to constant activity of carbonic anhydrase indicated that the enzyme was located in the plasma membrane of the isolated protoplasts. This research group also reported a more detailed study of the response of photosystem II to alteration of exogenous carbon dioxide (Ignatova *et al.*, 1993), using simultaneous measurements of oxygen evolution and chlorophyll fluorescence in the same protoplast system. These later experiments confirmed carbonic anhydrase to be associated with the plasma membrane, with a suggestion by the authors that the enzyme acts as a proton conductor.

As already discussed, protoplasts have provided ideal material for studies of cell membranes, sometimes linked to the photosynthetic capability of the cells. In this respect, increased dynamics of the plasma membrane and of chloroplast membranes were demonstrated to occur following treatment of pea mesophyll protoplasts with the long chain aliphatic alcohol I-triacontanol, with enhancement of the carbon dioxide fixation rate after exposure to the compound at 10^{-6} M (Ivanov and Angelov, 1997). The effect of triacontanol was a fluidization and/or disordering of the lipid matrix of membranes, the effect being dependent upon the incubation time and concentration of the aliphatic alcohol.

The ability to isolate protoplasts from pea leaf guard cells and from mesophyll tissues has permitted comparison of the physiology of protoplasts from these two sources. For example, working with these protoplasts from the pea cv. Arkel, Vani and Raghavendra (1994) reported that respiratory oxygen uptake was several fold higher in guard cell protoplasts than in mesophyll protoplasts, although the rates of photosynthetic oxygen evolution were similar in both types of protoplasts. In addition, enzymes of the tricarboxylic acid cycle were 2 to 5 fold higher in guard cell protoplasts, through greater mitochondrial activity. Respiratory inhibitors were used to assess the activity of the cyanide-sensitive cytochrome and alternative (cyanide resistant) pathways in guard cell and mesophyll cell protoplasts, with potassium cyanide or antimycin A inhibiting respiration more in guard cells than in mesophyll cell protoplasts. The cyanide-resistant pathway constituted only one-third of the total respiration in guard cell protoplasts, but two-thirds of respiration in the case of mesophyll protoplasts. These results indicated that the cyanide-resistant pathway was not completely functional in protoplasts of guard cells, but reached its full capacity in those of mesophyll cells.

Following these experiments, Padmasree and Raghavendra (1999) assessed the role of mitochondrial respiration in optimising photosynthesis by exposing pea mesophyll protoplasts to oligomycin (an inhibitor of oxidative phosphorylation), antimycin A (an inhibitor of the cytochrome pathway of electron transport) and salicylhydroxamic acid (an inhibitor of alternative oxidase). Such compounds reduced photosynthetic oxygen evolution with the authors concluding that mitochondrial metabolism through both the cytochrome and alternative pathways was essential for optimising photosynthesis at limiting, as well as at optimal, carbon dioxide concentrations. Mitochondrial oxidative electron transport through these pathways appeared to be more important than oxidative phosphorylation in optimising photosynthesis, particularly under conditions of carbon dioxide limitation. It is interesting to note that pea plants have been the source material for all of the investigations of photosynthesis and respiration in legume protoplasts. This choice of source material may reflect the fact that uniform supplies of seedlings are readily available, with high yields of viable protoplasts. Although the latter are demanding in their long-term culture, they provide an ideal single cell system for short-term, physiological experiments.

4.7. EFFECTS OF HERBICIDES ON CELL METABOLISM

Doulis *et al.* (1998) investigated the antioxidant responses of pea protoplasts of the cvs. Progress and Nugget to treatment with the herbicide paraquat. Protoplasts of the cv. Nugget were more resistant to paraquat in the micro-molar range, with a bleaching concentration of 10 μM inhibiting carbon dioxide-dependent oxygen evolution by about 50% during the first 40 minutes of exposure to the herbicide, which remained at that rate ("coping behaviour") for up to 100 minutes. However, protoplasts of the cv. Progress did not exhibit such coping behaviour. Since antioxidant enzyme activities in Nugget protoplasts did not alter throughout the experiments, the overall coping strategy of Nugget protoplasts could not be attributed to changes in activities of antioxidant enzymes. High concentrations of herbicide at 0.1 mM for 12 hours bleached the protoplasts, and total protein decreased to 30% of its pre-stress concentration, with glutathione reductase also declining. In contrast, superoxidase dismutase activity remained constant during the first 9 hours of herbicide treatment, increasing to about 150% of its original level by 12 hours of exposure. The authors indicated that their data agreed with earlier reports, and they hypothesised that superoxide dismutase activity increases in response to reactive oxygen species, even under conditions leading to cell death.

5. Protoplast-based studies of plant-pathogen interactions

Isolated protoplasts provide a suitable experimental system for studies of plant-pathogen interactions. Thus, Akpa and Archer (1993) used mesophyll protoplasts isolated from five cultivars of pea to study their interaction with five races of *Pseudomonas stringae* pv. *pisi*. Although it was not possible to detect any trend in the attachment of bacteria to the surfaces of compatible and incompatible host protoplasts, and the viability of protoplasts from compatible and incompatible host-pathogen interactions did not differ significantly, the authors concluded that pea protoplasts were a useful model system for investigating the early responses of pea to infection by *Pseudomonas*. A similar approach could also be used to study other plant-pathogen interactions.

Since protoplasts readily take up macromolecules, they also provide a convenient single cell system for studying plant-virus interactions. Purified particles of subterranean clover stunt virus, a single-stranded DNA virus, failed to infect their host plant when attempts were made to transmit the virus using the aphid vector, *Aphis craccivora*. However, following polyethylene glycol and electroporation-induced uptake of virus into clover and pea protoplasts, Chu *et al.* (1993) were able to detect *de novo* synthesis of viral particles using ELISA and nucleic acid hybridisation, demonstrating that purified viral particles are capable of replication in isolated protoplasts. These workers concluded that protoplast-based assays could be exploited to characterise new or altered segments or functional domains of the viral genome. They proposed that such investigations could provide the basis for the possible development of the subterranean clover stunt virus genome as a vector for foreign gene expression in plants. Other studies reported about the same time (Taliansky *et al.*, 1993) involved the use of cowpea protoplasts to study virion formation and the induction of tubular transport structures by the cowpea mutant virus mutant N123. Normally, this mutant fails to spread in intact plants because of mutations in its mRNA, which impair or completely abolish cell-to-cell transport. However, isolated protoplasts provide an experimental system which with to circumvent some of the difficulties associated with the use of intact plants.

Two more recent studies have focused attention on the movement within plant cells of particles of alfalfa mosaic virus (Zheng *et al.*, 1997) and of both alfalfa and brome mosaic viruses (Kasteel *et al.*, 1997). Using protoplasts isolated from cowpea, the latter authors employed immunofluorescence microscopy to demonstrate that the movement proteins of these viruses assemble into long tubular structures at the surface of infected protoplasts, with electron microscopy and immunogold analyses confirming the presence of virus particles in the tubes.

6. Conclusions and future prospects

This review of the exploitation of isolated legume protoplasts indicates that whilst most physiological and plant-pathogen studies have focused upon grain legumes, particularly protoplasts isolated from pea, most progress in terms of gene transfer has been achieved with protoplasts of forage legumes, with *Medicago sativa* featuring in the majority of successful somatic hybridisation attempts. This is probably a reflection of the fact that, in general, forage legume protoplasts are more amenable than those of grain legumes in terms of their basic tissue culture, since several protoplast-to-plant systems have been reported in the literature for forage legumes. Undoubtedly, in this respect, there is still a requirement for reproducible protoplast-to-plant systems to be developed and existing ones improved upon for the major grain legumes, if somatic cell approaches are to be fully exploited for gene transfer at the inter-generic and inter-specific levels. At present, there appears to be limited use for isolated legume protoplasts in the generation of transgenic plants, the main application being in short-term gene expression studies to monitor gene constructs.

References

Akpa A D and Archer S A (1993) Interaction between *Pseudomonas syringae* pv. *pisi* and isolated pea mesophyll protoplasts. *J. Phytopath.*, **139**: 110–118.

Ando S, Takahashi C, Ikumi K, Masuda S and Shimizu T (1997) Asymmetric hybridisation between alfalfa (*Medicago sativa* L.) cultivar and CMS line. *Breed. Sci.*, **47**: 195–201.

Arcioni S, Damiani F, Mariani A and Pupilli F (1997) Somatic hybridization and embryo rescue for the introduction of wild germplasm. In: *Biotechnology in Agriculture Series, Vol. 17, Biotechnology and the Improvement of Forage Legumes* (Eds McKersie B D and Brown D C W), Springer-Verlag, Heidelberg, 61–89.

Bhadra S K, Hammatt N and Davey M R (1991) Tissue and protoplast culture of rice bean (*Vigna umbellata* (Thunb.) Ohwi and Ohashi). *Trop. Agric.*, **68**: 344–348.

Bhadra S K, Hammatt N, Power J B and Davey M R (1994) A reproducible procedure for plant regeneration from seedling hypocotyl protoplasts of *Vigna sublobata* L. *Plant Cell Rep.*, **14**: 175–179.

Böhmer P, Meyer B and Jacobsen H J (1995) Thidiazuron induced high frequency of shoot induction and plant regeneration in protoplast derived pea callus. *Plant Cell Rep.*, **15**: 26–29.

Chowdhury J, Jain S and Jain R K (1993) Biotechnological approaches for developing salt tolerant field crops. *J. Plant Biochem. Biotechnol.*, **2**: 1–7.

Chu P W G, Qiu B S, Li Z Y and Larkin P (1993) Replication of subterranean clover stunt virus in pea and subterranean clover protoplasts. *Virus Res.*, **27**: 173–183.

Crea F, Calderini O, Nenz E, Cluster P D, Damiani F and Arcioni S (1997) Chromosomal and molecular rearrangements in somatic hybrids between tetraploid *Medicago sativa* and diploid *Medicago falcata. Theort. Appl. Genet.*, **95**: 1112–1118.

Diekmann W, Hedrich R, Raschke K and Robinson D G (1993) Osmocytosis and vacuolar fragmentation in guard cell protoplasts – their relevance to osmotically-induced volume changes in guard cells. *J. Exp. Bot.*, **44**: 1569–1577.

Dimitrova A P and Christov A M (1992) Electrically induced protoplast fusion using pulse electric fields for dielectrophoresis. *Plant Physiol.*, **100**: 2008–2012.

Doulis A G, Donahue J L and Alscher R G (1998) Differential responses to paraquat-induced oxidative injury in a pea (*Pisum sativum*) protoplast system. *Physiol. Plant.*, **102**: 461–471.

Durieu P and Ochatt S J (2000) Efficient intergeneric fusion of pea (*Pisum sativum* L.) and grass pea (*Lathyrus sativus* L.) protoplasts. *J. Exp. Bot.*, **51**: 1237–1242.

Elzenga J T M and Van Volkenburgh E (1997) Characterization of light-controlled anion channel in the plasma membrane of mesophyll cells of pea. *Plant Physiol.*, **113**: 1419–1426.

Estruch J J and Beltran J P (1991) Gibberellic acid stimulates acid invertase secretion in pea ovary protoplasts. *FEBS Letts.*, **279**: 303–306.

Fedina I S, Tsonev T D and Guleva E I (1994) ABA as a modulator of the response of *Pisum sativum* to salt stress. *J. Plant Physiol.*, **143**: 245–249.

Frearson E M, Power J B and Cocking E C (1973) The isolation, culture and regeneration of *Petunia* leaf protoplasts. *Dev. Biol.*, **33**: 130–137.

Gilmour D M, Golds T J and Davey M R (1989) *Medicago* protoplasts: fusion, culture and plant regeneration. In: *Biotechnology in Agriculture and Forestry, Vol. 8, Plant Protoplasts and Genetic Engineering I* (Ed Bajaj Y P S), Springer-Verlag, Heidelberg, pp. 370–388.

Gram T, Mattsson O and Joersbo M (1996) Division frequency of pea protoplasts in relation to starch accumulation. *Plant Cell, Tiss. Org. Cult.*, **45**: 179–183.

Hashimoto T, Yamada T, Tada A, Kawamata S, Tanaka Y, Sriprasertsak P, Ichinose Y, Kata H, Izutsu S, Shiraishi T, Oku H and Ohtsuki Y (1992) Transient expression in electroporated pea protoplasts-elicitor responsiveness of a phenylalanine ammonia-lyase promoter. *Plant Cell Rep.*, **11**: 183–187.

Ignatova L K, Moskvin O V, Ivanov B N and Romanova A K (1993) The effect of CO_2 uptake by pea protoplasts on O_2 evolution rate and parameters of chlorophyll fluorescence quenching. *Plant Physiol. Biochem.*, **31**: 295–301.

Ignatova L K and Romanova A K (1992) Involvement of carbonic anhydrase in inhibition of photosynthesis of pea protoplasts by CO_2 excess. *Soviet Plant Physiol.*, **39**: 461–465.

Iriawati H M, Miyake H and Taniguchi T (1994) Organogenesis in leaf protoplast-derived calli of mungbean. *Jpn. J. Crop Sci.*, **63**: 714–720.

Iriawati H M, Taniguchi T and Takeoka Y (1996a) Ultrastructural studies on the fusion product between protoplasts of mung bean and adzuki bean. *Jpn. J. Crop Sci.*, **65**: 330–339.

Iriawati H M, Taniguchi T and Takeoka Y (1996b) Effects of iodoacetamide on the fine structure of adzuki bean protoplast and its fusion product. *Jpn. J. Crop Sci.*, **65**: 532–539.

Ivanov A G and Angelov M N (1997) Photosynthesis response to triacontanol correlates with increased dynamics of mesophyll protoplast and chloroplast membranes. *Plant Growth Reg.*, **21**: 145–152.

Kaimori N, Senda M, Ishikawa R, Akada S, Harada T and Niizeki M (1998) Asymmetric somatic cell hybrids between alfalfa and birdsfoot trefoil. *Breed. Sci.*, **48**: 29–34.

Kalbin G, Strid A and Frohnmeyer H (1999) Transcriptional activation of the parsley chalcone synthase promoter in heterologous pea and yeast systems. *Plant Physiol. Biochem.*, **37**: 821–829.

Kao K N and Michayluk M R (1975) Nutritional requirements for growth of *Vicia hajastana* cells and protoplasts at a very low population density in liquid media. *Planta*, **126**: 105–110.

Kapros T, Bogre L, Nemeth K, Bako L, Gyorgyey J, Wu S C and Dudits D (1992) Differential expression of histone H3 gene variants during cell cycle and somatic embryogenesis in alfalfa. *Plant Physiol.*, **98**: 621–625.

Kapros T, Stefanov I, Magyar Z, Ocsovszky I and Dudits D (1993) A short histone H3 promoter from alfalfa specifies expression in S-phase cells and meristems. In Vitro *Cell Dev. Biol. – Plant*, **29**: 27–32.

Kasteel D T J, van der Wel N N, Jansen K A J, Goldbach R W and van Lent J W M (1997) Tubule-forming capacity of the movement proteins of alfalfa mosaic virus and brome mosaic virus. *J. Gen. Virol.*, **78**: 2089–2093.

Kumar V and Davey M R (1991) Genetic improvement of legumes using somatic cell and molecular techniques. *Euphytica*, **55**: 157–169.

Kurkdjian A, Leitz G, Manigault P, Harim A and Greulich K O (1993) Nonenzymatic access to the plasma membrane of *Medicago* root hairs by laser microsurgery. *J. Cell Sci.*, **105**: 263–268.

Lehminger-Mertens R and Jacobsen H J (1993) Regeneration of plants from protoplasts of pea (*Pisum sativum* L.). In: *Biotechnology in Agriculture and Forestry, Vol. 22, Plant Protoplasts and Genetic Engineering III* (Ed Bajaj YPS), Springer-Verlag, Heidelberg, pp. 97–104.

Li X B, Xu Z H and Wei Z M (1995) Plant regeneration from protoplasts of immature *Vigna sinensis* cotyledons via somatic embryogenesis. *Plant Cell Rep.*, **15**: 282–286.

Li Y G, Tanner G J, Delves A C and Larkin P J (1993) Asymmetric somatic hybrid plants between *Medicago sativa* L. (Alfalfa, Lucerne) and *Onobrychis viciifolia* Scop (Sainfoin). *Theort. Appl. Genet.*, **87**: 455–463.

Lu T G, Sun S C, Jian L C, Sun L H and Zhao G L (1992) Cell division, callus formation and plant regeneration from cell suspension derived protoplasts of *Onobrychis viciifolia* Scop. *J. Plant Physiol.*, **140**: 106–109.

Luo J P and Jia J F (1998) Plant regeneration from callus protoplasts of the forage legume *Astragalus adsurgens* Pall. *Plant Cell Rep.*, **17**: 313–317.

Mikaye I H, Taniguchi T and Takeoka Y (1996) Ultrastructural studies on the fusion product between protoplasts of mung bean and adzuki bean. *Jpn. J. Crop Sci.*, **65**: 330–339.

Murashige T and Skoog F (1962) A revised medium for rapid growth and bioassays with tobacco tissue cultures. *Physiol. Plant.*, **15**: 473–497.

Nenz E, Pupilli F, Damiani F and Arcioni S (1996) Somatic hybrid plants between the forage legumes *Medicago sativa* L. and *Medicago arborea* L. *Theort. Appl. Genet.*, **93**: 183–189.

Niizeki M, Nakajo S, Ishikawa R, Harada T and Saito K (1992) Behaviour of mitochondrial and chloroplast DNA in somatic hybrid calli between rice and alfalfa. *Jpn. J. Breed.*, **42**: 803–809.

Niizeki M and Saito K (1989) Callus formation from protoplast fusion between leguminous species of *Medicago sativa* and *Lotus corniculatus*. *Jpn. J. Breed.*, **39**: 373–377.

Ochatt S J, Mousset-Declas C and Rancillac M (2000) Fertile pea plants regenerate from protoplasts when calluses have not undergone endoreduplication. *Plant Sci.*, **156**: 177–183.

Olsson J, Nordstrom O, Nordstrom A C and Karlberg B (1998) Determination of ascorbic acid in isolated pea cells by capillary electrophoresis and amperometric detection. *J. Chromat.*, **826**: 227–233.

Opaskornkul C, Lindberg S and Tillberg J E (1999) Effects of ABA on the distribution of sucrose and protons across the plasmalemma of pea mesophyll protoplasts. Suggesting a sucrose/proton symport. *J. Plant Physiol.*, **154**: 447–453.

Opaskornkul V C, Greger M and Tillberg T E (1994) Effects of apoplastic sucrose on carbohydrate pools and sucrose efflux of mesophyll protoplasts of pea (*Pisum sativum*). *Physiol. Plant.*, **90**: 685–691.

Padmasree K and Raghavendra A S (1999) Importance of oxidative electron transport over oxidative phosphorylation in optimizing photosynthesis in mesophyll protoplasts of pea (*Pisum sativum* L.). *Physiol. Plant.*, **105**: 546–553.

Pladys D, Dimitrijevic L and Rigaud J (1991) Localization of a protease in protoplast preparations in infected cells of French bean nodules. *Plant Physiol.*, **97**: 1174–1180.

Puonti-Kaerlas J, Ottosson A and Eriksson T (1992) Survival and growth of pea protoplasts after transformation by electroporation. *Plant Cell, Tiss. Org. Cult.*, **30**: 141–148.

Pupilli F, Arcioni S and Damiani F (1991) Protoplast fusion in the genus *Medicago* and isozyme analysis of parental and somatic hybrid cell-lines. *Plant Breed.*, **106**: 122–131.

Pupilli F, Businelli S, Cacered M E, Damiani F and Arcioni S (1995) Molecular, cytological and morpho-agronomical characterisation of hexaploid somatic hybrids in *Medicago. Theort. Appl. Genet.*, **90**: 347–355.

Pupilli F, Scarpa G M, Damiani F and Arcioni S (1992) Production of interspecific somatic hybrid plants in the genus *Medicago* through protoplast fusion. *Theort. Appl. Genet.*, **84**: 792–797.

Rose R J and Nolan K E (1995) Regeneration of *Medicago truncatula* from protoplasts isolated from kanamycin sensitive and kanamycin resistant plants. *Plant Cell Rep.*, **14**: 349–353.

Rybczynski J J (1997) Plant regeneration from highly embryogenic callus, cell suspension and protoplast cultures of *Trifolium fragiferum. Plant Cell Tiss. Org. Cult.*, **51**: 159–170.

Saradadevi K, Padmasree K and Raghavendra A S (1996) Correlation between the inhibition of photosynthesis and the decrease in area of detached leaf discs or volume/absorbance of protoplasts under osmotic stress in pea (*Pisum sativum*). *Physiol. Plant.*, **96**: 395–400.

Saradadevi K and Raghavendra A S (1994) Inhibition of photosynthesis by osmotic stress in pea (*Pisum sativum*) mesophyll protoplasts is intensified by chilling or photoinhibitory light-intriguing responses of respiration. *Plant Cell Environ.*, **17**: 739–746.

Sarangi B K, Kuchuk N and Gleba Y Y (1992) Isolation and culture of protoplasts of pigeon pea (*Cajanus cajan* L.). *Plant Cell Rep.*, **11**: 462–465.

Sato T (1995) Basic study of biotechnology in Adzuki bean (*Vigna angularis* Ohwi and Ohashi). *Rep. Hokkaido Prefectural Agric. Expt. Stats.*, **87**: 1–68.

Sato T, Asaka D, Harada T and Matsukawa I (1993) Plant regeneration from protoplasts of adzuki bean (*Vigna angularis* Ohwi and Ohashi). *Jpn. J. Breed.*, **43**: 183–191.

Scarpa G M, Pupilli F, Damiani F and Arcioni S (1993) Plant regeneration from callus and protoplasts in *Medicago polymorpha. Plant Cell, Tiss. Org. Cult.*, **35**: 49–57.

Taliansky M E, De Jager C P, Wellink J, Van Lent J W M and Goldbach R W (1993) Defective cell-to-cell movement of cowpea mosaic virus mutant N123 is efficiently complemented by sunn-hemp mosaic virus. *J. Gen. Virol.*, **74**: 1895–1901.

Taylor I, Burbidge A and Thompson A J (2000) Control of abscisic acid synthesis. *J. Exp. Bot.*, **51**: 1563–1574.

Tegeder M, Gebhardt D, Scheider O and Pickardt T (1995) Thidiazuron-induced plant regeneration from protoplasts of *Vicia faba* cv. Mythos. *Plant Cell Rep.*, **15**: 164–169.

Tian D and Rose R J (1999) Asymmetric hybridisation between the annual legumes *Medicago trunculata* and *Medicago scutellata. Plant Cell Rep.*, **18**: 989–996.

Trossat C, Rathinasabapathi B, Weretilnyk E A, Shen T L, Huang Z H, Gage D A and Hanson A D (1998) Salinity promotes accumulation of 3-dimethylsulfoniopropionate and its precursor S-methylmethionine in chloroplasts. *Plant Physiol.*, **116**: 165–171.

Vani T and Raghavendra A S (1994) High mitochondrial activity but incomplete engagement of the cyanide-resistant alternative pathway in guard cell protoplasts of pea. *Plant Physiol.*, **105**: 1263–1268.

Willmer C M, Padmasree K and Raghavendra A S (1999) A novel method of measuring volume changes of mesophyll cell protoplasts and the effect of mercuric chloride on their osmotically-induced swelling. *J. Exp. Bot.*, **50**: 401–406.

Xu Z Q and Jia J F (1997) Regeneration of intergeneric somatic hybrids by protoplast fusion between *Onobrychis viciaefolia* and *Medicago sativa. Sci. China Series C-Life Sci.*, **40**: 363–370.

Zafar Y, Nenz E, Damiani F, Pupilli F and Arcioni S (1995) Plant regeneration from explant and protoplast-derived calluses of *Medicago littoralis. Plant Cell, Tiss. Org. Cult.*, **41**: 41–48.

Zheng H Q, Wang G L and Zhang L (1997) Alfalfa mosaic virus movement protein induces tubules in plant protoplasts. *Mol. Plant-Microbe Interact.*, **10**: 1010–1014.

WIDE HYBRIDIZATION IN IMPORTANT FOOD LEGUMES

NALINI MALLIKARJUNA
Genetic Resources and Enhancement Program,
International Crops Research Institute for Semi Arid Tropics (ICRISAT),
Patancheru P.O. 502 324, Andhra Pradesh, India
e-mail: n.mallikarjuna@cgiar.org

Abstract

Wild species are not only sources of resistance to important pests and diseases but are also sources of variability. Groundnut, chickpea and pigeonpea have a rich germplasm in the form of wild species. Interspecific hybridization to transfer resistance to important constraints is an on-going program at ICRISAT. Progress has been made to transfer foliar and viral disease resistance genes into interspecific derivatives of groundnut. Interspecific derivatives of pigeonpea obtained as a result of crossing *Cajanus platycarpus* with *C. cajan* have resistance to Phytophthora blight disease, a character inherited from the wild species. None of the cultivated pigeonpea have resistance to Phytophthora blight disease. One of the reasons for difficulty in accessing genes from the wild species is the barrier to hybridization. Anatomical studies of the cross between *C. platycarpus* and *C. cajan* revealed that one of the post zygotic barriers operating in the cross was the breakdown of the walls of the immature seed leading to the crushing of the hybrid embryo. The progress made in wide hybridization in groundnut, pigeonpea and chickpea are discussed in detail.

1. Introduction

Crop genetic resources are one of the important natural resources of the earth. With growing population and major changes in climate, there is increasing pressure on crop improvement programs to produce new and improved crop varieties. Many of the species from which the crop plants were selected continue to survive in the wild even today. Wild relatives of crop plants have evolved to survive droughts and floods, extreme heat and cold, and pest and disease attacks. This is the main reason why they are still so valuable to agriculture today.

High yielding cultivars of crop plants have grown in popularity and they have slowly replaced the land races that had coexisted with the wild relatives, and had sometimes

P.K. Jaiwal and R.P. Singh (eds.), Improvement Strategies for Leguminosae Biotechnology, 155–171.

crossbred with them. The new cultivars emerging from breeding programs are uniform and very few in number, compared to the immense diversity and vast number of land races. Genetic uniformity can make a crop vulnerable to epidemics of pests and diseases. When farmers grow the same variety, a pest or disease that strikes one plant quickly spreads over a large area. The dangers of genetic uniformity are well known in the case of Ireland's potato blight of 1845 when the Irish potato crop was wiped out leading to starvation and death. In 1943, brown spot disease aggravated by a typhoon destroyed the rice crop in India and started the great Bengal famine. In 1953 and 1954, wheat stem rust ravaged most of the hard wheat crop in the United States. During the summer of 1970, the southern leaf blight fungus killed corn plants in the United States in large areas. At the same time in the Philippines and Indonesia, the deadly tungro virus was spread by the leafhoppers on high yielding rice.

The most important use of wild species has been to breed crops that can resist pests and diseases. Tomato simply could not exist as a commercial crop without the genes from its wild relatives. Genes from numerous wild relatives of potato protect it from pests and diseases. Wild relatives of rice have played a key part in providing disease resistance to rice cultivars that feed most of Asia. *Oryza nivara*, which has resistance to grassy stunt virus, was collected in 1963 in Uttar Pradesh. Rice variety IR36 has the resistance gene to grassy stunt virus from *O. nivara*.

In spite of the demonstrated value and potential, wild relatives are usually considered by breeders to be the last resort because breeders in general are unfamiliar with the wild material and special techniques that are sometimes required to use it. To make wild relatives accessible and easy to use, germplasm enhancement programs prebreed them to remove some of their undesirable traits before crossing them with the crop. Through backcrossing with elite cultivars, prebreeding transfers useful genes from the wild relatives into a genetic environment in which plant breeders everywhere may more easily use them.

The semi-arid tropics (SAT) encompasses parts of 48 developing countries including most of India, parts of Southeast Asia, sub-Saharan Africa and southern and eastern Africa. Approximately one-sixth of the world's population lives in SAT, which is typified by unpredictable weather, limited and erratic rainfall and nutrient-poor soils. Groundnut, pigeonpea and chickpea constitute important legume crops of the SAT. An extensive array of unique constraints, some highly intractable, including climatic, edaphic, biological and social factors, severely limit the production of these crops.

2. Groundnut

Arachis hypogaea L. is a member of family *Leguminosae*, tribe *Aeschynomeneae* and subtribe *Stylosanthinae*. Krapovickas and Gregory (1994) divided the genus *Arachis* into nine sections. The genus *Arachis* was first discovered in, and is widely distributed in South America. ICRISAT gene bank holds 42 wild species representing eight sections and distributed in 452 accessions. Of the 27 known wild species under section *Arachis*, 20 are available at ICRISAT.

Based on crossability studies, genetic diversity in *Arachis* can be classified into three gene pools. The primary gene pool consists of landraces of *A. hypogaea* and its wild form

A. monticola. The secondary gene pool consists of diploid species from section *Arachis* and members from other sections, if any, that are cross compatible with *A. hypogaea*. The tertiary gene pool consists of the remaining wild species of *Arachis*, which are incompatible with *A. hypogaea*. Many of the wild species are resistant to diseases caused by fungi, viruses, and insect pests (Table 1).

Table 1. Resistance to foliar, viral and insect pests in wild Arachis *species*

Section and species	Rust	LLS	ELS	PSV	GRV	PMV	TSWV	PCV	THR	APH	Mit	JAS	Spod	Nem
Arachis annuae														
A. batizocoi	I													
A. duranensis	I			R					R			R		R
A. spagazzinii	I											R		
Arachis perennes														
A. helodes								T						R
A. correntina	I					R	R		R			R		
A. cardenasii	I	I		R		R	R		R			R		R
A. chacoensis	I	R	R/I		R	R	R		R	R			R	
A. stenosperma	R	R	R		R									R
A. kempf-mercadoi	R	R	R										R	
A. hoehnoi														
Caulorhizae														
A. repens		R	I	R	R				I			R		
Extranervosae														
A. vilosulicarpa	I	I	I											
Triseminalae														
A. pusilla	I					R	R		R			R		
Erectoides				R										
A. paraguariensis	I	R							I				R	
A. rigoni					R				R			R		
A. appressipila	I	R			R								R	
Rhizomatosae														
A. glabrata	R	I	R	R	R	R			R	R		R		
A. hagenbeckii	I	R	I									R		
Procumbentes														
A. kretschmeri	R													R
A. chiquitana														
A. magna														
Heteranthae														
A. sylvestris														R

Rust = rust, LLS = late leaf spot, ELS = early leaf spot, PSV = peanut stunt virus, GRV = groundnut rosette virus, PMV = peanut mottle virus, TSWV = tomato spotted wilt virus, PCV = peanut clump virus, THR = thrips, APH = aphids, Mit = mites, JAS = jassids, Spod = *spodoptera litura*, Nem = nematode (*Meloidogyne javanica*), R = resistant, I = immune, and T = tolerant.

2.1. INTERSPECIFIC HYBRIDIZATION

Arachis hypogaea, the cultivated groundnut, belongs to section *Arachis* with a tetraploid chromosome number of 2n = 40. It is readily crossable with the related tetraploid wild species *A. monticola* and produces fertile hybrids. *A. monticola* has been proposed as a wild subspecies of *A. hypogaea* (Smartt and Stalker, 1982; Singh and Moss, 1984). The variety "Spancross", developed in the USA from interspecific cross involving *A. monticola*, was released in Tanzania. Other compatible wild species in section *Arachis* are diploid with 2n = 20. Wild species in section *Arachis* have been assigned three genomes. Majority of the wild species in the section *Arachis* have A genome and *A. batizocoi* is the only wild species with the B genome. *A. spinaclava* is the wild species with the D genome. *A. hypogaea* has been successfully crossed with the wild species with both A and B genome. Hybrids have not been reported with *A. spinaclava*.

Gregory and Gregory (1979) crossed many of the wild species and reported interspecific hybrids. The diploid species of the section *Arachis* are cross compatible with *A. hypogaea*, but many a times require cytological intervention, growth regulator treatment, embryo germination and/or chromosome doubling to obtain fertile hybrids. Using backcross method of triploid, hexaploid and autotetraploid route, germplasm lines ICGV 86699, ICGV 87165 were developed at ICRISAT. ICGV 86699 is an interspecific derivative from the cross (*A. batizocoi* × *A. duranensis*) × *A. hypogaea*. It is a high-yielding germplasm with resistance/tolerance to rust, early and late leaf spots, peanut bud necrosis, peanut mottle virus and stem and pod rots. ICGV 87165 is an interspecific derivative between *A. hypogaea* and *A. cardenasii* and is high yielding with resistance to rust and late leaf spot. One of the interspecific derivative ICGV-SM 86715 was released as cultivar Veronica in Mauritius. Germplasm lines released for resistance against nematodes were derived from the (*A. cardenesaii* × *A. chacoense*) × *A. batizocoi* (Simpson *et al.*, 1993a). Similarly the registration of the line with resistance against late leaf spot was derived from the cross *A. hypogaea* × *A. cardenesii* (Stalker and Beute, 1993).

There are many reports of producing hybrids between *A. hypogaea* and members of the section *Arachis* (Stalker and Simpson, 1995). Reports on intersectional hybrids are few (Mallikarjuna and Sastri, 1985a, b; Shen *et al.*, 1995). There is no report of crossing wild species from section *Arachis* with *A. glabrata*, but it is now possible (N. Mallikarjuna, unpublished data). Crosses between *A. hypogaea* and members involving other eight sections require the intervention of *in vitro* techniques to obtain hybrid plants. *A. glabrata* belonging to the section *Rhizomatosae* has many desirable characters such as resistance to fungal and viral diseases, and insect pests. *A. glabrata* does not produce seeds and is propagated through rhizomes. Embryo rescue technique for groundnut (Mallikarjuna and Sastri, 1985) has been used to obtain hybrids from the cross *A. hypogaea* × *A. glabrata* (Fig. 1). The hybrid showed meiotic aberrations with univalents in metaphase. An important feature of the hybrid was the inheritance of disease resistance character from *A. glabrata* (Table 2). The BC_1 hybrid (*A. hypogaea* × *A. glabrata*) × *A. hypogaea* was meiotically more stable but abnormal telophase with micronuclei were observed (Fig. 2). Similarly, hybrids between *A. chacoense* and *A. glabrata*, and *A. duranensis* and *A. glabrata* were obtained by germinating hybrid embryos *in vitro* (Figs 3 and 4). The hybrids were screened for

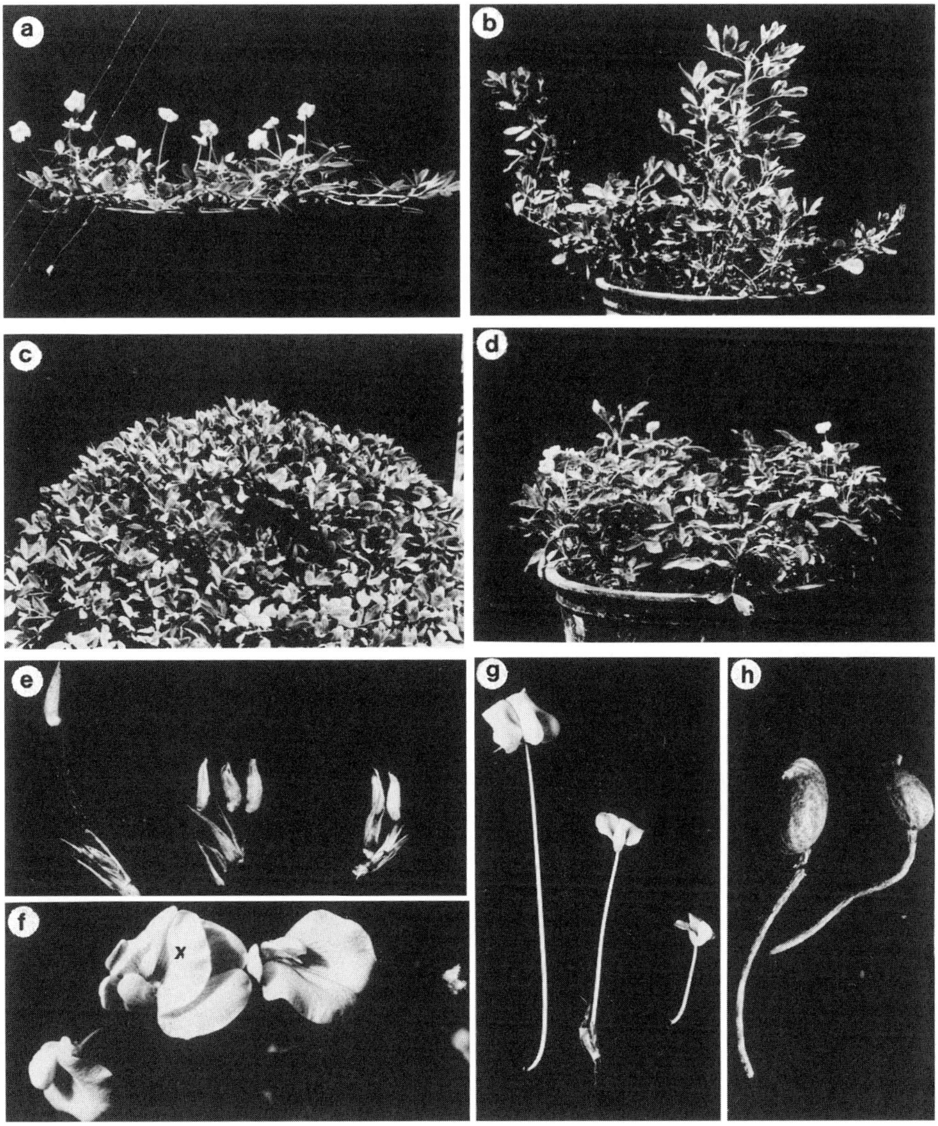

Figure 1. Interspecific hybrid between Arachis hypogaea *and* A. glabrata. *(a) Wild species* A. glabrata, *the male parent of the cross. (b) Cultivated species* A. hypogaea, *the female parent of the cross. (c) F_1 hybrid* (A. hypogaea × A. glabrata). *(d) BC_1 hybrid* (A. hypogaea × A. glabrata) × A. hypogaea. *(e) Flower bud morphology,* A. glabrata *and* A. hypogaea *on either sides with BC_1 buds at the center. (f) BC_1 flowers. The abnormal flower* (x) *rarely produced in BC_1 hybrid, which was a common feature of the F_1 hybrid. (g) Flower morphology,* A. glabrata *and* A. hypogaea *on the left and right with BC_1 flower at the center. (h) BC_1 pods. Note the presence of prominent beak, a wild species character.*

Table 2. Results of disease screening of the parents and hybrids of Arachis *species*

Identity	Rust	LLS	ELS	PMV	PSTV	TSWV
A. glabrata	R	R	R	R	R	R
A. hypogaea	S	S	S	S	S	S
A. duranensis	MR	R	MR	S	S	S
A. chacoensis	MR	R	R	R	S	R
RIMG	R	R	R	S	R	R
RIMD	R	S	R	NT	NT	NT
RIMC	R	S	NT	NT	NT	NT

RIMG = *A. hypogaea* × *A. glabrata*, RIMD = *A. duranensis* × *A. glabrata*, RIMC = *A. chacoense* × *A. glabrata*, R = resistant, MR = moderately resistant, S = susceptible, NT = not tested, LLS = late leaf spot, ELS = early leaf spot, PMV = peanut mottle virus disease, PSTV = peanut stripe disease, and TSWV = tomato spotted wilt virus disease.

three foliar diseases, i.e. rust, late leaf spot, and early leaf spot. It was observed that the hybrids had acquired the disease resistance character from *A. glabrata* (Table 2).

Apart from contributing genes to important constraints, crosses between wild and cultivated germplasm have been employed to decipher consanguinity between different gene pools and phylogenetic relationships. According to Stalker and Simpson (1995), members of the section *Rhizomatosae* have recently evolved and share one of their genomes with *A. hypogaea*. Mallikarjuna (2001) was able to prove that the A genome of *A. hypogaea* is closely related to R genome of *A. glabrata*. It was possible to cross *A. chacoense* and *A. duranensis* (A genome species) and produce fertile hybrids. Meiotic analysis of the hybrids between *A. hypogaea, A. duranensis* and *A. chacoense* with *A. glabrata* further confirmed that the two genomes are closely related. The BC_1 seeds obtained as a result of crossing the F_1 hybrids (*A. hypogaea* × *A. glabrata*), (*A. chacoense* × *A. glabrata*) and (*A. duranensis* × *A. glabrata*) were bold, but not completely mature. Embryos were germinated *in vitro* to obtain hybrid plants. It was not possible to cross *A. batizocoi* with *A. glabrata* (N. Mallikarjuna, unpublished data). With more than two decades of research on methods to overcome barriers to interspecific hybridization at ICRISAT, it is now possible to cross *A. hypogaea* with members from other sections of *Arachis* such as *Procumbentae, Heteranthae, Erectoides* and *Rhizomatosae* (N. Mallikarjuna, unpublished data).

3. Chickpea

The genus *Cicer* L. is a diploid with 2n = 16, and comprises of 33 perennial and nine annual species and one unspecified species (van der Maesen, 1987). Information is available on the presence of desirable genes in some wild species (Table 3). The primary gene pool comprises of *C. arietinum* and *C. reticulatum* which cross freely to produce fertile hybrids with normal chromosome segregation. The secondary gene pool comprises of only *C. echinospermum*. The tertiary gene pool consists of the remaining six annual and the perennial species. None of the wild species from the tertiary gene pool produce

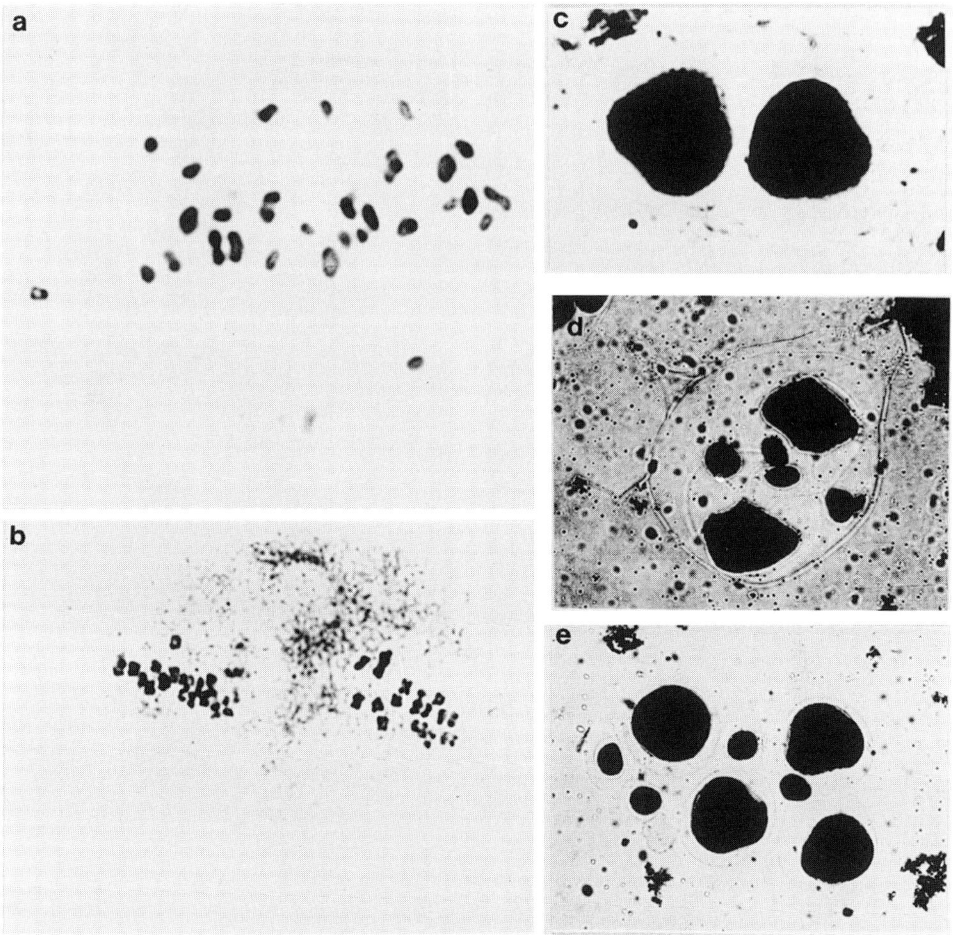

Figure 2. Meiosis in BC_1 hybrid (Arachis hypogaea × A. glabrata) × A. hypogaea. *(a) Meiosis in BC1 hybrid* {(A. hypogaea × A. glabrata) × A. hypogaea}. *Note the presence of a large number of univalents. (b) Anaphase in BC_1 hybrid with normal chromosome disjunction. (c) A diad. (d) A diad with 4 micronuclei. (e) A tetrad with 4 micronuclei.*

mature seeds when crossed with cultivated chickpea with the exception of the report by Verma *et al.* (1990), which needs to be confirmed.

3.1. INTERSPECIFIC HYBRIDIZATION

Ladizinsky and Alder (1976) produced fertile hybrids between *C. arietinum* and *C. reticulatum* with normal chromosome segregation. Although it was possible to successfully cross *C. arietinum* and *C. echinospermum* (Pundir and Mengesha, 1995), the hybrids had high pollen sterility with abnormal meiotic association. Singh *et al.* (1984) isolated early flowering segregants from the cross *C. arietinum* × *C. reticulatum*. It is presumed that

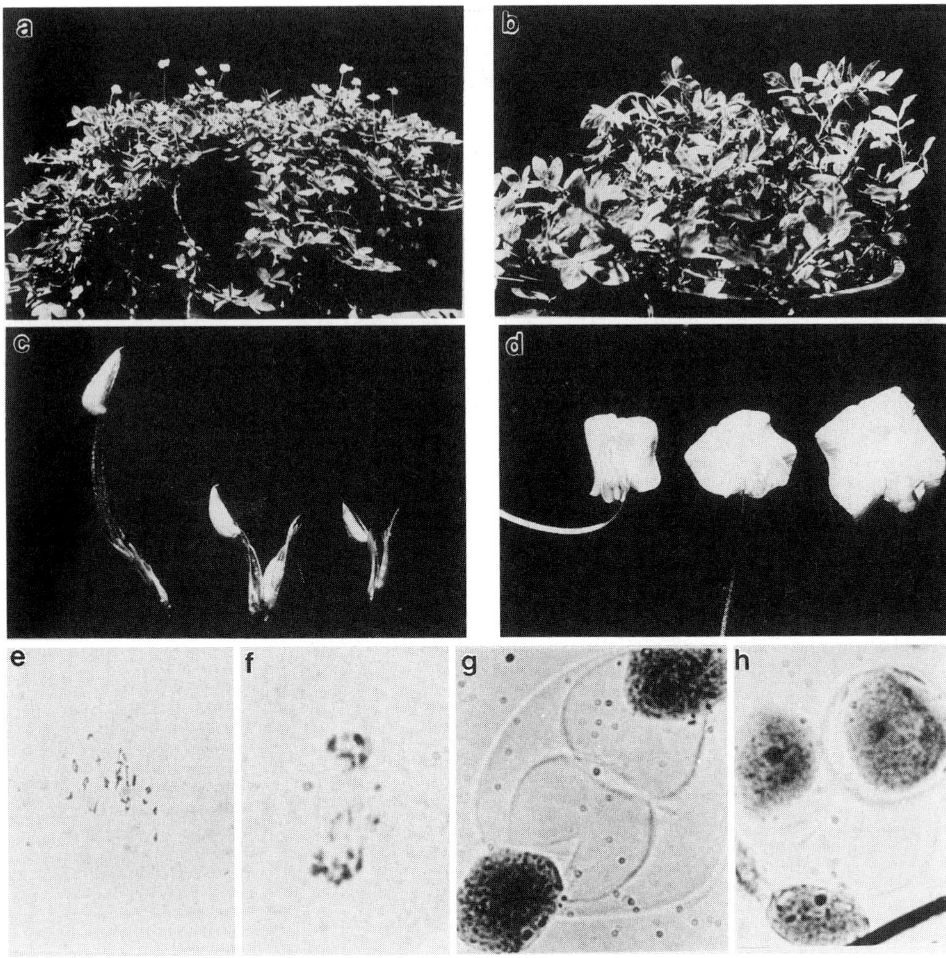

Figure 3. Interspecific hybrid between Arachis duranensis *and* A. glabrata. *(a) Wild species A. duranensis, used as the female parent. (b) F_1 hybrid* (A. duranensis × A. glabrata). *(c) Flower bud morphology of the parents* (A. glabrata *and* A. duranensis) *and the hybrid (center). (d) Flower morphology of the parents* (A. glabrata *and* A. duranensis) *and the hybrid (center). (e) Meiotic metaphase of the hybrid. Note the presence of univalents. (f) Anaphase with laggards. (g) A diad. (h) A triad.*

since the segregants flower early, they may escape heat and drought stress. Jaiswal *et al.* (1986, 1987) were able to obtain broad variation in yield and yield-related traits in F_2, F_3 and F_4 progenies from the cross *C. arietinum* × *C. reticulatum*.

Singh *et al.* (1994) isolated progenies from the cross *C. arietinum* × *C. echinospermum* phenotypically similar to the cultigen, with better agronomic attributes such as 100-seed weight, plant height and biological yield. The introgression of resistance to cyst nematode from *C. reticulatum* into chickpea (Di Vito *et al.*, 1996) is being evaluated in yield trials at the International Center for Agricultural Research in Dry Areas (ICARDA), Syria.

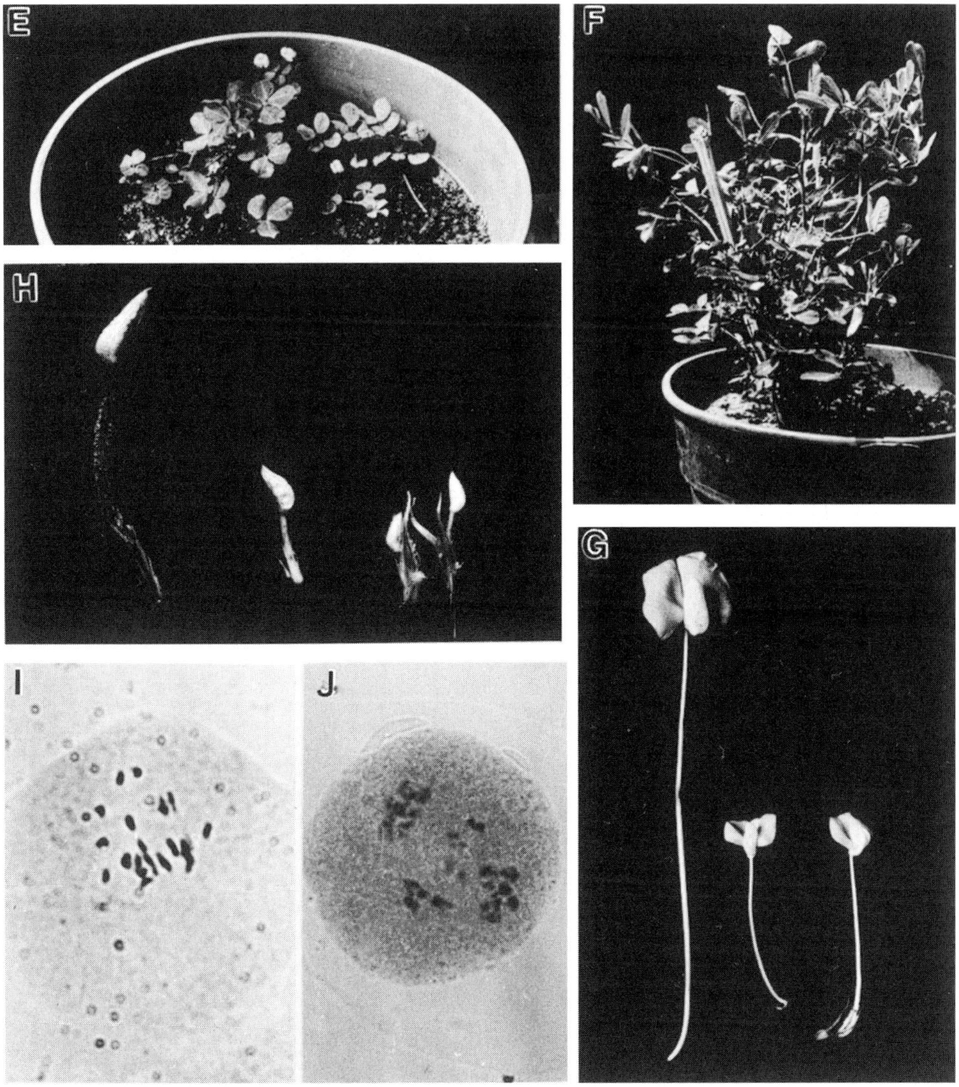

Figure 4. Interspecific hybrid between Arachis chacoensis *and* A. glabrata. *(E) Wild species* A. chacoensis, *the female parent of the cross (F)* F_1 *hybrid between* A. chacoensis *and* A. glabrata. *(G) Flower morphology of the parents and the hybrid (center). (H) Flower bud morphology of the parents and the hybrid (center). (I) Meiotic metaphase of the hybrid. Note the presence of univalents. (J) Hybrid anaphase. Note the presence of laggards.*

The wild species in the tertiary gene pool possess important and desirable genes for improving chickpea. Naik (1993) reported crossing *C. arietinum* with *C. pinnatifidum* (a member of the tertiary gene pool) and obtained shriveled seeds which upon germination gave rise to albino seedlings, that did not survive beyond 20 days of growth. The barriers

Table 3. Presence of useful traits in wild Cicer *species*

Species	Wilt	S.B. fungi	B. grey mold	Asco. blight	Cyst nema-tode	Leaf miner	Bru-chid	High prot-ein	Cold	Multi. seed	Twin pod
C. reticulatum	+				+	+		+	+		
C. echinospermum	+					+			+		+
C. pinnatifidum	+	+		+	+	+			+		+
C. bijugum	+	+	+	+	+	+	+	+	+		+
C. judaicum	+	+		+		+			+		
C. cuneatum	+	+				+				+	+
C. chorassanicum	+					+					
C. oxyden											+
C. canariense	+										
C. microphyllum											+
C. monbretii				+						+	
C. oxyden											
C. soongaricum											+
C. anatolicum											+

S.B. fungi = soil borne fungi, Asco. blight = Ascochyta blight, + = resistance, and B. grey mould = Botrytis grey mould.

to hybridization were studied and embryo rescue technique for chickpea was developed (Mallikarjuna, 1999). It was possible to save aborting hybrid embryos from the cross *C. arietinum* × *C. pinnatifidum* by *in-ovulo* embryo culture. Some of the hybrid shoots were pale yellow in color and scanning electron microscopy (SEM) studies showed that the chloroplasts were abnormal. It was observed that the use of a cytokinin in culture medium in combination with light helped the conversion of plastids to chloroplasts (Badami *et al.*, 1994). The hybrids between *C. arietinum* and *C. pinnatifidum* were fragile with the leaves resembling those of *C. pinnatifidum*. The color of the flower was pale violet resembling the violet color of the male parent. The hybrids were 100% pollen-sterile (Mallikarjuna, 1999). With the information on barriers to hybridization and the availability of a reliable embryo rescue technique, other annual and perennial wild species of *Cicer* (Mallikarjuna, 2001) may be amenable to gene transfer through wide-hybridization.

4. Pigeonpea

Pigeonpea, *Cajanus cajan* (L.) Millsp., belongs to the subtribe *Cajaninae*, tribe *Phaseoloideae*, subfamily *Papilionoideae* and family *Leguminosae*. Pigeonpea is a perennial shrub, but is usually grown as an annual crop. Pigeonpea originated in India and later spread to Africa. The revised genus *Cajanus* now comprises 32 species, with 18 species distributed in Asia, 15 in Australia and one in Africa. ICRISAT genebank holds 555 wild species accessions belonging to six genera and 20 species. Many of the wild species are a storehouse of useful genes (Table 4).

Table 4. Presence of useful traits in wild species of pigeonpea (Cajanus cajan)

Trait	*C. albicans*	*C. acutifolius*	*C. lineatus*	*C. platycarpus*	*C. scaraboides*	*C. sericeus*	*C. volubilis*	*R. rothii*
Early flowering				+				+
Photo insensitive				+				+
Drought resistance	+		+		+	+		+
SMV resistance	+		+	+		+	+	
Phytophthora blight				+				
High seed protein			+		+			
Pod borer resistance		+		+	+			
Salinity tolerance				+				

SMV = sterility mosiac virus.

Cajanus cajanifolius is the probable progenitor of cultivated pigeonpea. According to van der Maesen (1986), cultivated pigeonpea and *C. cajanifolius* belong to the primary gene pool. The only difference between the two species is the presence of strophiole in *C. cajanifolius*. The species readily crosses with cultivated pigeonpea. On the basis of the success of hybridization, the species in the secondary gene pool are interfertile with pigeonpea. Those species, which do not readily cross with pigeonpea are placed in the tertiary gene pool.

4.1. INTERSPECIFIC HYBRIDIZATION

An intergeneric cross between *C. cajan* and *Cajanus lineatus* contributed partial cleistogamic character to pigeonpea (Saxena *et al.*, 1992). Agronomically superior dwarf varieties were developed by the contribution of dwarfing genes from *Cajanus scaraboides* (Saxena *et al.*, 1987). High protein lines were isolated from the interspecific cross *C. cajan* × *C. scaraboides* (ICRISAT, 1987). *Cajanus sericius* and *C. scaraboides* have been used as female parents to develop cytoplasmic male sterile system in pigeonpea (Ariyanayagam *et al.*, 1995; Saxena and Kumar, 1999).

Several reports on the success of hybridization between pigeonpea and compatible wild species in the secondary gene pool are available (Dundas, 1990). However, except for *C. sericius, C. lineatus, C. albicans* and *C. scaraboides*, none of the other wild species in the secondary gene pool have been used in the improvement of pigeonpea. Some of the wild species in the tertiary gene pool of pigeonpea are the only source of disease resistance genes. One such example is *Cajanus platycarpus*. Resistance to phytophthora blight disease is only present in two accessions of *C. platycarpus* (Reddy *et al.*, 1996). *C. platycarpus* cannot be crossed with pigeonpea by conventional hybridization techniques. Barriers to hybridization were identified (Fig. 5) and techniques were developed to overcome them (Mallikarjuna and Moss, 1985; Mallikarjuna, 1999). Hybrids between *C. platycarpus* and *C. cajan* at F_2 and F_3 generations were screened for resistance to phytophthora blight disease (Fig. 6). It was observed that resistance present in *C. platycarpus* is governed by a recessive gene (N. Mallikarjuna, unpublished data). The embryo rescue and tissue culture techniques developed for pigeonpea have now opened new vistas to access genes from other wild species in the tertiary gene pool.

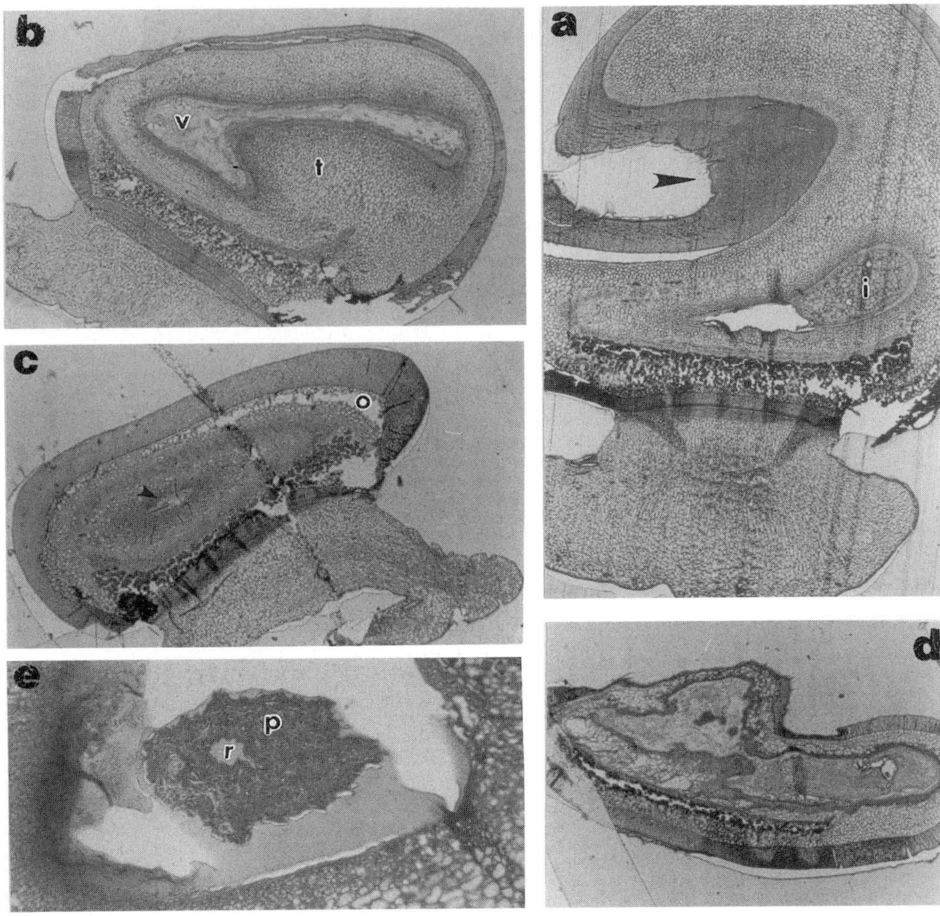

Figure 5. Hybrid embryo abortion. (a) Cross section of the ovule after 18 days after pollination (DAP) showing hybrid embryo (i) abortion. Arrow points towards the region of multilayer scleroid formation (×16). (b) Section of the ovule after 18 DAP, showing divisions in the inner integuement (t) resulting in narrowing of the embryo cavity (v) (×20). (c) Section of the ovule after 34 DAP showing crushed embryo cavity (arrow) and degeneration of the inner integuement (o) (×20). (d) Section of the ovule after 34 DAP showing total collapse of the ovule walls (×20). (e) Section of the ovule after 25 DAP showing abnormal growth and collapse of the inner integuement (p) and crushing the embryo cavity (r) (×46).

The major biotic constraint to pigeonpea production is pod borer (*Helicoverpa armigera*) attack, which causes up to 100% yield loss. In Andhra Pradesh, the districts of Warangal and Guntur have experienced the greatest havoc caused by pod borers in the farmers' fields of pigeonpea. The results of screening more than 10,000 lines have demonstrated that the limited amount of tolerance identified in cultivated pigeonpea towards *Helicoverpa* breaks down under intense insect pressure. Evaluation of the available wild germplasm of pigeonpea for resistance to *Helicoverpa* has shown

Figure 6. Phytophthora blight screening of the parents and the hybrids with P3 race. (a) Wild species Cajanus platycarpus, *the female parent of the cross. The plant is healthy after screening. (b) Cultivated pigeonpea* Cajanus cajan, *the male parent of the cross, has withered after screening. (c) Hybrid plants resistant to phytophthora blight screening. (d) Susceptible hybrids.*

that *C. scaraboides, C. platycarpus* and *C. acutifolius* have resistance to the insect pest. These wild species are rarely damaged, although the insect sometimes feeds on them under no-choice conditions. The nature of resistance present in *C. scaraboides* is largely due to the presence of non-glandular trichomes on the pod walls. The resistance present in *C. platycarpus* may be due to antibiosis against *Helicoverpa* as non-glandular hairs are not present on the pods of *C. platycarpus*.

5. Marker assisted gene introgression from wild species

Interspecific gene transfer, conferring resistance to pests and pathogens is a complex and lengthy procedure. The traits of interest are often difficult to identify in the segregating populations. It may not be possible to screen for the transfer of a gene from the wild species because the trait associated with the gene may not occur in that region. For example, rosette disease of groundnut does not occur in India, but is a major constraint in Africa. The development of alternate methods that would allow the characterization of the derivatives would be advantageous. The identification of genetic markers would improve the precision and speed of gene introgression. Molecular markers based on PCR technology would be an effective tool for early selection of desirable derivatives from interspecific hybridization.

Lanham *et al.* (1992) demonstrated the potential of RAPD markers in characterization of a synthetic amphiploid between *A. hypogaea* and the wild species hybrid *A. batizocoi* × *A. duranensis*. They were able to demonstrate the introgression of DNA from the hybrid *A. batizocoi* × *A. duranensis* into *A. hypogaea*. Halward *et al.* (1992) used RAPD markers to study evolutionary relationships of the wild species of through RFLP markers. Garcia *et al.* (1996) identified RAPD, SCAR and RFLP markers linked to nematode-resistance genes introgressed from *A. cardenasii* into *A. hypogaea*.

Mallikarjuna (2000) used RAPD markers to identify introgression of genes from wild species *A. glabrata* into *A. hypogaea*, and also to identify the success of the crosses between *A. chacoense* and *A. glabrata*, and *A. duranensis* and *A. glabrata* (N. Mallikarjuna, unpublished data). RAPD marker was used to identify introgression of DNA from *C. platycarpus* into *C. cajan* (Fig. 7). A linkage map was constructed using the cross *C. arietinum* × *C. reticulatum* because Simon and Muelhbauer *et al.* (1987) found that there was not enough polymorphism in the cultivated species. The application of molecular and bioinformatic tools for genetic enhancement will speed up the improvement of legumes.

Advances in molecular biology coupled with bioinformatics promise improved utilization and enhancement of wild species. Recent advances in information on entire genome of humans, microbes, model animal and plant species has had a great impact on plant breeding.

6. Critical gaps in wild species utilization

The gaps and problems in the utilization of wild species for crop improvement are the lack of information on useful traits present in them. Care must taken to prevent the extinction of

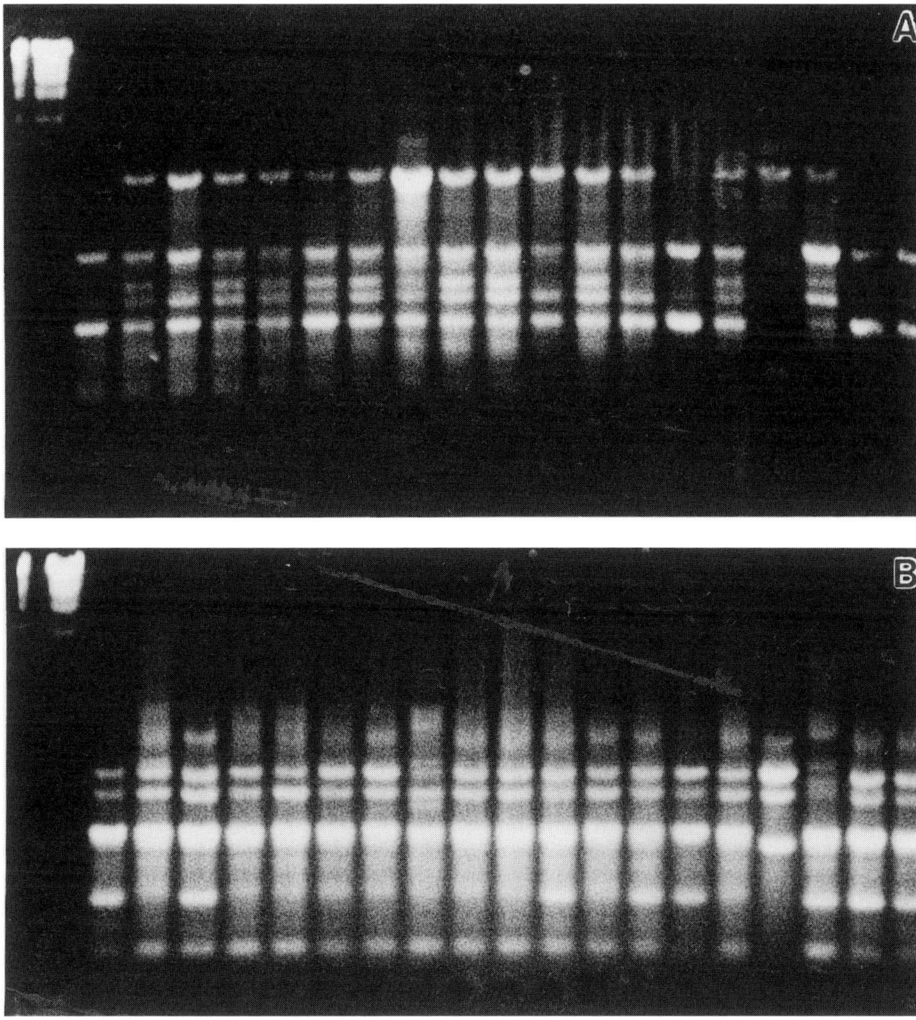

Figure 7. RAPD-DNA profile of the parents Cajanus platycarpus *and* C. cajan *and the hybrids* (C. platycarpus × C. cajan). *(A) RAPD-DNA profile with Operon primer OPN06 (lane 17 = female parent, lane 18 = male parent, lane 1 = molecular weight marker, rest of the lanes = hybrids). (B) RAPD-DNA profile with Operon primer OPN04 (lane 17 = female parent, lane 18 = male parent, lane 1 = molecular weight marker, rest of the lanes = hybrids).*

the wild species due to biotic or abiotic factors, or natural disasters. The nature of resistance present in the wild species, whether it is a physical factor or a chemical substance conferring resistance is another important factor in wild species utilization. The knowledge about barriers to hybridization plays a key role in the utilization of wild species to overcome biotic and abiotic constraints. Information on embryo rescue and tissue culture techniques will determine the success of using incompatible wild species. Restricted recombination and linkage drag can limit the use of wild species. Ploidy differences between the wild species and the crop of interest, and sometimes unknown causes hamper the success of utilization.

Gene transfer from wild species to cultivated crops is time consuming, but a rewarding program. Commitment in terms of financial as well as human resources are essential for the success of the program.

References

Ariyanayagam R P, Nageshwar Rao A and Zaveri P P (1995) Cytoplasmic genic male sterility in interspecific matings of pigeonpea. *Crop Sci.*, **35**: 981–985.

Badami P S, Mallikarjuna N and Moss J P (1997) Interspecific hybridization between *Cicer arietinum* and *C. pinnatifidum. Plant Breed.*, **116**: 393–395.

Di Vito M, Singh K B, Greco N and Saxena M C (1996) Sources of resistance to cyst nematode in cultivated and wild *Cicer* sps. *Genet. Resources Crop Evolut.*, **43**: 103–107.

Garcia G M, Stalker H T, Shroeder E and Kochert G (1996) Identification of RAPD, SCAR, and RFLP markers tightly linked to nematode resistance genes introgresses from *Arachis cardenasii* into *A. hypogaea. Genome*, **39**: 836–845.

Halward T A, Stalker S, Larue E and Kochert G (1992) Use of single primer DNA amplification in genetic studies of peanut (*Arachis hypogaea* L.). *Plant Mol. Biol.*, **18**: 315–325.

ICRISAT (International Crops Research Institute for the Semi-Arid Tropics) (1987) In: *Annual Report 1986*. ICRISAT, Patancheru, India, pp. 195–196.

Jaiswal H K, Singh B D and Singh R M (1987) Improvement of chickpea (*Cicer arietinum*) through introgression of genes from *C. reticulatum. Indian J. Agric. Sci.*, **57**: 880–883.

Jaiswal H K, Singh B D, Singh A K and Singh R M (1986) Introgression of genes for yield and yield traits from *C. reticulatum* to *C. arietinum. Int. Chickpea Newslett.*, **14**: 5–8.

Krapovikas A and Gregory W C (1994) Taxonomia del genero *Arachis* (Leguminosae). *Bonplandia*, **8**: 1–86.

Ladizinsky G and Alder A (1976) Genetic relationship among the annual species of *Cicer* L. *Proc. Indian Natl. Sci. Acad.*, **41**: 78–82.

Lanham P G, Fennel S, Moss J P and Powell W (1992) Detection of polymorphic loci in *Arachis* germplasm using random amplified polymorphic DNAs. *Genome*, **35**: 885–889.

Mallikarjuna N (1998) Ovule culture to rescue aborting embryos from pigeonpea (*Cajanus cajan* (L.) Millspaugh) wide crosses. *Indian J. Exp. Biol.*, **36**: 225–228.

Mallikarjuna N (1999) Ovule and embryo culture to obtain hybrids from interspecific incompatible pollinations in chickpea. *Euphytica*, **110**: 1–6.

Mallikarjuna N (2001) Use of markers for the detection of gene introgression from *Arachis glabrata* into *A. hypogaea, A. duranensis* and *A. diogoi. Euphytica* (in press).

Mallikarjuna N (2001) Prospects of using perennial wild species *C. canariense* for chickpea improvement. *Int. Chickpea Pigeonpea Newslett.*, **8** (accepted for publication).

Mallikarjuna N and Moss J P (1995) Production of hybrids between *C. platycarpus* and *C. cajan. Euphytica*, **83**: 43–46.

Mallikarjuna N and Sastri D C (1985a) *In vitro* culture of ovules and embryos from some incompatible interspecific crosses in the genus *Arachis* L. In: *Cytogenetics of* Arachis. *Proceedings of International Workshop on Cytogenetics of* Arachis (Ed Moss J P), International Crops Research Institute for Semi-Arid Tropics, Patancheru India, pp. 135–138.

Mallikarjuna N and Sastri D C (1985b) Utilization of incompatible species in *Arachis*: Sequential hormone applications. In: *Cytogenetics of* Arachis*: Proceedings of International Workshop on Cytogenetics of* Arachis (Ed Moss J P), International Crops Research Institute for the Semi-Arid Tropics, Patancheru, India, pp. 144–152.

Naik S V (1993) Interspecific hybridization and characterization of hybrids in the genus *Cicer* L. M.Sc. thesis, Indira Gandhi Krishi Vishwavidhyalaya, Raipur, Madhya Pradesh, India.

Pundir R P S and Mengesha M H (1995). Cross compatibility between chickpea and its wild relative *C. echinispermum* Davis. *Euphytica*, **83**: 241–245.

Reddy M V, Raju T N and Sheila V K (1996) Phytophthora blight disease in wild pigeonpea. *Int. Chickpea Pigeonpea Newslett.*, **3**: 52–53.

Saxena K B, Ariyanayagam R P and Reddy L J (1992) Genetics of a high selfing trait in pigeonpea. *Euphytica*, **59**: 125–127.

Saxena K B and Kumar R V (1999) Development of cytoplasmic male sterility in pigeonpea. Progress Report. Genetic Resources and Enhancement Program, International Crops Research Institute for Semi-Arid Tropics, Patancheru, India.

Saxena K B, Singh M D and Singh U (1987) Can seed size and protein content in pigeonpea be increased simultaneously? *Int. Pigeonpea Newslett.*, **6**: 29–31.

Shen F W, Wang C T and Duan S F (1995) Aseptic culture of gynophores to obtain peanut intersectional hybrids. *Euphytica*, **81**: 245–249.

Simon C J and Muehlbauer F J (1997) Construction of a chickpea linkage map and its comparison with maps of pea and lentil. *J. Hered.*, **88**: 115–119.

Singh A K and Moss J P (1984) Utilization of wild relatives in genetic improvement of *Arachis hypogaea* L.V. Genome analysis in section *Arachis* and its implications in gene transfer. *Theort. Appl. Genet.*, **68**: 1–10.

Singh B D, Jaiswal H K, Singh R M and Singh A K (1984) Isolation of early-flowering recombinants from interspecific cross between *Cicer arietinum* and *C. reticulatum. Int. Chickpea Newslett.*, **11**: 14–16.

Smartt T and Stalker H T (1982) Speciation and cytogenetics in *Arachis*. In: *Peanut Science and Technology* (Eds Pattee H E and Young C T), American Peanut Research and Education Society, Yoakum, Texas, USA, pp. 21–49.

Stalker H T and Simpson C E (1995) Germplasm Resources in *Arachis*. In: *Advances in Peanut Science* (Eds Patte H E and Stalker H T), American Peanut Research and Education Society, Inc., Stillwater, OK, USA, pp. 14–53.

Van der Maesen L J G (1986) *Cajanus* DC. and *Atylosia* W. and A. (*Leguminosae*). Agricultural University Wageningen papers 85-4 (1985). Wageningen, The Netherlands Agricultural University, p. 225.

Van der Maesen L J G (1987) Origin, history and taxonomy of chickpea. In: *The Chickpea* (Eds Saxena M C and Singh K B), CAB International Publishers, Wallingford, Oxon, UK, pp. 11–34.

Verma M M, Sandhu J S, Brar H S and Brar J S (1990) Crossability studies in different species of *Cicer* (L.). *Crop Improv.*, **17**: 179–181.

PART II
NITROGEN METABOLISM

PERCEPTION AND PROCESSING OF NOD FACTOR SIGNALS

HUBERT H. FELLE
Botanisches Institut I, Senckenbergstraße 17,
Justus-Liebig-Universität, D-35390 Gießen, Germany
e-mail: Hubert.Felle@bio.uni-giessen.de

Abstract

Plant responses to Nod factors comprise a variety of processes that affect membrane transporters, the apoplastic milieu, metabolism, phytohormone balance and gene expression. While some of these responses occur in seconds after the first encounter of the root (hair) plasma membrane with the appropriate Nod factor, others need minutes, hours or days before becoming detectable. Following Nod factor perception by one or more putative plasma membrane receptors, activation of Ca^{2+} channels is the earliest response observed so far. The resulting Ca^{2+} influx stimulates anion channels which, giving rise to Cl^- release, depolarizes the cells and, by releasing organic anions, alkalizes the external space. Although these effects have not been found with all such investigated plant species, a function with respect to permitting the symbiotic rhizobial bacteria access to the root hair symplast appears likely. The conspicuous root hair deformation seems to be caused by a perturbation in apical free Ca^{2+} which causes both a rapid breakdown of the root hair cytoskeleton followed by its rearrangement. Ca^{2+} spiking in the nucleous area, that occurs with a 10 min delay, may be connected with gene activation leading to the early nodulin gene expression in connection with plant development and nodule formation. Although Nod factors (without rhizobia being present) induce the formation of nodules in certain legumes, they may not be the primary signal. As, due to an imbalance of phytohormones (auxin and cytokinin), nodules may spontaneously form, gene products like ENOD40 are very likely the trigger or regulator of cortical cell divisions.

1. Introduction

Species of *Rhizobium*, *Bradyrhizobium* and *Azorhizobium* (collectively known as rhizobia) enter into a mutualistic association with legumes in which the bacteria provide reduced nitrogen to the plant and the plant provides carbon and energy to the rhizobia. This association is initiated when plant flavonoids from the legume root and seed coat serve as signals to induce expression of the nodulation genes of rhizobia in the immediate

P.K. Jaiwal and R.P. Singh (eds.), Improvement Strategies for Leguminosae Biotechnology, 175–189.

rhizosphere. In turn the rhizobia synthesize a class of lipochito-oligosaccharides which interact with the host legume, commonly known as Nod factors. These signal molecules are tetra- and pentalipo-oligosaccharides of N-acetylglucosamine and induce specific responses in the epidermis as well as in the cortex of legume roots at nano- or even at picomolar concentrations. Although Nod factor signalling is rather complex, evidence is beginning to emerge that Nod factors trigger a variety of responses which become effective at different stages of the symbiotic interaction between rhizobia and plant. Apart from the rather diverse effects on the tissue, cellular or subcellular levels, the Nod factor responses occur at distinctly different time intervals. This raises the questions to what extent Nod factors are primary signal molecules and what are the transducing elements? Nod factor responses may be classed into basic processes:

- Changes in plasma membrane properties, which include activation changes of a number of membrane transporters, probably serve to enable the symbiotic bacteria to cross the membrane barrier.
- Root hair curling, leading to the formation of pockets, serves to harbour the symbiotic bacteria.
- Gene activation.
- Induction of cortical cell divisions, leading to the formation of nodules in which nitrogen is fixed by the bacteroids.

2. Early events

Here, early events will be referred to responses that occur within seconds or at most within a few minutes following encounter of the root hair surface either with the bacteria or with the appropriate Nod factor alone. Following Nod factor perception these are changes in membrane potential, ion fluxes across the plasma membrane, cytosolic changes in free [Ca^{2+}], pH and [Cl^-] and other intracellular events that process the Nod factor signal, such as the activation of G-proteins, protein kinases and protein phosphatases, as summarized in Fig. 1.

3. Nod factor perception

Although chitin fragments from fungal cell walls are known elicitors of defense reactions in plants, molecules with a lipid chain attached to an oligosaccharide backbone have not been conclusively shown either to be endogenous signalling molecules in plants or to be produced by other interacting organisms (Hahn, 1996). Nod factors are amphiphilic molecules and seemingly exist as monomers in aqueous solution at least up to 10^{-8} M, rather than forming micelles through their hydrophobic acyl chains. These molecules will rapidly insert into membranes, but do not flip-flop between membrane leaflets and therefore will not spontaneously enter the cell without specific perception and transport mechanism. Thus, their initial perception is at or outside the plasma membrane and might involve recognition either from the aqueous rhizosphere or directly from bacterial membranes attached to the root hair. In the latter case the definition of a concentration is difficult, but has been calculated to be around 5 nM (Goedhart *et al.*, 2000). Moreover, the capacity of cell wall

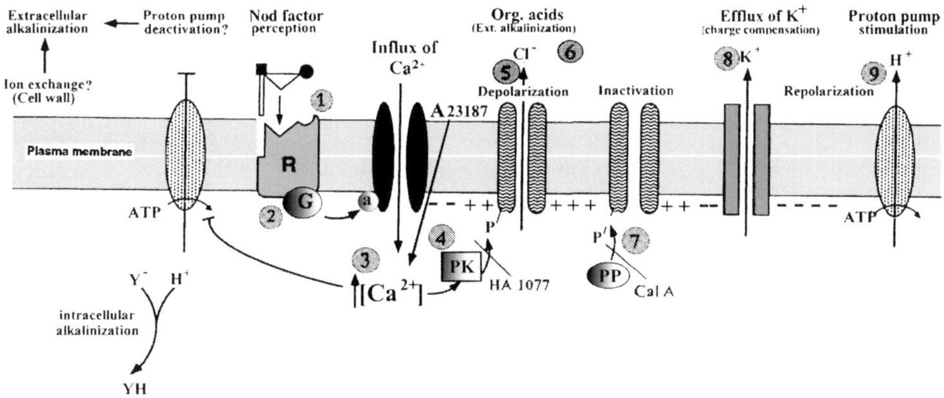

Figure 1. Model of the early steps in Nod factor signalling in alfalfa. (1) Nod factor perception; (2) activation of G-protein (G); (3) activation of Ca^{2+} channel and elevation of cytosolic free $[Ca^{2+}]$; (4) activation of protein kinase (PK) and anion channel phosphorylation (P), inhibited by HA 1077; (5) efflux of Cl^- and depolarization; (6) release of organic acid anions and external alkalinization; (7) inactivation of anion channel through dephosphorylation by protein phosphatases (PP); (8) efflux of K^+ and charge compensation; (9) pump stimulation and repolarization.

binding of Nod factors seems rather high, and could result in their accumulation up to 50 times over the Nod factor concentration of the surrounding media. This binding could be important for increasing the efficiency of Nod factor perception in case of low ambient concentrations. In fact, it has been shown that the lag phase after which Nod factor-induced membrane depolarization (Ehrhardt *et al.*, 1992) occurs is roughly constant at 10 to 15 s in the presence of Nod factor concentrations between 10^{-7} and 10^{-9} M, but increases rapidly with decreasing concentrations (Felle *et al.*, 1995). A variety of approaches are presently being used to identify the molecular components of Nod factor perception. Perhaps the most direct chemical evidence for the existence of a Nod factor receptor in legumes is the characterization of two different Nod factor binding sites (NFBS1 and NFBS2) in *Medicago truncatula* (Bono *et al.*, 1995; Gresent *et al.*, 1999; Cullimore *et al.*, 2001). NFBS2 reveals an affinity of 4 nM to Nod factors which would be in good aggreement with the value of 1 to 2 nM observed in *Medicago sativa* depolarization tests (Felle *et al.*, 1995), the latter of which, however, would have to rate as functional perception.

3.1. FUNCTIONAL PERCEPTION

In want of unambiguously identified Nod factor receptors at the plasma membrane, perception studies so far have largely relied on a functional analysis, i.e. Nod factor perception turns on a signal chain, and one of the responses following within or at the end of the chain(s) is recorded and brought into connection with the initial stimulus. Clearly, this approach implies some caveats which renders the interpretation of actual Nod factor binding to the putative receptor(s) difficult. For instance, the depolarization response occurs typically 15 to 20 s after adding the Nod factor, a delay which in terms of signalling represents a long time during which a series of processes must have taken place. In spite of these uncertainties, apparent correlations of both Nod factor concentration and Nod factor decoration with the depolarization response have been demonstrated revealing specificities, which indicate a quantitative processing of the binding/recognition signal (see below). The concentration dependent lag-phase, however, cannot be attributed to Nod factor/ receptor interaction, but may either indicate a pool-dependent processing of the signal or, as mentioned above, a concentration step-up through accumulation of the Nod factor within the cell wall prior to plasma membrane binding by a receptor.

Although because of the above mentioned reasons Nod factor binding analysis is still in its infancy, one can show that oligochitins, structurally related to Nod factors, very likely do not bind to the same receptor, but nevertheless interfere with the Nod factor responses, depending on the mode of addition (jointly or successively). When added successively, responses (depolarization, external pH changes, Cl^--efflux) are inhibited to some extent, but are enhanced, when the two compounds are added at the same time (Felle *et al.*, 2000). Whereas the latter effect clearly rules out a joint perception site for Nod factors and oligochitins, the interferences point to a physiological "pin hole" through which both signals may have to be transduced.

3.2. DIFFERENT RECEPTORS FOR ENTRY AND FOR SIGNALLING?

Purified Nod factors are able to induce the formation of preinfection threads, but fail in the formation of complete infection threads. *R. meliloti* mutants (nodL$^-$ or nodFE$^-$) that produce Nod factors with structural deficiencies are able to induce root hair deformation and cortical cell activation as scored by the presence of starch grains, but are affected in their efficiency to initiate infection sites (Ardourel *et al.*, 1994). Based on this, it has been proposed that at the epidermis two types of Nod factor receptors should be present – a signalling receptor and an entry receptor. By this model, the signalling receptor would be less selective in its recognition of Nod factors, leading to initiation of inner cortical cell divisions, root hair tip growth, and root hair deformation, whereas the entry receptor would only recognize fully decorated Nod factors. Only upon triggering of the entry receptor are infection threads formed that will guide the bacteria towards the nodule primordium.

That different receptors may in fact be operational comes from studies of Nod factor-induced alkalinization. Felle *et al.* (1996) demonstrated that application of both the sulfated and the non-sulfated factor induced cytosolic alkalinization. In order to fully appreciate the importance of this finding, it should be mentioned that the sulfated factor strongly depolarizes while the non-sulfated factor up to 10^{-7} M had no effect. When both factors were added successively, both induced an alkalinization, regardless in which order of addition, indicating that the alkalinization responses were not just a consequence of differential binding of the Nod factors to the same receptor, but in fact of binding to different receptors.

3.3. SPECIFICITY OF NOD FACTOR RECOGNITION: ARE THE FAST RESPONSES REQUIRED FOR NODULATION?

Specificities of Nod factor recognition, functionally analyzed, have been found with several investigated measuring parameters: membrane depolarization (Felle *et al.*, 1995;

Kurkdjian, 1995), cytosolic alkalinization (Felle *et al.*, 1996), ion fluxes within the root hair space (Felle *et al.*, 1998a), and free Ca^{2+} (Erhardt *et al.*, 1996; Felle *et al.*, 1999a). In most cases (with the exception of cytosolic alkalinization) the responses follow a common pattern, namely that Nod factors which induce larger early responses also cause higher nodulation. As for alfalfa and their effective Nod factors, the hierarchy would be NodRm-IV(C16:2,S) = NodRm-IV(C16:2,S,Ac) > NodRm-IV(C16:0,S) > NodRm-V (C16:2,S) > NodRm-IV(C16:2).

Such an observation is amazing because perception and nodulation, with respect to signalling time, are "light years" apart from each other. Although there is a solid data base for alfalfa, such experiments have rarely been carried out for other symbiotic partners. In spite of the observation that in *Medicago sativa* these early responses seem somehow related to nodulation, preliminary experiments carried out on *Medicago truncatula* show less clear causal links.

Whereas *M. truncatula* bv. R108 essentially follows the pattern of alfalfa (although somewhat less pronounced), *M. truncatula* bv. *jemalong* and other *truncatula* biovars respond rather sluggishly to Nod factors, resisting so far a rigorous analysis and thus indicate that the apparently clear correlation of early responses and nodulation observed in alfalfa may not be a general principle (Felle *et al.*, 1998b). Still, it would not be justified to declass the observed correlations as unimportant "side effects" or even as "physiological artifacts". These effects are far too specific not to have a meaning in Nod factor signalling. The weak correlations with *Medicago truncatula* could mean that our instrumentary of investigation is too crude to visualize the molecular events, which for some reason appear more strongly bundled in *M. sativa* than in *M. truncatula*. The general approach of testing Nod factors, namely to add them to the entire root system, which is certainly not what happens in nature, may be one clue to that problem. It could be that in *M. truncatula* bv. *jemalong* not as many cells are susceptible to Nod factors than in alfalfa. In that case, ion fluxes within the root hair space in *M. truncatula jemalong* would not be measured as massive as observed in alfalfa. It could also be that ion fluxes are rapidly masked by charge compensation. Basically, however, it is not the measurable quantity of the Nod factor response that matters for Nod factor signalling, it is the local perturbance at a small membrane area on the root hair which probably triggers the subsequent signalling sequences.

4. Processing of the signal

It appears that the Nod factor signal is processed through different pathways, as also indicated by the different responses in cytosolic free [Ca^{2+}] (see below). There is some evidence that activation of G-proteins takes place; an involvement of the phosphoinositide pathway is currently discussed. Obviously, part of the information following Nod factor perception goes to the activation of plasma membrane transporters, leading to the early responses described and discussed above. More information goes to the genome to activate specific genes involved in the expression of products which regulate the installment of symbiotic specific changes in root morphology and physiology. Apart from the Ca^{2+}-spiking which may be involved in such processes, not much is known about the mechanisms by which the Nod factor signal is transferred from the site of perception to the formation of the nodules.

4.1. G-PROTEIN ACTIVATION/PHOSPHOINOSITIDE PATHWAY

Relatively little information is available with regard to the involvement of G-proteins in Nod factor signalling. The only solid evidence in favour of G-protein action comes from Pingret *et al.* (1998) who investigated Nod factor signal transduction in the *Medicago* root epidermis by using a pharmacological approach in conjunction with transgenic plants expressing the Nod factor-responsive reporter construct pMtENOD12-GUS. Evidence for the participitation of heterotrimeric G-proteins in Nod factor signalling comes from three complementary observations: (1) the amphiphilic peptides mastoparan and Mas7, known G-protein agonists, are able to mimic Nod factor-induced epidermal MtENOD12 expression; (2) growth of plants in nodulation-inhibiting conditions (i.e. 10 mM NH_4NO_3) leads to a dramatic reduction in both Nod factor- and mastoparan-elicited gene expression; (3) bacterial pertussis toxin, a well characterized G-protein antagonist, blocks the activities of both Nod factor and mastoparan. Evidence from the same laboratory has been provided for phosphoinositide signalling. Antagonists that interfere with phospholipase C acivity (neomycin and U73122) block Nod factor/mastoparan activity. These results are supported by preliminary data from tests on *Medicago sativa*, where it was shown that mastoparan, added prior to the Nod factor, effectively blocked the Nod factor-induced external alkalinization (H. H. Felle and coworkers, unpublished data).

4.2. PROTEIN KINASES/MAP-KINASES

Although it has been demonstrated that elicitor signalling seemingly involves protein phosphorylation (Grosskopf *et al.*, 1991; Felix *et al.*, 1991; Dietrich *et al.*, 1990; Viard *et al.*, 1994) and MAP-kinase activation (Cardinale *et al.*, 2000), not much information exists with respect to Nod factors. Recently, we could demonstrate that the protein kinase inhibitors K252a and HA1077 most effectively inhibited Nod factor-induced depolarization and external alkalinization (H. H. Felle and coworkers, unpublished data), but so far were unable to prove that these effects were in fact direct links and not due to disturbances somewhere else in the cell.

4.3. CALCIUM

Calcium obviously plays an important role in Nod factor signalling. Although changes in free cytosolic [Ca^{2+}] in response to externally applicated Nod factors have been reported from different laboratories, the diversity of the data creates great problems in their interpretation as well as in formulating a general theory. Erhardt *et al.* (1996), using Calcium-green as fluorescent probe, report Ca^{2+} spiking in alfalfa which occurs mainly in the nucleus area roughly 10 min after Nod factor addition. The effect is specific and does not occur in Nod mutants or nonactive Nod factors. The calcium spiking was extended recently to do a genetic analysis on nodulation mutants of *Medicago truncatula* (Wais *et al.*, 2000). It was shown that mutants which are completely Nod$^-$ (blocked at early stages of the symbiosis) show a strongly reduced Ca^{2+} response to Nod factors only downstream of *DMI1* and *DMI2*, required in common for early steps of infection and nodulation and for Ca^{2+} spiking, but upstream of *DMI3*. Since all three genes apparently are responsible for the root hair swelling, an unequivocal conclusion was not formulated. Nevertheless, the approach to use calcium spiking for the investigation of Nod signalling is promising and was also used in pea. Walker *et al.* (2000) demonstrated that root hairs of pea responded to Nod factors with Ca^{2+} spiking that was similar to the one reported from the Long-group. Within 1 to 2 min after Nod factor addition, an increase in fluorescence was observed, followed about 10 min later by spikes at the rate of 1 spike per min. As in *Medicago* the spikes were localized around the nuclear region. Mutations affecting loci *sym8*, *sym10*, *sym19* abolished Nod factor-induced calcium spiking, whereas a normal response was observed in peas carrying alleles of *sym2^A*, *sym7*, *sym9* and *sym30*. Chitin oligomers of four or five N-acetylglucosamine residues could also induce calcium spiking, although the response was qualitatively different from that induced by Nod factors.

De Ruijter *et al.* (1998), using acid loading of indo-1, showed that the apical free [Ca^{2+}] increased about 6-fold in root hairs of *Vicia sativa* after treatment with Nod factors. Since the studies were carried out on mature root hairs in which tip growth was re-initiated by the Nod factors, the increased tip Ca^{2+} was brought into connection with this. Unfortunately, since the first measurements of Ca^{2+} were made about 70 min after adding the Nod factor, conclusive signalling events within the time of Nod factor application and response could not be formulated.

On *Phaseolus vulgaris* root hairs, Cárdenas *et al.* (1999) demonstrate that Nod factors increase tip [Ca^{2+}] from 400 nM to 1500 or 1800 nM within 5 to 10 min after application and observe oscillatory Ca^{2+} changes around the nucleus 10 to 15 min after the external application of active Nod factors corresponding to the findings from the Long- and Downie-groups. They also report that in the apical region the Ca^{2+} hot spots travel, i.e. can disappear in one location and reappear in another location.

In electrophysiological studies on alfalfa root hairs, Felle *et al.* (1999a, b) further contribute to the rather complex view of the role of Ca^{2+} in Nod factor action. They demonstrate that within 1 to 2 min after Nod factor addition the free [Ca^{2+}] drops in growing root hairs, but increases in non-growing root hairs (which form no tip-to-base [Ca^{2+}] gradient) as well as in general outside of the tip region. These effects have been shown to be Nod factor specific, essentially corresponding with the depolarization experiments. It was demonstrated in the same group that the ion-fluxes, discussed above, crucially depended on an increase in cytosolic free [Ca^{2+}], but that this would not have to involve Nod factors specifically, i.e. any increase in cytosolic [Ca^{2+}] will probably lead to (anion-)channel activation. As such, it has been demonstrated that Nod factor action can be mimicked through bypassing Nod factor perception using the Ca^{2+} ionophore A23187 to force Ca^{2+} into the cytosol, and also by using BHQ, a potent inhibitor of endomembrane Ca^{2+} ATPases in mammalian cells, which mobilizes Ca^{2+} from IP3-sensitive stores.

It appears that there are at least three different Ca^{2+} effects in Nod factor signalling:

(1) One that is involved in tip growth, probably also connected with cytoskeletal changes, finally leading to the root hair deformation. There we find changes in cytosolic free Ca^{2+} to depend on the current physiological state of the root hair, i.e. whether Nod factors are perceived by growing or non-growing, young or mature root hairs. Since it has been demonstrated in tip growing cells in general and in root hairs in particular that tip growth is causally

linked to elevated apical free [Ca^{2+}], root hair deformation is very likely caused by locally differing cytosolic [Ca^{2+}], i.e. growth would be stimulated where [Ca^{2+}] is elevated, on the other hand would be stopped by lower [Ca^{2+}] on the far side. Such a differential effect on the growing root hair will lead to curling. *Rhizobia*, which attach to the root hair tip through excreting the Nod factors, might cause a local decrease in cytosolic [Ca^{2+}] which stops growth, whereas the far side remains unaffected and continues growing. Thus, the root hair would form a pocket to harbour the bacteria. On the other hand, Nod factors added to the medium, will affect the entire root hair with a different outcome. Since the direction of the stimulus is missing the root hairs deform at random without following a special pattern.

(2) Alterations in the membrane conductivity and permeability are required to permit the bacteria to overcome the membrane barrier. And thus some of the early ion fluxes are not side products, but may represent a necessary prerequisite for the interaction of symbiont and host. The induction of ion fluxes across the plasma membrane requires cytosolic [Ca^{2+}] to increase. This has been demonstrated in alfalfa by placing ion-selective microelectrodes in the root hair space. Felle *et al.* (1998a) observed Ca^{2+}-influx seconds after Nod factor application and underscored this finding by using the Ca^{2+} ionophore A23187, which produced an effect nearly identical to the application of the Nod factor. The idea that Nod factors induce a Ca^{2+} influx was supported by Cárdenas *et al.* (1999, 2000), who also demonstrated such Ca^{2+} influx using vibrating probe techniques. Further support for this idea comes from studies on soybean protoplasts (Yokoyama *et al.*, 2000), and from the rapid Ca^{2+} increases measured in *Vigna unguiculata* root hairs by Gehring *et al.* (1997).

(3) Since the Ca^{2+} spiking occurs around the nucleus mainly occurring with a much longer lag phase than the other Ca^{2+} responses, its origin must be different. Since external measurements did not reveal a rythmic Ca^{2+} decrease, Ca^{2+} spiking appears an entirely internal process. It is likely that this event is linked to gene activation. Although first screenings of mutants have been carried out, a clear attribution of the spiking to certain events in the symbiotic signal chain is yet to come (Walker *et al.*, 2000; Wais *et al.*, 2000).

4.4. EXTRACELLULAR AND INTRACELLULAR pH

An interesting Nod factor response is the extracellular alkalinization, which occurs transiently and is just as Nod factor-specific as for instance the membrane depolarization (Felle *et al.*, 1996). Interestingly, very similar effects have also been demonstrated in tomato suspension cells using both Nod factors or chitotetraose (Felix *et al.*, 1993). This indicates that the alkalinization response is not a typical symbiotic response, but could rather be typical of a defense reaction. This argumentation is weakened, however, by the observation that oligochitin elicitors alkalize the extracellular space of *Medicago sativa* to a minor extent (less than 0.1 pH unit) only, before a more substantial external acidification (0.2 pH units or more) takes place (Felle *et al.*, 2000). Baier *et al.* (1999) also contribute to the complex pH behaviour, in that they compared the responses of alfalfa and tobacco cell cultures to chitin oligosaccharides and Nod factors. They demonstrate that the pH change to these compounds is never transient and that the response to Nod factors may be an acidification, depending on the applied concentration. Taken together, the diversity of these data is not surprising because an (external) pH change may have many causes, such as changes in proton pump activity, movement of different (strong) ions in and out of the investigated compartment across a membrane or ion exchange processes within the cell wall. This makes the pH response rather complex and thus difficult to interpret from case to case. As such, the external pH change would not be a signal, but rather the result of underlying (transport) processes.

It is a widespread opinion that extracellular alkalinization should go along with an intracellular acidification. This, however, only holds if one looks at a single transmembrane process which involves protons, but not when metabolism or several intracellular compartments contribute to the cytosolic pH change. As such, the observation that Nod factors alkalize the cytosol, while at the same time they also alkalize the extracellular space (Felle *et al.*, 1996), is by no means a paradox, but reflects the fact that both alkalinizations come from entirely different sources. Whereas one can be confident that most of the external alkalinization comes from the efflux of organic acid anions (Felle *et al.*, 2000), it is uncertain where the cytosolic alkalinization arises from, but one could hypothesize that metabolic shifts due to the far-reaching symbiotic interactions may be the cause. Recently, it has been demonstrated that defense-related cytosolic pH changes in alfalfa were acidifications (Felle *et al.*, 2000) and, since there is evidence that cytosolic acidification is one of the prerequisites for activation of defense genes, it was argued that the Nod factor-induced alkalinization was an efficient block to prevent defense gene activation. Moreover, the simultaneous addition of Nod factors and oligochitins provided a highly interesting result, in that the first response was a cytosolic alkalinization, followed by a 10 to 15 min delayed acidification. It appears that there is a hierarchy in signalling: for as long as the cytosol remains more alkaline, defense genes are not activated, which buys time for the host to establish the symbiotic pathway(s).

4.5. THE NOD O PROTEIN

R. leguminosarium bv. *viciae* has a nodulation gene (*nodO*) that encodes a secreted protein involved in pea and vetch nodulation. Although mutation of *nodO* does not block nodulation, it is evident that in the absence of *nodE* gene, *nodO* is required for the nodulation of pea or vetch. The *nodE* gene product is a key determinant of host specificity and introduces functional specificity into the Nod factor by causing the production of the C18:4 acyl group. It appears that the role of NodO must be complementary to that of the Nod factors (Downie, 1994; Downie and Walker, 1999). The primary target of NodO is most likely to be at the plasma membrane of plant cells. NodO forms cation selective channels in planar lipid bilayers (Sutton *et al.*, 1994) and it was proposed that it may enhance nodulation by forming pores (Dalla Serra *et al.*, 1999). As such it may be an important auxiliary protein to facilitate the intrusion of the bacteria into the host symplast (Sutton *et al.*, 1994).

5. Intermediate responses

5.1. CYTOSKELETON REARRANGEMENTS AND ROOT HAIR DEFORMATION

Within 3 h after addition of Nod factor to legume roots, root hair deformation takes place in a narrow zone, the so-called susceptible zone. It can be demonstrated that root hair deformation can be inhibited by actinomycin D and cycloheximide, suggesting that the deformation process depends on plant DNA-specified RNA and protein synthesis. Evidence is accumulating that modulation of apical cytosolic free [Ca^{2+}] is an important factor in tip growth. Since Ca^{2+} concentrations within set margins are indispensable for a stable cytoskeleton, Nod factors cause dramatic changes in the cytoskeleton of root hairs. As the cytoskeleton is one of the determinants of the cell form, a breakdown of this structure could initiate root hair curling, which appears to arise from a disturbance of normal unidirectional root hair growth. Ridge (1992), using *Rhizobium* sp. bacteria to elicit a response, showed a diffuse fluorescence in the deformed region of curled root hairs of *Vicia*; no actin bundles were evident in the treated hairs. It was proposed that this area was a region in which the actin could be fragmented in response to the presence of bacteria. Allen *et al.* (1994), using purified Nod factors from *Rhizobium meliloti*, showed the presence of actin foci in curled root hairs of alfalfa.

The causal interactions of the effects of Nod factors on the cytoskeleton were investigated on root hairs of *Phaseolus vulgaris* (Cárdenas *et al.*, 1998). The actin cytoskeleton, usually arranged in long bundles and oriented parallel to the long axis of the root hair, extending into its apex, within 5 to 10 min responded to exposure to *R. etli* Nod factors with a rapid breakdown of the actin bundles. In the continued presence of the Nod factor the filamentous actin partially recovered, i.e. after about 1 h the filamentous actin had partially recovered forming long bundles. It is discussed that this rearrangement could be a prelude for the formation of infection threads. The appearance of short filaments appears to be indicative of an elevation of cytosolic Ca^{2+}. The most dramatic effect thereof was detected at the tip, which was visualized as a region of abundant fluorescence, suggesting the presence of short actin filaments. It has been proposed that an increase in cytosolic free [Ca^{2+}], triggered by the Nod factors, could be the cause of cytoskeleton modulation (Sánches *et al.*, 1991; Allen *et al.*, 1994). In fact, it was demonstrated in pollen tubes (also tip-growing) that elevated levels of Ca^{2+} induce fragmentation of F-actin and arrest cytoplasmic streaming (Kohno and Shimmen, 1987).

5.2. EARLY NODULIN GENE EXPRESSION

Two genes have been identified in which expression is induced in the root epidermis within 2 to 3 h following addition of Nod factors. These are *ENOD12*, encoding a Pro-rich protein, and the peroxidase encoding *MtRip1* (Journet *et al.*, 1994; Cook *et al.*, 1995). Both genes are expressed in a specific zone of the root epidermis extending from the root meristem through the area of root hair differentiation. Based on the timing and site of expression, it is unlikely that these genes are involved in root hair deformation, but they may play a role in the next step to occur, the infection. In this context, it is discussed that peroxidases may contribute to the infection process by altering root hair wall structure at sites of incipient penetration (Salzwedel and Dazzo, 1993).

The early nodulin gene *ENOD40* encodes a peptide of 10–13 amino acids that might influence primordium formation (Van den Sande and Bisseling, 1997). This gene has been found in legumes as well as in non-legumes and apparently plays a role in regulating plant development. Studies with tobacco protoplasts show that the externally added peptide is recognized by these cells, indicating a perception or uptake mechanism at their surface. This makes *ENOD40* an interesting candidate for cell-to-cell signalling. Since

ENOD40 transcription in the root pericycle precedes cortical cell divisions, it may be that ENOD40 can diffuse to the inner cortex, where it might be involved in local changes in the auxin/cytokinin balance and trigger the onset of cell division.

6. Late responses: Root cortical cell divisions and nodule formation

The formation of a root nodule primordium requires the de-differentiation of fully differentiated cortical cells. Which root cortical cells divide depends on the type of nodule a particular host forms. In general, two types of nodules can be distinguished in leguminous plants: a determinate and an indeterminate form. In indeterminate nodules the cells of the inner cortex are reactivated to divide and a nodule with a persistent apical meristem is formed, e.g. in pea, vetch and alfalfa. In the determinate nodule type the primordium is formed in the outer cortex and the formed nodule has a determinate growth pattern, in which the mitotic activity ceases during development, and cell expansion, rather than cell division, is responsible for the increase in nodule size, e.g. soybean and common bean.

6.1. IMBALANCE OF PHYTOHORMONES OR NOD FACTOR SIGNAL

It has been reported that purified Nod factors (without rhizobia being present) can induce genuine nodules in some but not all legumes (Spaink *et al.*, 1991; Stokkermans and Peters, 1994; Vijn *et al.*, 1995). Since the usual approach is to add the Nod factor to the entire root and not just to the root hair tip, it remains very unclear whether in fact the Nod factor-induced signals are carried through the root hair into the cortex. In alfalfa a certain proportion of wild-type plants can spontaneously develop genuine root nodule structures in the absence of *Rhizobium* bacteria or Nod factor. As the number of root nodules as well as their position on the root are indistinguishable from those observed in the infected situation, this indicates that the plant itself is able to trigger the genes involved in the nodule formation in the same or similar way as rhizobia do. Since analysis shows that in roots where nodules develop spontaneously, a phytohormone (auxin, cytokinin, ethylene) imbalance exists, this raises the question whether the Nod factor signal itself or signals induced by Nod factors in fact just feed into preformed pathways, open new paths or induce phytohormone imbalance and, by activating certain genes, coordinate nodule formation. It apears that rhizobia convert the process of arbitrarily occurring nodule formation into a well-directed and controlled process that occurs at a certain time after infection.

6.2. PHYTOHORMONE SIGNALLING

The involvement of ethylene in root nodule formation is indicated by legume mutants. The ethylene-insensitive *Medicago truncatula* mutant “sickle” forms many more nodules in the so-called susceptible zone than wild-type plants (Penmetsa and Cook, 1997), whereas nodulation of low-nodulating pea sym5 mutants can be restored by the addition of Ag^+, an inhibitor of ethylene functioning. Heidstra *et al.* (1997) showed that the expression of ACC oxidase, which catalyses the last step of ethylene biosynthesis, is confined to the pericycle in the phloem sectors of the root, which leads to a localized production of ethylene in the susceptible zone. The blocking of ethylene production leads to an increased number of nodules and the formation of nodule primordia opposite the phloem poles. Thus, ethylene obviously acts as a signalling compound providing positional information for nodule formation and also for controlling nodule number.

It is still uncertain whether Nod factors themselves are perceived by the cortical cells or whether a primary signal generated in the epidermis is involved. Auxin and cytokinin are involved in cell division, and two types of experiments indicate that changes in the phytohormone balance are required for the induction of nodule primordia. To initiate cell divisions in the inner cortex, which are required to form an indeterminate nodule, Nod factors with a highly unsaturated acyl chain must be used. For instance, on *Vicia sativa* Nod factors should carry a C18:4 acyl chain to induce cell divisions, whereas factors containing a C18:1 group are unable to do so (Spaink *et al.*, 1991). In contrast, Nod factors produced by rhizobia that nodulate on plants that form determinate nodules do not have specific saturation requirements. Thus, the presence or absence of highly unsaturated acyl chains in Nod factors is correlated with the place in the cortex where cell divisions will be induced. Accordingly, it has been proposed that highly unsaturated acyl chains may contribute to facilitate transport into the inner cortical cell layers (Spaink *et al.*, 1995).

On alfalfa roots, cytokinin can induce the formation of nodules, resembling rhizobial nodules: cell divisions are induced in the inner cortex, and lead to the formation of a nodule with peripheral vascular tissue, a central region surrounded by a cortex, and one or more meristem-like regions. The expression of early nodulin genes, such as *ENOD12*, is induced in these structures. Such mimicking of the effects of Nod factors by cytokinin suggests that Nod factors might induce an increased cytokinin level in the root, which subsequently triggers cell division, in line with the general role of cytokinin in plant development.

7. Conclusions and future prospects

Nod factor signalling is far from being cleared up. Whereas we have a fair idea of some of the early processes concerning the activation of ion channels at the plasma membrane level, in long-term signalling, i.e. signal transduction from the site of perception to the cortex, huge caveats exist. There is some hope that within the next years the chemical nature of the Nod factor-perceiving membrane constituents might be cleared up; with respect to the elucidation of the processes leading to nodule formation one might be less optimistic, as this involves the interaction of phytohormones. Our present understanding of phytohormone action in general is still rudimentary, and this ignorance is potentiated where more than one phytohormone is effective, as it seems to be the case here. The observation that Nod factors, without rhizobia being present, may lead to the formation of nodules is puzzling and would point to the Nod factor as the triggering molecule not only at the root hair level. Reports claiming that Nod factors can also affect cortical cells directly, raise the question how this could happen in nature. We recall, that the bacteria enter the symplast at the root hair tips and form an infection thread to reach the cortical cells. Is there information that on this stage of infection Nod factors are excreted and transported to the cortex? And if not, experiments will have to be performed where Nod factors only get into contact with the root hair tips. Only when nodules are formed under such conditions, can we move to the next step and ask what kind of information travels and how is it carried to the cortex.

References

Allen N S, Bennet M N, Cox D N, Shipley A, Erhardt D W and Long S R (1994) Effects of Nod factors on alfalfa root hair Ca^{2+} and H^+ currents and on cytoskeletal behaviour. In: *Advances in Molecular Genetics of Plant-Microbe Interactions, Vol. 3* (Eds Daniels M J, Downie J A and Osbourn A E), Kluwer Academic Publishers, Dordrecht, The Netherlands, pp. 107–113.

Ardourel M, Demont N, Debellé F, Maillet F, de Billy F, Promé J-C, Dénarié J and Truchet G (1994) *Rhizobium meliloti* lipooligosaccharide nodulation factors: different structural requirements for bacterial entry into target root hair cells and induction of plant symbiotic developmental responses. *Plant Cell*, **6**: 1357–1374.

Baier R, Schiene K, Kohring B, Flaschel E and Niehaus K (1999) Alfalfa and tobacco cells react differently to chitin oligosaccharides and *Sinorhizobium meliloti* nodulation factors. *Planta*, **210**: 157–164.

Bono J J, Riond J, Nicolaou K C, Bockovich N J, Estevez V A, Cullimore J V and Ranjeva R (1995) Characterization of a binding site for lipo-oligosaccharidic NodRm factors in particulate fractions prepared from roots. *Plant J.*, **7**: 253–260.

Cárdenas L, Feijo J A, Kunkel J G, Sánches F, Holdaway-Clarke T L, Hepler P K and Quinto C (1999) *Rhizobium* Nod factors induce increases in intracellular free calcium and extracellular calcium influxes in bean root hairs. *Plant J.*, **19**: 347–352.

Cárdenas L, Holdaway-Clarke T L, Sanches F, Quinto C, Feijo J A, Kunkel J G and Hepler P K (2000) Ion changes in legume root hairs responding to Nod factors. *Plant Physiol.*, **123**: 443–451.

Cárdenas L, Vidali L, Domínguez J, Pérez H, Sánchez F, Hepler P K and Quinto C (1998) Rearrangement of actin microfilaments in plant root hairs responding to *Rhizobium etli* nodulation signals. *Plant Physiol.*, **116**: 871–877.

Cardinale F, Jonak C, Ligterink W, Niehaus K, Boller T and Hirt H (2000) Differential activation of four specific MAPK pathways by distinct elicitors. *J. Biol. Chem.*, **275**: 36734–36740.

Cook D, Dreyer D, Bonnet D, Howell M, Nony E and Van den Bosch K (1995) Transient induction of a peroxidase gene in *Medicago truncatula* precedes infection by *Rhizobium meliloti*. *Plant Cell*, **7**: 43–55.

Cullimore J V, Ranjeva R and Bono J J (2001) Perception of lipo-chitooligosaccharidic Nod factors in legumes. *Trends Plant Sci.*, **6**: 24–30.

Dalla Serra M, Sutton J M, Höper F, Downie J A and Menestrina G (1999) Effects of calcium and protons on the secondary structure of the nodulation protein NodO from *Rhizobium leguminorarum* biovar *viciae*. *Biochem. Biophys. Res. Commun.*, **263**: 516–522.

De Ruijter N C A, Rook M B, Bisseling T and Emons A (1998) Lipochito-oligosaccharides re-initiate root hair tip growth in *Vicia sativa* with high calcium and spectrin-like antigen at the tip. *Plant J.*, **13**: 341–350.

Dietrich A, Mayer J E and Halbrock K (1990) Fungal elicitor triggers rapid, transient and specific protein phosphorylation in parsley cell suspension cultures. *J. Biol. Chem.*, **265**: 6360–6368.

Downie J A (1994) Signalling strategies for nodulation of legumes by rhizobia. *Trends Microbiol.*, **2**: 318–323.

Downie J A and Walker S A (1999) Plant responses to nodulation factors. *Curr. Op. Plant Biol.*, **2**: 483–489.

Erhardt D W, Atkinson E M and Long S R (1992) Depolarization of alfalfa root hair membrane potential by *Rhizobium meliloti* Nod factors. *Science*, **256**: 998–1000.

Erhardt D W, Wais R and Long S R (1996) Calcium spiking in plant root hairs responding to *Rhizobium* nodulation signals. *Cell*, **85**: 673–681.

Felix G, Grosskopf D G, Regenass M and Boller T (1991) Rapid changes of protein phosphorylation are involved in transduction of the elicitor signal in plant cells. *Proc. Natl. Acad. Sci. USA*, **88**: 8831–8834.

Felix G, Regenass M and Boller T (1993) Specific perception of subnanomolar concentrations of chitin fragments by tomato cells: Induction of extracellular alkalinization, changes in protein phosphorylation, and establishment of a refractory state. *Plant J.*, **4**: 307–316.

Felle H H, Kondorosi É, Kondorosi Á and Schultze M (1995) Nod signal-induced plasma membrane potential changes in alfalfa root hairs are differentially sensitive to structural modifications of the lipochitooligosaccharide. *Plant J.*, **10**: 295–301.

Felle H H, Kondorosi É, Kondorosi Á and Schultze M (1996) Rapid alkalinization in alfalfa root hairs in response to rhizobial lipochitooligosaccharide signals. *Plant J.*, **10**: 295–301.

Felle H H, Kondorosi É, Kondorosi Á and Schultze M (1998a) The role of ion fluxes in Nod factor signalling in *Medicago sativa. Plant J.*, **13**: 455–463.

Felle H H, Kondorosi É, Kondorosi Á and Schultze M (1998b) Analysis of early events in Nod factor signalling in root hairs of *Medicago*. In: *Biological Nitrogen Fixation for the 21st Century* (Ed Elmerich C), Kluwer Academic Publishers, Dordrecht, The Netherlands, pp. 253–253.

Felle H H, Kondorosi É, Kondorosi Á and Schultze M (1999a) Nod factors modulate the cytosolic [Ca^{2+}] differently in growing and non-growing root hairs of *Medicago sativa. Planta*, **209**: 207–212.

Felle H H, Kondorosi É, Kondorosi Á and Schultze M (1999b) Elevation of the cytosolic free [Ca^{2+}] is indispensable for the transduction of the Nod factor signal in alfalfa. *Plant Physiol.*, **121**: 273–279.

Felle H H, Kondorosi É, Kondorosi Á and Schultze M (2000) How alfalfa root hairs discriminate between Nod factors and oligochitin elicitors. *Plant Physiol.*, **124**: 1373–1380.

Gehring C A, Irving H R, Kabarra A A, Parish R W, Boukli N M and Broughton W J (1997) Rapid, plateau-like increases in intracellular free calcium are associated with Nod-factor-induced root hair deformation. *Mol. Plant Microbe Interact.*, **7**: 791–802.

Geurts R and Franssen H (1996) Signal transduction in *Rhizobium*-induced nodule formation. *Plant Physiol.*, **112**: 447–453.

Goedhart J *et al.* (2000) *In vivo* fluorescence correlation microscopy (FCM) reveals accumulation and immobilization of Nod factors in root hair cell walls. *Plant J.*, **21**: 109–119.

Gressent F (1999) Ligand specificity of a high-affinity binding site for lipochito-oligosaccharidic Nod factors in *Medicago* cell suspension cultures. *Proc. Natl. Acad. Sci. USA*, **96**: 4704–4709.

Grosskopf D G, Felix G and Boller T (1991) K-252a inhibits the response of tomato cells to fungal elicitors *in vivo* and their microsomal protein kinase *in vitro. FEBS Lett.*, **275**: 177–180.

Hahn M G (1996) Microbial elicitors and their receptors in plants. *Annu. Rev. Phytopathol.*, **34**: 387–412.

Heidstra R, Yang W C, Yalcin Y, Peck S, Emons A, van Kammen A and Bisseling T (1997) Ethylene provides positional information on cortical cell division but is not involved in Nod factor-induced root hair tip growth in *Rhizobium*-legume interaction. *Development*, **124**: 1781–1787.

Journet E P, Pichon M, Dedieu A, De Billy F, Truchet, G and Barker D G (1994) *Rhizobium meliloti* Nod factors elicit cell-specific transcription of the *ENOD12* gene in transgenic alfalfa. *Plant J.*, **6**: 241–249.

Kohno T and Shimmen T (1988) Mechanism of Ca^{2+} inhibition of cytoplasmic streaming in lily pollen tubes. *J. Cell Sci.*, **91**: 501–509.

Kurkdjian A C (1995) Role of the differentiation of root epidermal cells in Nod factor (from *Rhizobium meliloti*)-induced root hair depolarization of *Medicago sativa. Plant Physiol.*, **107**: 783–790.

Penmetsa R V and Cook D R (1997) A legume ethylene-insensitive mutant hyperinfected by its rhizobial mutant. *Science*, **275**: 527–530.

Pingret J L, Journet E P and Barker D G (1998) *Rhizobium* Nod factor signaling: evidence for a G protein-mediated transduction mechanism. *Plant Cell*, **10**: 659–671.

Ridge R W (1992) A model of legume root hair growth and *Rhizobium* infection. *Symbiosis*, **14**: 359–373.

Salzwedel J L and Dazzo F B (1993) pSym nod gene influence on elicitation of peroxidase activity from white clover and pea roots by rhizobia and their cell-free supernatants. *Mol. Plant Microbe Interact.*, **6**: 127–134.

Sánches F, Padilla J E, Pérez H E and Lara M (1991) Control of nodulin genes in root-nodule development and metabolism. *Annu. Rev. Plant Physiol. Plant Mol. Biol.*, **42**: 507–528.

Spaink H P, Sheeley D M, van Brussel A A N, Glushka J, York W S, Tak T, Geiger O, Kennedy E P, Reinhold V N and Lugtenberg B J J (1991) A novel highly unsaturated fatty acid moiety of lipo-oligosaccharide signals determines host specificity of *Rhizobium. Nature*, **354**: 125–130.

Stokkermans T J W and Peters N K (1994) *Bradyrhizobium elkanii* lipo-oligosaccharide signals induce complete nodule structures on *Glycine soja* Siebold et Zucc. *Planta*, **193**: 413–420.

Sutton J M, Lea E J A and Downie J A (1994) The nodulation-signalling protein Nod O from *Rhizobium leguminosarium* biovar *viciae* forms ion channels in membranes. *Proc. Natl. Acad. Sci. USA*, **91**: 9990–9994.

Sutton J M, Peart J, Dean G and Downie J A (1996) Analysis of the C-terminal secretion signal of the *Rhizobium leguminosarium* nodulation protein NodO: A potential system for the secretion of heterologous proteins during nodule invasion. *Mol. Plant Microbe Interact.*, **9**: 671–680.

Van de Sande K and Bisseling T (1997) Signalling in symbiotic root nodule formation. *Ess. Biochem.*, **32**: 127–142.

Viard M P, Martin F, Pugin A, Ricci P and Blein J P (1994) Protein phosphorylation is induced in tobacco cells by the elicitor cryptogein. *Plant Physiol.*, **104**: 1245–1249.

Vijn I, Martinez-Abarca F, Yang W C, das Neves L, van Brussel A, van Kammen A and Bisseling T (1995) Early nodulin gene expression during Nod factor-induced processes in *Vicia faba*. *Plant J.*, **8**: 111–119.

Wais R J, Galera C, Oldroyd G, Catoria R, Penmetsa R V, Cook D, Gough C, Dénarié J and Long S R (2000) *Proc. Natl. Acad. Sci. USA*, **97**: 13407–13412.

Walker S A, Viprey V and Downie J A (2000) Dissection of nodulation signaling using pea mutants defective for calcium spiking induced by Nod factors and chitin oligomers. *Proc. Natl. Acad. Sci. USA*, **97**: 13413–13418.

Yokoyama T, Kobayashi N, Kouchi H, Minamisawa K, Kaku H and Tsuchiya K (2000) A lipochito-oligosaccharide, Nod factor, induces transient calcium influx in soybean suspension-cultured cells. *Plant J.*, **22**: 71–78.

NITRATE ASSIMILATION

NIHARIKA SHANKAR AND H.S. SRIVASTAVA
Department of Plant Science, MJP Rohilkhand University,
Bareilly – 243006, India
e-mail: niharikashanker@rediffmail.com

Abstract

Although most legumes obtain nutrient nitrogen through symbiotic fixation of atmospheric nitrogen to ammonia, uptake and assimilation of nitrate from soil is also a significant process and often it is possible to modulate growth and productivity of nitrogen fixers as well as non-fixer legumes through the nitrate assimilation process. The process of nitrate reduction in legumes is well studied and the characterization of nitrate reductase, the rate limiting enzyme in the process, is almost similar to that in non-legumes. It is possible to manipulate process through the endogenous levels of carbohydrates and other photosynthates, nitrate supply, growth regulators and some of the environmental factors. The molecular genetics of nitrate assimilation process is now better understood than earlier. Transgenic tobacco (a non-legume) with altered levels of nitrate reductase activity have been produced, although it has not been possible to increase growth and productivity by increasing the expression of nitrate reductase gene in the transgenics. Clearly, the process is quite complex and many inter-linked metabolic aspects are also to be manipulated to achieve the desired effects.

1. Introduction

Legumes belonging to the family Leguminosae are important for their high nutritional value both for humans and cattle, primarily on account of high protein content in their seeds and also in the leaves. Morphologically the leguminous species range from small annual herbs such as pea and peanuts to large perennial trees such as tamarinds. The nitrogen requirement of many of the legumes is satisfied by the symbiotically associated microorganisms, which are present in specialised root nodules. However even in these nitrogen fixing species all the nitrogen requirement is not fulfilled by nitrogen fixation (Sinha, 1978) and they have to absorb and assimilate inorganic nitrogen from the soil to achieve optimum growth and productivity (Asthana and Srivastava, 1979). The soil nitrogen, which is usually in the form of nitrate is assimilated in to amino acids and other organic nitrogenous molecules through its reduction to nitrite and then to ammonia, by

P.K. Jaiwal and R.P. Singh (eds.), Improvement Strategies for Leguminosae Biotechnology, 191–199.

the sequential action of the enzymes nitrate reductase and nitrite reductase. The ammonia is incorporated primarily to glutamic acid and glutamine. The enzyme nitrate reductase (NR, E.C.1.6.6.1-2) is considered to be the rate limiting enzyme in the pathway and its activity is often correlated with the protein and total organic nitrogen content of the tissue (Srivastava, 1980). A positive correlation between NR activity and growth had been observed in suspension cultures of soybean also (Bayley *et al.*, 1972). Further, the activity of the enzyme is maximum during flowering and pod formation, when the requirement for nitrate assimilation for the synthesis of protein in the grains and over all pod development may be very high (Liu *et al.*, 1997). Perhaps the nitrogen fixer legumes have developed nitrate assimilation potential as an evolutionary strategy to make maximum use of whatever nitrogen is available in the environment. When nitrate is available as a source of nitrogen the process of nodulation and the activity of the enzyme nitrogenase, which is responsible for N_2 fixation in symbiotically associated legumes, are inhibited (Streeter, 1988). This inhibition is apparently due to resistance in O_2 diffusion in short term exposure to nitrate and linked to nitrate metabolism in long term effects (Kaiser *et al.*, 1997). Thus, nitrate reduction and assimilation is a significant process in the nitrogen nutrition of nitrogen fixer legumes also and it is possible to modulate growth and productivity of legumes through nitrate assimilation process.

2. Nitrate reduction in legumes

As mentioned earlier, both nodulating and non-nodulating legumes have the potential of absorbing and assimilating nitrate derived nitrogen. The nitrate absorbed from the soil is assimilated either in the roots or/and in shoots. Green leaves are undoubtedly the most active centres of active nitrate assimilation as they are for most other activities. However, the process depends upon the availability of nitrate in the leaves. For example, it has been demonstrated in soybean that nitrate reduction in the leaves is determined by the nitrate flux from the xylem. In other words, the export of nitrate from roots determines the nitrate reducing potential of the leaves (Gojon *et al.*, 1991). However, for short term requirement the nitrate from the storage pool in the leaves is mobilized to the metabolic pool, where it is reduced and assimilated (Gojon *et al.*, 1991). Stem also plays an important role in nitrate assimilation in herbaceous legumes of tropical origin (Andrews *et al.*, 1984). The legumes of the temperate region have very little nitrate reduction in the stem and they reduce nitrate mostly in the leaves (Sprent, 1980). It is apparent from these accounts that the factors affecting the growth and metabolic status of the leaves will have a significant effect on nitrate assimilation process.

The distribution of nitrate reduction in roots or in leaves varies and is affected by the nitrate supply. At low nitrate concentrations, NR activity is localised primarily in the roots, and contribution by the leaf increases with the increasing level of nitrate supply to the plant (Glaab and Kaiser, 1993). Apparently, the supply of nitrate from roots to shoots increases with the increase in external nitrate concentration. As indicated in the earlier paragraph, the distribution of NR activity also varies according to the species. In legumes belonging to the *Phaseolae* (*Glycine max*, *Phaseolus coccineus*, *P. vulgaris*, *Vigna radiata* and *V. unguiculata*), the enzyme activity is substantially lower (15% of the total) in roots than in shoots, while in legumes of *Viciae* tribe (*Cicer arietinum*, *Pisum sativum* and *Vicia faba*), *Genistae* (*Lupinus albus*) and in *Trifolieae* (*Medicago sativa* and *M. truncatula*), 30–60 of the total activity was found in roots (Wallace, 1986). In trees species, such as in *Leucaena leucocephala*, the enzyme activity is located primarily in the roots (Jaenicke *et al.*, 1996).

3. The characterization on nitrate reductase

Nitrate reductase has been extensively studied for its enzymic characteristics and regulation in legumes such as in soybean and in pea. Evans and Nason (1953) originally isolated and purified NR from soybean leaves and demonstrated that the enzyme could use either NADH or NADPH as reductant. The presence of enzyme has been demonstrated in nearly all parts of the plants, although roots or leaves appear to be the major sites of the enzyme activity. However, in many herbaceous species of tropical origin, stem also contains a major portion of total enzyme activity (Andrews *et al.*, 1984). The proportion of stem NR activity to total NR activity varies according to the species. It is highest in *Vicia faba* and the lowest in *Cajanus cajan* (Andrews *et al.*, 1984). The enzyme is present in root nodules also, which are the sites of symbiotic nitrogen fixation, although the increase in enzyme activity in nodules in response to nitrate supply is not as pronounced as in roots and leaves (Caba *et al.*, 1995; Kanayama *et al.*, 1999). However, variation from this general trend has also been reported. For example, in *Medicago sativa* inoculated with *Rhizobium meliloti*, about 17% of the total NR activity is in nodules, which is increased by almost 5-fold by the supply of nitrate (Deroche and Bablar, 1987). On the other hand, the enzyme activity in leaves or stem is not affected by nitrate supply in this system. In nodules, the activity is detected in cytosol and it is absent from the bacteroids as reported in faba bean and pea (Caba *et al.*, 1990; Chaliflour and Nelson, 1988) and in soybean (Kanayama *et al.*, 1999). However, in some studies, when the enzyme was assayed by *in vivo* method, the activity was demonstrated in bacteroids also (Stephens and Neyra, 1983; Becana *et al.*, 1989). The variation in the presence or absence of NR activity in the bacteroid also appears to be species dependent. In a study by Becana *et al.* (1989), no enzyme activity was detected in the bacteroids from *Phaseolus vulgaris* or *Pisum sativum* and very little activity was detected in *Glycine max* bacteroids. However, *Vigna radiata* and *V. unguiculata* bacteroids showed high NR activity both in the presence or absence of nitrate.

Three forms of NR, namely constitutive NADH : NR, constitutive NAD(P)H : NR and inducible NADH : NR, have been purified and identified from soybean leaves (Streit *et al.*, 1987). However, by using immunochemical characterization procedures, it was demonstrated that the constitutive NADH : NR and the constitutive NAD(P)H : NR were the same (Robin *et al.*, 1985). The constitutive or the inducible forms of the enzyme seem to be regulated in independent manner. The NADH:NR from the nodule cytosol of *Glycine max* has sub-unit Mr of 100 kDa and has a Km nitrate 160 mM and optimum pH 7.5 (Kanayama *et al.*, 1999). In this respect, it is similar to NR from non-legume (Srivastava *et al.*, 1995).

Low nitrate reductase (NADH : NR) mutants have been produced in soybean by using chemical mutagens such as nitrosomethyl urea (Streit and Harper, 1986). However, the mutation had no effect on the level of NADPH : NR activity. NR deficient mutants of pea have also been produced (Walsh and Carroll, 1992; Kaiser *et al.*, 1997), which have provided some useful information about the characters and role of NR in plants. Pea mutants with altered nitrate reduction potential have also been detected, which may be useful in the development of cultivars with desired qualities (Shelp *et al.*, 1998).

4. Molecular aspects of nitrate reductase and transgenics

Nitrate reductase gene has been identified and characterized in soybean and common bean. Two nitrate reductase genes *iNR1* and *iNR2* have been cloned (cDNA) (Wu *et al.*, 1995). *iNR1* has 2661 bp and codes for 99.7 kDa protein containing 886 amino acids and *iNR2* has 2673 bp coding for a 99.9 kDa protein of 891 amino acids. Two NR genes isolated from common bean, *NR1* (Hoff *et al.*, 1991) and *NR2* (Jensen *et al.*, 1994) show differences in developmental and tissue specific expression. The 5′ flanking promoter sequence of the two NR genes have also been isolated (Jensen *et al.*, 1996).

Attempts have been made to under-or over-express NR activity through genetic transformation of the organisms. In crop plants, it is possible to manipulate nitrate assimilating potential and growth using such an approach. Transgenic *Nicotiana plumbaginifolia* plants that express either a 5-fold increase or a 21-fold decrease in NR activity were used to study relationship between carbon and nitrogen assimilation (Foyer *et al.*, 1994). In this study, under maximum irradiance, the maximum rate of photosynthesis per unit surface area, was decreased in the low NR expressers but was relatively unchanged in high NR expressers compared with the wild type controls. In another study, transformed tobacco plants, constitutively expressing a chimaeric nitrate reductase gene (*nia2* gene fused to the constitutive CaMV 35S promoter) were grown in greenhouse (Quillere *et al.*, 1994). These plants had 25 to 150% higher NR activity and 32 to 47% lower nitrate content in the leaves than the wild type plants, although the total nitrogen content, protein content and dry matter production were similar. Thus, increasing NR activity could not support higher organic nitrogenous molecules and growth in the transgenics. However, this strategy could be employed for obtaining low nitrate content leafy vegetables such as spinach, lettuce, cabbage, etc.

Such experiments have not been conducted with legumes, but as with other plants, it would be possible to manipulate nitrate assimilation process through genetic transformation in legumes also.

5. Improvement of nitrate assimilation

Nitrate reductase activity is sensitive to quite a few chemical and environmental factors (Srivastava, 1980) and the process of nitrate reduction may be modified through induction of changes in enzyme activity by these factors.

5.1. CARBOHYDRATE METABOLISM

Assimilation of nitrate requires ATP and reductants (NADH or NADPH) which can be supplied by the light reactions of photosynthesis in green tissues. In non-chlorophyllous tissues, it may be derived from respiratory metabolism. The carbon skeleton required for the fixation of ammonia derived from the reduction of nitrate also comes from the respiratory metabolism of carbohydrates synthesized in dark reactions of photosynthesis. The net reaction of the synthesis of glutamate molecules from nitrate via nitrite, ammonia and glutamine, showing the requirements of reductants and ATP is as follows:

$$NO_3^- + ATP + 2\text{-oxoglutarate} + 10e^- + 9H^+ \rightarrow \text{Glutamate} + ADP + Pi$$

Thus, the rate of nitrate assimilation is closely linked with the carbohydrate metabolism. Studies related to the link between photosynthesis, nitrogen nutrition and nitrogen use efficiency have been conducted in a few leguminous species. In pea leaves, about 75 to 80% of total nitrogen is accounted in chloroplasts (Makino and Osmond, 1991). Further, the response of photosynthesis to nitrogen supply is much greater than that of dark respiration.

The NR activity in legumes and other species is upregulated by light and downregulated by darkness. Like in other species, the light-dark regulation of NR activity in legumes, such as in pea roots and leaves also involves reversible dephosphorylation-phosphorylation mechanisms (Glaab and Kaiser, 1993). Experiments conducted with non-legumes have shown some link between light activation of NR and photosynthesis. In barley, when photosynthesis is inhibited by the electron transport inhibitor DCMU, the NR activity drops considerably (Lillo *et al.*, 1996). Further, excluding CO_2 from the atmosphere surrounding the leaves also causes a declined in NR activity (Kaiser and Brendle-Behnich, 1991), and a high rate of CO_2 assimilation favours a high rate of nitrogen assimilation (Ferrario *et al.*, 1995). Apparently, a signal from the photosynthetic process triggers dephosphorylation of NR. Provan and Lillo (1999) have suggested that Calvin cycle intermediates were involved in the activation of NR in light and that there were multiple signal transduction pathways inducing the activation state of NR. Among others, sucrose and glucose may be possible signal transducers, as they are known to prevent dark induced decline in NR activity (De Cires *et al.*, 1993). They also induce activation of NR from barley (Provan and Lillo, 1999) and external supply of sugars, especially sucrose, increases NR activity in some species including bean leaves (Puranik and Srivastava, 1983). Thus, at physiological level, it is possible to improve nitrate assimilation process through endogenously elevated levels of sugars.

5.2. NITRATE SUPPLY

Increasing nitrate supply may increase NR activity and overall nitrate assimilation and growth potential in most species. However, this may not be a desirable strategy in improving overall productivity, because the accumulation of nitrate in the plant also increases with increasing nitrate supply (Nambiar *et al.*, 1988; Hervas *et al.*, 1991) and this will depreciate the food and feed values of the legumes. Further, nitrate supply may inhibit nodule formation and nitrogenase activity in nodulating legumes (Srivastava and Ormrod, 1986; Kaiser *et al.*, 1997). Nevertheless, nitrate supply increases plant growth as measured by root and shoot dry weight and total nitrogen content in both nodulating (Srivastava and Ormrod, 1986) and non-nodulating (Srivastava *et al.*, 1990) common bean (*Phaseolus vulgaris*) plants. Similarly, in *Phaseolus lunatus* also, nitrate supply increases plant growth, although it causes a decline in nodulation (Da Silveira *et al.*, 1998). Thus, given a choice, the nodulated legumes also would prefer nitrate over nitrogen fixation for fulfilling the requirement of nitrogen nutrition. This is apparently because the nitrogen nutrition through nitrogen fixation process is much more energy demanding than through nitrate assimilation.

5.3. GROWTH REGULATORS

Both natural as well as synthetic growth regulators are often used in agriculture for the qualitative as well as quantitative improvement of crop plants. The nitrate assimilation process in legumes and also in non-legumes is responsive to many growth regulators and thus there is a possibility of modifying nitrate acquisition and assimilation process in legumes through the growth regulators. In some relevant studies, cytokinins benzyl adenine and kinetin are known to increase NR activity and total organic nitrogen content in the cotyledons of *Leucaena leucocephala* (Pandey and Srivastava, 1995). Increased NR activity and total nitrogen content has been demonstrated in common bean also (Srivastava *et al.*, 1994). The supply of polyamine spermidine also increases *in vivo* NR activity, total nitrogen and growth in the roots and shoots of *Leucaena leucocephala* seedlings (Pandey and Srivastava, 1995). Increased nitrate assimilation potential in response to spermine and spermidine has been reported in the leaves of common bean also (Srivastava *et al.*, 1995).

5.4. ENVIRONMENTAL FACTORS

Heavy metals, which often contaminate agricultural soils in sub-urban and industrial areas often inhibit nitrate assimilation by inhibiting NR activity. The inhibitory effect of Pb on NR activity has been recorded in species such as in bean (Jain and Gadre, 1996) and mungbean (Singh *et al.*, 1997). The inhibition in bean by 0.1 mM Pb acetate could be overcome by the supply of inorganic nitrogen in the form of NH_4Cl or NH_4NO_3 (Jain and Gadre, 1996). Similarly, in mungbean also supply of nitrate or of cytokinin benzyl amino purine could overcome the inhibitory effect of Pb on root NR activity (Singh *et al.*, 1997). Thus, nitrate fertilization in polluted soils may be an appropriate strategy to ameliorate the toxic effects of heavy metals on nitrate utilization.

Several other environmental factors also affect the nitrate assimilation process in legumes. In *Phaseolus vulgaris* (Castrillo *et al.*, 1990) and in *Cicer arietinum* (Wery *et al.*, 1988), drought stress inhibits NR activity, although inhibition is not as pronounced as that of nitrogen fixation. Thus, nitrate reduction process may assume significance in nitrogen in drought stressed legumes.

6. Conclusions and future prospects

Legumes are the principal source of dietary protein for vegetarian population of the world. Thus, there is a need for quantitative as well as qualitative improvement of legumes, especially the grain legumes. Conventional breeding methods have met with only limited success, so far. The molecular biology and the genetic engineering methods are being used for changing the amino acid composition of seed proteins and also for increasing the total protein and nitrogen contents. The enzyme NR, which is the rate limiting enzyme in nitrate assimilation process, has been over or under expressed in tobacco, a non-legume. But, the expected result, that is a correlated increase in protein content and ultimately the growth of the plants, has not been achieved. Apparently, the process is complex and other aspects of metabolism linked to nitrate assimilation process are also to be manipulated.

References

Andrews M, Sutherland J M, Thomas R J and Sprent J I (1984) Distribution of nitrate reductase activity in six legumes: The importance of the stem. *New Phytol.*, **98**: 301–310.

Asthana J S and Srivastava H S (1979) Nitrate reductase activity, grain yield and nitrogen content of pea seeds as affected by nitrate supply and *Rhizobium* inoculation. *Proc. Natl. Acad. Sci. India*, **49**: 163–165.

Bayley J M, King J and Gamborg O L (1972) The ability of amino compounds and conditioned medium to alleviate the reduced nitrogen requirement of soybean cells grown in suspension culture. *Planta*, **105**: 25–32.

Becana M, Minchin F R and Sprent J I (1989) Short term inhibition of legume N2 fixation by nitrate I: Nitrate effects reductase activities of bacteroids and nodule cytosol. *Planta*, **180**: 40–45.

Caba J M, Hervas A, Lluch C and Ligero F (1990) Nitrate metabolism in roots and nodules of *Vicia faba* in responses to exogenous nitrate. *Physiol. Plant*, **79**: 531–539.

Caba J M, Lluch C and Ligero F (1995) Distribution of nitrate reductase activity in *Vicia faba*: Effect of nitrate and plant genotype. *Physiol. Plant*, **93**: 667–672.

Castrillo M, Fernandez P, Molina B and Kazandjian A (1990) Nitrogen metabolism in *Phaseolus vulgaris* L. under water deficit. *Turrialba*, **40**: 515–519.

Chalifour F P and Nelson L M (1988) Short-term effects of nitrate on nitrate reductase activity and symboidic dinitrogen fixation in faba bean and pea. *Can. J. Bot.*, **66**: 1639–1645.

Da Silviera J A G, Matos J C D, Ceccatto V M, Sampaio A H and Da Costa R C L (1998) Induction of nitrate reductase activity and nitrogen fixation in two *Phaseolus* species in response to exogenous nitrate levels. *Physiol. Mol. Biol. Plants*, **4**: 181–188.

De Cires A, De la Tore A, Delgado B and Lara C (1993) Role of light and CO_2 fixation in the control of nitrate reductase activity in barley leaves. *Planta*, **190**: 277–283.

Deroche Me and Babalar M (1987) *In vivo* nitrate reductase activity in different organs on lucerne. Colloques de I. *INRA*, **37**: 121–133.

Evans H J and Nason A (1953) Purine-nucleotide nitrate reductase from extracts of higher plants. *Plant Physiol.*, **28**: 233–254.

Foyer C H, Hescure J C, Lefebvre C, Morot-Gaudry J F, Vincetz M and Vaucheret H (1994) Adaptation of photosynthetic electron transport, carbon assimilation and carbon partitioning in transgenic *Nicotiana plumbaginifolia* plants to changes in nitrate reductase activity. *Plant Physiol.*, **109**: 171–178.

Glaab J and Kaiser W M (1993) Rapid modulation of nitrate reductase in pea leaves. *Planta*, **191**: 173–179.

Gojon A, Wakrim R, Passama L and Robin P (1991) Regulation of NO_3 assimilation by anion availability in excised soybean leaves. *Plant Physiol.*, **96**: 398–405.

Haenicke H, Lips H S and Ullrich W R (1996) Growth, ion distribution, potassium and nitrate uptake of *Leucaena leucocephala*, and effects of NaCl. *Plant Physiol. Biochem.*, **34**: 743–751.

Hain M and Gadre (Puranik) R (1996) Inhibition of nitrate reductase activity by lead in excised bean leaf segments. *Proc. Natl. Acad. Aci. India*, **66**: 343–348.

Hervas A, Ligero F and Lluch C (1991) Nitrate reduction in pea plants: Effects of nitrate application and *Rhizobium* strains. *Soil Biol. Biochem.*, **23**: 695–699.

Hoff T, Stumann B M and Henningsen K W (1991) Cloning and expression of a gene encoding a root specific nitrate reductase in bean (*Phaseolus vulgaris*). *Physiol. Plant*, **82**: 197–204.

Jensen P E, Hoff T, Moller M G, Stumann B M and Henningsen K W (1994) Identification of characterization of a nitrate reductase gene from bean (*Phaseolus vulgaris*) containing four introns. *Physiol. Plant*, **92**: 613–623.

Jensen P E, Hoff T, Stumann B M and Henningsen K W (1996) Functional analysis of two bean nitrate reductase promoters in transgenic tobacco. *Physiol Plant*, **96**: 351–358.

Kaiser B N, Layzell D B and Shelp B J (1997) Role of oxygen limitation and nitrate metabolism in the nitrate inhibition of nitrogen fixation by pea. *Physiol. Plant*, **101**: 45–50.

Kaiser W M and Brendle-Behnisch E (1991) Rapid modulation of spinach leaf nitrate reductase activity by photosynthesis. I. Modulation *in vivo* by CO_2 availability. *Plant Physiol.*, **96**: 363–367.

Kanayama Y, Kimura K, Kakamura Y and Lke T (1999) Purification and characterization of nitrate reductase from nodule cytosol of soybean plants. *Physiol. Plant*, **105**: 396–401.

Lillo C, Smith L H, Nimmo H G and Wilkins M B (1996) Regulation of nitrate reductase and phosphoenol pyruvate carboxylase activities in barley leaf protoplasts. *Planta*, **200**: 181–185.

Liu H C, Guan P C and Chen R Y (1997) Effect of N, K nutrition on yield, quality and activity of some enzymes in asparagus bean. *J. South China Agric. Univ.*, **18**: 31–35.

Makino A and Osmond B (1991) Effect on nitrogen nutrition on nitrogen partitioning between chloroplasts and mitochondria in pea and wheat. *Plant Physiol.*, **96**: 355–362.

Nambiar P T C, Rego T J and Rao B V S (1988) Nitrate concentration and nitrate reductase activity in the leaves of legumes and three cereals. *Ann. Appl. Biol.*, **112**: 547–553.

Pandey S and Srivastava H S (1995) Stimulation of growth and nitrate assimilation on *Leucaena leucocephala* seedlings. *Biol. Plant*, **37**: 153–157.

Provan F and Lillo C (1999) Photosynthetic post translational activation of nitrate reductase. *J. Plant Physiol.*, **154**: 605–609.

Purianik R and Srivastava H S (1983) Increase in nitrate reductase activity in the presence of sucrose in bean leaf segments. *Phytochemistry*, **22**: 2383–2387.

Quillere I, Dufosse C, Roux Y, Foyer C H, Caboche M and Morto-Gaudry J F (1994) The effects of deregulation of NR gene expression on growth and nitrogen metabolism of *Nicotiana plumbaginfolia* plants. *J. Exp. Bot.*, **45**: 1205–1211.

Robin P, Streit L, Campbell W H and Harper J E (1985) Immunochemical characterization of nitrate reductase forms from wild type (cv Williams) and nrl mutant soybean. *Plant Physiol.*, **77**: 232–236.

Santucci D M, Haas B and Smarelli J Jr. (1995) Regulation of the inducible soybean nitrate reductase isoform in mutants lacking constitutive isoforms(s). *Biochim. Biophys. Acta*, **1247**: 46–50.

Shelp B J, Kaiser B N and Deschesne A M (1998) Registration of five near isogenic genetic stocks of "Juneaur" pea with altered modulation and nitrate reductase activity. A 3171, nod 31, A 317 nod 31, E 1351 and R251. *Crop Sci.*, **38**: 554.

Singh R P, Dabas S, Chaoudhary A and Maheshwari P (1997) Effect of lead on nitrate reductase activity and alleviation of lead toxicity by inorganic salts and 6-benzylaminopurine. *Biol. Plant*, **40**: 399–404.

Sprent J I (1980) Root nodule anatomy, type of export product and evolutionary origin in some *Leguminosae*. *Plant Cell Environ.*, **3**: 35–43.

Srivastava H S (1980) Regulation of nitrate reductase activity in higher plants. *Phytochemistry*, **17**: 725–733.

Srivastava H S (1995) Nitrate reductase. In: *Nitrogen Assimilation in Higher Plants* (Eds Srivastava H S and Singh R P), Associated Publ. Co., New Delhi, India, pp. 45–69.

Srivastava H S and Ormrod D P (1986) Effects of nitrogen dioxide and nitrate nutrition on nodulation, nitrogenase activity, growth and nitrogen content of bean. *Plant Physiol.*, **81**: 737–741.

Srivastava H S, Ormrod D P and Hale-Marie B (1990) Photosynthetic photon flux effects on bean response to nitrogen dioxide. *Environ. Exp. Bot.*, **14**: 463–367.

Srivastava H S, Ormrod D P and Hale B A (1994) Cytokinins affect the responses of growing and green bean leaves to nitrogen dioxide and nutrient nitrate supply. *J. Plant Physiol.*, **144**: 156–160.

Srivastava H S, Ormrod D P and Hale B A (1995) Polyamine mediated modifications of bean leaf responses to nitrogen dioxide. *J. Plant Physiol.*, **146**: 313–317.

Stephens B D and Neyra C A (1983) Nitrate and nitrite reduction in relation to nitrogenase activity in soybean nodules and *Rhizobium japonicum* bacteroids. *Plant Physiol.*, **71**: 731–735.

Streeter J G (1988) Inhibition of legume nodule formation and N_2 fixation of nitrate. *Crit. Rev. Plant Sci.*, **7**: 1–23.

Streit L and Harper J E (1986) Biochemical characterization of soybean mutants lacking constitutive NADH: Nitrate reductase. *Plant Physiol.*, **81**: 593–596.

Streit L, Martin B A and Harper J E (1987) A method for the separation and partial purification of the three forms of nitrate reductase present in wild type soybean leaves. *Plant Physiol.*, **84**: 654–657.

Wallace W (1986) Distribution of nitrate assimilation between the root and shoot of legumes and a comparison with wheat. *Physiol. Plant.*, **66**: 630–636.

Walsh K B and Carroll B J (1992) A reappraisal of nitrate inhibition of nitrogenase in A317, a nitrate reductase deficient mutant of pea (*Pisum sativum*). *Physiol. Plant*, **84**: 283–291.

Wery J, Deschamps M and Leger-Cresson N (1988) Influence of some agroclimatic factors and agronomic practices on nitrogen nutrition of chickpea (*Cicer arietinum* L.). In: *Nitrogen Fixation in Legumes in Mediterranean Agriculture* (Eds Beach D P and Materon L A), Martinus-Nijhoff, Dordrecht, The Netherlands, pp. 287–301.

Williams W M, Hoh C H, Lenz F and Broughton W J (1988) *Rhizobia* in tropical legumes: Environmental factors and the reduction of nitrogen. *Soil Biol. Biochem.*, **20**: 667–675.

Wu S, Lu Q, Kriz A L and Harper J E (1995) Identification of cDNA clones corresponding to two inducible nitrate reductase genes in soybean: Analysis in wild type and nr1 mutant. *Plant Mol. Biol.*, **29**: 491–500.

IMPROVEMENT OF NITROGEN UTILIZATION

BERTRAND HIREL[1], JUDITH HARRISON[2] AND ANIS LIMAMI[1]
[1] *Unité de Nutrition Azotee des Plantes, INRA, Route de St-Cyr, F-78026 Versailles Cedex, France*
[2] *Laboratoire de Biologie Vegetale et Microbiologie, CNRS ESA 6169, Université de Nice Sophia-Antipolis, Parc Valrose, 06108 Nice cedex 2, France*
e-mail: hirel@versailles.inra.fr

Abstract

In this chapter we discuss the ways in which our understanding of the nature of the molecular controls of nitrogen assimilation have been increased by the use of legumes with modified capacities for ammonium assimilation. These modifications have been achieved through genetic engineering and breeding. An improved understanding of nitrogen assimilation will be vital if improvements in legume crop nitrogen use efficiency are to be made to reduce the need for excessive input of fertilizers. In this review we present an overall view of past work and more recent studies on this topic. Using *Medicago* sp. and *Lotus* sp. as model plants, glutamine synthetase activity have been altered by stimulating or inhibiting, in an organ or tissue-specific manner, the expression of the corresponding genes. The physiological impact of these genetic manipulations has been studied on plants grown under different nitrogen regimes. More recently a genetic approach using either different ecotypes of *Medicago truncatula* or recombinant inbred lines of *Lotus japonicus* have been used to identify key enzymes involved in the genetic variation of nitrogen use efficiency as well as various traits such as plant biomass production. The possible involvement of these genes in plant nitrogen use efficiency and plant productivity is discussed.

1. Introduction

Nitrogen is one of the major limiting factors for plant growth. However, an excessive external supply of nitrogen causes major problems in agriculture and the environment, mostly due to nitrate leaching into underground water (Benes *et al.*, 1989) not only polluting the aquatic environment but also resulting in a high nitrate content in food and drinks (Moller *et al.*, 1990). Paradoxically, dinitrogen is 80% of the atmosphere but it can only be assimilated by prokaryotic symbiotic or free-living diazotrophic organisms possessing

P.K. Jaiwal and R.P. Singh (eds.), Improvement Strategies for Leguminosae Biotechnology, 201–220.

nitrogenase activity (Dilworth, 1974). In higher plants, nitrate and ammonium are the two major inorganic nitrogen compounds which can be directly assimilated. They are provided either artificially by the external supply of fertilizers, bacterial nitrification, atmospheric reduction of dinitrogen during lighting or naturally through biological nitrogen fixation. Therefore, regulation of inorganic nitrogen assimilation and incorporation of inorganic nitrogen into organic matter is of major importance for both maintaining a sustainable agriculture and protecting the environment.

Three major reactions are involved in inorganic nitrogen assimilation in plants (see review by Lea and Ireland, 1999). The first reaction is catalyzed by the enzyme nitrate reductase (NR)(E.C.1.6.1.1), which reduces nitrate to nitrite. Nitrites are further reduced into ammonia in a reaction catalyzed by the enzyme nitrite reductase (NIR)(E.C.1.7.99.3). Ammonia is then incorporated into an organic molecule, by the enzyme glutamine synthetase (GS)(E.C.6.3.1.2) to form glutamine. Thus, glutamine is a key metabolite that acts as an amine donor to the various transamination reactions involved in amino acid, nucleotide and polyamine biosynthesis (Joy, 1988). The transfer of the amino group is primarily catalyzed by the enzyme glutamate synthase (GOGAT)(E.C.1.4.7.1) in order to recycle a molecule of glutamate in a GS-GOGAT-coupled reaction. The GS-GOGAT pathway is now considered as the major route for inorganic nitrogen assimilation in plants (Miflin and Lea, 1980).

In the past few years, attention has been particularly focused on glutamine synthetase because of its central role in nitrogen metabolism and its diverse metabolic and developmental regulation in different plant species and organs (Cren and Hirel, 1999). There are two major isoforms of the GS enzyme: cytosolic GS1 occurring in the cytoplasm of leaves and non photosynthetic organs, and plastidic GS2, present in the chloroplasts of photosynthetic tissues and the plastids of roots or etiolated plants. The relative proportions of the cytosolic and plastidial GS activity may vary within different organs of the same plant or within different plant species depending on their photosynthetic type (McNally *et al.*, 1983). Altogether, these observations led to the proposal that each GS isoenzyme has a specific function in assimilating or reassimilating ammonia derived from a variety of processes such as nitrate reduction, photorespiration or nitrogen recycling (Cren and Hirel, 1999). In addition, a number of studies have shown that GS is encoded by a multigene family and one or several gene products (Forde and Cullimore, 1989) may constitute each isoenzyme. The expression of each gene appears to be tissue-specific or regulated by external factors such as light, the nitrogen source and the symbiotic association with *Rhizobium* (Cren and Hirel, 1999). Moreover, it has also been recently demonstrated that in addition to organ- or tissue-specific expression, a number of GS genes can be differentially regulated by several metabolites including amino acids and soluble carbohydrates (Oliveira and Coruzzi, 1999). These recent findings suggest that the metabolic status of individual organs and/or tissues may be as important as a developmental regulation to efficiently control the pathway of ammonia assimilation in the plant kingdom. The pattern of expression of the various cytosolic GSs has been most thoroughly investigated in legumes where GS is actively synthesised to assimilate the large supply of ammonium produced by the symbiotic nitrogen-fixing bacteroids located inside specialized and complex cellular structures called root nodules (Robertson *et al.*, 1975). Although different from one legume species to the other (temperate legumes versus tropical legumes forming either determinate or indeterminate nodules, Pate and Atkins, 1983; Corby *et al.*, 1983), it has been clearly shown that metabolic and/or developmental events control the expression of some members of the nodule GS multigene family in a tissue-specific manner. Moreover, each of these genes is differentially transcribed and translated to allow the synthesis of a polypeptide component of the native GS enzymes (Forde and Cullimore, 1989). In all legume species studied so far, it was found that compared to roots or shoots higher cytosolic GS activity was always present in nodules where ammonia resulting from N_2 fixation is diffused out from the bacteroids to the host cytoplasm. However, the exact role of various GS isoenzymes identified in the different nodule cell types is still not fully understood. In particular, recent findings have shown that final nodule GS activity is highly regulated and subject to a multiple step co-ordinated process, starting at the transcriptional level up to the control of the holoenzyme turnover (Ortega *et al.*, 1999). This complexity seems to be consistent with the fact that most legumes have a greater potential to assimilate major amounts of nitrogen in their root system whether or not they are under atmospheric nitrogen fixing conditions. However, in these species, the efficiency of root nitrogen assimilation has been questioned because of a possible competition with shoot inorganic nitrogen assimilation (Oaks, 1992). This particularity is likely to be the result of an evolutionary adaptation to the symbiosis with *Rhizobium* bacteria.

Therefore, considering both the economical and ecological importance of atmospheric nitrogen fixing symbiosis (Pawlowski, 1997; Shantharam and Mattoo, 1997) and the unique developmental and molecular events associated with nodule development, a number of studies have been performed to unravel the regulatory mechanisms controlling reduced nitrogen transfer from the bacterium to the host (Waters *et al.*, 1998). However, despite the significant progress made in understanding the physiological and molecular mechanisms involved during the establishment of the symbiosis, not much success has been obtained in either enhancing biological nitrogen fixation in legumes or transferring important biological nitrogen fixation traits to non-nitrogen fixing plants (Shantharam and Mattoo, 1997). Consequently, alternative physiological approaches have been proposed for improving mobilization, redistribution and utilization of N fixed and stored N reserves within the host plant rather than N_2 fixation itself. This prompted a number of groups to modify ammonia assimilation in legumes by the means of genetic manipulation. This approach provides at the same time a means of assessing the role of the different nodule GS genes and isoenzymes in response to the massive supply of ammonia resulting from N_2 fixation. In addition, the impact of such genetic manipulation was examined to determine if the reaction catalysed by GS was one of the limiting factors in terms of nitrogen use efficiency (NUE) and yield. By the means of selected examples, we will present a general view on the recent attempts to alter ammonia assimilation in legumes using genetic engineering.

However, this approach only allows the identification of the role of a single gene and does not account for the variation of complex traits (such as NUE) often found in agronomical applications. Furthermore, the public still poorly accepts the use of improved transgenic crop plants. Therefore, a parallel complementary strategy is necessary to understand in a more integrated manner the physiological and molecular basis of NUE in legumes. In the past, both conventional breeding techniques and mutagenesis (Herridge and Danso, 1995) have been developed to improve legume productivity in relation to nitrogen fixation and nitrogen metabolism. Although these approaches were, in a few cases, successful in terms of yield enhancement, there were no real attempts to identify the regulatory control mechanisms responsible for either a more efficient symbiotic association or leading to improved NUE. More recently the development of quantitative genetic approaches has opened new perspectives to identify and characterize quantitative trait loci (QTLs) involved in the genetic variation of NUE in crop plants including legumes. The aim of this new strategy and its potential to plant breeders in the carrying out of marker assisted crop selection for improved NUE in relation to yield will be also introduced and discussed in this chapter.

2. Modification of ammonia assimilation in legumes by genetic engineering: impact on plant physiology and development

2.1. IN ALFALFA

The original idea of modulating GS activity in alfalfa arose from work published by Knight and Langston-Unkefer (1988) in which nodulated plants were infested with *Pseudomonas syringae* pv. *Tabaci*, bacteria living at the root surface of many plants and releasing tabtoxin-B-lactam (TBL), an irreversible inhibitor of GS. Following the infection, an approximate doubling of plant growth, total nitrogen, nodulation and overall dinitrogen fixation was seen. This spectacular effect on plant biomass production was explained by the selective action of the inhibitor on the root-specific cytosolic GS isoform and although the nodule-specific GS was not affected, total GS activity in the nodules was reduced by half. Although detailed physiological analysis of the infested plants were not presented, the authors proposed that alternative routes of ammonia assimilation may be more efficient in assimilating the ammonium produced as the result of nitrogen fixation. In addition, the important decrease in the glutamine pool (a potent feed-back inhibitor of nitrogenase activity) in nodules of infested plants, could explain the enhancement of nitrogen fixation, thus being beneficial to the overall nitrogen metabolism.

However, this work was a matter of debate for several years, because it was not reproducible (B. Hirel and coworkers, unpublished results). This prompted Sengupta-Gopalan's group to develop an antisense RNA strategy to reduce GS activity in roots and root nodules of alfalfa and to examine if a decrease in the enzyme activity would be beneficial for plant biomass production.

Before the genetic manipulation of GS1 levels in an organ-specific manner, preliminary studies were performed by this group to characterise the pattern of GS1 gene and protein expression in alfalfa. In this species, two genes encoding cytosolic glutamine synthetase (pGS13 and pGS100) have been identified. They have shown that both genes are expressed in all organs tested, although at varying levels. Genes represented by pGS13 are expressed at a level approximately 20-fold higher in nodules than in roots. The expression level of pGS13 genes in leaves is approximately equivalent to that found in roots, while genes represented by pGS100 are expressed at a slightly elevated level in nodules compared to roots and at a much lower level in leaves (Temple *et al.*, 1995).

Following the characterization of the two members of the alfalfa GS multigene family, Temple *et al.* (1994) showed that when alfalfa plants were transformed with a construct containing pGS100 cDNA in the antisense orientation behind the CaMV 35S promoter, no phenotypic variations were observed in the transformants compared to untransformed control plants. Western blot analysis of several independent transformants showed no significant decrease in the level of GS1 subunits. They hypothesized that the down regulation of GS was lethal, likely to be the result of ammonia toxicity or shutdown in the availability of reduced nitrogen necessary for plant growth and development. This could be due to the lack of specificity of the GS1 antisense RNA whose expression was not targeted to either roots or nodules. Also the high degree of homology between plastidic and cytosolic gene sequence may result in a reduction of GS2 expression as well, thus provoking a toxic built up of photorespiratory intermediates known to be lethal for plants grown under ambient atmosphere (Morris *et al.*, 1989).

Two other strategies were then developed by the same group using either a root-specific promoter or specific sequences of the two GS1 mRNAs. In a first series of experiments the alfalfa GS cDNA pGS100 was placed in the antisense orientation behind the acidic chitinase promoter from *Arabidopsis* (Temple *et al.*, 1994). In the few plants that resulted from the transformation, a decrease in the steady state level of pGS100 transcripts in all organs was observed. The amount of GS1 subunits dropped both in roots and nodules and there was a reasonably good correlation with the changes in GS activity. Once again, due to the limited number of plants obtained, the authors suspected that a fully active antisense RNA was lethal. Unfortunately, the effect of reduced GS activity on the physiology and the development of the survivors was not presented.

In the second set of experiments, aiming to block the expression of the two genes encoding GS1, the specific 3′-untranslated regions of pGS13 and pGS100 were utilized to make antisense constructs and introduced individually into alfalfa (Temple *et al.*, 1998a). The two constructs were effective in lowering the level of the corresponding transcripts in the various organs of alfalfa. However, transgenic plants with up to 80% reduction in the transcript level corresponding to each gene class showed no reduction in GS activity and GS1 polypeptides.

Altogether, these two series of experiments revealed that in root and root nodules of alfalfa, cytosolic GS levels are controlled at both the transcriptional and translational/post translational level, thus allowing the enzyme activity to be maintained at a fairly constant level. This may not be surprising considering the key role of GS in temperate legumes known to assimilate ammonia mostly in roots (Oaks, 1992). Supposing that a complete inhibition of the corresponding transcripts is lethal, this renders very difficult in the use of antisense RNA strategies to down regulate GS activity to a certain threshold.

An alternative would be to use cell-specific homologous promoters to target more efficiently the antisense inhibition in tissues that could cope with impaired ammonia assimilation, although we have no idea if such modification would be beneficial in terms of plant performance. However, this strategy merits further attention if we consider that in a number of legume species, each member of the GS multigene family is differentially expressed in specialized cell types such as the vascular tissue, the cortex or the uninfected cells, which are not directly connected with the site of nitrogen fixation taking place in the infected cells (Streeter, 1991).

Another possibility would be the use of inducible promoters (Gatz and Lenk, 1998) allowing the synthesis of antisense RNAs at specific stages of root or root-nodule development, thus avoiding a toxic build up of ammonia and/or a nitrogen shortage in the early stages of plant growth. The finding that in a *Lotus japonicus* ecotype, Funakura, the increase in biomass production is concomitant to a decrease in GS activity only at a certain stage of plant development supports this idea (Limami *et al.*, 1999). Whether this phenomenon can be generalized to other legume species and follows the same timing during the plant life cycle requires further investigation.

Attempts were also made to overexpress GS in roots of alfalfa using the acidic chitinase promoter fused to pGS100 (Temple *et al.*, 1994). Although in the transgenic plants a significant increase in GS1 subunit concentration was observed in roots, stems and leaves an important decrease in nodule GS protein, attributed to a co-suppression phenomenon, was surprisingly observed. Since no physiological analyses were performed, it was impossible to tell if the expression of the transgene had any impact on the overall metabolism although these authors mentioned that no change in the plant phenotype was apparent.

2.2. IN *LOTUS* SPECIES

Another model plant currently used to study the expression of genes expressed during the establishment of the *Rhizobium*-legume symbiosis is *Lotus*. Originally most of the studies used *Lotus corniculatus* because it was easy to transform and relatively fast to regenerate over a three-month period. However, the transformation procedure was only possible *via Agrobacterium rhizogenes* and regeneration of plantlets from hairy roots. In addition, *L. corniculatus* is not self-fertile and T_1 progeny are only obtained following manual cross-pollination. Today most of the studies are conducted with *L. japonicus* for which a transformation protocol was developed using *A. tumefaciens* as the bacterial vector. Because this diploid species with a relatively small genome is also self-fertile it was adopted by a number of groups for both genetic and biochemical studies (Handberg and Stougaard, 1992).

As for alfalfa, attempts were made to modify the pathway of ammonia assimilation by modulating the expression of GS in various organs of *Lotus* although only one gene encoding cytosolic GS has been characterized in this species (Thykjaer *et al.*, 1997). A preliminary study was conducted using *L. corniculatus* in which a soybean gene encoding cytosolic GS (Hirel *et al.*, 1992) fused to the constitutive 35S CaMV promoter was introduced. The aim of this study was to examine if a constitutive over-expression of GS in the cytosol of roots, shoots and nodules had any effect on plant growth and development. When these plants were grown under non-nodulating conditions they had a 50–80% increase in total shoot GS activity, whereas root GS activity was unexpectedly unchanged. Detailed analysis of the transgene expression in roots revealed that, although mRNAs corresponding to the introduced soybean GS gene were highly transcribed, neither the GS protein content nor the corresponding activity were modified. As already found in alfalfa, this result suggests that root GS expression is controlled at the post-transcriptional level (Temple *et al.*, 1998a). These plants were grown under different nitrogen regimes (4 or 12 mM NH_4^+ or symbiotic N_2 fixation) and aspects of carbon and nitrogen metabolism were examined. No changes in ammonium or carbohydrate metabolism were observed when plants were grown at 4 mM NH_4^+. However, when grown at 12 mM NH_4^+ an increase in free amino acids and ammonium in roots was accompanied by a decrease in soluble carbohydrates in transformed plants in comparison to the wild type grown under the same conditions. Labelling experiments using $^{15}NH_4^+$, which allows the monitoring of ammonium influx and its subsequent incorporation into amino acids, showed that both ammonium uptake into the roots and the subsequent translocation of amino acids to the shoots were lower in plants overexpressing GS. Thus, it was concluded that the increase in both ammonium and amino acid concentration in the roots was the result of shoot protein degradation. In addition, early floral development was also observed in the transformed plants. Considering that the changes observed in transformed plants are typically observed in senescing plants it is suggested that expression of cytosolic GS in the shoots may accelerate plant development, leading to early senescence and premature flowering when plants are grown on ammonium-rich medium (Vincent *et al.*, 1997). This result confirmed that cytosolic GS, found to be induced during leaf senescence, plays a major role during the remobilization of leaf nitrogen (Masclaux *et al.*, 2000). Since *L. corniculatus* is not self-fertile it was not possible to investigate if accelerating leaf nitrogen remobilization was beneficial for both seed nitrogen content and seed yield. This aspect of plant nitrogen metabolism is currently under investigation using *L. japonicus* and other legumes and non-legumes that assimilate substantial amounts of nitrogen for further export into the seeds.

Another source of reduced inorganic nitrogen is the ammonia produced in the nodules in the process of dinitrogen fixation. Therefore, further studies were performed using nodulated plants overexpressing constitutively the soybean cytosolic GS gene. In transgenic plants grown under nitrogen fixing conditions, although the soybean GS transgene was expressed at a very high level, GS activity was surprisingly 40% lower in nodules compared to untransformed control plants. However, the decrease in GS activity was accompanied by a 50% increase in shoot, root and nodule biomass. A large increase in amino acids (mostly asparagine) concentration concomitant to a decrease in the soluble carbohydrate content revealed a more efficient ammonium assimilation in the nodules of transgenic plants. A $^{15}N/^{14}N$ dilution experiment was performed to monitor the partitioning of the newly fixed atmospheric nitrogen between nodules, roots and shoots. Results from this experiment showed that the increase in plant biomass followed the onset of symbiotic nitrogen fixation. The newly fixed nitrogen was used primarily to develop new nodules, and then for root and shoot growth. The rate of nitrogen fixation on a dry weight basis remained similar in the transgenic plants compared to the untransformed controls. However, since nodule biomass was 50% higher in transgenic plants it was concluded that nitrogen fixation per plant was enhanced. Following these observations, it was proposed that regulatory mechanisms, as yet unknown, control the balance between nodule GS activity and nodule biomass for optimal NUE and plant growth (Hirel *et al.*, 1997). Whether these regulatory mechanisms are acting in specific nodule cell types or, as in roots, are subjected to a post transcriptional control still needs further investigations. For further experimental work, one suggestion would be to avoid post translational controls by using GS transgenes originating from other eucaryotic or prokaryotic origins. Moreover, the lack of specificity of the CaMV 35S promoter could be circumvented by using promoters driving the expression of the transgene in specific nodule cell types including infected cells, uninfected cells or cortical cells (Forde *et al.*, 1989; Terce-Laforgue *et al.*, 1999). The combination of these two approaches appears now to be a prerequisite to a better understanding of the complex regulation of nodule GS expression, if we want to improve in a more targeted manner nitrogen assimilation in N-fixing legumes.

To achieve this, a first experiment was conducted to direct over-expression of GS specifically in the nodule infected cells. The soybean gene encoding cytosolic GS was fused to the leghemoglobin promoter (*LBC3*) and introduced into *Lotus japonicus* (Harrison *et al.*, 2000). In transgenic plants, nodule GS activity was 2-to 3-fold higher compared to untransformed control plants. After 4 weeks following inoculation by *Rhizobium*, plant biomass was severely decreased in transformed lines showing the strongest increase in GS activity although nodule biomass was unaffected. No differences were seen in the concentration of sugars or free ammonium, except that the free ammonium content of the nodules was reduced by approximately 50%. The amino acid content of the nodules was found to be severely decreased. When the individual amino acids were analyzed low amounts of asparagine were found in the nodules. Asparagine is one of the major long-distance nitrogen transport compounds, especially in nitrogen-fixing legumes and this loss would explain the severely decreased biomass of the shoots. Although nitrogen fixation was not affected, these findings suggest that the up-regulation of nodule GS activity has a detrimental effect on the overall nitrogen metabolism of the nodule. This could be in support of the hypothesis proposed by Oaks (1992) who suggested that there was an inefficient competition between root and nodule glutamine synthetases. Both the negative effect of nodule GS over-expression in *Lotus* and the positive effect of nodule GS inhibition in alfalfa (Knight *et al.*, 1988) and *Lotus* (Vincent *et al.*, 1997) on plant growth allows the conclusion that the enzyme plays a major role in the control of plant productivity. These results also suggest an optimal flux of nitrogen through the ammonia assimilatory pathway is necessary for maximal growth.

GS activity in temperate legumes is a key enzyme controlling also the flux of reduced nitrogen in roots because these species, when they are not under nitrogen fixing conditions, reduce nitrogen mostly in the roots. Experiments were therefore conducted to up-regulate the enzyme specifically in roots by introducing into *L. japonicus* a construct composed of the soybean gene encoding cytosolic GS fused to the *RolD* promoter (Limami *et al.*, 1999). In transgenic plants, a significant increase in root GS activity (25–30%) was observed compared to the non-transformed plants. The most striking result was a 30 to 50% decrease in both shoot and root biomass production of transgenic plants when grown on a nutrient solution containing either nitrate or ammonia as nitrogen source. The shoot and root nitrate content was also significantly modified in transgenic plants. Nitrate concentration in shoots of transgenic plants was approximately twice as high as the control plants, whereas a 37% decrease in the root nitrate content was observed. This suggests that nitrogen metabolism was markedly affected. Free ammonium and carbohydrate content of the roots in the transformed plants showed a 50% decrease, however the amino acid pool was similar to that in the non-transformed plants. Feeding plants with $^{15}NO_3$ revealed that NO_3 uptake was significantly reduced in transformed plants compared to the non-transformed plants. Moreover, in the transformed lines, more of the newly absorbed N was transported to the shoots where it could not be reduced due to the lack of nitrate reductase activity in *Lotus* leaves (Limami *et al.*, 1999). It was therefore hypothesised that by increasing root GS activity the flux of amino acids derived from glutamine and/or glutamate is increased, leading to a decreased nitrate uptake and assimilation by feedback inhibition.

These different attempts to modify ammonia assimilation in shoots, roots and nodules of *Lotus* indicate that the reaction catalysed by the cytosolic GS is of major importance in controlling plant growth and development. In addition, these studies strongly suggest that a fine balance exists between the flux of nitrogen taken up from the soil or produced as the result of biological nitrogen fixation and its optimal utilization by the plant and that GS activity in the cytosol may be one of the main components controlling this balance.

3. Conventional plant breeding and mutagenesis: another way of improving nitrogen use efficiency in legumes

In this section, we will first summarize the results obtained from research programs to define genetic variation in N_2 fixation and to enhance N_2 fixation through selection and breeding with a particular emphasis on those criteria used to define the most metabolically efficient and productive plants. Future prospects will also be presented in light of the results obtained with a preliminary quantitative genetic approach using *Lotus japonicus* as a model plant.

Research to select and breed for enhanced N_2 fixation was mostly performed with common bean and soybean, two tropical legumes using ureides as nitrogen transport molecules (Herridge and Danso, 1995) or with alfalfa and pea, two temperate legumes exporting amides (Pate and Atkins, 1983). Selection criteria allowing the discrimination between high- and low-fixing lines consisted of a choice of morphological, physiological and agronomic traits which facilitated in a precise quantification of the traits of interest (Bliss, 1993). Among those traits, the number or weight of the nodules in bean was found to be inadequate, indicating that a more direct and useful selection trait would be N_2 fixation or the proportion of plant N derived from N_2 fixation (Pfix). This latter trait could be considered for use either for commercial release of new genotypes or as a selection of high fixing parents in breeding programs. In addition, it has been noted that in soybean a strong positive association exists between seed dry weight and fixed N content of the seed, suggesting that selection for high seed yield will probably result in the identification of plants with high rate of nitrogen fixation (Ronis *et al.*, 1985). Similarly in alfalfa, it was shown that the rate of nitrogen fixation in field grown plants mirrored growth rates (Viands *et al.*, 1981).

The capacity of the plant to nodulate and fix N_2 in the presence of soil nitrate, known to inhibit both nodulation and N_2 fixation, appeared to be also one of the most effective selection criteria particularly in soybean. A number of nitrate tolerant and supernodulating mutants of soybean and pea have been isolated and evaluated for their agronomic performance (Gresshoff, 1993; Sagan *et al.*, 1993). Although a number of field trials gave contradictory results, it was generally found that these mutants did not yield as well as the wild type. These results indicate a complex environmental and genetic control of their agricultural performance, which remains still not clearly understood.

Increasing N_2 fixation by optimizing the number and effectiveness of rhizobia in the rooting zone through strain selection and inoculation technique in a given genotype is another possible way to improve plant productivity. For example, Hardason *et al.* (1982) showed that plant selection programs in alfalfa should be conducted with the most effective and competitive strains of indigenous *Rhizobium*.

Despite the occurrence of significant genetic variability observed for the different traits described above, the attempts to select cultivars with improved capacity for

N_2 fixation were rather disappointing (Bliss *et al.*, 1989). One of the two major reasons for this lack of success was the difficulty in combining a single desirable trait like N_2 fixation with other agronomic and yield traits. The difficulty of accurately measuring N_2 fixation in the field was the other reason for this lack of success. Nevertheless, it was generally well established that N_2 fixation is variable in different cultivars (Zhang *et al.*, 1997). One can therefore extrapolate that the subsequent steps involved in ammonia assimilation and nitrogen redistribution and management are also variable in different genotypes or ecotypes of a given legume species.

To illustrate the later assumption and investigate more deeply the physiological and genetic basis of NUE in legumes, four different ecotypes of the model legume *Medicago truncatula* (Barker *et al.*, 1999) were analysed with respect to their potential to fix atmospheric N_2, to assimilate ammonia and to produce biomass (Fig. 1). Two of these ecotypes

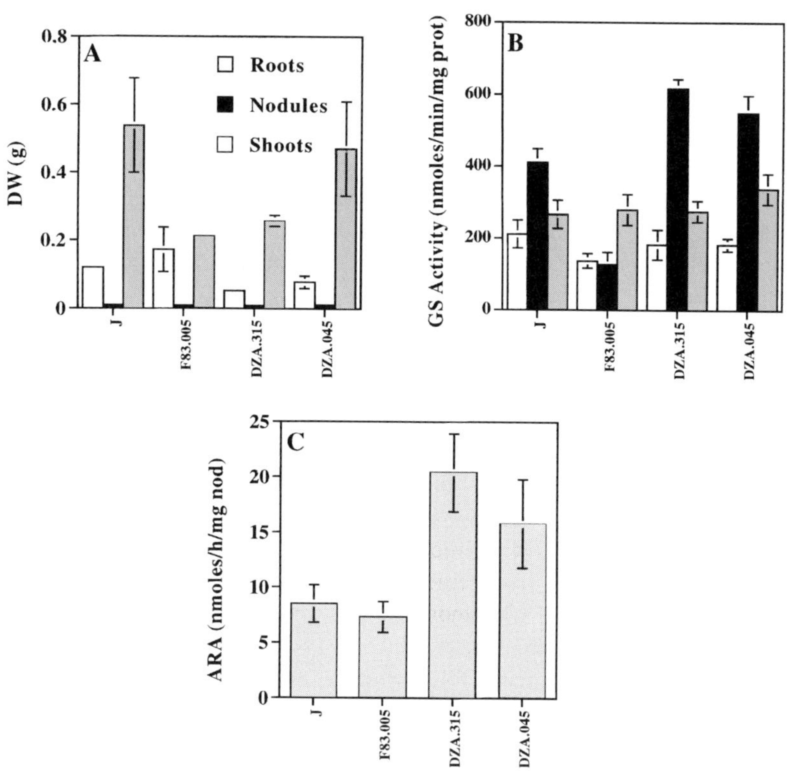

Figure 1. Some characteristics of four Medicago truncatula *ecotypes (J = Jemalong, F83.005, DZA.315 and DZA.045) grown under nitrogen fixing conditions. (A) Dry weight (DW), (B) glutamine synthetase (GS) activity and (C) nitrogen fixation measured using the acetylene reduction assay (ARA) of plants grown on a nitrogen free solution in an aeroponic system over a 3 weeks period following inoculation with* Rhizobium *(Deroche and Babalar, 1987). GS activity was measured according to the protocol described by O'Neal and Joy (1973). Nitrogen fixation was measured as (ARA) described by (Deroche and Babalar, 1987).*

DZA.315 and DZA.045 were collected from the natural population of Algeria. The two others Jemalong and F83.005 also originating from arid areas were collected in Australia and in the south of France respectively (Prosperi, 1993). It was found that acetylene reduction activity (ARA) (Fig. 1B) in the four ecotypes was variable and not necessarily correlated with plant biomass production (Fig. 1A). Interestingly, an increase in ARA capacity was not the result of increased nodule biomass, which suggests that the host may genetically control a more efficient reduction of N_2. In contrast, nodule GS activity (Fig. 1C) matched exactly ARA, indicating that the level of the enzyme activity is a good biomarker for the capacity of the plant to fix atmospheric nitrogen. However, it is likely that these biological characteristics, subjected to complex regulatory mechanisms, are species-specific since it was found that soybean cultivars which are better able to fix N_2 usually have a higher nodule biomass (Zhang *et al.*, 1997). Whether they are under the control of specific developmental and/or metabolic signals originating from the host remains to be determined.

Today, with the application of new techniques such as quantitative genetics, it has become possible to obtain more information on the genetic basis of nitrogen use efficiency in legumes. This powerful tool is based on the use of the genetics of quantitative physiological function and agronomic traits associated to DNA markers (Prioul *et al.*, 1997). The principle consists in using segregating lines derived from two homozygote parents and mapping them with as many RFLP markers so as to link each marker to at least another one (forming a saturated map), after which a statistical relationship between the quantitative trait value and the genotype at each marker is searched for QTL (Quantitative Trait Loci). A significant link means that in the vicinity of the markers there exists at least one gene controlling part of the trait variability. Moreover the use of a RFLP map consisting of a large number of probes for genes of known function allows the search for co-locations with a QTL related to the same function. In such cases, the gene becomes a candidate gene related for the same function. Although successful in identifying QTLs for phenotypic traits in model plants (Alonso-Blanco *et al.*, 1998) or for carbon metabolism in plants of agronomic interest (Pelleschi *et al.*, 1999), to our knowledge, this approach has never been applied to biochemical and agronomic traits related to NUE in legumes. Nevertheless, preliminary studies performed on *Lotus japonicus* RIL have revealed the great potential of using such an approach to identify key enzymes controlling nitrogen assimilation and management.

Considering the key role of GS in inorganic nitrogen assimilation and management and its direct connection with N_2 fixation in legumes it is likely that this enzyme is of major importance in controlling complex traits such as plant growth and development. GS was therefore chosen as a potential candidate gene influencing NUE in legumes. In addition, it is now clearly established that nitrogen metabolites, besides being a source of reduced or organic nitrogen, play a signalling role during the plant life cycle. For example, it has been shown that NO_3^-, ammonia and glutamine are all metabolite signals influencing both plant metabolism and plant development (Trewavas, 1983; Scheible *et al.*, 1997).

In order to reveal a genetic diversity for GS activity and examine the contribution of the enzyme to plant biomass production, a series of 50 recombinant inbred lines were produced (Jiang and Gresshoff, 1997), following the crossing of two *Lotus japonicus* ecotypes, Gifu and Funakura. These two ecotypes, easily distinguishable by their phenotype (Fig. 2), were

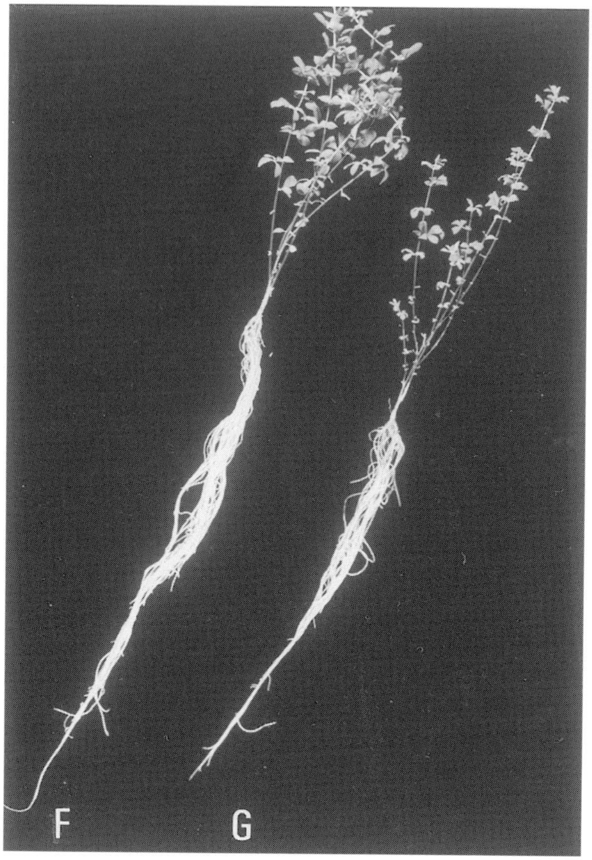

Figure 2. The phenotype of the two Lotus japonicus *ecotypes Gifu (G) and Funakura (F) used to produce the recombinant inbred lines (Jiang and Gresshoff, 1997).*

found to exhibit very different characteristics in terms of nitrate and ammonia assimilation. In particular, when the two *Lotus* ecotypes were grown on a nutrient solution containing nitrate as sole nitrogen source a negative correlation between plant biomass production and root GS activity was clearly observed. After 48 days of culture, Funakura produced 40% more biomass than Gifu but its root GS activity was 10 times lower (Limami *et al.*, 1999). This observation tends to confirm the hypothesis of Oaks (1992) who suggested that there is a competition between roots and shoots in terms of ammonia assimilation, the latter being less efficient. Therefore, a lower root GS activity in the ecotype Funakura could result in more efficient shoot assimilation leading to higher biomass production.

In order to strengthen the conclusion that root cytosolic GS activity is one of the important factors controlling plant biomass production, the distribution of root GS in a recombinant inbred line population derived from a cross between Gifu and Funakura was analysed. Statistical analyses revealed genetic differences between individuals for plant growth and cytosolic root GS activity. These traits were normally distributed (Gaussian distribution) in the F_6 progeny exhibiting a significant negative correlation ($r = -0.38$) which was the same under the two different culture conditions used (Limami *et al.*, 1999). This result indicates that GS activity is a polygenic trait likely to share common QTLs with growth traits and that root cytosolic GS activity is one of the limiting factors that controls plant growth. Following statistical variance analysis, the contribution of root GS activity to plant biomass production was estimated to be around 15% ($R^2 = 0.15$) (Limami *et al.*, 1999). Furthermore, the negative correlation between root GS activity and plant biomass production is consistent with the results obtained with transgenic plants in which a higher root GS activity led to a decrease in growth. Although the physiological explanation for this detrimental effect needs further investigation, it seems probable that GS (in root nitrogen assimilators such as a number of temperate legumes) controls the flux of reduced nitrogen between roots and shoots. Thus, when GS is more active there would be a greater flux of glutamine or derivatives to the shoots, which would lead to a feedback inhibition of inorganic nitrogen absorption and/or reduction to balance the C/N ratio. Plant growth is then adjusted according to an optimum C/N ratio.

The other question arising from this preliminary analysis is the genetic basis of the continuous variation of GS activity in the RILs. As mentioned earlier, several cytosolic GS isoforms occur in plants. In *Phaseolus vulgaris* nodules, nine GS isoforms were identified, differing in the relative proportion of the two subunits β and γ components of the octameric holoenzyme (Cai and Wong, 1989). Although the exact polypeptide composition of GS in roots of *Lotus* is not known, it is logical to think that the relative proportions of the different subunits corresponding to the translation products of different polymorphic genes may account, in part, for the variability in GS activity. Except in the case where there would be only one polymorphic structural gene for GS (in this case the individuals of the progeny would have either the isoform β8 or the isoform γ8 in the proportion 1 : 1), the combination of monomers encoded by unlinked polymorphic genes would result in a number of different octamers (de Vienne and Rodolphe, 1985), their relative abundance differing from one line to the other. In addition, it is well known that within the GS multigene family some members are constitutively expressed while others are subjected to developmental or metabolic regulation (Cren and Hirel, 1999). Therefore the relative abundance of each glutamine synthetase isoform, corresponding to the translation products of differentially regulated polymorphic loci, would influence the protein abundance and turnover and thus the final enzyme activity in a given genotype. Therefore, the occurrence of GS polymorphic loci would have an adaptive significance to naturally select plants suited for a particular environment with respect to carbon and nitrogen nutrition.

As a result of the variation seen between the two ecotypes of *Lotus japonicus* in root GS activity and plant biomass, further studies have been conducted to unravel the changes in nitrate and ammonia assimilation that occur. This work was focused on root nitrate absorption and reduction because *Lotus* reduces inorganic nitrogen mostly in roots as NR activity is lacking in shoots (Mellor and Shread, 1971). Preliminary studies had shown that within the RIL of *Lotus japonicus* described earlier, four different relationships could be found in terms of nitrate content and root or shoot biomass. Representative RILs were selected and along with the two parental lines were grown hydroponically on 10 mM nitrate as the sole nitrogen source for 4 weeks after which various enzymes and metabolites involved in N assimilation were examined. The differences between the lines in terms of the nitrate

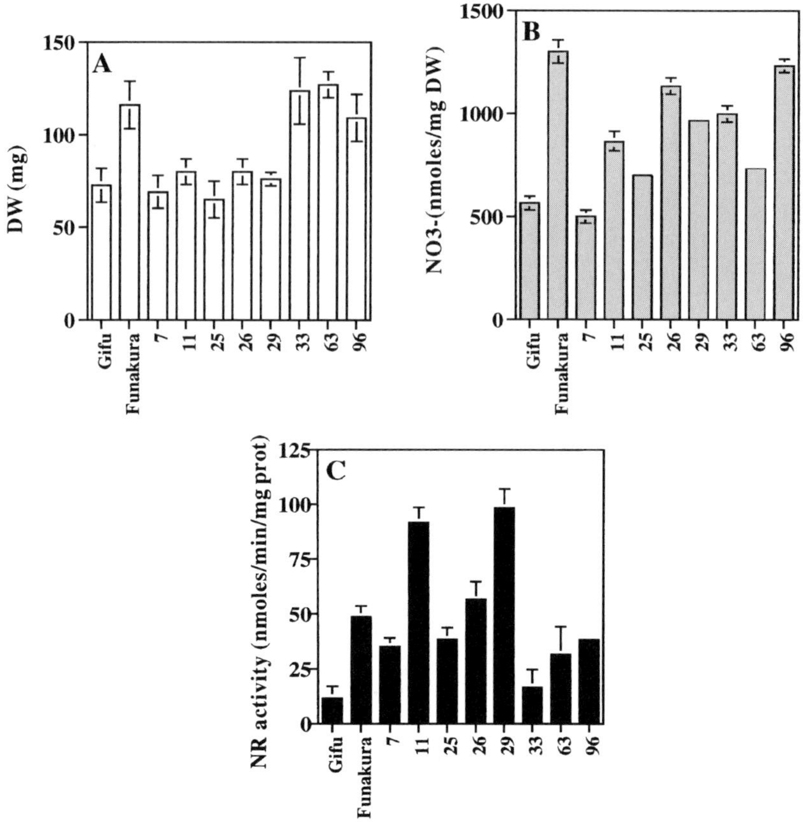

Figure 3. Root dry weight (DW) (A), nitrate content (B) and nitrate reductase (NR) activity (C) in two ecotypes of Lotus japonicus *and some of their RILs after growing for 4 weeks before harvest on 10 mM NO_3^- as the sole nitrogen source.*

content and dry weight of roots (similar differences were obtained for shoot dry weight) are shown in Fig. 3. The parental line Gifu and RILs 7, 11, 25, 26 and 29 were all smaller in biomass than the other parental line Funakura and RILs 33, 63 and 96 (Fig. 3A). However, nitrate contents of roots were highest in Funakura, 11, 26, 29, 33 and 96 (Fig. 3B). Thus, lines 11, 26 and 63 showed inverse relationships between their root nitrate content and root biomass. A Northern analysis (Fig. 4A) revealed that changes in expression of the high affinity nitrate transporter corresponded well with the nitrate content in all lines except line 63, with the lowest level of expression in Gifu and the highest in Funakura and the RIL showing varying intermediate levels of expression. Therefore, differences in root (or whole plant) growth were not always correlated with the uptake of nitrate. In addition, the activity of nitrate reductase was examined in order to give an indication of whether the capacity for nitrate reduction was similar in all lines. Again, differences were seen between the lines (Fig. 3C), in that the lower nitrate containing plants generally had the lowest NR activity, although this was not true for all lines. Furthermore, expression levels of nitrate

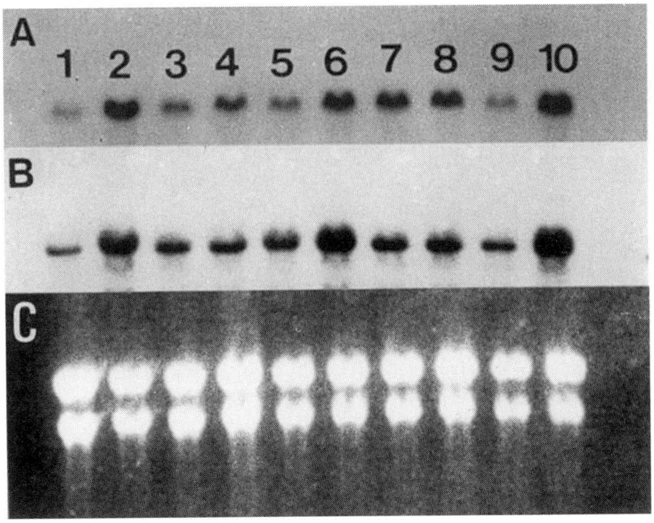

Figure 4. Northern blot analysis of (A) high affinity nitrate transporter and (B) nitrate reductase expression levels in roots of two ecotypes of Lotus japonicus *and some of their RILs. Total RNA was extracted from: Lane 1 – Gifu, lane 2 – Funakura, lane 3 – RIL 7, lane 4 – RIL 11, lane 5 – RIL 25, lane 6 – RIL 26, lane 7 – RIL 29, lane 8 – RIL 33, lane 9 – RIL 63 and lane 10 – RIL 96. 10 μg of total RNA was loaded in each well. Probes for nitrate transporter and nitrate reductase were isolated from* L. japonicus *and provided by Dr. Brian Forde (Rothamsted Experimental Station, Harpenden, UK) and Prof. David Clarckson (Long-Ashton Experimental Station, Bristol, UK) respectively. (C) represents an ethidium bromide stained gel showing equal loading in all lanes.*

reductase (Fig. 4B) did not always correlate with enzyme activity (either potential or actual), which may indicate a post-translational modification. Thus, although NR activity appears to be primarily controlled by nitrate content in some lines, complex interaction between light, sugars and nitrate may well be determining NR activity in others.

These results show that within *Lotus* a significant amount of variation can be found in terms of nitrogen metabolism, and there are many points of control. This variation, likely to be a result of adaptation to growth under differing environmental conditions (Gifu originates from mountainous areas of Japan whilst Funakura is found on the, possibly more nitrate rich, plains), is worth exploiting. Not only could it be useful as a means of gaining a greater understanding of the regulation of nitrogen metabolism in legumes but also, more directly, in selection programs for higher yielding lines.

4. Conclusions and future prospects

From both the field and laboratory experiments performed in the last two decades, it is evident that there are a number of limitations on the ability to genetically improve symbiotic nitrogen fixation and thus NUE in legumes. One of the main reasons for this failure remains the lack of understanding of the fine regulatory mechanisms controlling both the flux of ammonia resulting from atmospheric nitrogen fixation and the management of reduced nitrogen during plant growth and development. In addition, the species-specific particularities in terms of nitrogen metabolites used for export (ureides or amides; Pate and Atkins, 1983) as well as nodule structure and function (determinate and indeterminate; Corby *et al.*, 1983) within the legume family, does not facilitate the identification of common regulatory models depicting the metabolic regulation of nitrogen assimilation and management. Consequently, it is necessary to develop two main approaches, one exploiting the basic knowledge which will be gained from model species (Cook *et al.*, 1997) and the other more focused on the identification of species-specific adaptive regulatory mechanisms (Doyle, 1998). This will allow us to identify why some legumes were selected for their agronomic performances and why they are adapted to particular environmental conditions. This also implies that both of these approaches will have to be conducted both on the host and the bacterial side.

As a result of the idea to transfer fixation abilities to non-legumes, a large number of studies were undertaken to identify the mechanisms controlling the plant bacterial interactions (Denarie *et al.*, 1996) and the developmental program controlling nodule formation and effectiveness (Nap and Bisseling, 1990). In contrast, although most legumes are used for their high nitrogen content in either vegetative organs or seeds much less attention has been given towards the understanding of the regulatory mechanisms controlling reduced nitrogen acquisition and its subsequent translocation and reallocation during the plant life cycle (Tegeder *et al.*, 2000).

The use of transgenic plants and/or mutants altered in the pathway of ammonia assimilation and its subsequent incorporation into organic molecules is of great importance in identifying which enzymes of the pathway may be limiting. Preliminary work performed on the two temperate legumes *Lotus* and alfalfa suggested that GS might be one of these. However, further work is necessary to determine if other enzymes of the pathway such as GOGAT, acting at the crossroad between carbon and nitrogen metabolites (Temple *et al.*, 1998b) play a major role in controlling the fluxes of carbohydrates *via* α-keto glutarate and reduced nitrogen *via* glutamine and glutamate. Since these compounds are thought to play an important role as signalling molecules controlling both the availability of carbon skeletons (Sheible *et al.*, 1997) and the synthesis of a number of amino acids (Ferrario-Mery *et al.*, 2000), modification of their fluxes will certainly be one way to unravel the metabolic control of N assimilation either at the cellular or at the whole plant level.

How these metabolic changes are sensed remains, however, largely unknown in higher plants in general and legumes in particular, mainly because we still do not have any idea of the nature of the receptor molecule capable of sensing the C and N status of the cell. The isolation of regulatory mutants or the discovery of plant homologues to bacterial or yeast regulatory proteins (Halford *et al.*, 1999; Hsieh *et al.*, 1998) may aid in the identification of some of these elements.

Another important aspect of plant nitrogen metabolism is the control and efficiency of nitrogen remobilization. Although it is known that leaf proteins are the main source of nitrogen use to develop new vegetative or reproductive organs during senescence (Feller and Fisher, 1994; Masclaux *et al.*, 2000), our knowledge on the metabolic pathways involved and their regulation is still fragmentary. In addition the molecular mechanisms controlling the relationship between source and sink organs and the export of assimilates are even more obscure. As for primary nitrogen assimilation, both the use of transgenic technology and molecular genetic techniques will certainly help to identify the enzymatic reactions and transport systems (Rentsch and Frommer, 1996) which are limiting, as well as the regulatory mechanisms involved.

The exploitation of natural or targeted genetic variability appears to be a good alternative by which more performant genotypes can be selected for their ability to assimilate or remobilize nitrogen more efficiently. Classical plant breeding was performed empirically by selecting the most appropriate characters in terms of yield or technological characteristics. However, it was very difficult to approach the fine regulatory mechanisms controlling the expression of these characters due to the large number of structural or regulatory genes involved. Now, the exponential development of quantitative genetics originally developed for marker assisted selection (Ribaud and Hoisington, 1998) has open new perspectives toward the understanding of how these genes are regulated and how the regulatory mechanisms interact. Concerning nitrogen assimilation, the preliminary work performed on model legumes is encouraging, as we were able to demonstrate that root or root nodule GS activity, at least in temperate legume species, may be a good marker to select genotypes producing more biomass or having a better capacity to assimilate fixed nitrogen. Further investigations are, however, necessary to identify QTLs associated to the expression of the two quantitative characters and these will be possible when a saturated genetic map is available. Furthermore, the recent development of genome sequencing and mapping projects in a number of model and crop legumes will be a valuable tool to locate precisely the QTLs associated to the desired characters. In addition genetic dissection of the identified QTLs through sequence analysis will certainly allow the identification of possible structural or regulatory genes controlling NUE both during plant growth and according to different environmental conditions.

References

Alonso-Blanco C, El Assal S E D, Coupland G and Koornneef M (1998) Analysis of natural allelic variation at flowering time loci in the landsberg *erecta* and Cape Verde Islands ecotypes of *Arabidopsis thaliana. Genetics*, **149**: 749–764.

Barker D G, Bianchi S, Blondon F, Dattée Y, Duc G, Flament P, Gallusci P, Génier P, Guy P, Muel X, Tourneur J, Dénarié J and Huguet T (1999) *Medicago truncatula*, a model plant for studying the molecular genetics of the *Rhizobium*-legume symbiosis. *Plant Mol. Biol. Rep.*, **8**: 40–49.

Benes V, Pekny V, Skorepa J and Vrba J (1989) Impact of diffuse nitrate pollution sources on groundwater quality – some examples from Czechoslovakia. *Environ. Health Perspect.*, **83**: 5–24.

Bliss F A (1993) Breeding common bean for improved biological nitrogen fixation. *Plant Soil.*, **152**: 71–79.

Bliss F A, Pereira P A A, Araujo R S, Henson R A, Kmiecik K A, McFerson J R, Texeira M G and da Silva C C (1989) Registration of five high nitrogen fixing common bean germplasm lines. *Crop Sci.*, **29**: 240–241.

Cai X and Wong P P (1989) Subunit composition of glutamine synthetase isozymes from root nodules of bean (*Phaseolus vulgaris* L.). *Plant Physiol.*, **91**: 1056–1062.

Cook D R, VandenBosch K, de Bruijn F and Huguet T (1997) Model legumes get the nod. *Plant Cell*, **9**: 275–281.

Corby H D L, Polhill R M and Sprent J I (1983) Taxonomy. In: *Nitrogen fixation, Vol. 3, Legumes* (Ed Broughton W J), Clarendon Press, Oxford, pp. 1–35.

Cren M and Hirel B (1999) Glutamine synthetase in higher plants: Regulation of gene and protein expression from the organ to the cell. *Plant Cell Physiol*, **40**: 1187–1193.

Denarie J, Debelle F and Prome J C (1996) *Rhizobium* lipo-chitooligo-saccharide nodulation factors: Signaling molecules mediating recognition and morphogenesis. *Ann. Rev. Biochem.*, **65**: 503–535.

Devienne D and Rodolphe F (1985) Biochemical and genetic properties of oligomeric structures: A general approach. *J. Theort. Biol.*, **116**: 527–568.

Dilworth M J (1974) Dinitrogen fixation. *Ann. Rev. Plant Physiol.*, **25**: 81–114.

Doyle J J (1998) Phylogenetic perspectives on nodulation: Evolving views of plants and symbiotic bacteria. *Trends Plant Sci.*, **3**: 473–478.

Feller U and Fischer A (1994) Nitrogen metabolism in senescing leaves. *Crit. Rev. Plant Sci.*, **13**: 241–273.

Ferrario-Mery S, Suzuki A, Kunz C, Valadier M H, Roux Y, Hirel B and Foyer C (2001) Modulation of amino acid metabolism in transformed tobacco plants deficient in Fd-GOGAT. *Plant Soil* (in press).

Forde B G and Cullimore J V (1989) The molecular biology of glutamine synthetase in higher plants. *Oxford Surveys of Plant Mol. Cell Biol.*, **6**: 247–296.

Forde B G, Day H M, Turton J F, Wen-jun S, Cullimore J V and Oliver J E (1989) Two glutamine synthetase genes from *Phaseolus vulgaris* L. display contrasting developmental and spatial patterns of expression in transgenic *Lotus corniculatus* plants. *Plant Cell*, **1**: 391–401.

Gatz C and Lenk P (1998) Promoters that respond to chemical inducers. *Trends Plant Sci.*, **3**: 352–358.

Gresshoff P M (1993) Molecular genetics of nodulation in soybean. *Plant Breed. Rev.*, **11**: 275–318.

Halford N G, Purcell P C and Hardie D G (1999) Is hexokinase really a sugar sensor in plants? *Trends Plant Sci.*, **4**: 117–120.

Handberg K and Stougaard J (1992) *Lotus japonicus*, an autogamous, diploid legume species for classical and molecular genetics. *Plant J.*, **2**: 487–496.

Hardarson G, Heichel G H, Barnes D K and Vance C P (1982) Rhizobial strain preference of alfalfa populations selected for characteristics associated with N_2 fixation. *Crop Sci.*, **22**: 55–58.

Harrison J, Brugiere N, Phillipson B, Ferrario-Mery S, Becker T, Limami A and Hirel B (1999). Manipulating the pathway of ammonia assimilation through genetic manipulation and breeding. Consequences on plant physiology and development. *Plant Soil*, **221**: 81–93.

Herridge D F and Danso S K A (1995) Enhancing crop legume N_2 fixation through selection and breeding. *Plant Soil*, **174**: 51–82.

Hirel B, Marsolier M C, Hoarau A, Hoarau J, Branjeon J, Schafer R and Verma D P S (1992) Forcing expression of a soybean root glutamine synthetase gene in tobacco leaves induces a native gene encoding cytosolic enzyme. *Plant Mol. Biol.*, **20**: 207–218.

Hirel B, Phillipson B, Murchie E, Suzuki A, Kunz C, Ferrario S, Limami A, Chaillou S, Deleens E, Brugiere N, Chaumont-Bonnet M, Foyer C H and Morot-Gaudry J F (1997) Manipulating the pathway of ammonia assimilation in transgenic non-legumes and legumes. *Z. Pflanzenernhr. Bodenk.*, **160**: 283–290.

Hsieh M H, Lam H M, van der Loo F and Coruzzi G (1998) A pII-like protein in *Arabidopsis*: Putative role in nitrogen sensing. *Proc. Natl. Acad. Sci. USA*, **95**: 13965–13970.

Jiang Q and Gresshoff P M (1997) Classical and molecular genetics of the model legume *Lotus japonicus. Mol. Plant Microbe Interact.*, **10**: 59–68.

Joy K W (1988) Ammonia, glutamine, and asparagine: a carbon-nitrogen interface. *Can. J. Bot.*, **66**: 2103–2109.

Knight T J and Langston-Unkeffer P J (1988) Enhancement of symbiotic dinitrogen fixation by a toxin-releasing plant pathogen. *Science*, **241**: 951–954.

Lea P J and Ireland R J (1999) Nitrogen metabolism in higher plants. In: *Plant Amino Acids* (Ed Singh B K), Marcel Dekker, Inc., NewYork, Basel, Hong Kong, pp. 1–47.

Limami A, Phillipson B, Ameziane R, Pernollet N, Jiang Q, Roy R, Deleens E, Chaumont-Bonnet M, Gresshoff P M and Hirel B (1999) Does root glutamine synthetase control plant biomass production in *Lotus japonicus* L.? *Planta*, **209**: 495–502.

Masclaux C, Valadier M H, Brugiëre N, Morot-Gaudry J F and Hirel B (2000) Characterization of the sink/source transition in tobacco (*Nicotiana tabacum* L.) shoots in relation to nitrogen management and leaf senescence. *Planta*, **211**: 519–527.

McNally S F, Hirel B, Gadal P, Mann A F and Stewart G R (1983) Glutamine synthetase in higher plants. Evidence for a specific isoform content related to their possible physiological role and their compartmentation within the leaf. *Plant Physiol.*, **72**: 22–25.

Mellor G E and Sheard R W (1971) Comparison of the nitrate metabolism in orchardgrass and birdfoot trefoil. *Can. J. Plant Sci.*, **51**: 399–404.
Miflin B J and Lea P J (1980) Ammonia assimilation. In: *The Biochemistry of Plants* (Ed Miflin B J), Academic Press, New York, pp. 169–202.
Moller H, Landt J, Pederson E, Jensen P, Autrup H and Jensem O M (1990) Endogenous nitrosation in relation to nitrate exposure from drinking water and diet in a Danish rural population. *Cancer Res.*, **49**: 3117–3121.
Morris P F, Layzell D B and Canvin D T (1989) Photorespiratory ammonium does not inhibit photosynthesis in glutamate synthase mutants of *Arabidopsis. Plant Physiol.*, **89**: 498–500.
Nap J P and Bisseling T (1990) Nodulin function and nodulin gene regulation in root nodule development. In: *Molecular Biology of Symbiotic Nitrogen Fixation* (Ed Gresshoff P M), CRC Press Inc. Boca Raton, FL, pp. 181–229.
Oaks A (1992) A re-evaluation of nitrogen assimilation in roots. *Bioscience*, **12**: 103–110.
Oliveira I C and Coruzzi G M (1999) Carbon and amino acids reciprocally modulate the expression of glutamine synthetase in *Arabidopsis. Plant Physiol.*, **121**: 301–309.
Ortega J L, Roche D and Sengupta-Gopalan C (1999) Oxidative turnover of soybean root glutamine synthetase. *In vitro* and *in vivo* studies. *Plant Physiol.*, **119**: 1483–1495.
Pate J S and Atkins (1983) Nitrogen uptake, transport and utilization. In: *Nitrogen Fixation, Vol. 3, Legumes* (Ed Broughton W J), Clarendon Press, Oxford, pp. 1–35.
Pawlowski K (1997) Nodule-specific gene expression. *Physiol. Plant.*, **99**: 617–631.
Pelleschi S, Guy S, Kim J Y, Pointe C, Mahe A, Barthes L, Leonardi A and Prioul J L (1999) *Ivr2*, a candidate gene for a QTL of vacuolar invertase activity in maize leaves: Gene-specific expression under water stress. *Plant Mol. Biol.*, **39**: 373–380.
Prioul J, Quarrie S, Causse M, and de Vienne D (1997) Dissecting complex physiological functions into elementary components through the use of molecular quantitative genetics. *J. Exp. Bot.*, **48**: 1151–1163.
Properi J M (1993) Selection of annual medics for French Mediterranean regions. In: *Workshop on Introducing the Ley Farming Systems to the Mediterranean Basin* (Eds Christiansen S, Materson L, Farcinelli M and Cock P), Perugia, Italy, pp. 173–191.
Rentsch D and Frommer W B (1996) Molecular approaches towards understanding of loading and unloading of assimilates in higher plants. *J. Exp. Bot.*, **47**: 1199–1204.
Ribaud J M and Hoisington D (1998) Marker assisted selection: new tools and strategies. *Trends Plant Sci.*, **3**: 236–239.
Robertson J G, Farnden K J F, Warburton M P and Banks J A M (1975) Induction of glutamine synthetase during nodule development in lupin. *Aust. J. Plant Physiol.*, **2**: 265–272.
Ronis D H, Sammons D J, Kenworthy W J and Mesinger (1985) Heritability of total and fixed N content of the seed in two soybean populations. *Crop Sci.*, **25**: 1–8.
Sagan M, Ney B and Duc G (1993) Plant symbiotic mutants as a tool to analyse nitrogen nutrition and yield relationship in field grown peas (*Pisum sativum* L.). *Plant Soil*, **153**: 33–45.
Scheible W R, Gonzalez-Fontes A, Lauerer M, Miller-Rôber B, Caboche M and Stitt M (1997) Nitrate acts as a signal to induce organic acid metabolism and repress starch metabolism in tobacco. *Plant Cell*, **9**: 783–798.
Shantharam S and Mattoo A K (1997) Enhancing biological nitrogen fixation: An appraisal of current and alternative technologies for N input into plants. *Plant Soil*, **194**: 205–216.
Streeter J G (1991) Transport and metabolism of carbon and nitrogen in legume nodules. *Advances Bot. Res.*, **18**: 129–187.
Tegeder M, Offler C E, Frommer W B and Patrick J W (2000) Amino acids transporters are localized to tranfer cells of developing pea seeds. *Plant Physiol.*, **122**: 319–325.
Temple S J, Bagga S and Sengupta-Gopalan C (1994) Can glutamine synthetase activity be modulated in transgenic plants by the use of recombinant DNA technology? *Biochem. Soc. Trans.*, **22**: 915–920.
Temple S J, Bagga S and Sengupta-Gopalan C (1998a) Down-regulation of specific members of the glutamine synthetase gene family in alfalfa by antisense RNA technology. *Plant Mol. Biol.*, **37**: 535–547.
Temple S J, Vance C P and Gantt J S (1998b) Glutamate synthase and nitrogen assimilation. *Trends Plant Sci.*, **3**: 51–56.
Temple S J, Heard J, Ganter G, Dunn K and Sengupta-Gopalan S (1995) Characterization of nodule-enhanced glutamine synthetase from alfalfa: Nucleotide sequence, *in situ* localization, and transcript analysis. *Mol. Plant Microbe Intract.*, **8**: 218–227.

Terce-Laforgue T, Carrayol E, Cren M, Desbrosses G, Hecht V and Hirel B (1999) A strong constitutive positive element is essential for the ammonium-regulated expression of a soybean gene encoding cytosolic glutamine synthetase. *Plant Mol. Biol.*, **39**: 551–564.

Thykjaer T, Danielsen D, She Q and Stougaard J (1997) Organization and expression of genes in the genomic region surrounding the glutamine synthetase gene *gln 1* from *Lotus japonicus. Mol. Gen. Genet.*, **255**: 628–636.

Trewavas A J (1983) Nitrate as a plant hormone. In: *Interactions between Nitrogen and Growth Regulators in the Control of Plant Development* (Ed Jackson M B), British Plant Growth Regulator Group, Wantage, Oxfordshire, pp. 97–110.

Viands D R, Barnes D K and Heichel (1981) Nitrogen fixation in alfalfa. Responses to bi-directional selection for associated characteristics. *U.S. Dep. Agricul. Tech. Bull.*, **1643**.

Vincent R, Fraisier V, Chaillou S, Limami M A, Deleens E, Phillipson B, Douat C, Boutin J P and Hirel B (1997). Overexpression of a soybean gene encoding cytosolic glutamine synthetase in shoots of transgenic *Lotus corniculatus* L. plants triggers changes in ammonium assimilation and plant development. *Planta*, **201**: 424–433.

Waters J K, Hughes II B L, Purcell L C, Gerhardt K O, Mawhinney T P and Emerich D W (1998) Alanine, not ammonia, is excreted from N_2-fixing soybean nodule bacteroids. *Proc. Natl. Acad. Sci. USA*, **95**: 12038–12042.

Zhang F, Dijak M, Smith D L, Lin J, Walsh K, Voldeng H, Macdowell F and Layzell D B (1997) Nitrogen fixation and nitrate metabolism of six diverse soybean (*Glycine max* (L.) Merr) genotypes under low temperature stress. *Environ. Exp. Bot.*, **38**: 49–60.

PART III
ABIOTIC AND BIOTIC STRESS TOLERANCE

GENETIC ENGINEERING FOR ENHANCING ABIOTIC STRESS TOLERANCE

RANA P. SINGH, USHA, S.M.H. RIZVI, SONIA AND PAWAN K. JAIWAL
Department of Biosciences, M.D. University, Rohtak – 124 001, India
e-mail: rana_psingh@rediffmail.com

Abstract

A major factor, which reduces the potential yield of legume crops very significantly, is their sensitivity to abiotic stresses, e.g. drought, flooding, heavy metals, high and low temperature, salinity, alkalinity and acidity of the soil, ozone and UV-irradiations, etc. Stress signals are perceived by the plants and their transduction within the cellular compartments can modulate the gene expression leading to the synthesis and build up of the strategic biomolecules which help the plants to develop stress tolerance. The signal perception involves membrane receptors. The perceived signal is further transduced into the cells by a change in the levels of intracellular signaling molecules known as secondary messengers. Certain protein-modifiers, scaffolds and adapters, etc. are also known to be involved in regulating the abiotic stress signals in plants as the signaling partners, which provide physical supports for the certain signaling events. Many stress responsive genes, promoters and transcription factors have been identified in the plants including the legumes. The transcription factors modulate the expression of a cascade of stress inducible genes to impart tolerance to the stressed plants. Several stress related proteins have been characterized in plants, which are involved in conferring the stress tolerance by various ways. Compatible solutes provide osmoprotection and scavenge free radicals generated during the stress. Antioxidative defense system is also induced during the various abiotic stresses, which provide a quick and efficient removal of toxic free radicals, e.g. reactive oxygen species (ROI). Recent studies indicate that the manipulation of specific key regulatory steps, singly or in combinations or by a cascade of genes using transcription factors through genetic engineering, can confer the stress tolerance in transgenic plants. Most of the studies, however, are related to model plants like *Arabidopsis* and *Nicotiana* which have many differences with legumes. Direct studies with legumes or a leguminous model plant like *Lotus japonicus* or *Medicago truncatula* can provide better insight to understand the complex interactions of abiotic stress signaling and tolerance mechanisms and strategic target points can be characterized to achieve stable stress tolerance to the abiotic stresses in legumes.

P.K. Jaiwal and R.P. Singh (eds.), Improvement Strategies for Leguminosae Biotechnology, 223–243.

1. Introduction

The present century is experiencing the burgeoning population to exert pressure for increase in agricultural productivity. The legume crops are at the centre of concern, because of increase in population with a habit of consuming vegetarian dietary proteins, mainly through legume pulses and vegetables. Also there is an upsurge in vegetarian protein diet throughout the world as a consequence of health consciousness. On the other hand, the production of legumes has been stagnant during the last four decades and the so-called green revolution has been limited to the cereals, e.g. wheat, rice and maize, etc. The field crops experience many stress conditions, i.e. biotic and abiotic stresses, which reduce their potential yield very significantly in the stress-prone ecosystems (Grover *et al.*, 1998). The abiotic stresses include drought, flooding, heavy metals, high and low temperatures, salinity, alkalinity and acidity of the soil, ozone, and UV-irradiations, etc. Yield losses in cool season food legumes due to drought ranges from 30 to 60%, depending upon the geographical locations, and the climatic conditions, during the crop season (Saxena *et al.*, 1993). About 20% of the world's cultivated lands and nearly half of all irrigated lands are affected by salinity; moreover, the soil salinity is increasing every year in the irrigated agriculture in several areas (see Rhoades and Loveday, 1990; Gulati and Jaiwal, 1997; Zhu, 2001; Xiong and Zhu, 2001). India is the major producer of chickpea, pigeonpea and mungbean, where these legumes are grown in marginal lands and suffer very significantly due to the abiotic stresses. Most of the cultivated grain legumes are sensitive to such stresses and their production, despite the attempts made for breeding high yielding cultivars and increase in the cultivation area, remained stagnant during the last decades. The wild relatives, on the other hand, generally show a better stress tolerance as compared to their cultivated counterparts.

The extent of damage by the abiotic stresses varies with the plant species and the developmental stage of the plants. A wide variety of plants, moreover, have common responses for the stress perception and signaling and adaptations to the stresses. Stress signals up- or down-regulate the genes coding for specific stress related proteins. The genes for some of these proteins have been isolated, characterized and cloned (see Grover *et al.*, 1998, 2001; Zhu, 2001; Xiong and Zhu, 2001). Most of the stresses also affect the water status of the plants resulting in the impairment of major vital metabolic processes (Bohnert *et al.*, 1995). Certain stress related metabolites, e.g. proline, glycine betaine, polyamines, polyols, oligosaccharides, etc. have been reported to accumulate under the stress conditions (Bohnert *et al.*, 1995; McNeil *et al.*, 2000; Zhu, 2001). Much of the injury to plants caused by the stress exposure is associated with the oxidative damage at the cellular level (Allen, 1995; Zhu, 2001). Antioxidant defense system of plants includes enzymes, such as superoxide dismutase, ascorbate peroxidase and glutathione reductase, etc., and certain low molecular weight metabolites, e.g. ascorbic acid and glutathione, etc, which are also induced under the stress perception (Allen, 1995; Wellburn *et al.*, 1998; Jimenez *et al.*, 1998; Knorzer *et al.*, 1999; Velikova *et al.*, 2000; Zhu, 2001; Grover *et al.*, 2001).

The present chapter analyzes the current status and state of art for various molecular strategies adapted to modulate the stress tolerance in major grain legumes using the biotechnological tools and techniques. The key regulatory steps involved in the abiotic stress tolerance in legumes and attempts to obtain the stress tolerant genotypes by plant genetic engineering have been discussed.

2. Perception and transduction of stress signal

In an ecosystem the abiotic stresses to the plants may be lethal, sub-lethal or a sub-lethal stress preceded by a lethal stress (see Grover *et al.*, 2001). Plant responses to stress starts from the successful perception of the stress signals, its transduction within the cellular compartments and modulation of gene expression leading to the synthesis and build up of the strategic biomolecules. In case of direct lethal stress, whole plant senescence starts and plant death occurs. Plants selectively tolerate the sub-lethal stress, even those preceded by a lethal one by evolving various strategies at physiological, biochemical and molecular levels.

Occurrence of the cross-tolerance in plants, i.e. protection against the other stresses on exposure to one stress indicates the existence of common signal transduction pathway for the various stresses (Schoffl *et al.*, 1998; Sabehat *et al.*, 1998; Zhu, 2001; Xiong and Zhu, 2001; Shanker and Srivastava, 2002). Recently it has been demonstrated that a tobacco stress induced gene *Tsi1* acts as a positive transacting factor, which might be involved in two separate signal transduction pathways under abiotic and biotic stresses (Park *et al.*, 2001). However, looking at the recent evidence for a complex cross-talk signaling system in the plant cell, there seems to be specific signal perception and transduction pathways too, in addition to the common ones, which can identify different kinds of stress signals. For example, one of the C-repeat dehydration responsive element, *CBF1*, functions primarily in cold response, whereas other binding protein, DREB2, is involved in drought response (Liu *et al.*, 1998; Kasuga *et al.*, 1999).

Plant signal transduction mechanism is almost similar to that in the animals and involves membrane receptors, e.g. protein kinases, G-proteins, GTP-binding proteins, cyclic nucleotides, and Ca^{+2}/calmodulin-phospholipase, etc. (Grover *et al.*, 2001). The most important steps have been suggesed as receptor-coupled phosphorelay, phosphoinositol-induced Ca^{+2} changes, mitogen-activated protein kinase (MAP-kinase) cascade and transcriptional activation of stress responsive genes (Xiong and Zhu, 2001). In addition, Xiong and Zhu (2001) have suggested the involvement of protein post translational modifications and adapter and scaffold-mediated protein-protein interactions in abiotic stress signal transduction. The extra-cellular stress signal binds to a transmembrane receptor which in turn activates GTP-binding proteins to regulate either a cascade of kinases (e.g. MAP-kinases) or a G-protein effector (e.g. adenylate cyclase, cAMP, phoshodiesterase, phospholipase-C, ion channels, etc). A transmembrane hybrid-type histidine kinase (ATHK1) functions as an "osmosensor", and has been reported in *Arabidopsis* (Urao *et al.*, 1999). This novel histidine kinase transmits the stress signal to a downstream MAP kinase cascade. Similarly, Reggiani and Laoreti (2000) have demonstrated the involvement of phospholipase C in the anaerobic signal transduction in rice.

It has been demonstrated that membranes are involved in various signal transduction pathways (see Shinozaki and Yamaguchi-Shinozaki, 1997). The role of membrane fluidity in temperature perception and transduction of signals in higher plants have been studied. However, the molecular mechanisms elucidating the nature of sensor proteins that perceive changes in membrane fluidity and changes in their own configuration after perception of the stress signals are yet to be understood. It has recently been suggested by Xiong and Zhu (2001) that G-protein associated receptors may be participating in the perception of environmental signals in plants, though it seems that plants do not use a large number of hormone and neurotransmitter receptors of wide range as extensively used in animals. Phospholipid metabolism is involved in hyperosmotic stress responses in plants. After perceiving the signals through the membrane receptors, cells often utilize multiple phosphoprotein cascades to transduce and amplify the information for downward transmission (Xiong and Zhu, 2001). Protein phosphorylation and dephosphorylation involving many protein kinases and phosphatases perform such signaling events (see Xiong and Zhu, 2001). These steps regulate a wide range of cellular processes such as enzyme activation, assembly of macromolecules, protein localization and degradation, etc.

Recently, Takahashi *et al.* (2001), have demonstrated a rapid and transient increases in inositol 1,4,5-triphosphate (IP_3) independent of abscisic acid in (ABA) *Arabidopsis* cell cultures induced by hyperosmotic stress. The transient IP_3 production occurs due to the activation of phosphoinositide-specific phospholipase-C (PI-PLC) in response to hypertonic stress, and the involvement of PI-PLC and IP_3, in an ABA-independent hyperosmotic stress signal transduction pathway has been suggested in higher plants (Takahashi *et al.*, 2001). De Wald *et al.* (2001) have recently suggested that in response to salt and osmotic stresses, *Arabidopsis* uses a signaling pathway in which a small but significant portion of phosphoinositide phosphatidyl-inositol 4,5-biphosphate [Ptd Ins (4,5) P_2] is hydrolyzed to IP_3. It has been demonstrated that the accumulation of IP_3 occurs during a time frame similar to that observed for stress-induced calcium mobilization. It appears, therefore, that the majority of the Ptd Ins (4,5) P_2 synthesized in response to salt and osmotic stress may be utilized for cellular signalling events distinct from the canonical IP_3 signalling pathway (De Wald *et al.*, 2001). Sanchez and Chua (2001) have proposed that phospholipase C_1 (encoded by *ALPLC1)* plays a role in secondary responses to ABA signals. Hyperosmotic stress induces the rapid phosphorylation of a *Glycine max* phosphatidylinositol transfer protein (PITP), homolog through activation of the protein kinases SPK1 and SPK2 (Monks *et al.*, 2001). The participation of Ssh1p, a soybean PITP-like protein, in the early events of osmosensory signal transduction in plants, a function not attributed previously to animal or yeast PITPs, throws new light on signal transduction systems in plants. Exposure of plant tissues to hyperosmotic stress led to the rapid phosphorylation of Ssh1p, a modification that decreased its ability to associate with the membranes. This study reveals that the activation of SPK1 and/or SPK2 and the subsequent phosphorylation of Ssh1p act as two early successive events in a hyperosmotic stress induced signaling cascade in plants. Furthermore, Ssh1p is shown to enhance the activities of a plant phosphatidylinositol 3-kinase and phosphidylinositol 4-kinase. This observation suggests that the ultimate function of Ssh1p in cellular signaling is to alter the plant's capacity to synthesize phosphoinositides during the period of hyperosmotic stress (Monks *et al.*, 2001). A membrane associated 48-kDa phospholipase A_2 (PLA_2), isolated from the leaves of *Vicia faba*, has been suggested to play an important role in an elicitor-induced defense response (Jung and Kim, 2000). Expression of some calcium dependent protein kinases (CDPKs) has been shown to be induced by physical stress, salt stress and $CaCl_2$ in *Vigna radiata* (Botella *et al.*, 1996). The role of such signalling molecules in perception of the abiotic stress responses in legumes needs attention to explore the sequential events involved.

The perceived stress signal is further transduced into the cell by a change in the levels of intracellular signaling molecules known as secondary messengers, such as cAMP, cGMP, protein kinase-C, Ca^{+2} dependent kinases and calmodulin-dependent kinases, etc. (Redhead and Palme, 1996; Murata and Los, 1997; Shinozaki and Yamaguchi-Shinozaki, 1997; Knight *et al*., 1997; McAinsh and Hetherington, 1998; Grover *et al*., 2001; Xiong and Zhu, 2001; Zhu, 2001). In addition, many other molecules are also known to act as messenger signal molecules in the transmission of the stress signals, e.g. hydrogen peroxide (H_2O_2), salicyclic acid (SA), abscisic acid (ABA), ethylene, jasmonic acid, polyamine, nitric oxides, etc. (Kato-Naguchi, 1999; Shanker and Srivastava, 2002). The evidences for the involvement of these molecules are based on an increase in their cellular levels during the stress, and the protection of plants against the stress, when these molecules are supplied exogenously along with the simulated stress conditions. For example, in bush bean plants, 75 mM NaCl caused a 3-fold increase in the ABA content within 2 h (Montero *et al*., 1997).

In recent years, the role of Ca^{+2} as secondary messenger, in transduction of the stress signals, has been documented extensively (Knight *et al*., 1997, 1998, 1999; McAinsh and Hetherington, 1998; Sanders *et al*., 1999; Clark *et al*., 2001; Thomashow, 2001; Xiang and Zhu, 2001). The Ca^{+2}-mediated stress signal transduction involves Ca^{+2}-transporters, calmodulin-(CaM) and CaM-dependent protein kinases, etc. (see Grover *et al*., 2001; Xiang and Zhu, 2001). In *Arabidopsis* and *Medicago sativa*, the cytoplasmic Ca^{+2} levels increase rapidly in response to the low temperature (Monroy *et al*., 1993, 1998; Thomashow, 2001). This increase in Ca^{+2} influx induces at least some C-repeat dehydration responsive elements (*CRT*/*DRE*) regulated genes that impart responsiveness to low temperature and dehydration. The gene encoding Ca^{+2}-dependent protein kinases have been isolated from many plants including legumes. *Vigna radiata* and *Glycine max* have been reported to express them under the environmental stresses (see Ellard-Ivery *et al*., 1999). Little is known, however, about the steps between Ca^{+2} influx and the activation of the gene expression, but it appears that protein phosphorylation may be involved as an underlying mechanism (Monroy *et al*., 1998). The kinetics and magnitude of Ca^{+2} signal differ between different stimuli and are thought to contribute to the specificity of the end response (Knight *et al*., 1998). The altered Ca^{+2} response may encode a "memory" of the previous stress encounters, and thus may perhaps be involved in acclimation to the environmental stresses. Recently, Knorpp and coworkers (Galvis *et al*., 2001) have suggested that the mitochondrial diphosphate kinase (mtNDPK), which catalyzes transfer of the phosphoryl group from GTP to ATP, is involved in the heat stress response in *Pisum sativum*, possibly as a modulator of the novel 86-kDa heat shock proteins. Two genes, *HOS1* and *HOS2* (high expression of osmotically responsive genes), which seem to encode negative regulators of the low temperature signal transduction, have been reported (see Thomashow, 2001).

It has been suggested that certain protein-modifiers, scaffolds and adapters are also involved in regulating the abiotic stress signals in plants as the signaling partners (see Xiong and Zhu, 2001 for a recent review). These signaling partners are protein modifiers (other than protein kinases/phosphatases) and scaffolds or adaptors that provide various physical supports for the certain signaling events. The common proteins modifications include phosphorylation, acetylation, methylation, ADP-ribosylation, glycosylation, myristoylation and isopenylation, etc. The cascade of events involved in the signal perception and transduction under the abiotic stress in legumes is still in infancy. Most of the studies have been done in the model plant systems, e.g. *Arabidopsis* and *Nicotiana*, which do not have N_2-fixing systems. Studies in the model legumes, *Lotus japonicus* and *Medicago truncatula*, may provide an insight to understand the gaps in the knowledge on signal transduction pathways in legumes. Transgenic plants that modify the expression of stress responsive genes, and mutants having stress sensitive and tolerant phenotypes, may provide more insights in understanding of the molecular mechanisms of the stress responses to these plants. Manipulation in key steps in signal perception and signal transduction to reduce the stress sensitivity of the exposed cells, i.e. low level of constitutive expression of the stress perception gene(s), are considered as a possible way to bring out enhanced stress tolerance (see Grover *et al.*, 1998). On the other hand, overexpression of *DREB1* in transgenic plants has been shown to enhance tolerance to drought, salt and freezing (Kasuga *et al.*, 1999).

3. Stress responsive genes, promoters and transcription factors

Abiotic stresses regulate the expression of stress inducible genes and studies on their regulation and function will provide new insights into the understanding of stress tolerance mechanisms (Chai *et al.*, 1998; Sengupta *et al.*, 1999; Raymond *et al.*, 2000; Grover *et al.*, 2001; Knight and Knight, 2001; Xiong and Zhu, 2001; Zhu, 2001). A large number of stress inducible genes are becoming available in cloned form and many studies have been followed by analysis of their promoters to determine *cis*-acting elements involved in the regulated expression of these genes. Most of the stress related promoters contain an array of stress specific *cis*-acting elements that are recognized by the requisite transcription factors (see Grover *et al.*, 2001; Xiong and Zhu, 2001; Thomashow *et al.*, 2001). Many genes that are induced by cold are also induced by drought or ABA. For example, the drought inducible genes have been isolated from *Vigna unguiculata* and one of these genes, *VuNCED 1*, encodes a *cis* epoxy carotenoid dioxygenase that catalyzes the key steps in ABA biosynthesis under drought stress (Iuchi *et al.*, 2000). Several genes are induced by drought, salt and cold in *aba* (ABA deficient) or *abi* (ABA insensitive) *Arabidopsis* mutant. This suggests that these genes do not require ABA for their expression under cold or drought conditions but do respond to exogenous ABA. Two separate regulatory systems function in gene expression during drought and cold stresses, one is ABA independent and the other is ABA dependent. The *rd29A* is one of the many cold and drought inducible genes that has been found to contain a novel *cis*-acting element, dehydration responsive element (*DRE*) or C-repeat (*CRT*) consisting of 9 bp conserved sequences, TACCGACAT, in addition to *ABRE* (abscisic acid responsive elements). The *DRE* is involved in the first rapid responses of *rd29a* to conditions of dehydration, high salt and low temperature but not in ABA responsive slow expression in which *ABRE* is involved. The *DRE* related motif has been reported in the promoter region of several cold and drought-inducible genes. A similar motif (C-repeat TGGCCGAC) has also been reported in the promoter region of cold inducible gene *cor15a* in *Arabidopsis* (Thomashow, 1990; Thomashow *et al.*, 2001). The CCGAC core sequence has been found in the promoter region of the cold inducible gene *bn115* and designated as low temperature responsive element (*LTRE*). The regulation of heat shock genes is mediated by the core heat shock element (*HSE*), located in the promoter region of *hsp* genes (Grover *et al.*, 2001). *Cis*-acting HSE sequence is present in almost all *hsp* genes sequenced so far. The presence of at least 3–5 bp modules (nGAAn) arranged as continuous repeats – nGAAnnTTCnnGAAn – is the key feature of HSEs.

Genes such as *rd29a* contain both *DRE* and *ABRE* elements in their promoters and can be activated by ABA dependent and ABA independent pathways. A single *ABRE* element cannot function independently, for instance, *rd29b* has two copies (see Thomashow *et al.*, 2001). The *rd29b* has both *ABRE* and *DRE*, suggesting that *ABRE* require the *DRE* for ABA induced expression. The promoter region of anaerobic inducible genes, *adh1*, contain a string of specific nucleotide bases called the "Anoxia Responsive Element" with a consensus sequence of its core element as TGGTTT, which is required for anaerobic induction of transcription.

This *cis*-acting promoter sequences interact with specific protein for their activation. Such proteins are generally termed as transcription factors. For the regulation of cold and drought induced gene *rd29a*, two groups of transcription factors *DREB1* and *DREB2* bind to *DRE* sequences in the promoter region of the *rd29a* gene. The *DREB1* and *DREB2* genes encode structurally different proteins that are induced specifically by the low temperature and by salt or drought respectively. Overexpression of the *DREB1*, a cDNA in *A. thaliana* under the constitutive promoter 35S, not only induced a strong expression of the target gene (stress inducible gene), but also caused dwarf phenotypes, which showed tolerance to freezing, drought and salinity stresses (Liu *et al.*, 1998). However, overexpression of *DREB2* under the stress induced promoter from *rd29A* gene caused minimal negative effect on plant growth and greater tolerance to stress was observed (Kasuga *et al.*, 1999).

On the other hand, overexpression of tobacco *Tsi1* gene encoding *EREBP/AP$_2$* type transcription factor enhances resistance against pathogen attack and osmotic stress in tobacco, suggestingt that *Tsi1* might be involved as a positive transacting factor in two separate signal transduction pathways under abiotic and biotic stresses (Park *et al.*, 2001). Winicow and Bastola (1999) identified Alfin1 cDNA, which encodes a putative transcription factor that binds in the promoter region of a NaCl-inducible gene *MsPRP$_2$*. The function of Alfin1 was tested in transgenic alfalfa under the control of 35S promoter in sense and antisense orientation with *MsPRP$_2$* as a reporter gene. Transgenic calli or plants overexpressing Alfin1 in the sense orientation showed not only enhanced level of *MsPRP$_2$* mRNA accumulation but also found resistance to growth inhibition by 171 mM NaCl. In contrast, the antisense transgenic plants grew poorly under NaCl conditions. This suggests that Alfin1 activates probably all the genes associated with salt tolerance in alfalfa.

For the regulation of HS promoter, heat shock transcription factor (*HSF*) genes have been identified and cloned from plants like *A. thaliana, Zea mays* and *Glycine max* that bind specifically to *HSEs* (see Grover *et al.*, 2001). Overexpression of *HSF1* and *HSF3* causes plant to constitutively express at least some HSPs and produce somewhat higher basal thermotolerance (Lee *et al.*, 1995; Prandl *et al.*, 1998). Recently another *HSF, HSP101* has been identified whose overexpression plays a pivotal role in not only basal but also in acquired thermotolerance in *Arabidopsis* (Queitsch *et al.*, 2000) and *Phaseotus lunatus* (Keeler *et al.*, 2000).

The regulation of cold inducible gene (*cor*) expression is controlled by the *CBF* family of transcriptional activators (Gilmoure *et al.*, 1998, 2000), or also known as *DREB1* protein (Liu *et al.*, 1998), which binds to *CRT/DRE* sequences in promoter region of *cor* genes. *CBF1, CBF2* and *CBF3* (also known as *DREB1b, DREB1c* and *DREB1a* respectively), are the *CRT/DRE* binding factors discovered for overexpression of *CBF1* and *CBF3* in *Arabidopsis*. They were subsequently known to activate expression of the entire battery of known *CRT/DRE* regulated *cor* genes and to enhance whole plant freezing survival without a low temperature stimulus (Jaglo-Ottosen *et al.*, 1998; Thomashow *et al.*, 2001).

Several different DNA binding proteins interact with *adh1* gene promoter of maize. According to the sequence of their binding sites, the *adh1* gene promoter interacting proteins are of following types. (a) those that have a 5′ GTGG 3′ within their binding sites and (b) those that have 5′ GCCCC 3′ sequences in the same. It is speculated that GTGG binding protein may represent a group of general transcription factors while the proteins that interact with the GCCCC sequences are uniquely a part of *ARE* (see Grover *et al.*, 2001). A protein complex (ARF-B2) specifically binds to part of the anaerobic responsive element of maize *adh1*.

Engineering overexpression of transgenic factor genes induced stress tolerance by altering levels of several target genes at the same time. This would definitely enhance the resistance levels as compared to single gene manipulations. However, a very few transcription factors have been isolated to date. Isolation and characterization of more transcription factors require urgent attention.

4. Abiotic stress inducible proteins in legumes

In response to the abiotic stresses crop plants produce a large number of proteins which show up/down regulation due to stresses as documented by one/two dimensional protein gel analysis (see Grover *et al.*, 1998, 2001; Thomashow, 2001; Zhu, 2001; Shankar and Srivastava, 2002 for recent reviews). The proteins which are up-regulated by the stress conditions are referred as stress proteins. Several proteins have been characterized in the crop plants including legumes in response to high temperature, cold, salinity, drought and heavy metals.

One group of the most extensively characterized stress proteins is the heat shock proteins (HSPs), which are mostly encoded by nuclear genes, but are localized in different cell compartments, e.g. cytoplasm, mitochondria, chloroplast and endoplasmic reticulum, etc. (see Grover *et al.*, 2001). These proteins are generally produced in response to high temperature stress, however, the synthesis and accumulation of such proteins in plants have also been documented in the presence of water stress, abscisic acid, salinity, chilling and anoxia, etc. apart from high temperature stress (see Shankar and Srivastava, 2002 for a recent review).

Many legumes, e.g. *Glycine max*, *Phaseolus vulgaris*, *P. aureus*, *Pisum sativum* and *Vigna radiata* have been studied for the synthesis and accumulation of HSPs of various molecular weights (see Grover *et al.*, 1998). Recently, Keeler and coworkers (Keeler *et al.*, 2000) have demonstrated that an acquired thermotolerance in Lima bean

(*Phaseolus lunatus*) was associated with the expression of HSP 100/clp B proteins. A cultivar specific 22 kDa heat shock protein has been characterized from mitochondria of *Pisum sativum* (Wood *et al.*, 1998). Knorpp and coworkers (Galvis *et al.*, 2001) have recently suggested that heat stress response in pea involves interaction of a mitochondrial nucleoside diphosphate kinase (mtNDPK), which possibly act as a modulator of 86 kDa HSP.

Hamilton, III and Heckathorn (2001) have demonstrated that complex I electron transport in mitochondria is protected by antioxidants and small HSPs in maize under the high soil sodium stress, whereas complex II is protected by proline and betaine. This study indicates a correlated action of various components of stress tolerance mechanisms to confer the tolerance to the plants. At the molecular level, the heat-shock response is a transient reprogramming of cellular activities evident by the synthesis of HSPs, concomitant with accessation of normal protein synthesis (see Schoffl *et al.*, 1998). The HSPs accumulate in a close dependent manner and are suggested to act as molecular chaperons that are defined by their capacity to recognize and to bind substrate proteins that are in an unstable, inactive state.

Well-characterized HSP proteins responsive to low temperature, drought, salt stress and heavy metal stress, etc. have also been reported in legumes. Low temperature responsive proteins have been characterized in *Glycine max*, *Medicago sativa* (see Grover *et al.*, 1998, 2001 and references therein) and *Arachis hypogea* (Dave and Mitra, 1998). Recently, Thomashow (2001) and Xiong and Zhu (2001) have suggested that the CBF/DREB1 transcriptional activators regulate the cold acclimation of plants and the CPF/DREB1 proteins regulate expression of freezing tolerance genes. Water stress and salt stress responsive genes and proteins are other stress proteins plants synthesize and accumulate under such stresses. Water and salt responsive genes/proteins have been reported in legumes, e.g. *Glycine max*, *Pisum sativum*, *Vigna radiata* and *Medicago sativa* (Grover *et al.*, 1998, 2001). A tissue specific modulation of salt inducible gene expression in callus and whole plant in *Medicago sativa* have been shown to be correlated to salt stress (Winicov and Shirzadegan, 1997). The cell wall proteins of 22–36 kDa have been reported to get induced by water deficit in *Phaseolus vulgaris* (Covarrubias *et al.*, 1995). A new LEA protein Pvlea18 has been reported to get accumulated during water stress in bean seedlings (Colmenero-Flores *et al.*, 1999). Sinha *et al.* (1999) have demonstrated synthesis of stress induced polypeptides in *Lathyrus sativus* in presence of salt stress, mannitol and ABA treatments.

5. Compatible solutes in the abiotic stress tolerance

Plants exposed to the various environmental stresses accumulate certain compatible solutes (osmoprotectants) in the cytoplasm for the osmotic adjustment. The major stress related osmoprotectants are glycine betaine, proline, γ-aminobutyric acid, polyols, sugars, and polyamines, etc.

Glycine betaine affords osmoprotection in bacteria, plants and animals (Hung *et al.*, 2000). Trigonelline (nicotinic acid betaine) and glycine betaine and choline have been reported to accumulate in salt stressed *Glycine max* leaves (Cho *et al.*, 1999), and *Medicago* plants (Baburina and Shevyukova, 1998). Overexpression of bacterial choline oxidase (*codA*) gene isolated from *Arthrobacter globiformis* and *A. passens* in transgenic plants have been reported to increase synthesis of glycine betaine, which confers higher tolerance to the abiotic stresses, e.g. high salt and low and high temperature (Huang *et al.*, 2000; McNeil *et al.*, 2000; Sakamoto and Murata, 2000, 2001). It has been suggested that the plants, which do not synthesize glycine betaine in nature, can also be made stress tolerant by engineering *codA* gene in those plants (see Sakamoto and Murata, 2001). Such studies, however, are not available for the legume improvement and need an urgent attention.

Proline is another important osmoprotectant, which has been reported to accumulate in elevated level in large number of plants subjected to stresses (Delauney and Verma, 1993). A high level of endogenous proline has been estimated in *Vigna radiata* cultivars grown under salt stress and Cd-supplementation (R.P. Singh and coworkers, unpublished data). High proline accumulating cultivars show a higher level of tolerance to the stress as compared to the low accumulating ones. Hydroxyproline resistant cell lines of mungbean cultivars which accumulated 14 times higher level of free proline were found tolerant to salt stress *in vitro* (Gulati and Jaiwal, 1997). The exogenous applications of proline protect plant tissue from stress conditions (see Dubey, 1997). Increased accumulation of proline helps the cells in osmoprotection as well as in regulating their redox potential, scavenging hydroxyl radicals and in protection against denaturation of various macromolecules. Overexpression of *Vigna aconitifolia* Δ^1 pyrroline-5-carboxylate synthetase (*P5CS*) gene or a modified *P5CS* 129a gene impart tolerance to the osmotic stress in tobacco (Kavi Kishore *et al.*, 1995; Hong *et al.*, 2000).

Hyashi *et al.* (2000) have demonstrated recently that expression of genes for P5CS and proline dehydrogenase (ProDH), the enzymes involved in proline biosynthesis and degradation respectively in *Arabidopsis thaliana* involve light irradiation and change in cellular water status, and accumulated proline keeps the osmotic balance between cells and the outer environment. Recently accumulation of proline under the stresses has been shown to suppress the RUBISCO carboxylase activity by dissociating small subunits from holoenzyme of RUBISCO isolated from *Sesbania sesban* leaflets (Shivakumar *et al.*, 1998, 2000). The activity of enzymes of NH_4^+ assimilation, i.e. NADH-glutamate dehydrogenase (Singh and Srivastava, 1983) and glutamine synthetase (R. P. Singh and coworkers unpublished data) were also suppressed by the exogenously supplied or endogenously accumulated proline in maize and mungbean seedlings respectively. On the other hand, proline has been reported to be involved in protecting thylakoid membranes, photoinhibitory loss of photochemical activity, and alleviates salt stress induced enhancement in RUBISCO oxygenase activity (Alia *et al.*, 1991, 1992, 1993, 1997; Shivakumar *et al.*, 2001). Such observations caution the fruitfulness of overproducing proline under the stress even if it supports the initial survival of the stressed plants. More attention is needed to understand the effects of such stress induced accumulation of the osmoprotectants in regulating the vital metabolic processes of the cell.

γ-aminobutyric acid (GABA) is a non-protein amino acid known to accumulate in a variety of plant tissues under the stresses (Serraj *et al.*, 1998). It is produced primarily from the α-decarboxylation of L-glutamate or by the oxidative degradation of polyamines (SatyaNaryan and Nair, 1990; Tuin and Shelp, 1994). A high GABA accumulation has been reported in nodules of a number of legumes, and its accumulation has increased in the various parts of plants under the stresses, e.g. in nodulated soybean in response to drought stress (Serraj *et al.*, 1998). It has been suggested that GABA may be involved in transduction of the stress signals, however, its precise role in plants is not yet known. The accumulation of plant polyols, either straight chain metabolites such as mannitol or sorbitol or cyclic polyols such as myo-inositol and its methylated derivatives ononitol and pinitol, has been correlated with salinity and drought tolerance (Bohnert and Jensen, 1996). The polyols maintain osmotic potential of cell by retention of water and sequestration of sodium to the vacuoles or apoplast. They protect membrane protein complexes or enzymes by scavenging active oxygen (Bohnert *et al.*, 1995). Many legumes such as *Vigna* species, *Cajanus cajan, Microptilium* species have been reported to accumulate cyclic polyols such as D-pinitol (1-D-3-o-methyl-chiro-inositol) or D-ononitol (1-D-4-o-methyl-myo-inositol) in response to water deficit (Ford, 1982, 1984; Sheriff *et al.*, 1986; Keller and Ludlow, 1993; Wanek and Richter, 1997). The bacterial mannitol phosphate dehydrogenase (*mtlD*) gene has been transferred into tobacco (Tarczynski *et al.*, 1993), *Arabidopsis* (Thomas *et al.*, 1995) and eggplant (Rajam *et al.*, 1998) to develop transgenic plants tolerant to salt stress.

Polyamines are ubiquitous polycations, which accumulate in plants during various abiotic stresses such as mineral deficiency, drought, salinity, low and high temperature, low pH, etc. (see Rajam, 1997). Recently, Choudhary and Singh (2000) have shown accumulation of higher polyamine levels in *Vigna radiata* seedlings under the heavy metal stress. Engineering polyamine biosynthesis pathways by introducing arginine decarboxylase (*adc*) and ornithine decarboxylase (*odc*), S-adenosyl methionine decarboxylase (*samdc*) genes, the PA content of the plants can be manipulated to increase the abiotic stress tolerance (see Rajam *et al.*, 1998; Roosens *et al.*, 1998).

It is evident from the various studies that the manipulation of certain metabolic pathways of these compatible solutes can confer the stress tolerance in transgenic plants. However, only one of the genes in the particular biosynthetic pathway has not always resulted in accumulation of desired solute and thus the tolerance could not be imparted to the plants in all the cases. There is a need to transfer more than one gene of biosynthetic pathway or genes of different pathways responsible for the synthesis of more than one solute or a suitable transcription factor for more coordinated responses.

6. Antioxidative defense system

The plants exposed to various abiotic stresses produce excess reactive oxygen species, i.e. superoxide radical (O_2^-), hydrogen peroxide (H_2O_2) and the hydroxyl radical (OH^-), which cause oxidative damage at cellular levels (Allen *et al.*, 1994; Foyer *et al.*, 1994; Rao and Ormrod, 1995; Allen, 1995; Jimenez *et al.*, 1998; Mckersie *et al.*, 1999; Knorzer *et al.*, 1999; Burkey *et al.*, 2000; Roxas *et al.*, 2000; Xiong *et al.*, 2001). Plants are endowed with an array of enzymes and low molecular weight compounds with antioxidant properties to efficiently metabolize oxygen free radicals, which enhance their tolerance under such stresses. The superoxide dismutase (SOD) converts O_2^- to H_2O_2, which is quickly scavenged by ascorbate peroxidase (APX) and subsequently an ascorbate-glutathione cycle, which is maintained by the regeneration of reduced glutathione produced by glutathione reductase (GR), reduces the oxidative damage caused by the reactive oxygen species (ROI)

(Allen, 1995). Physiological and genetic evidences suggest that the ROI scavenging systems of plants are an important component of the stress protective mechanisms. The ability to manipulate the expression levels of specific enzymes involved in this process by genetic engineering provides insight into the specific functions of the various components of this system and raises hopes to improve the stress tolerance of the target plants.

Antioxidative defense system of mitochondria and peroxisomes get depressed during the senescence of pea leaves (Jimenez *et al.*, 1998). Knorzer *et al.* (1999) have reported that salicylic acid and BTH (benzo (1,2,3) thidiazole-7-carbothioic acid s-methylester) elicit soybean cells from herbicide induced lipid peroxidation. In bean leaves lipid peroxidation was induced by acid rain treatment with the polyamines spermidine and spermine (Velicova *et al.*, 2000). The impact of PAs could be attributed to their acid neutralizing and antioxidative effects, as well as to their ability to stabilize membrane by associating with negatively charged phospholipids.

Gonzalez *et al.* (1998) have demonstrated that the activities of mitochondrial, cytosolic and chloroplastic SOD isoforms and APX increased in *Phaseolus vulgaris* treated with high light intensity and excess manganese, however, GR activity was not affected by the stress. Copper imposed oxidative stress, as well as the antioxidative defense response in the leaves of bean plants, appear to be biphasic (Cuypers *et al.*, 2000). An indirect preventive effect on the antioxidative defense system was observed during the first phase before the leaf copper content increased. A root to shoot signaling system appears to be involved. Direct oxidation of reduced cell metabolites occurred during the second phase by copper when the leaf copper content was enhanced. A very significant increase in peroxidase and SOD activities in the cotyledons, root, stem and leaves of *Vigna radiata* have been observed under the toxicity of cadmium and nickle (R.P. Singh and co-workers, unpublished data). Ruegsegger *et al.* (1990) have reported cadmium induced high glutathione synthesis in roots and shoots of *Pisum sativum*.

The SODs are a family of metalloenzymes with isoforms (Cu/Zn-SOD, Mn-SOD and Fe-SOD) located in various compartments of plant cells. A full length Fe-SOD cDNA from soybean has been characterized with a chloroplast-targeting domain (Crowell and Amasino, 1991). Expression of *Pisum sativum* chloroplastic Cu/Zn-SOD in tobacco (cv Xanthi) resulted in a 3-fold increase in total SOD activity, which led to significant increase in the resistance to low levels of light mediated methyl viologen (MV)-induced membrane damage (Sengupta *et al.*, 1993a, b; Allen, 1995). At high MV-concentration, however, this SOD was inactivated by H_2O_2 as Cu/Zn-SOD is sensitive to H_2O_2 generated high under the stresses. It appears, therefore that increased expression of chloroplastic Cu/Zn-SOD alone may be insufficient to confer visible tolerance to the stress, unless some additional stress protection is supplemented. A chloroplastic Mn-SOD chimeric gene construct developed by Bowler *et al.* (1991) in which the native mitochondrial transit peptide of *Nicotiana plumbaginifolia* was replaced by a chloroplastic transit peptide from the *Arabidopsis RuBPcase* gene transferred in *Medicago sativa* has shown a significant increase in resistance to the herbicide actifluorofen and to freezing (Mckersie *et al.*, 1993). A novel study by Schake (1995) in which *Mn-SOD cDNA* (transgene) from pea with Cu/Zn-SOD chloroplast targeting sequence and Cu/Zn-SOD construct from the same source was expressed in tobacco demonstrate clearly that at low MV concentration both the isoforms of SOD could confer protection to the oxidative damage, but at higher

MV levels (2.4 μM), protection was maintained only in Mn-SOD plants and not in Cu/Zn-SOD plants (see Allen, 1995).

Overexpression of *sod* in chloroplasts provides increased protection from oxidative stress. Elevated activity of H_2O_2 resistant Mn-SOD in the chloroplast stroma apparently provides superior protection to the oxidative stress caused by chemical exposure (MV, acti-fluorofen, ozone) but this isoenzyme is much less effective than Cu/Zn SOD in providing protection from the photo-oxidation under photoinhibitory conditions. These differences show that simply elevating SOD activity in the chloroplast stroma may not be sufficient. Rather, the type of SOD that is increased is also a critical factor. Since the Cu/Zn SOD is associated with the surface of the thylakoid membranes in close association with PSI, it seems likely that different protective activities of Cu/Zn and Mn-SOD in chloroplast are due to the differences in the biochemical characteristics and sub-organellar localization of these enzymes.

Overexpression of Mn-*sod* in transgenic alfalfa plants often improves the winter survival and subsequent herbage yield of the crop, but in some independent transgenic plants, the winter survival and subsequent yield were actually lessened (Mckersie *et al.*, 1996, 1999). There was no apparent difference in primary freezing injury in wild and transgenic plants, and therefore, it is suggested that the trait is not associated with a change in the primary site of freezing injury.

Using transgenic tobacco expressing the bacterial *gor* gene in the cytosol and the *Arabidopsis Fe-sod* gene in chloroplast on four lines of poplar (*Populas tremula* × *P. alba*) Tyystjarvi *et al.* (1999) have demonstrated that the activities of GR and Fe-SOD increased by many fold in transgenic plants overexpressing these genes. They found that the ability to convert O_2^- to H_2O_2 is not a limiting factor in protection against photoinhibition or in the repair of photoinhibitory damage or that the site of O_2^- production is not accessible to the transgene production. The repair of photoinhibitory damage was not enhanced in plants overproducing GR in the cytosol too.

Another important group of enzymes of antioxidant defense system are ascorbate peroxidases (APXs) which have high substrate specificity for ascorbate and are the primary H_2O_2 scavenging enzymes in the chloroplasts and cytosol of plant cells (Asada, 1992; Allen, 1995). Although catalase is an alternative enzyme for H_2O_2 scavenging, its location in peroxisomes and its relatively low K_m limit its ability to keep H_2O_2 level low enough to prevent chloroplast damage. Cytosolic APXs are encoded by nuclear genes and a number of cDNA for cytosolic APXs have been isolated and characterized from plants including pea (Mittler and Zilinskas, 1991) and *Arabidopsis* (Kubo *et al.*, 1992). Distinct thylakoid bound and stromal forms of APX also apparently exist in chloroplasts (Miyake *et al.*, 1993). Transgenic tobacco plants that overexpress cytosolic *apx* have shown an increased tolerance to MV damage (Pitcher *et al.*, 1994; Allen, 1995). Recently, Wang *et al.* (1999) have demonstrated that overexpression of *Arabidopsis* gene *apx3* encoded a peroxisomal membrane bound APX3 isoform provided protection against the oxidative stress originated from peroxisomes and not from that originated from chloroplasts. Gene construct that encode chimeric chloroplastic APX isoform that consists of a 5′ chloroplast transit peptide sequence fused to cytosolic APX cDNA was developed and successfully expressed in tobacco plants, but this form could not increase tolerance (Pitcher *et al.*, 1994). A 3-fold increase in APX activity and mRNA in transgenic tobacco plants that overexpress chloroplastic *cu/zn-sod* indicates expression of other co-ordinated genes of antioxidative defense system simultaneously in certain cases (SenGupta *et al.*, 1993b; Allen, 1995).

Glutathione reductase (GR) is involved in the regeneration of oxidized ascorbate through reducing glutathione (see Creissen *et al.*, 1994). In pea leaves GR is present in chloroplast, mitochondria and cytosolic cellular compartments (Edwards *et al.*, 1990). The various GR isoforms are encoded by nuclear genes. GR-encoding cDNA from pea and tobacco have been isolated and sequenced (Allen, 1995). The derived amino acid sequence from these cDNA have substantial homology with bacterial and mammalian GRs, but they include putative amino-terminal chloroplast transit peptide sequences.

Attempts to develop transgenic plants overexpressing bacterial *GR* gene under the control of the CaMV 35S promoter in tobacco increased specific GR activity by 3.5-fold, but due to lack of transit peptide for chloroplast it was presumably confined to the cytosol (see Allen, 1995). A partial success in maintaining a high reduced ascorbate pool and tolerance towards MV and SO_2 but not ozone could be achieved through trangenic tobacco plants overexpressing cytosolic *GR* and by targeting of the *E. coli GR* to chloroplasts (Aono *et al.*, 1991). Transgenic tobacco plants that overexpress pea *GR* have also been developed (Creissen *et al.*, 1994).

Tolerance to atmospheric ozone in transgenic tobacco overexpressing glutathione synthetase in plastids have been reported which causes reduced ethylene emissions and increased or unchanged total pigment concentrations in transgenic plants (Wellburn *et al.*, 1998). Roxas *et al.* (2000) have recently demonstrated that overexpression of a tobacco glutathione s-transferase with glutathione peroxidase activity (GST/GPX) in transgenic tobacco seedlings provide increased glutathione dependent peroxidase scavenging and alteration in glutathione and ascorbate metabolism that lead to reduced oxidative damage caused by the various stresses, e.g. thermal or salt stress. This protective effect is primarily responsible for the ability of GST + seedlings to maintain growth under stressful conditions.

It has been suggested by Burkey *et al.* (2000) that the screening for ozone tolerance based on antioxidant content is not a reliable approach in *Phaseolus vulgaris* as ozone sensitive and tolerant plants were not clearly distinguished by differences in leaf antioxidant contents, i.e. ascorbic acid, glutathione and α-tocopherol (vitamin E). An increased resistance to oxidative stress in transgenic tobacco plants has been achieved by targeting mannitol biosynthesis to chloroplast (Shen *et al.*, 1977). It has been suggested that the chloroplast location of mannitol can supplement endogenous free radical scavenging mechanisms and reduce oxidative damage of cells by hydroxyl radicals.

The antioxidative defense system with its variable organ specific enzymatic components and redox active molecules plays an important role in protecting plants from the oxidative damage which is evident from various physiological, biochemical, molecular and genetic studies. The emergence of genetic engineering as an effective tool to elucidate the actual significance of a metabolic system for plant growth, stress tolerance and yield, and developing tolerance in the plants by introducing a target transgene, has also shown promising results at least in model systems. It appears that in addition to broad range antioxidative defense system, some stress specific and organelle specific genes and gene products are involved in the tolerance mechanism. The key regulatory steps or transcription factors that induce the cascade of events related to antioxidative defense mechanism may be characterized to find out more precise manipulations of this potential of the stressed plants.

7. Conclusions and future prospects

The abiotic stress responses to the plants seems to involve many kinds of molecular dynamics, starting from the perception and transduction of the stress signals, expression of stress genes and biosynthesis and accumulation of stress proteins, osmoprotection and free radical scavenging by osmoregulatory compatible solutes and antioxidative defense enzymes and molecules, etc. In most cases, recent studies indicate the involvement of various kinds of stress related molecules and mechanisms at a time to protect plants from the stresses, and therefore modifications with one gene may not be very significantly effective. A lot of exciting information is expected in the coming years from the studies related to *Arabidopsis* genome, however, legumes have a different kind of metabolic system, e.g. N_2-fixation and high protein synthesis and accumulation.The studies on the general and specific roles of various stress related molecules are incomplete in plants and in an infancy stage in legumes. Identification and characterization of stress related genes and their regulation in response to the specific/common stresses and long/short term stresses need attention. Manipulation in key steps in signal perception and signal transduction to reduce the stress sensitivity of the exposed cells, i.e. low level of constitutive expression of the stress perception gene(s), are considered as a possible way to bring out enhanced stress tolerance. Engineering over-expression of gene, promoters and transcription factor genes can induce stress tolerance by altering levels of several target genes and proteins at the same time. This would definitely enhance the resistance levels as compared to single gene manipulations. However, a very few transcription factors have been isolated to date. Isolation and characterization of more transcription factors and identification of key regulatory steps for general and specific stresses require urgent attention. The understanding of actual share of different strategies adopted for stress tolerance in plants may provide insight to select the target genes for the significant improvement in abiotic stress tolerance. The stability of transgenic plants in the changing and dynamic ecosystems is another key point, which needs to be assessed for the success of transgenic technology for abiotic stress tolerance in legumes.

Acknowledgements

Financial support from the Department of Science and Technology, Department of Biotechnology and University Grant Commission, New Delhi as research projects to RPS and PKJ to carry out research work on legume biotechnology is gratefully acknowledged.

References

Alia A, Pardha Saradhi P and Mohanty P (1991) Proline enhances primary photochemical activities in isolated thylakoid membranes of *Brassica juncea* by arresting photoinhibitory damage. *Biochem. Biophys. Res. Commun.*, **181**: 1238–1244.

Alia A, Pardha Saradhi P and Mohanty P (1992) Enhancement of photosystem-II photoreaction and high pH stability in thylakoid from cotyledonary leaves of *Brassica juncea* raised under sodium chloride stress. *Physiol. Plant.*, **86**: 189–196.

Alia A, Pardha Saradhi P and Mohanty P (1993) Proline in relation to free radical production in seedlings of *Brassica juncea* raised under sodium chloride stress. *Plant Soil*, **156**: 497–500.

Alia A, Pardha Saradhi P and Mohanty P (1997) Involvement of proline in protecting thylakoid membranes against free radical-induced photodamage. *J. Photochem. Photobiol.*, **38**: 253–257.

Allen R D (1995) Dissection of oxidative stress tolerance using transgenic plants. *Plant Physiol.*, **107**: 1049–1054.

Allen R D, Sen Gupta A, Webb R P and Holaday A S (1994) Protection of plants from oxidative stress using *sod* transgenes: Interactions with endogenous enzymes. In: *Frontiers of Reactive Oxygen Species in Biology and Medicine* (Eds Asada K and Yoshikawa A), Excerpta Medica, Amsterdam, pp. 321–322.

Aono M, Kubo A, Saji H, Natori T, Tanaka K and Kondo N (1991) Resistance to active oxygen toxicity of transgenic *Nicotiana tabaccum* that express the gene for glutathione reductase from *Escherichia coli. Plant Cell Physiol.*, **32**: 691–697.

Asada K (1992) Ascorbate peroxidase, a hydrogen peroxide scavenging enzyme in plants. *Physiol. Plant*, **85**: 235–241.

Baburina O K and Sheryakora N I (1988) Salt tolerance and contents of quarternary ammonium compounds in wild spp. of *Medicago. Fiziol Rastenii*, **34**: 887–893.

Bohnert H J and Jensen R G (1996) Strategies for engineering water stress tolerance in plants. *TIBTECH*, **14**: 89–87.

Bohnert H J, Nelson D E and Jensen R G (1995) Adaptations to environmental stresses. *Plant Cell*, **7**: 1099–1111.

Botella J R, Arteca J M, Somodevilla M and Arteca R N (1996) Calcium-dependent protein kinases gene expression in response to physical and chemical stimuli in mungbean (*Vigna radiata*). *Plant Mol. Biol.*, **30**: 1129–1137.

Bowler C, Slooten L, Vandenbranden S, De Rycke R, Botterman J, Sybesma C, Van Montague M and Inze D (1991) Manganese superoxide dismutase can reduce cellular damage mediated by oxygen radicals in transgenic plants. *EMBO J.*, **10**: 1723–1732.

Burkey O K, Wei C, Easan G, Ghosh P and Fenner G (2000) Antioxidant metabolite levels in ozone sensitive and tolerant genotypes of snap bean. *Physiol. Plant.*, **110**: 195–200.

Chai T Y, Zhang Y X and Burkard G (1998) Heavy metal responsive genes in kidney bean: cloning of cDNA and gene expression analysis. *Acta Phytophysiol. Sinica*, **24**: 399–404.

Cho Y, Lightfoot D A and Weed A J (1999) Trigonelline concentrations in salt stressed leaves of cultivated *Glycine max. Phytochemistry*, **52**: 1235–1238.

Cho M H and Spalding E P (1996) An anion channel in *Arabidopsis* hypocotyls activated by blue light. *Proc. Natl. Sci. Acad. USA*, **93**: 8134–8138.

Choudhary A and Singh R P (2000) Cadmium induced changes in diamine oxidase activity and polyamines levels in *Vigna radiata* L. Wilczek seedlings. *J. Plant Physiol.*, **156**: 704–710.

Clark G B, Thompson Jr G and Roux S J (2001) Signal transduction mechanism in plants; An overview. *Curr. Sci.*, **80**: 170–177.

Colmenero-Flores J M, Moreno L P, Smith C E and Covarrubias A A (1999) Pvlea-18, a member of a new late embryogenesis abundant protein family that accumulates during water stress and in the growing region of well irrigated bean seedlings. *Plant Physiol.*, **120**: 93–103.

Covarrubias A A, Ayala J W, Reyes J L, Hernandez M and Garciarrubio A (1995) Cell wall proteins induced by water deficit in bean (*Phaseolus vulgaris*) seedlings. *Plant Physiol.*, **107**: 1119–1128.

Creissen G P, Edwards A and Mullineaux P M (1994) Glutathione reductase and ascorbate peroxidase. In: *Causes of Photooxidative Stress and Amelioration of Defence Systems in Plants* (Eds Foyer C H and Mullineaux P M), CRC Press, Boca Raton, FL, pp. 343–364.

Crowell D N and Amasino R M (1991) Nucleotide sequence of an iron superoxide dismutase cDNA from soybean. *Plant Physiol.*, **96**: 1393–1394.

Cuypers A, Vangronsveld J and Clijsters H (2000). Biospheric effect of copper on the ascorbate-glutathione pathway in primary leaves of *Phaseolus vulgaris* seedlings during the early stages of metal assimilation. *Physiol. Plant*, **110**: 512–517.

Dave R S and Mitra R K (1998) A low temperature induced apoplastic protein isolated from *Arachis hypogaea*. *Phytochemistry*, **49**: 2207–2213.

Delauney A, Hu C, Kavikishore P B and Verma D P S (1993) Cloning of ornithin-d-aminotransferase cDNA by trans complementation in *E. coli* and regulation of proline biosynthesis. *J. Biol. Chem.*, **268**: 18673–18678.

DeWald D B, Torabinejad J, Jones C A, Shope J C, Cangelosi A R, Thompson J E, Prestwich G D and Hama H (2001) Rapid accumulation of phoshatidylinositol 4,5-biphosphate and inositol 1,4,5-triphosphate correlates with calcium mobilization in salt stressed *Arabidopsis*. *Plant Physiol.*, **126**: 759–769.

Dubey R S (1997) Nitrogen metabolism in plants under salt stress. In: *Stategies for Improving Salt Tolerance in Higher Plants* (Eds Jaiwal P K, Singh R P and Gulati A), Science Publishers, Enfield, USA/Oxford and IBH Pub. Co. Pvt Ltd., Delhi, India, pp. 129–138.

Edwards A, Rawsthorne S and Mullineaux P M (1990) Subcellular distribution of multiple isoforms of glutathione reductase in leaves of pea (*Pisum sativum* L.). *Planta*, **180**: 278–284.

Ellard-Ivery M, Hopkins R B, White T J and Lomax T L (1999) Cloning, expression and N-terminal nuristoylation of CPCK1, a calcium dependent protein kinase from Zucchini (*Cucurbita pepo* L.). *Plant Mol. Biol.*, **39**: 208.

Ford C W (1982) Accumulation of O-methyl ionositols in water stressed *Vigna* spp. *Phytochemistry*, **21**: 1149–1151.

Ford C W (1984) Accumulation of low molecular weight solutes in water stressed tropical legumes. *Phytochemistry*, **23**: 1007–1015.

Foyer C H, Descourvieres P and Kunert K J (1994) Protection against oxygen radicals: an important defense mechanism studied in transgenic plants. *Plant Cell Environ.*, **17**: 507–523.

Galvis M L E, Marttila S, Hakansson G, Forsberg J and Knorpp C (2001) Heat stress responses in pea involves interaction of mitochondrial nucleoside di phosphate kinase with a novel 86-kilodalton protein. *Plant Physiol.*, **126**: 69–77.

Gilmour S J, Seboly A M, Salazar M P, Everard J D and Thomashow M F (2000) Overexpression of the *Arabidopsis* CBF3 transcriptional activator mimics multiple biochemical changes associated with cold acclimatization. *Plant Physiol.*, **124**: 1854–1865.

Gilmour S J, Zarka D G, Stockinger E J, Salazar M P, Houghton J M and Thomashow M F (1998) Low temperature regulation of the *Arabidopsis* CBF family of AP_2. *Plant J.*, **16**: 433–442.

Gonzalez A, Steffen K L and Lynch J P (1998). Light and excess manganese: Implications for oxidative stress in common bean. *Plant Physiol.*, **118**: 493–503.

Grover A, Kapoor A, Satya Lakshmi O, Agarwal S, Sahi C, Katiyar-Agarwal S, Agarwal M and Dubey H (2001) Understanding molecular alphabets of the plant abiotic stress responses. *Curr. Sci.*, **80**: 206–216.

Grover A, Pareek A, Singla S L, Minhas D, Katiyar S, Ghawana S, Dubey H, Agarwal M, Rao G U, Rathee J and Gover A (1998) Engineering crops for tolerance against abiotic stresses through gene manipulation. *Curr. Sci.*, **75**: 689–696.

Gulati A and Jaiwal P K (1997) The potential of plant tissue culture and related techniques for the improvement of salt tolerance in higher plants. In: *Strategies for Improving Salt Tolerance in Higher Plants* (Eds Jaiwal P K, Singh R P and Gulati A), Science Publishers Inc. Enfield, USA/Oxford and IBH Pub. Co. Pvt. Ltd, New Delhi, India.

Guy C L (1990) Cold acclimation and freezing stress tolerance: role of protein metabolism. *Annu. Rev. Plant Physiol. Plant Mol. Biol.*, **41**: 187–223.

Hamilton III E W and Heckathorn S A (2001) Mitochondrial adaptation to NaCl complexI is protected by antioxidants and small heat shock proteins, whereas complexII is protected by proline and betaine. *Plant Physiol.*, **126**: 1266–1274.

Hayashi F, Ichino T, Osanai M and Wada K (2000). Oscillation and regulation of proline content by P5CS and ProDH gene expressions in the light/dark cycles in *Arabidopsis thaliana* L. *Plant Cell Physiol.*, **41**: 1096–1101.

Hong Z, Lakkineni K, Zhang Z and Verma D P S (2000) Removal of feedback inhibition of pyrroline-5-carboxylate synthetase results in increased proline accumulation and protection of plants from osmotic stress. *Plant Physiol.*, **122**: 1129–1136.

Huang C H, Tate B F, Crain R C and Cote C G (1995) Multiple phosphoinositol specific phospholipase in oat roots: characterization and partial purification. *Plant J.*, **8**: 257–267.

Huang J, Hiriji R, Adam L, Rozwadowski L K, Hammerlindl K J, Keller A W, Selveraj G (2000) Genetic engineering of glycine betaine production towards enhancing stress tolerance in plants: metabolic limitation. *Plant Physiol.*, **122**: 747–756.

Iuchi S, Kobayashi M, Yamaguchi-Shinozaki K and Shinozaki K (2000). A stress inducible gene for 9-cis epoxycaratenoid dioxygenase invoved in abscisic acid biosynthesis under water stress in drought tolerant cowpea. *Plant Physiol.*, **123**: 553–562.

Jaglo-Ottosen K R, Gilmour S J, Zarka D G, Schabenberger O and Thomashow M F (1998) *Arabidopsis* CBF1 overexpression induces *cor* genes and enhances freezing tolerance. *Science*, **280**: 104–106.

Jimenez A, Hernandez J A, Pustori G, del Rio L A and Sevilla F (1998) Role of ascorbate-glutathione cycle of mitochondria and peroxisomes in the senescence of pea leaves. *Plant Physiol.*, **118**: 1327–1335.

Jung K M and Kim D K (2000) Purification and characterization of a membrane-associated 48-kilodalton phospholipase A_2 in leaves of broad bean. *Plant Phsiol.*, **123**: 1057–1067.

Kasuga M, Liu Q, Miura S, Yamaguci-Shinozaki K and Shinozaki K (1999) Improving plant drought, salt and freezing tolerance by gene transfer of a signal stress-inducible transcription factor. *Nature Biotech.*, **17**: 287–291.

Kato-Naguchi H (1999). Abscisic acid increase anaerobic tolerance in alfalfa seedlings. *Physiol. Plant.*, **107**: 395–398.

Kavikishor P B, Hong Z, Guo-Hua M, Hu Chein A N A and Verma D P S (1995) Overproduction of D^1 pyrroline-5-carboxylate synthetase increase proline production and confer osmotolerance in transgenic plants. *Plant Physiol.*, **108**: 1387–1394.

Keeler S J, Boettger C M, Haynes J G, Kuches K A, Johnson M M, Thureen D L, Keeler C L and Kitto S L (2000) Acquired thermotolerance and expression of the HSP100/ *ClpB* genes of Limebean. *Plant Physiol.*, **123**: 1121–1132.

Keller F and Ludlow M W (1993) Carbohydrates metabolism in drought-stressed leaves of pigeonpeas (*Cajanus cajan*). *J. Exp. Bot.*, **44**: 1351–1359.

Knight H, Brandt S and Knight M R (1998) A history of stress alters drought calcium signaling pathways in *Arabidopsis*. *Plant J.*, **16**: 681–687.

Knight H and Knight M R (2001) Abiotic stress signaling pathways: specificity and cross talk. *Trends Plant Sci.*, **6**: 262–267.

Knight H, Trewas A J and Knight M R (1996) Cold calcium signaling in *Arabidopsis* involves two cellular pools and a change in calcium signature after acclimation. *Plant Cell*, **8**: 489–503.

Knight H, Trewas A J and Knight M R (1997) Calcium signaling in *Arabidopsis thaliana* responding to drought and salinity. *Plant J.*, **12**: 1067–1078.

Knight H, Veale E L, Warren G J and Knight M R (1999) The *sfr6* mutation in *Arabidopsis* suppresses low temperature induction of genes dependent on the CRT/DRE sequence motif. *Plant Cell*, **11**: 875–886.

Knight M R, Campbell A K, Smith S M and Trewas A J (1991) Transgenic plant aequorin reports the effect of touch and cold shock and elicitors and cytoplasmic calcium. *Nature*, **352**: 524–526.

Knorzer O C, Lederer B, Durner J and Boger P (1999) Antioxidative defense activation in soybean cells. *Physiol. Plant.*, **107**: 294–302.

Kubo A, Saji H, Tanaka K and Kondo N (1992) Cloning and sequencing of a cDNA encoding ascorbate peroxidase from *Arabidopsis thaliana*. *Plant Mol. Biol.*, **18**: 691–701.

Lee J H, Habel A and Schoffl F (1995) Derepression of the activity of genetically engineered heat shock factor causes constitutive synthesis of heat shock proteins and increased thermotolerance in transgenic *Arabidopsis*. *Plant J.*, **8**: 603–612.

Liu Q, Kasuga M, Sakuma Y, Abe H, Miura S, Yamaguch-Shinozaki K and Shinozaki K (1998). Two transcription factors, *DREB 1* and *DREB 2*, with an EREBP/AP2 DNA binding domain separate two cellular signal transduction pathways in drought and low temparature responsive gene expression, respectively in *Arabidopsis*. *Plant Cell*, **10**: 1391–1406.

McAinsh M R and Hetherington A M (1998) Encoding specificity in Ca^{2+} signaling systems. *Trends Plant Sci.*, **3**: 32–36.

Mckersie B D, Bowley S R and Jones K S (1999) Winter survival of transgenic alfalfa over expressing superoxide dismutase. *Plant Physiol.*, **119**: 839–847.

Mckersie B D, Bowley S R, Harjanto E and Leprince O (1996) Water deficit tolerance and field performance of transgenic alfalfa overexpressing superoxide dismutase. *Plant Physiol.*, **111**: 1177–1181.

Mckersie B D, Chen Y, De Beus M, Bowley S R and Bowler C (1993) Superoxide dismutase enhances tolerance of freezing stress in transgenic alfalfa (*Medicago sativa* L.) *Plant Physiol.*, **103**: 1155–1163.

McNeil S D, Rhodes D, Russell B L, Nuccio M L, Shachar-Hill Y and Hanson A D (2000) Metabolic modeling identifies key constraints on an engineered glycine betaine synthesis pathway in tobacco. *Plant Physiol.*, **124**: 153–162.

Mittler R and Zilinskas B A (1991) Molecular cloning and nucleotide sequence analysis of a cDNA encoding pea cytosolic ascorbate peroxidase. *FEBS Lett.*, **289**: 257–259.

Miyake C, Cao W H and Asada K (1993) Purification and molecular properties of the thylakoid bound ascorbate peroxidase in spinach chloroplasts. *Plant Cell Physiol.*, **34**: 881–889.

Monks D E, Aghoram K, Courtney P D, DeWald D B and Dewey R E (2001) Hyperosmotic stress induces the rapid phosphorylation of through activation of the protein kinases SPK1 and SPK2. *Plant Cell*, **13**: 1205–1219.

Monroy A F, Sahan F and Dhindsa R S (1993) Cold induced changes in freezing tolerance protein phosphorylation and gene expression. Evidence for a role of calcium. *Plant Physiol.*, **102**: 1227–1235.

Monroy A F, Sangwan V and Dhindsa R S (1998) Low temperature signal transduction during cold acclimitization: Protein phosphatase 2A as an early target for cold inactivation. *Plant J.*, **13**: 653–660.

Montero E, Cabot C, Barcela J and Poscheri R C (1997) Endogenous abscisic acid levels are linked to decreased growth of bushbean plants treated with NaCl. *Physiol. Plant*, **101**: 1722–1730.

Muratta N and Los D A (1997) Membrane fluidity and temperature perception. *Plant Physiol.*, **115**: 875–879.

Park J M, Park C J, Lee S B, Ham B K, Shin R and Paek K H (2001) Overexpression of the tobacco *Tsi1* gene encoding an EREBP/AP_2 type transcription factor enhances resistance against pathogen attack and osmotic stress in tobacco. *Plant Cell*, **13**: 1035–1046.

Pitcher L H, Repetti P and Zilinskas B A (1994) Overproduction of ascorbate protects transgenic tobacco plants against oxidative stress. *Plant Physiol.*, **105**: S-169.

Prandl R, Hinderhofer K and Eggers-Schoffl F (1998) HSF3, a new heat shock factor from *Arabidopsis thaliana* derepresses the heat shock response and confers thermotolerance when over expressed in transgenic plants. *Mol. Gen. Genet.*, **258**: 269–278.

Queitsch C, Hong S H, Vierling E and Lindquist S (2000) Heatshock protein 101 plays a crucial role in thermotolerance in *Arabidopsis. Plant Cell*, **12**: 479–492.

Rajam M V (1997) Polyamines in plant ecophysiology. In: *Plant Ecophysiology* (Ed Prasad M N V). John Wiley, New York, USA, pp. 343–374.

Rajam M V, Dagar S, Waie B, Yadav J S, Kumar P A, Shoeb F and Kumaria R (1998) Genetic engineering of polyamine and carbohydrate metabolism for osmotic stress tolerance higher plant. *J. Biosci.*, **23**: 473–482.

Rao M V and Ormrod D P (1995) Antioxidative enzymes, the missing links between physiological, biochemical and molecular response. *Physiol. Mol. Biol. Plants*, **1**: 99–100.

Raymond P, Weber H, Damond M and Farmer F E (2000) Differential gene expression in response to mechanical wounding and insect feeding in *Arabidopsis. Plant Cell*, **12**: 707–719.

Redhead C R and Palme K (1996) The genes of plant signal transduction. *Crit. Rev. Plant Sci.*, **15**: 425–454.

Reggiani R and Laoreti P (2000) Evidences for the involvement of phospholipase C in the anaerobic signal transduction. *Plant Cell Physiol.*, **41**: 1392–1396.

Rhoades J D and LoveDay J (1990) Salinity in irrigated agriculture. In: *American Society of Civil Engineers; Irrigation of Agriculture Crops* (Eds Steward B A and Nielsen D R), American Society of Agronomists, USA.

Roosens H C J N, Thu T, Iskander M H and Jacobs M (1998) Isolation of ornithine aminotransferase cDNA and effect of salt stress on its expression in *Arabidopsis thaliana. Plant Physiol.*, **117**: 263–271.

Roxas V P, Lodhi S A, Garret D K, Mahan J R and Allen R D (2000) Stress tolerance in transgenic tobacco seedlings that overexpress glutathione-S-transferase/glutathione peroxidase. *Plant Cell Physiol.*, **41**: 1229–1234.

Ruegsegger A, Schmutz D and Brunold C (1990) Regulation of glutathione synthesis by cadmium in *Pisum sativum* L. *Plant Physiol.*, **93**: 1579–1584.

Sabehat A, Weiss D and Lurie C (1998) Heat shock proteins and cross tolerance in plants. *Physiol Plant.*, **103**: 437–441.

Sakamoto A and Murata N (2000) Genetic engineering of glycine betaine synthesis in plants: Current status and implications for enhancement of stress tolerance. *J. Exp. Bot.*, **51**: 81–88.

Sakamoto A and Murata N (2001) The use of bacterial choline oxidase, a glycine betaine synthesizing enzyme to create stress resistant transgenic plants. *Plant Physiol.*, **125**: 180–188.

Sanchez J P and Chua N H (2001) *Arabidopsis* PLC1 is required for secondary responses to abscisic acid signals. *Plant Cell*, **13**: 1143–1154.

Sanders D, Brownlee C and Harper J F (1999) Communicating with calcium. *Plant Cell*, **11**: 691–706.

SatyaNarain V and Nair P M (1990) Metabolism enzymology and possible roles of 4-amino butyrate in higher plants. *Phytochemistry*, **29**: 367–375.

Saxena N P, Johanson C, Saxena M C and Silim S N (1993). Selection for drought and salinity tolerance in cool season food legumes. In: *Breeding for Stress Tolerance in Cool Season Food Legumes* (Eds Singh K B and Saxena M C), John Wiley and Sons, New York, pp. 245–270.

Schake S A (1995) Analysis of pea chloroplastic Mn-SOD overexpressed in tobacco. MS thesis, Texas Tech University, Lubbock, TX.

Schoffl F, Prandl R and Reindl A (1998). Regulation of heat shock response. *Plant Physiol.*, **117**: 1135–1144.

Sen Gupta A, Heinen J, Holaday A S, Burke J J and Allen R D (1993a) Increased resistance to oxidative stress in transgenic plants that overexpress chloroplastic Cu/Zn superoxide dismutase. *Proc. Natl. Acad. Sci. USA*, **90**: 1629–1633.

Sen Gupta A, Webb R P, Holaday A S and Allen R D (1993b) Overexpression of superoxide dismutase protects plants from oxidative stress. *Plant Physiol.*, **103**: 1067–1073.

Sengupta D N, Banarjee K and Rouf A (1999) Abiotic stress responsive elements, their binding factors and their utilization for the improvement of tolerance to abiotic stress in plants. *J. Plant Biol.*, **26**: 179–185.

Serraj R, Shelp J B and Sinclair R T (1998) Accumulation of γ-aminobutyric acid in nodulated soybean in response to drought stress. *Physiol. Plant*, **102**: 79–86.

Shankar N and Srivastava H S (2002) 'Abiotic stress induced proteins and signal transduction'. In: *Plant Genetic Engineering. Vol. 1. Applications and Limitations* (Eds Singh R P and Jaiwal P K), Sci-Tech Pub. LCC, Houstan, USA (in press).

Shen B, Jensen R G and Bohnert H J (1997) Increased resistance to oxidative stress in transgenic plants by targeting mannitol biosynthesis to chloroplasts. *Plant Physiol.*, **118**: 1177–1183.

Sheriff D W, Fisher M J, Rusitzka G and Ford C W (1986) Physiological reaction to an imposed drought by two twining pasture legumes: *Macroptilium atropurpureum* (desiccation sensitive) and *Galactia striata* (desiccation insensitive). *Aust. J. Plant Physiol.*, **13**: 431–435.

Shinozaki K and Yamaguchi-Shinozaki K (1997) Gene expression and signal transduction in water stressed response. *Plant Physiol.*, **115**: 327–334.

Singh R P and Srivastava H S (1983) Regulation of glutamate dehydrogenase activity by aminoacids in maize seedlings. *Physiol. Plant.*, **57**: 549–564.

Sinha K M, Dachdev A, Johari R P and Mehta S L (1999) Stress induced polypeptides in *Lathyurs sativus. Plant Biochem. Biotech.*, **8**: 47–51.

Sivakumar P, Sharmila P and Pardha Saradhi P (1998) Proline suppresses Rubisco activity in higher plants. *Biochem. Biophys. Res. Comm.*, **252**: 428–432.

Sivakumar P, Sharmila P and Pardha Saradhi P (2000) Proline alleviates salt stress-induced enhancement in Ribulose-1,5-bisphosphate oxygenase activity. *Biochem. Biophys. Res. Comm.*, **279**: 512–515.

Sivakumar P, Sharmila P and Pardha Saradhi P (2001) Proline suppresses Rubisco activity by dissociating small subunits from holoenzyme. *Biochem. Biophys. Res. Comm.*, **282**: 236–241.

Sugihara K, Hanagata N, Dubinsky Z, Baba S and Karube I (2000) Molecular characterization of cDNA encoding oxygen evolving enhancer protein1 increased by salt treatment in the Mangrove plant, *Bruguiera gymnorrhiza. Plant Cell Physiol.*, **41**: 1279–1285.

Takahashi S, Katagiri T, Hirayama T, Yamaguchi-Shinozaki K and Shinozaki K (2001) Hyperosmotic stress induces a rapid and transient increase in inositol 1,4,5-triphosphate independent of abscisic acid in *Arabidopsis* cell culture. *Plant Cell Physiol.*, **42**: 214–222.

Tarczynski M C, Jensen R G and Bohnert H J (1993) Stress protection of transgenic tobacco by producing the osmolytes mannitol. *Science*, **259**: 508–510.

Thomas J C, Sepahi M and Bonhert H J (1995) Enhancement of seed germination in high salinity by engineering mannitol-expression in *Arabidopsis thaliana. Plant Cell Environ.*, **18**: 801–806.

Thomashow M F (1990) Molecular genetics of cold acclimation in higher plants. *Adv. Genet.*, **28**: 99–131.

Thomashow M F (1998) Role of cold responsive gene in plant freezing tolerance. *Plant Physiol.*, **118**: 1–7.

Thomashow M F (2001) So what's new in the field of plant cold acclimation? Lots. *Plant Physiol.*, **125**: 89–93.

Thomashow M F, Gilmour S J, Stockinger E J, Jaglo-Ottosen K R and Zarka D G (2001) Role of the *Arabidopsis* CBI transcriptional activators in cold acclimation. *Physiol. Plant*, **112**: 171–175.

Tuin L G and Shelp B J (1994) *In situ* [^{14}C]-glutamate metabolism by developing soybean cotyledons: metabolic routes. *J. Plant Physiol.*, **143**: 1–7.

Tyystjarvi E, Riikonen M, Arisi A M, Ketturen R, Jouanin L and Foyer C H (1999) Photoinhibition of photosystem-II in tobacco plants overexpressing glutathione reductase and poplars overexpressing superoxide dismutase. *Physiol. Plant*, **105**: 409–416.

Urao T, Yakubov B, Satoh R, Yamaguchi-Shinozaki K, Seki M, Hirayama T and Shinozaki K A (1999) Transmembrane hybrid-type histidine kinase in *Arabidopsis* functions as an osmosensor. *Plant Cell*, **11**: 1743–1754.

Velikov V, Yordanov I and Edreva A (2000) Oxidative stress and some antioxidant systems in acid rain treated bean plants protective role of exogenous polyamines. *Plant Sci.*, **151**: 59–66.

Wanek W and Ritcher A (1997) Biosynthesis and accumulation of D-ononitol in *Vigna umbellata* in response to drought stress. *Physiol. Plant*, **101**: 416–424.

Wang J, Zhang H and Allen R D (1999) Overexpression of an *Arabidopsis* peroxisomal ascorbate peroxidase gene in tobacco increase protection against oxidative stress. *Plant Cell Physiol.*, **40**: 725–732.

Wellburn F A M, Creissen G P, Lake J A, Mullineaux P M and Wellburn A R (1998) Tolerance to atmospheric ozone in transgenic tobacco overexpressing glutathione synthetase in plastids. *Physiol. Plant*, **104**: 623–629.

Winicov I and Bastola D R (1999) Transgenic overexpression of the transcription factor *Alfin1* enhances expression of the endogenous *MsPRP2* genes in Alfalfa and improves salinity tolerance of the plants. *Plant Physiol.*, **120**: 473–480.

Winicov I and Shirzadegan M (1997) Tissue specific modulation of salt inducible gene expression: Callus versus whole plant response in salt tolerant alfalfa. *Physiol. Plant*, **100**: 314–319.

Wood C K, Pratt J R and Moore A L (1998) Identification and characterization of cultivar-specific 22 KDa heat shock proteins from mitochondria of *Pisum sativum*. *Physiol. Plant*, **103**: 369–376.

Xiong C, Werner B L, Christensen E M and Oliver D J (2001) The biological functions of glutathione revisited in *Arabidopsis* transgenic plants with altered glutathione levels. *Physiol Plant*, **126**: 564–574.

Xiong L and Zhu J K (2001) Abiotic stress signal transduction in plants: Molecular and genetic perspectives. *Physiol. Plant*, **112**: 152–166.

Zenoff A M, Hital M, Galo M and Moreno H (1994). Changes in roots lipid composition and inhibition of the extrusion of protons during salt stress in two genotype of soybean resistant or susceptible to stress varietal differences. *Plant Cell Physiol.*, **35**: 729–735.

Zhu J K (2001) Plant salt tolerance. *Trends Plant Sci.*, **6**: 66–76.

GENETIC ENGINEERING FOR INSECT RESISTANCE

RAJANI JAISWAL AND P. ANAND KUMAR
National Research Centre for Plant Biotechnology,
Indian Agricultural Research Institute, New Delhi – 110 012
e-mail: polumetla@hotmail.com

Abstract

Legume crops are the most important source of vegetable proteins in the diet of people in the developing world, especially the Indian subcontinent. These crops are also responsible for nitrogen enrichment of soil through symbiosis with nitrogen fixing bacteria. Legumes have been qualitatively and quantitatively improved by conventional breeding over the past five decades. However resistance to several pests and diseases still remains elusive which is the major limiting factor for productivity. Excessive use of synthetic, organic insecticides resulted in the degradation of environment, adverse effect on human health and development of resistant insects. Recent advances in genetic engineering have clearly demonstrated the possibility of incorporating foreign genes for desired characters while preserving the existing traits of improved genotypes. Reduction in the consumption of insecticides is only possible by introducing the genes encoding insecticidal proteins such as δ-endotoxins of *Bacillus thuringiensis* (Bt), protease inhibitors, alpha-amylase inhibitors, lectins, enzymes such as chitinase and peroxidase. Genetically modified legume crops carrying insect resistance genes have been shown to exhibit considerable protection against the target insects. It is possible to incorporate the resistance against more than one insect pest in the transgenic plants through genetic engineering. This can be achieved by introducing fusion genes or multiple genes in combination encoding the insecticidal proteins which are toxic against a wide range of target pests. Development of resistance in insects is a possibility which can be circumvented or prevented by deploying suitable resistance management strategies. Such approaches include the usage of multiple toxins, crop rotation practices, high or ultra dosage of toxins and refugia. Use of δ-endotoxins of *Bacillus thuringiensis* in conjunction with other bioinsecticides in an IPM mode can drastically reduce the consumption of chemical pesticides and pave the way for safe and sustainable agriculture.

1. Introduction

Legumes or pulses constitute an important part of the Indian diet which is predominantly vegetarian in nature. Legumes are rich and a comparatively cheaper source of proteins.

P.K. Jaiwal and R.P. Singh (eds.), Improvement Strategies for Leguminosae Biotechnology, 245–253.

However, considering the present area (24 million hectares) and yield (552 kg/ha), the production of legumes in India is inadequate to meet the needs of the country (ICAR, 1997). In near future, the burgeoning population will make additional demands for further increase in the yield and productivity of pulse crops.

Over the past few decades, plant breeders have rendered yeoman service to agriculture by developing high yielding hybrids and cultivars of different pulse crops. Another quantum jump in the productivity of these crops can only be realized by adopting the novel methods and techniques of modern biology. Fortunately, advancements made in the recent years in the areas of molecular biology, genetic engineering and plant tissue culture provide an altogether new dimension to crop improvement. It is now possible to harness economically important genes from any biological source and introduce into a plant of our choice with precision. Using such tools many important crop plants have been genetically engineered over the past one decade to confer agronomically valuable traits such as herbicide tolerance, viral resistance, insect resistance, fungal resistance, protein and oil quality, metabolic changes, production of biodegradable plastics, edible vaccines etc. (Dale *et al.*, 1993). Many leguminous crops such as soybean, pea, peanut, bean, etc. have been genetically transformed for important characters like herbicide tolerance, viral resistance and pest resistance. Some of these crops are already under commercial cultivation (James and Krattiger, 1996). In the present account, we describe the progress made so far in the genetic engineering of leguminous crops *vis-à-vis* insect resistance and discuss the prospects for the future.

2. Insect pests of legume crops

India is predominantly tropical and sub-tropical in its climate. Because of the warm temperatures prevailing all over the country the insect prevalence is permanent and the biodiversity of the insect pests is astounding. Pulse crops in such a situation are damaged by a plethora of insect pests. At times, the losses due to insect pests reach alarming levels (Atwal, 1986). The most damaging insects are usually lepidopterans followed by coleopterans and dipterans. Some of the major insect pests of pulse crops of India are enumerated in Table 1. In addition to these major pests, insects such as aphids spread debilitating viral diseases (e.g. mung bean yellow mosaic virus). Thus, insect pest management in pulse crops is of paramount importance. Over the past few decades, usage of synthetic, organic insecticides has resulted in considerable control of pests. However, the consequences of extensive and very often indiscriminate usage of pesticides have been the adverse effects on human health, degradation of environment, extermination of the beneficial insects (pollinators and predators) and development of resistant insects. Hence, a drastic reduction in the consumption of insecticides in agriculture is imperative to safeguard the environment and human health in addition to the primary objective of productivity enhancement.

3. Biotechnology and insect pest management

A durable and eco-friendly alternative for insect pest management in legumes is to genetically transform the elite genotypes and breeding lines of pulse crops using genes

Table 1. Major insect pests of legume crops of India

	Crops	Insect Pest	
1.	Chickpea	Gram podborer	*Helicoverpa armigera*
2.	Pigeonpea	Gram podborer	*Helicoverpa armigera*
		Podfly	*Melanagromyza obtusa*
		Maruca podborer	*Maruca testulalis*
3.	Mung bean	Maruca podborer	*Maruca testulalis*
4.	Soybean	Cutworm	*Agrotis ipsilon*
5.	Groundnut	Leaf miner	*Aproaerema modicella*
		Tobacco caterpillar	*Spodoptera litura*
6.	Cowpea	Bean fly	*Ophiomyza phaseoli*
7.	Pea	Pea aphid	*Macrosiphum pisi*
		Pod borer	*Helicoverpa armigera*
		Pea weevil	*Bruchus pisorum*
8.	Lentil	Podborer	*Etiella zinckenella*

that encode insecticidal proteins. Such insecticidal proteins should be innocuous to all other organisms except the target pests and can be taken from any biological source. The transgenic plants carrying foreign genes for insect resistance have many advantages which include decreased pesticide usage, environment-friendly nature, decreased input costs to the farmer, season-long protection independent of weather conditions, effective control of burrowing insects difficult to reach with sprays and control at all stages of insect development (Kumar *et al.*, 1994). Nature has in its bounty many insecticidal molecules and proteins which can be safely expressed in transgenic plants for insect control. Some examples are protease inhibitors, alpha-amylase inhibitors, lectins, enzymes such as chitinase and peroxidase and δ-endotoxins of *Bacillus thuringiensis* (Bt). A list of the insecticidal protein genes and transgenic legume crops harbouring these genes is given in Table 2. A brief description of these insecticidal proteins is given below.

3.1. INSECTICIDAL PROTEINS OF *BACILLUS THURINGIENSIS*

The leading biorational pesticide, *Bacillus thuringiensis* (Bt), is a ubiquitous gram-positive, spore-forming bacterium that forms a parasporal crystal during the stationary phase of its growth cycle. Bt has a very wide insecticidal spectrum ranging from Lepidoptera, Diptera, Coleoptera, Hymenoptera, Homoptera, Orthoptera, Mallophaga and extending up to nematodes, mites and protozoa (Kumar *et al.*, 1996; Schnepf *et al.*, 1998). Bt produces two different types of insecticidal which are agronomically important, the most widely known one being called δ-endotoxin or insecticidal crystal protein (ICP).

The ICPs usually exist as protoxins of high molecular weight (135–138 kDa). Upon ingestion by insect larvae these are proteolytically processed into small molecular weight toxins in the highly alkaline midgut of the larvae. The toxin molecules bind to specific receptors present in the membranes of midgut epithelium and cause pore formation. This causes disruption of the electrical, K+ and pH gradients across the membrane leading to the death of the larvae (Kumar *et al.*, 1996). The presence of specific receptors in the insects and insect classes is what determines the lack of activity of Bt δ-endotoxins towards mammals and other organisms including beneficial insects. Genetic engineering has facilitated stable expression of Bt genes in crop plants with considerable success (Schuler *et al.*, 1998). Bt-transgenic pulse crops are dealt with in section 4.

3.2. PROTEASE INHIBITORS

Currently, there are two major groups of plant-derived genes used to confer insect resistance on crops: inhibitors of insect digestive enzymes (proteinase and α-amylase inhibitors) and lectins. Plant protease/proteinase inhibitors are polypeptides or proteins that occur naturally in a wide range of plants and are a part of the plant's natural defence system against herbivores. However, these proteins are expressed at very low levels in their native state. Hyperexpression of proteinase inhibitors in transgenic plants would lead to significant levels of insect protection. Fourteen different plant proteinase-inhibitor genes have been introduced into crop plants. Table 2 presents a list of the legume crop species transformed with different proteinase inhibitor genes. The most active inhibitor identified to date is the cowpea trypsin inhibitor (CpTi), which has been transferred to ten different crop species. Experiments with transgenic plants and artificial diets have shown that CpTi affects a wide range of lepidopteran and coleopteran species (Gatehouse and Hilder, 1994). The serine-proteinase inhibitors (from soybean) when expressed in transgenic tobacco and potato resulted in considerable larval mortality of *Spodoptera littoralis*. In addition to serine-proteinase inhibitors, one cysteine-proteinase inhibitor from rice has been introduced into several other crops. Recently, a gene encoding multi-domain

Table 2. Insect resistance legume crops

Crop species	Target pest	Reference
Bt insecticidal proteins		
Soybean (Cry1Ac)	*Helicoverpa zea*	Stewart *et al.*, 1996
Peanut (Cry1Ac)	Lesser cornstalk borer	Singsit *et al.*, 1997
Chickpea (Cry1Ac)	*Helicoverpa armigera*	Kar *et al.*, 1996
Alfalfa (Cry1C)	*Spodoptera littoralis*	Strizhov *et al.*, 1996
Alfalfa (Cry6A)	*Spodoptera littoralis*	Schuler *et al.*, 1998
Proteinase inhibitors		
Pea	*Helicoverpa armigera*	Charity *et al.*, 1999
Alfalfa	*Spodoptera littoralis*	Narviez-Vasquez *et al.*, 1992
Alfalfa	*Frankliniella* spp. (Thrips)	Thomas *et al.*, 1994
Bean-amylase inhibitor		
Pea	*Callosobruchus maculatus*	Shade *et al.*, 1994
Pea	*Bruchus pisorum*	Schroeder *et al.*, 1995
Azuki bean	*Callosobruchus maculatus*	Ishimoto *et al.*, 1996

proteinase inhibitor precursor was expressed in transgenic pea under the control of Rubisco small subunit promoter (Charity *et al.*, 1999). Insect feeding trials have shown that the mortality of *Helicoverpa armigera* larvae was high compared to controls. Protease inhibitors from insects have also been expressed in plants. In *Manduca sexta* (Tobacco hornworm), several protease inhibitors are present in hemolymph. One of these proteins inhibits the activity of the enzyme elastase. Expression of this inhibitor in transgenic alfalfa, a forage legume, has resulted in reduced thrips (*Frankliniella* spp.) infestation.

3.3. AMYLASE INHIBITORS

Inhibitors of α-amylases are the second type of enzyme inhibitors used to modify crop plants. α-Amylase inhibitor from the common bean (*Phaseolus vulgaris*) forms a complex with and inhibits α-amylases in the midgut of coleopteran and storage pests of the genera *Callosobruchus* and *Bruchus* and blocks larval development (Ishimoto *et al.*, 1996). Genes for three α-amylase inhibitors have been expressed in pea and Azuki bean (Table 2). Shade *et al.* (1994) have expressed the gene encoding α-amylase inhibitor under the control of a seed-specific promoter in transgenic pea and observed significant level of insect protection. Similar results were obtained when Azuki bean was transformed with α-amylase inhibitor gene and tested for protection against bruchid beetles (Ishimoto *et al.*, 1996). It would be very useful if α-amylase inhibitor genes are expressed in tropical legumes such as chickpea and pigeonpea which suffer from losses due to a variety of storage pests.

3.4. LECTINS

Lectins are carbohydrate binding proteins, some of which are toxic to insects. Various lectins have shown some toxicity against species of the insect orders, Homoptera, Coleoptera, Lepidoptera and Diptera. The mode of action of lectins against insects remains unclear, but it has been shown that at least some bind to insect midgut epithelial cells (Czapla, 1997). However, some insecticidal lectins also show significant mammalian toxicity, including lectins from *P. vulgaris*, winged bean, soybean and wheat germ. Other lectins, for example those from pea and snowdrop, have demonstrated insecticidal activity and are innocuous to mammals (Gatehouse and Hilder, 1994).

Lectin from snowdrop (*Galanthus nivali*) has been shown to be very effective against aphids and rice brown planthopper (Powell *et al.*, 1995). It has been expressed in nine different crops including potato and tomato. Laboratory tests with engineered potatoes showed that snowdrop lectin did not increase the mortality or development time of potato aphid but considerably reduced fecundity (Down *et al.*, 1996). Results of experiments with the potato peach aphid were similar, but in addition, the establishment of aphids on transgenic potatoes was reduced (Gatehouse *et al.*, 1996). Snowdrop lectin also enhanced the resistance of potato to larvae of tomato moth (*Lacanobia oleracea*). Gatehouse and others concluded that the effect of snowdrop lectin is antifeedant rather than insecticidal (Gatehouse *et al.*, 1997). To date, no legume crop has been transformed with snowdrop lectin gene.

4. Bt-transgenic legume crops

Although many insecticidal genes have been transferred to different crop species, the most satisfactory system so far in terms of field resistance is the one based on Bt. Bt toxins have been transferred and expressed in at least thirty different plant species (Schuler *et al.*, 1998). However, the level of resistance they confer will depend on whether native (wild-type) or truncated, modified genes have been used (Kumar *et al.*, 1996). The prokaryotic codon usage in Bt genes needs to be modified towards that of higher plant genes. In addition, features that can destabilize the transcripts in higher plant cells need to be removed. A pioneering study of this kind has been made by the scientists at Monsanto, who modified the Bt genes and expressed in cotton with considerable success (Perlak *et al.*, 1990, 1991). The performance of cotton plants transgenic for *cry1Ab or cry1Ac* against cotton bollworm and pink bollworm was very satisfactory. This was followed by the expression of a coleopteran-specific gene *cry3Aa* from Bt subspecies *tenebrionis* in transgenic potato plants (Perlak *et al.*, 1993). The plants were resistant to Colorado potato beetle (CPB; *Leptinotarsa decemlineata*) under high levels of natural field infestation. Improved plant expression of the gene *cry3Aa* was achieved by increaseing its overall G/C content from 36 to 49%. As of now, three Bt-transgenic crops are under commercial cultivation (De Maagd *et al.*, 1998). Bt insecticidal proteins have been expressed in soybean, alfalfa and peanut for resistance to their respective pests (Table 2). A native *cry1Ac* gene was expressed in chickpea to confer protection against podborer, *Helicoverpa armigera* (Kar *et al.*, 1997). Development of podborer larvae was affected when fed on transgenic tissues. A synthetic gene encoding Cry1Ac toxin was introduced in soybean by particle bombardment and the transgenic plants were observed to be resistant to corn earworm (*Helicoverpa zea*), soybean looper (*Psuedoplusia includens*) and velvetbean caterpillar (*Anticarsia gemmatalis*) (Stewart *et al.*, 1996). Similarly, a synthetic *cry1C* gene was transferred to alfalfa for resistance to *Spodoptera littoralis* (Strizhov *et al.*, 1996). The transgenic plants produced Bt-Cry1C to the extent of 0.01–0.2% of total soluble protein and were resistant to cotton leaf worm and beet army worm. Transformation of peanut by a synthetic *cry1Ac* gene resulted in various levels of resistance to the lesser cornstalk borer, from complete larval mortality to a 66% reduction in larval weight (Singsit *et al.*, 1997).

5. Resistance management

The fundamental purpose of the deployment of resistance genes in transgenic plants is to manage the insect pest population and to prevent the development of resistance in insects. Resistance management strategies try to prevent or diminish the selection of rare individuals carrying resistance genes and hence to keep the frequency of resistance genes sufficiently low for insect control. Strategy development generally relies on theoretical assumptions and on computer models simulating insect population growth under various conditions (Tabashnik, 1994). Proposed strategies include the use of multiple toxins, crop rotation, high or ultra high dosages, and spatial or temporal refugia. The most promising and currently practical strategy is that of using refugia. The strategy would reduce the possibility of resistant insects from mating with other resistant insects, thereby preventing the creation of a resistant population. This is achieved by ensuring that there are always plenty of susceptible insects nearby for the few resistant ones to mate with. The basic principle of high dose strategy is to deploy plants with high levels of expression of the toxin with the expectation that it would take a long time for insects to overcome the toxin. It assumes that most or all resistance is recessive and that most resistance carriers would be heterozygous. A viable complementary strategy that is best adopted with the above two strategies is the deployment of multiple resistance or pyramiding of resistance genes. This strategy requires more than one resistant gene with different modes of action. It could be achieved either with additional *cry/vip* genes or with novel methods of insect resistance, but requires the use of refugia (Gould, 1998). Targeted expression is also complementary to the above-described strategies and will become possible in the near future. A toxin gene is expressed only specifically in a certain vulnerable tissue/part of the plant or is expressed both in a certain part of the plant as well as at a particularly critical time in the development of the plant. This strategy would allow plenty of susceptible insects to breed normally, thus increasing their predator and parasitic populations, while at the same time be prevented from causing damage to the critical plant parts or life cycles.

One of the most important tools of resistance management is to apply integrated pest management (IPM) principles in transgenic crop cultivation. Use of biological control methods (predators, viruses, fungi, etc.), botanical pesticides (Neem and Pyrethrum), crop rotation and sanitation, traditional methods coupled with minimal application of chemical insecticides will prolong the "life" of transgenic crops.

6. Conclusions and future prospects

Though legumes are considered to be recalcitrant to tissue culture, many species such as soybean, peanut, chickpea, alfalfa have been successfully transformed with a variety of insecticidal protein genes. Apart from the most important category of Bt toxin genes legumes were also transformed with genes encoding protease inhibitors, alpha amylase inhibitor and lectins for insect resistance. However, novel insecticidal proteins and respective genes need to be identified and used in conjunction with Bt to prevent the development of resistant insects. In addition, integrated pest management will have to play a central role in sustainable agriculture. Synthetic insecticides will continue to have a minimal role in combating minor pests and those that cannot be managed by biotechnological tools. Many economically important grain legumes such as pigeonpea and *Vigna* species are yet to be transformed with insecticidal protein genes though Bt δ-endotoxins/genes are available. Some of the legume crops are relatively recalcitrant in tissue culture and procedures to improve their regeneration capacity need to be developed.

References

Atwal A S (1986) *Agricultural Pests of India and South-East Asia*, Kalyani Publishers, New Delhi.

Boulter D (1993) Insect pest control by copying nature using genetically engineered crops. *Phytochemistry*, **34**: 1453–1466.

Charity J A, Anderson M A, Bittisnich D J, Whitecross M and Higgins T J V (1999) Transgenic tobacco and pea expressing a proteinase inhibitor from *Nicotiana alata* have increased insect resistance. *Mol. Breed.*, **5**: 357–365.

Czapla T M (1997) Plant lectins as insect control proteins in transgenic plants. In: *Advances in Insect Control: The Role of Transgenic Plants* (Eds Carozzi N and Koziel M), Taylor and Francis, London, pp. 123–138.

Dale P J, Irwin J A and Scheffler J A (1993) The experimental and commercial release of transgenic crop plants. *Plant Breed.*, **111**: 1–22.

De Maagd R A, Bosch D and Stiekema W (1999) *Bacillus thuringiensis* toxin-mediated insect resistance in plants. *Trends Plant Sci.*, **4**: 9–13.

Down R E, Gatehouse A M R, Hamilton W D O and Gatehouse J A (1996) Snowdrop lectin inhibits development and decreases fecundity of the glasshouse potato aphid (*Aulacorthum solani*) when administered *in vitro* and via transgenic plants both in laboratory and glass house trials. *J. Insect Physiol.*, **42**: 1035–1045.

Gatehouse A M R, Down R E, Powell K S, Sauvion N, Rahbe Y, Newell C A, Merryweather A, Hamilton W D O and Gatehouse J A (1996) Transgenic potato plants with enhanced resistance to the peach-potato aphid, *Myzus persicae*. *Entomol. Exp. Appl.*, **79**: 295–307.

Gatehouse A M R, Davison G M, Newell C A, Merryweather A, Hamilton, W D O, Burgess E P J, Gilbert R J C and Gatehouse J A (1997) Transgenic potato plants with enhanced resistance to the tomato moth, *Lacanobia oleracea*: Growth room trials. *Mol. Breed.*, **3**: 49–63.

Gatehouse A M R and Hilder V A (1994) Genetic manipulation of crops for insect resistance. In: *Molecular Perspectives: Crop Protection* (Eds Marshall G and Walters D), Chapman and Hall, London, pp. 177–201.

Gould F (1998) Sustainability of transgenic insecticidal cultivars: Integrating pest genetics and ecology. *Annu. Rev. Entomol.*, **43**: 701–726.

Indian Council of Agricultural Research (1997) *Agricultural Research Data Book*, ICAR, New Delhi.

Ishimoto M, Sato T, Chrispeels M J and Kitamura K (1996) Bruchid resistance of transgenic azuki bean expressing seed amylase inhibitor of common bean. *Entomol. Exp. Appl.*, **79**: 309–315.

James C and Krattiger A F (1996) *Global Review of Field Testing and Commercialization of Transgenic Plants*. ISAAA Briefs No. 1, ISAAA, Ithaca, USA.

Kar S, Basu D, Das S, Ramkrishnan N A, Mukharjee P, Nayak P and Sen S (1997) Expression of *cryIA(c)* gene of *Bacillus thuringiensis* in transgenic chickpea plants inhibits development of podborer (*Heliothis armigera*) larvae. *Transgenic Res.*, **6**: 177–182.

Krattiger A F (1997) *Insect Resistance in Crops: A Case Study of Bacillus thuringiensis*. ISAAA, Ithaca, NY.

Kumar P A and Sharma R P (1994) Genetic engineering of insect-resistant crop plants with *Bacillus thuringiensis* crystal protein genes. *J. Plant Biochem. Biotech.*, **3**: 3–8.

Kumar P A, Sharma R P and Malik V S (1996) Insecticidal proteins of *Bacillus thuringiensis*. *Adv. Appl. Microbiol.*, **42**: 1–43.

Narviez-Vasquez J, Orozco-Cardenas M L and Ryan C A (1992) Differential expression of a chimeric CaMV tomato proteinase inhibitor I gene in leaves of transformed nightshade, tobacco and alfalfa plants. *Plant Mol. Biol.*, **20**: 1149–1157.

Perlak F J, Deaton R W, Armstrong T A, Fuchs R L, Sims S R, Greenplate J T and Fishhoff D A (1990) Insect resistant cotton plants. *Bio/Technology*, **8**: 939–943.

Perlak F J, Fuchs R L, Dean D A, McPherson S and Fischhoff D A (1991) Modification of the coding sequence enhances plant expression of insect control protein genes. *Proc. Natl. Acad. Sci. USA*, **88**: 3324–3328.

Perlak F J, Stone T B, Muskopf Y M, Petersen L J, Parker G B, McPherson S A, Wyman J, Love S, Reed G, Biever D and Fischhoff D A (1993) Genetically engineered potatoes: Protection from damage by Colorado potato beetles. *Plant Mol. Biol.*, **22**: 313–321.

Powell K S, Gatehouse A M R, Hilder V A and Gatehouse J A (1995) Antifeedant effect of plant lectins and an enzyme on the adult stage of the rice brown planthopper *Nilaparvata lugens*. *Entomol. Exp. Appl.*, **75**: 51–59.

Shade R E, Schroeder H E, Pueyo J J, Tabe L M, Murdock L L, Higgins T J V and Chrispeels M J (1994) Transgenic pea seeds expressing the alpha-amylase inhibitor of the common bean are resistant to bruchid beetles. *Bio/Technology*, **12**: 793–796.

Schroeder H E, Gollasch S, Moore A, Tabe L M, Craig S, Hardie D C, Chrispeels M J, Spencer D and Higgins T J V (1995) Bean alpha-amylase inhibitor confers resistance to the pea weevil (*Bruchus pisorum*). *Plant Physiol.*, **107**: 1233–1239.

Schnepf E, Crickmore N, Van Rie J, Lereclus D, Baum J, Feitelson J, Zeigler D R and Dean D H (1998) *Bacillus thuringiensis* and its pesticidal crystal proteins. *Microbiol. Mol. Biol. Rev.*, **62**: 775–806.

Schuler T H, Poppy G M, Kerry B R and Denholm I (1998) Insect-resistant transgenic plants. *Trends Biotech.*, **16**: 168–175.

Singsit C, Adang M J, Lynch R E, Anderson W F, Wang A, Cardineau G and Ozias-Akins P (1997) Expression of a *Bacillus thuringiensis cry1Ac* gene in transgenic peanut plants and its efficacy against lesser cornstalk borer. *Transgenic Res.*, **6**: 169–176.

Stewart C N, Adang M J, All J N, Boerma H R, Cardineau G, Tucker D and Parrot A (1996) Genetic transformation, recovery and characterisation of fertile soybean transgenic for a synthetic *Bacillus thuringiensis cryIAc* gene. *Plant Physiol.*, **112**: 121–129.

Strizhov N, Keller M, Mathur J, Koncz-Kalman Z, Bosch D, Prudovsky E, Schell J, Sneh B, Koncz C and Zilberstein A (1996) A synthetic *cry1C* gene, encoding a *Bacillus thuringiensis* δ-endotoxin confers *Spodoptera* resistance in alfalfa and tobacco. *Proc. Natl. Acad. Sci. USA*, **93**: 15012–15017.

Tabashnik B E (1994) Evolution of resistance to *Bacillus thuringiensis. Annu. Rev. Entomol.*, **39**: 47–79.

Thomas J C, Wasmann C C, Echt C, Dunn R L, Bohnert H J and McCoy T J (1994) Introduction and expression of an insect proteinase inhibitor in alfalfa (*Medicago sativa* L.). *Plant Cell Rep.*, **14**: 34–36.

GENETIC ENGINEERING FOR VIRUS RESISTANCE

R. USHA AND K. VELUTHAMBI
Department of Plant Biotechnology, School of Biotechnology, Madurai Kamaraj University, Madurai – 625 021, India
e-mail: veluthambi@mrna.tn.nic.in

Abstract

A wide range of RNA viruses, belonging to 19 families, infect legumes. DNA viruses of the families *Geminiviridae* and *Caulimoviridae* and a new genus, *Nanovirus* cause serious diseases in legumes. Since host resistance genes for virus resistance are rare in pulses, genetic engineering with pathogen-derived genes is pursued as a promising approach in many legumes. Transgenic alfalfa plants with Alfalfa mosaic virus coat protein gene and a mutant P2 replicase gene have been generated. Genetic engineering of pea with Pea enation mosaic virus coat protein and Pea seed-borne mosaic potyvirus replicase gene (NIb) has been accomplished. Transformation of peanut with antisense nucleocapsid gene of Tomato spotted wilt virus and engineering of soybean with Bean pod mottle Comovirus have been reported. In many RNA viruses infecting legumes, the feasibility of pathogen-derived resistance has been demonstrated in a heterologous host, tobacco. A range of geminiviruses that infect bean and mungbean are reviewed. Bean plants transformed with Bean golden mosaic virus antisense genes of Rep, TrAP, REn and BC1 exhibited resistance against the virus. The antisense RNA approach holds promise for the generation of geminivirus resistance in important pulse crops. Progress in genetic engineering of pulses has been slow due to lack of efficient plant regeneration systems and due to a lower transformation efficiency with *Agrobacterium*. These limitations are being addressed in many research laboratories.

1. Introduction

Plant viruses can cause an extensive reduction in the yield of crops. Therefore, various methods for controlling them have been developed. Conventional methods include the use of virus-free germplasm and the eradication of infected plants, timing of planting so as not to coincide with the presence of the vectors, and the use of chemicals to control such vectors, especially insects. However, these approaches suffer from a number of disadvantages. The use of virus-free starting material and eradication of infected plants means that one must be able to detect viruses at an early stage. Moreover, these methods

P.K. Jaiwal and R.P. Singh (eds.), Improvement Strategies for Leguminosae Biotechnology, 255–268.

are expensive and the chemicals to control the virus vectors can pollute the environment. Another method is the breeding for virus resistance. However, there are few natural sources of virus resistance and, especially for RNA viruses, which constitute the majority of the plant viruses and have a high frequency of recombination, virulent strains might arise quickly to overcome the resistance. In addition, some virus resistance genes are recessive or inherited in a complex manner that further complicates breeding programs.

The first virus-resistant transgenic plants were obtained 15 years ago (Powell-Abel *et al.*, 1986). They represent the first application of the concept of pathogen-derived resistance (Sanford and Johnston, 1985). Since then, numerous transgenic crops tolerant or resistant to a wide range of viruses have been developed (Beachy, 1997). Different genomic sequences were used, including viral genes encoding coat proteins (CP), replicases, defective movement proteins, proteases or helper components (for reviews see: Lomonossoff, 1995; Palukaitis and Zaitlin, 1997). CP-mediated resistance, which is the most frequently used strategy, has been used to confer resistance to viruses in at least 13 RNA virus genera, including 23 distinct virus species (Grumet, 1995). Virus-resistant transgenic plants offer numerous potential agronomic and ecological benefits. This is clearly the case when no corresponding natural host resistance genes have been identified, and also when introgression of such a natural resistance gene into commercial crops would be too lengthy a process. Virus-resistant transgenic plants should also allow reduction of the amount of pesticide used to eliminate virus vectors.

2. RNA viruses infecting legumes

Legumes are susceptible to a large number of RNA viruses. Table 1 lists the members of nineteen families of RNA viruses that infect legumes. The important legumes considered here are: Alfalfa, bean, clover, cowpea, lupin, pea, peanut, pigeonpea and soybean. The various approaches to genetically engineer some of these legumes are discussed in this review. Due to the difficulties in transforming and regenerating legumes, only a few legume plants have been made transgenic with pathogen-derived genes. These are alfalfa, pea, peanut and soybean. For some of the other legume-infecting RNA viruses, virus-resistance has been demonstrated in the model plant, tobacco. These are Cowpea mosaic virus, Cucumber mosaic virus, Pea early browning virus and Peanut stripe virus.

2.1. ALFALFA

2.1.1. Alfalfa mosaic virus

Hill *et al.* (1991) have generated transgenic alfalfa, expressing the coat protein of Alfalfa mosaic virus. The strongest expressers accumulated nearly 500 ng of coat protein per mg of soluble leaf protein. Transgenic alfalfa plants containing large amounts of coat protein were shown to be resistant to the virus. These plants did not develop systemic infections following inoculations with up to 50 μg/mL Alfalfa mosaic virus, while inoculated control plants developed systemic infections following inoculation with as little as 10 μg/mL of the virus. The coat protein of Alfalfa mosaic virus has also been used as a carrier molecule for the expression of antigenic peptides from rabies virus (Modelska *et al.*, 1998).

Table 1. RNA viruses infecting legumes

Family	Genus	Virus
Bromoviridae	Alfamoviruses	Alfalfa mosaic virus
	Bromoviruses	Cowpea chlorotic mottle virus
		Broad bean mottle virus
		Cassia yellow blotch virus
		Springbeauty latent virus
	Cucumoviruses	Peanut stunt virus
		Cucumber mosaic virus
	Ilarviruses	Tobacco streak virus
		Prunus necrotic ring spot virus
		Sunflower ringspot virus
Bunyaviridae	Tospoviruses	Peanut yellow spot virus
		Groundnut ring spot virus
		Tomato spotted wilt virus
Carlaviruses		Cowpea mild mottle virus
		Pea streak virus
		Red clover vein mosaic virus
		Elderberry virus
		Poplar mosaic virus
Closteroviruses		Broad bean severe chlorosis virus
		Clover yellows virus
		Cassia severe mosaic virus
Comoviridae	Comoviruses	Cowpea mosaic virus
		Bean pod mottle virus
		Bean rugose mosaic virus
		Broad bean stain virus
		Broad bean true mosaic virus
		Cowpea severe mosaic virus
		Glycine mosaic virus
		Pea green mottle virus
		Pea mild mosaic virus
		Quail pea mosaic virus
		Red clover mottle virus
	Fabaviruses	Broad bean wilt virus
	Nepoviruses	Crimson clover latent virus
		Lucerne Australian latent virus
		Tomato black ring virus
		Tomato ring spot virus
		Mulberry ring spot virus
		Dogwood mosaic virus
		Cacao necrosis virus
		Tobacco ring spot virus
Dianthoviruses		Red clover necrotic mosaic virus
		Sweet clover necrotic mosaic virus
Enamoviruses		Pea enation mosaic virus
Furoviruses		Broad bean necrosis virus
		Peanut clump virus
Luteoviruses		Bean leaf roll virus

Table 1. Continued

Family	Genus	Virus
		Groundnut rosette virus Soybean dwarf virus Soybean Indonesian dwarf virus Subterranean clover red leaf virus Cowpea stunt virus Beet western yellows virus
Potexviruses		Clover yellow mosaic virus White clover mosaic virus Clitoria mosaic virus Crotalaria yellow mosaic virus Groundnut chlorotic spot virus Foxtail mosaic virus Narcissus mosaic virus Cymbidium mosaic virus Commelina X virus Asparagus 3 virus
Potyviridae	Potyviruses	Artichoke latent virus Bean common mosaic virus Bean common mosaic necrosis virus Bean yellow mosaic virus Bidens mosaic virus Cassia yellow spot virus Chickpea bushy dwarf virus Chickpea distortion mosaic virus Clover yellow vein virus Cowpea green vein banding virus Cowpea aphid-borne mosaic virus Cowpea rugose mosaic virus Groundnut eyespot virus Pea mosaic virus Pea seed-borne mosaic virus Peanut green mosaic virus Peanut mottle virus Soybean mosaic virus Swordbean distortion mosaic virus Passiflora ring spot virus Passionfruit woodiness virus Turnip mosaic virus Tobacco etch virus
Reoviridae	Phytoreoviruses	Clover wound tumour virus
Rhabdoviridae	Cytorhabdoviruses	Broad bean yellow vein virus Pisum yellow blotch virus
Rhabdoviridae	Nucleorhabdovirus	Clover enation virus Lucerne enation virus Red clover mosaic virus Sweet clover latent virus
Sobemoviruses		Southern bean mosaic virus Lucerne transient streak virus Subterranean clover mottle virus

Table 1. Continued

Family	Genus	Virus
Tenuiviruses		Pigeon pea sterility mosaic virus
Tobamoviruses		Odontoglossum ring spot virus
		Sunn-hemp mosaic virus
Tobraviruses		Pea early browning virus
		Tobacco rattle virus
Tombusviridae virus	Aureusviruses	Pigeonpea isolate of pothos latent
	Carmoviruses	Bean mild mosaic virus
		Blackgram mottle virus
		Cowpea mottle virus
		Glycine mottle virus
		Galinsoga mosaic virus
	Tombusviruses	Cymbidium ring spot virus
Tymoviruses		Clitoria yellow vein virus
		Okra mosaic virus
		Abelia latent virus
		Turnip yellow mosaic virus
		Kennedya yellow mosaic virus
		Ononis yellow mosaic virus
		Plantago mottle virus
Umbraviruses		Bean yellow vein banding virus
		Groundnut rosette virus

References: For all except the pigeonpea viruses: Brunt A A, Crabtree K, Dallwitz M J, Gibbs A J, Watson L and Zurcher E J (Eds) (1996). Plant Viruses Online: Descriptions and Lists from the VIDE Database. URL: http://biology.anu.edu.au/Groups/MES/vide. For the pigeonpea viruses: Kumar *et al.* (2000) and Kumar *et al.* (2001).

Brederode *et al.* (1995) have demonstrated that when the GDD motif in the full-length P2 replicase protein of Alfalfa mosaic virus was changed into GGD (5 lines), GVD (15 lines), or DDD (13 lines), 20 to 30% of the transgenic lines showed a high level of resistance to Alfalfa mosaic virus infection. This resistance was effective to inoculum concentrations of 10 to 25 μg/mL of virus causing severe necrosis of control plants. For all transgenic lines, the expression of the transgenes was analyzed at the RNA level. With the GGD, GVD, and DDD mutants, resistance was generally observed in plants with a relatively high expression level. This indicates that the resistance is due to the mutant replicase rather than to an RNA-mediated co-suppression phenomenon.

2.2. PEA

2.2.1. Pea enation mosaic enamovirus

Chowrira *et al.* (1998) have reported the coat protein-mediated resistance to Pea enation mosaic virus in transgenic *Pisum sativum* L. Transgenic R_2, R_3 and R_4 plants displayed delayed or transient pea enation mosaic virus multiplication and attenuated symptoms as compared to control inoculated plants.

2.2.2. Pea seed-borne mosaic potyvirus

Transgenic pea lines carrying the replicase (NIb) gene of Pea seed-borne mosaic potyvirus showed a highly resistant state upon infection by the homologous isolate (Jones *et al.*, 1998). Resistance was observed in plants in either the homozygous or hemizygous condition and resulted in no overall yield loss despite the initial infection. Resistance was associated with a loss of both viral and transgene RNA, which is indicative of a mechanism based on post-transcriptional gene silencing.

2.3. PEANUT

2.3.1. Tomato spotted wilt virus

Magbanua *et al.* (2000) have generated peanut lines transgenic for the antisense nucleocapsid (N) gene of a Tomato spotted wilt virus strain isolated from peanut by microprojectile-mediated transformation of somatic embryos of cultivars VC1 and AT120. The selectable marker (hygromycin resistance) and the N gene were on separate plasmids. A total of 207 VC1 and 120 AT120 hygromycin-resistant lines were produced. Of all the VC1 plants recovered, 71% were co-transformed with the N gene, but all plants were sterile. For AT120, 48 of the transgenic cell lines converted into plants. PCR screening showed 15 of the lines were transgenic for the N gene and two of these lines were fertile. A field test was conducted using seeds from each fertile line along with non-transgenic controls. Plants from four randomly selected field plots were examined for symptoms and analyzed by double-antibody sandwich ELISA and PCR at 10 and 14 weeks after planting. At 14 weeks, 76% of the transgenic plants were symptomless, while 2% were severely symptomatic or dead. In contrast, only 42% of the plants lacking the N gene were symptomless and 50% were severely symptomatic or dead. Northern blot analysis of field-resistant plants detected transgene RNA and the transcript level appeared undiminished after viral exposure.

2.4. SOYBEAN

2.4.1. Bean pod mottle virus

Di *et al.* (1996) have developed transgenic soybean lines expressing the Bean pod mottle comovirus coat protein precursor gene. Soybean (*Glycine max*, Merrill. cv. Fayette) cotyledonary nodes were transformed with Bean pod mottle virus coat protein precursor gene via *Agrobacterium*-mediated transformation. The transformation rate was low and only five primary transformants derived from five different explants were obtained from 400 original cotyledonary nodes. Southern blot hybridization verified the integration of the Bean pod mottle virus coat protein precursor gene. Inheritance and expression of this gene in R_1 plants were also demonstrated. About 30% of R_2 plants derived from one transgenic line showed complete resistance to Bean pod mottle virus infection, assessed by symptomatology and ELISA, suggesting that homozygous, but not hemizygous plants exhibit the resistant phenotype.

2.5. PIGEONPEA

Virus diseases have emerged as important constraints in pigeonpea cultivation in India. Therefore, a brief review is presented on viruses that infect pigeonpea.

2.5.1. Pigeonpea sterility mosaic virus

Sterility mosaic disease results in annual losses exceeding 205,000 tons of grain in India alone. Although this disease was reported in 1931, the causal agent has only been recently identified in a major breakthrough by Kumar *et al.* (2000). They have detected tenui-like virus particles of asymmetric morphology from the diseased pigeonpea plants and have confirmed the mite-transmission of the virus.

2.5.2. Pigeonpea isolate of Pothos latent virus

Kumar *et al.* (2001) have also characterized a new virus from pigeonpea. The genome of this virus showed 93% homology to that of Pothos latent virus (PoLV), a recently described (Martelli *et al.*, 1998) virus in the genus *Aureusvirus*, family *Tombusviridae* and was indistinguishable from PoLV in gel double diffusion serological tests. This virus is, therefore, regarded as a pigeonpea isolate of PoLV (PoLV-PP).

2.6. LEGUME-INFECTING RNA VIRUSES FOR WHICH STUDIES TOWARDS PATHOGEN-DERIVED RESISTANCE HAS BEEN CONDUCTED IN TOBACCO

2.6.1. Cucumber mosaic cucumovirus

Nakajima *et al.* (1993) developed transgenic tobacco expressing the coat protein (CP) of Cucumber mosaic virus strain O (CMV-O) which showed a significant level of protection against CMV strains O and Y. When inoculum concentrations were increased, the transformants showed a stronger level of protection against CMV-O than against CMV-Y. The substitution of an amino acid residue between CMV-O and CMV-Y, which is presumed to change the conformation of CP, may reflect the difference in susceptibility to these viruses. Furthermore, the transgenic tobacco plants showed a significant level of protection against Chrysanthemum mild mottle virus, a member of the cucumovirus group but with no serological relationship to CMV.

Anderson *et al.* (1992) obtained transgenic *Nicotiana tabacum* plants by transformation with a modified and truncated replicase gene encoded by RNA-2 of cucumber mosaic virus. The transgenic plants were resistant to virus disease when challenged with either Cucumber mosaic virus virions or RNA at concentrations up to 500 μg/mL or 50 μg/mL, respectively, the highest concentrations tested. This resistance was absolute, as neither symptoms nor virus could be detected in uninoculated leaves, even after prolonged incubation (120 days after inoculation).

2.6.2. Cowpea mosaic comovirus

Sijen *et al.* (1996) have demonstrated that resistance to Cowpea mosaic virus (CPMV) in transgenic *Nicotiana benthamiana* plants is RNA mediated. In resistant CPMV movement protein (MP) gene-transformed lines, transgene steady state mRNA levels were low, whereas nuclear transcription rates were high, implying that a post-transcriptional gene-silencing mechanism is at the base of the resistance. The silencing mechanism was shown to affect Potato virus X RNAs when they contained CPMV MP gene sequences from the 3′ part, even as small as 60 nucleotides. Therefore, they conclude that the 3′ part of the MP transgene mRNA is the initial target of the silencing mechanism. The arrangement of the transgenes in the plant genome was found to play an important role in establishing resistance because the frequency of resistant lines increased from 20 to 60% when transformed with a transgene containing a direct repeat of MP sequences rather than a single MP transgene. They have also detected strong methylation in all of the plants containing the directly repeated MP sequences. In sensitive lines, only the promoter region was found to be heavily methylated, whereas in resistant lines, only the transcribed region was strongly methylated. Genetic engineering experiments on CPMV, in addition to providing important insight into the mechanism of resistance in virus resistant transgenic plants, have also paved the way towards the development of plant-based edible vaccines. CPMV was the first plant virus to be developed as a stable system for the expression of a number of foreign epitopes from different human and animal viruses like the foot and mouth disease virus, common cold virus, human immunodeficiency virus and mink enteritis virus (MEV) (Usha *et al.*, 1993; Porta *et al.*, 1994; Dalsgaard *et al.*, 1997). The three-dimensional structure of CPMV (Stauffacher *et al.*, 1987) revealed the most suitable positions on the surface of the virus particle for the insertion of the epitopes. There are several advantages in using CPMV as an epitope-presentation system. The virus is extremely stable, grows to very high titres and the purification of the virus is easy and quick. The chimaeric virus particles expressing epitopes from HIV-1 or MEV have been shown to be capable of raising neutralizing and protective antibodies, respectively, when injected into experimental animals (McLain *et al.*, 1995, 1996; Dalsgaard *et al.*, 1997). Thus, the structural and molecular biological studies on CPMV have made the virus an ideal system for the design of stable and inexpensive vaccines for human and animal diseases.

The coat protein precursor (CP-P) of Cowpea mosaic virus (CPMV) has been expressed in transgenic tobacco plants (Nida *et al.*, 1992). However, the infectivity experiments did not differentiate between the CP-P expressors and the vector-transformed plants.

2.6.3. Pea early browning tobravirus

MacFarlane *et al.* (1992) have shown that tobacco plants transformed with a region of the 201 kDa replicase gene of Pea early browning virus are resistant to virus infection.

2.6.4. Peanut stripe potyvirus

Cassidy and Nelson (1995) have engineered *Nicotiana benthamiana* plants with the coat protein gene (both ATG plus and ATG minus) of Peanut stripe potyvirus. They have shown that the plants with the CP-ATG plus transgene displayed the recovery phenotype whereas the CP-ATG minus transgenic plants lacked the recovery phenotype. So, they conclude that it is important for an engineered ATG to be present to obtain a recovery phenotype.

3. DNA viruses infecting legumes

Plant viruses with DNA genomes are classified into two families, *Geminiviridae* and *Caulimoviridae* and in a recently proposed genus, *Nanovirus*. Geminiviruses and nanoviruses have single-stranded genomes, whereas caulimoviruses have double-stranded genomes.

3.1. GEMINIVIRIDAE

Many important viruses that infect pulses belong to the family *Geminiviridae*. Geminiviruses have small single-stranded genomes of 2.5 kb to 3.0 kb, encapsidated as twin icosahedral, geminate particles (Lazarowitz, 1992). Geminiviruses are classified into three genera, *Mastrevirus*, *Curtovirus* and *Begomovirus* based on their genome organization, hosts infected and insect vectors (Mayo and Pringle, 1998). *Begomovirus* comprises bipartite (rarely monopartite) geminiviruses that infect dicotyledonous plants and are transmitted by whiteflies. Most legume-infecting geminiviruses fall into this group. Monopartite, dicot-infecting geminiviruses transmitted by leafhoppers are included in the genus *Curtovirus*. Monopartite, monocot-infecting geminiviruses transmitted by leafhoppers are included in the genus, *Mastrevirus*.

3.1.1. Begomovirus

Bean golden mosaic virus (BGMV). This is the type member of the genus *Begomovirus*. The genome of BGMV (Brazil isolate, BGMV-BR), which comprises DNA A of 2,647 nucleotides and DNA B of 2,587 nucleotides, was first sequenced by Howarth *et al.* (1985). The sequence information from BGMV helped in elucidating the salient features of genome organization in bipartite geminiviruses (Fig. 1).

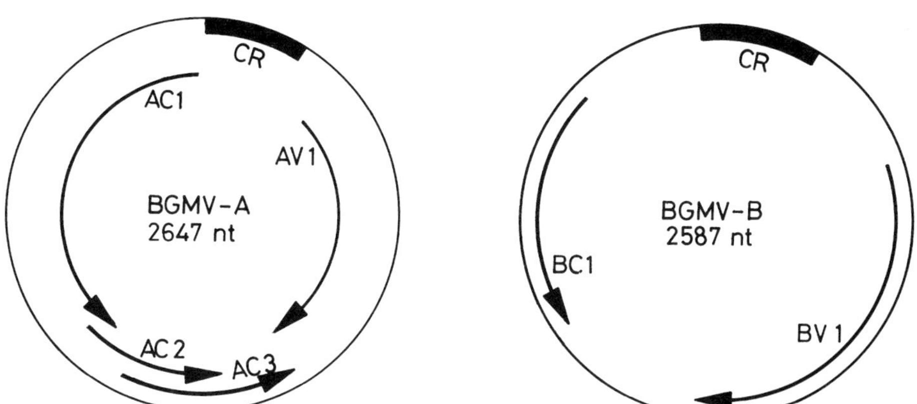

Figure 1. Genome organization of Bean golden mosaic virus (BGMV-BR). AV1 and BV1 are ORFs on the virion-sense strand. AC1, AC2, AC3 and BC1 are ORFs on the complementary-sense strand. CR, Common region. (Maps drawn based on Howarth et al., *1985.)*

DNA A and DNA B sequences of BGMV differed except in a 205-nt region named as common region (CR), which showed almost 100% sequence identity. A set of 12-nt GC-rich invert repeats with an intervening 12-nt region was proposed to form a stable stem-loop structure. A nonanucleotide sequence, TAATATTAC, found in the loop of the stem-loop structure was later found to be conserved in all geminiviruses studied subsequently (Timmermans *et al.*, 1994). The stem-loop structure constitutes the origin of rolling circle replication using which single-stranded virion DNA is replicated from a double-stranded replicative form (Arguello-Astorga *et al.*, 1994). The tandem (direct) repeats found in the CR were later found to constitute the iterons that serve as replicase (Rep)-binding domains (Hanley-Bowdoin *et al.*, 1999).

Howarth *et al.* (1985) reported six ORFs, four in DNA A and two in DNA B. The functions of proteins coded by the six ORFs have been delineated (Hanley-Bowdoin *et al.*, 1999). The complementary-sense strand of DNA A carries three ORFs. AC1 encodes replicase (Rep) that initiates the rolling circle replication of the double-stranded replicative form (RF). AC2 encodes a transactivating protein (TrAP) that enchances transcription from AV1 and BV1 promoters. AC3 encodes a replication-enhancing protein (REn), which greatly enhances viral DNA accumulation.

Bean golden yellow mosaic virus (BGYMV). BGMV-BR is not sap-transmissible and was classified as Type-I BGMV. Many sap-transmissible isolates that cause bean golden mosaic were isolated later and were initially classified as Type-II isolates of BGMV. The nomenclature of these viruses has been revised recently (Fauquet *et al.*, 2000). The sap-transmissible BGMV are now called as Bean golden yellow mosaic virus (BGYMV). Examples of this group are BGYMV-GA (Gautemala), BGYMV-PR (Puerto Rico), BGYMV-DR (Dominican Republic) and BGYMV-MX (Mexico) (Garrido-Ramirez *et al.*, 2000a).

Bean dwarf mosaic virus (BDMV). BDMV is serologically related to BGMV but distinguishable on the basis of symptoms produced and by hybridization to type-specific DNA probes (Morales *et al.*, 1990). BDMV is a typical bipartite *Begomovirus* (Hidayat *et al.*, 1993). BDMV has served as a good experimental system to understand virus movement functions encoded in DNA B. Microinjection of *E. coli*-expressed BV1 and BC1 movement proteins of BDMV has established that BV1 mediates export of viral DNA from nucleus to cytoplasm whereas BC1 mediates cell to cell movement of the virus by increasing the size-exclusion limit of plasmodesmata (Noueiry *et al.*, 1994). Rojas *et al.* (1998) showed that BDMV BV1 and BC1 interact with viral DNA on the basis of form and size rather than the sequence. The role of BV1 as a determinant of hypersensitive response was also demonstrated in BDMV (Garrido-Ramirez *et al.*, 2000b).

Bean yellow dwarf virus (BeYDV). BeYDV is an atypical *Mastrevirus* that is adapted to infect dicotyledonous species (Liu *et al.*, 1997). Adaptation of BeYDV to dicotyledonous hosts was linked to the evolution of viral genes associated with viral DNA replication and virus movement (Liu *et al.*, 1998).

Mungbean yellow mosaic virus (MYMV). A Thailand isolate of yellow mosaic virus infecting mungbean (*Vigna radiata*) was cloned (Morinaga *et al.*, 1990) and sequenced (Morinaga *et al.*, 1993). A blackgram (*V. mungo*) isolate of MYMV was cloned from South

India and its DNA A and DNA B were used to analyse several weeds that were proposed to serve as alternative hosts of MYMV (Vanitharani *et al.*, 1996). A second blackgram isolate of MYMV was cloned from North India and was shown to be agroinfectious on blackgram (Mandal *et al.*, 1997). The names of yellow mosaic viruses listed above have been recently revised (Fauquet *et al.*, 2000) as Mungbean yellow mosaic virus-Thailand (MYMV-TH), Mungbean yellow mosaic – *Vigna* (MYMV-Vi) and Mungbean yellow mosaic Indian virus (MYMIV), respectively. In the *Begomovirus* genus, MYMV strains are clustered into a group designated as 'legume begomovirus'. A related yellow mosaic virus, Horsegram yellow mosaic virus infecting *Dolichos biflorus* has been studied serologically (Swanson *et al.*, 1992).

Many other geminiviruses infecting pulses have been reported but have not been subjected to detailed analysis (Fauquet *et al.*, 2000). These include, Bean calico mosaic virus (AF110189, AF110190), Cowpea golden mosaic virus-[Nigeria] (AF029217) and Cowpea golden mosaic virus-[Brazil] (AF188708).

4. Nanovirus

This is a complex group of plant viruses with circular ssDNA genomes. Each ssDNA is about 1 kb long. An example of a pulse-infecting nanovirus is Faba bean necrotic yellows virus (FBNYV; Katul *et al.*, 1993). FBNYV was reported to have six ssDNA components, each principally encoding one ORF (Katul *et al.*, 1997). All ssDNA components carried a common region (CR) that can potentially form a stem-loop structure, a feature shared with geminiviruses.

5. Caulimovirus

This genus, coming under the family *Caulimoviridae*, comprises plant pararetroviruses. They contain a circular dsDNA of 7.8 to 8.2 kb size, with one or more single-strand discontinuities on either strand. A pulse-infecting *Caulimovirus*, Soybean chlorotic mottle virus (SbCMV) has been reported (see Mathews, 1991).

6. Genetic engineering of legumes for developing resistance against DNA viruses

Progress in genetic engineering of pulses has been very slow due to two major constraints: (1) Tissue culture conditions for shoot regeneration have not been developed in most of the pulses and (2) Tissues of pulses from which multiple shoots are generated are not highly amenable to *Agrobacterium*-based transformation. Therefore, microprojectile bombardment-based direct transformation approaches have been explored for transformation of pulses.

Aragão *et al.* (1998) reported engineering of beans (*Phaseolus vulgaris*) for developing resistance against Bean golden mosaic virus (BGMV-BR). They constructed the first antisense unit by cloning the Rep (AC1), TrAP (AC2) and REn (AC3) coding sequences under the CaMV35S promoter in antisense orientation. The second antisense unit was constructed by cloning the BC1 coding sequence under the CaMV35S promoter. Both antisense units were placed in one plasmid which was used to transform the shoot apex of embryonic axes by microprojectile bombardment. Cobombardment was carried out with a *gus-neo* gene. GUS-positive multiple shoots were excised, rooted and grown into plants. Of the four transgenic lines tested in R_3 and R_4 generation, two lines exhibited delayed and attenuated viral symptoms. This antisense RNA approach holds good promise in generating resistance against other geminiviruses that infect important pulse crops.

The effectiveness of Bean dwarf mosaic virus (BDMV) movement proteins in conferring broad spectrum resistance against geminiviruses has been tested in a heterologous system, tomato, by Hou *et al.* (2000). The BV1 and BC1 genes of BDMV were transferred to tomato using *Agrobacterium* and the transgenic plants were tested for resistance against Tomato mottle virus (ToMoV), a *Begomovirus*. Several R_0 transgenic tomato plants expressing either wild-type or mutated BV1 or BC1 proteins showed significant delay in ToMoV infection, compared with untransformed plants. The approaches involving the use of movement proteins offer an excellent scope for developing broad spectrum resistance against a wide range of related geminiviruses. This approach has not been extended to pulses so far.

References

Anderson J M, Palukaitis P and Zaitlin M (1992) A defective replicase gene induces resistance to Cucumber mosaic virus in transgenic tobacco plants. *Proc. Natl. Acad. Sci. USA*, **89**: 8759–8763.

Aragão F J L, Ribeiro S G, Barros L M G, Brasileiro A C M, Maxwell D P, Rech E L and Faria J C (1998) Transgenic beans (*Phaseolus vulgaris* L.) engineered to express viral antisense RNAs show delayed and attenuated symptoms of Bean golden mosaic geminivirus. *Mol. Breed.*, **4**: 491–499.

Arguello-Astorga G R, Herrera-Estrella L and Rivera-Bustamante R (1994) Experimental and theoretical definition of geminivirus origin of replication. *Plant Mol. Biol.*, **26**: 553–556.

Beachy R N (1997) Mechanisms and applications of pathogen-derived resistance in transgenic plants. *Curr. Opin. Biotech.*, **8**: 215–220.

Brederode F T, Taschner P E, Posthumus E and Bol J F (1995) Replicase-mediated resistance to Alfalfa mosaic virus. *Virology*, **207**: 467–474.

Cassidy B G and Nelson R S (1995) Differences in protection phenotypes in tobacco plants expressing coat protein genes from Peanut stripe potyvirus with or without an engineered ATG. *Mol. Plant-Microbe Interact.*, **8**: 357–365.

Chowrira G M, Cavileer T D, Gupta S K, Lurquin P F and Berger P H (1998) Coat protein-mediated resistance to Pea enation mosaic virus in transgenic *Pisum sativum* L. *Transgenic Res.*, **7**: 265–271.

Dalsgaard K, Uttenthal X, Jones T D, Xu F, Merryweather A, Hamilton W D O, Langeveld J P M, Boshuizen R S, Kamstrup S, Lomonossoff G P, Porta C, Vela C, Casal J L, Meloen R H and Rodgers P B (1997) Plant-derived vaccine protects target animals against a virus disease. *Nature Biotech.*, **15**: 248–252.

Fauquet C M, Maxwell D P, Gronenborn B and Stanley J (2000) Revised proposal for naming geminiviruses. *Arch. Virol.*, **145**: 1733–1761.

Garrido-Ramirez E R, Sudarshana M R and Gilbertson R L (2000a) Bean golden yellow mosaic virus from Chiapas, Mexico: Characterization, pseudorecombination with other bean-infecting geminiviruses and germplasm screening. *Phytopathology*, **90**: 1224–1232.

Garrido-Ramirez E R, Sudharsana M R, Lucas W J and Gilbertson R L (2000b) Bean dwarf mosaic virus BV1 protein is a determinant of the hypersensitive response and avirulence in *Phaseolus vulgaris*. *Mol. Plant-Microbe Interact.*, **13**: 1184–1194.

Grumet R (1995) Genetic engineering for crop virus resistance. *HortScience*, **30**: 449–456.
Hanley-Bowdoin L, Settlage S B, Orozco B M, Nagar S and Robertson D (1999) Geminiviruses: Models for plant DNA replication, transcription and cell cycle regulation. *Crit. Rev. Plant Sci.*, **18**: 71–106.
Hidayat S H, Gilbertson R L, Hanson S F, Morales F J, Ahlquist P, Russel D R and Maxwell D P (1993) Complete nucleotide sequence of the infectious cloned DNAs of Bean dwarf mosaic geminivirus. *Phytopathology*, **83**: 181–187.
Hill K K, Jarvis-Eagan N, Halk E L, Krahn K J, Liao L W, Mathewson R S, Merlo S E, Rashka K E and Loesh-Fries L S (1991) The development of virus-resistant alfalfa, *Medicago sativa* L. *Bio/Technology*, **9**: 373–377.
Hou Y, Sanders R, Ursin V M and Gilbertson R L (2000) Transgenic plants expressing geminivirus movement proteins: Abnormal phenotypes and delayed infection by Tomato mottle virus in transgenic tomatoes expressing the Bean dwarf mosaic virus BV1 or BC1 proteins. *Mol. Plant-Micr. Interact.*, **13**: 297–308.
Howarth A J, Caton J, Bossert M and Goodman R M (1985) Nucleotide sequence of Bean golden mosaic virus and a model for gene regulation in geminiviruses. *Proc. Natl. Acad. Sci. USA*, **82**: 3572–3576.
Jones A L, Johansen I E, Bean S J, Bach I and Maule A J (1998) Specificity of resistance to Pea seed-borne mosaic potyvirus in transgenic peas expressing the viral replicase (NIb) gene. *J. Gen. Virol.*, **79**: 3129–3137.
Katul L, Maiss E, Morozov S Y and Vetten H J (1997) Analysis of six DNA components of the Faba bean necrotic yellows virus genome and their structural affinity to related plant virus genomes. *Virology*, **233**: 247–259.
Katul L, Vetten H J and Mauss E (1993) Characterization and serology of virus-like particles associated with Faba bean necrotic yellows. *Ann. Appl. Biol.*, **123**: 629–647.
Kumar P L, Jones A T, Srinivasulu P and Reddy D V R (2000) Breakthrough in the identification of the causal agent of Pigeonpea sterility mosaic disease. *J. Mycol. Plant Pathol.*, **30**: 249.
Kumar P L, Jones A T, Sreenivasulu P, Fenton B and Reddy D V R (2001) Characterization of a virus from pigeonpea with affinities to species in the genus *Aureusvirus*, family *Tombusviridae. Plant Disease*, **85**: 216–219.
Lazarowitz S G (1992) Geminiviruses: Genome structure and gene function. *Crit. Rev. Plant Sci*, **11**: 327–349.
Liu L, Davies J W and Stanley J (1998) Mutational anlaysis of Bean yellow dwarf virus, a geminivirus of the genus *Mastrevirus* is adapted to dicotyledonous plants. *J. Gen. Virol.*, **79**: 2265–2274.
Liu L, Van Tonder T, Pieterson G, Davies J W and Stanley J (1997) Molecular characterization of a subgroup I geminivirus from a legume in South Africa. *J. Gen. Virol.*, **78**: 2113–2117.
Lomonossoff G P (1995) Pathogen-derived resistance to plant viruses. *Annu. Rev. Phytopathol.*, **33**: 323–343.
MacFarlane S A and Davies J W (1992) Plants transformed with a region of the 201-kiloDalton replicase gene of Pea early browning virus are resistant to virus infection. *Proc. Natl. Acad. Sci. USA*, **89**: 5829–5833.
Magbanua Z V, Wilde H D, Roberts J K, Chowdhury K, Abad J, Moyer J W, Wetzstein H Y and Parrott W A (2000) Field resistance to Tomato spotted wilt virus in transgenic peanut (*Arachis hypogaea* L.) expressing an antisense nucleocapsid gene sequence. *Mol. Breed.*, **6**: 227–236.
Mandal B, Varma A and Malathi V G (1997) Systemic infection of *Vigna mungo* using the cloned DNAs of the blackgram isolate of Mungbean yellow mosaic geminivirus through agroinfection and the transmission of the progeny virus by whiteflies. *J. Phytopath.*, **145**: 505–510.
Martelli G P, Russo M, Rubino L and Sabanadzovic S (1998) *Aureusvirus*, a novel genus in the family Tombusviridae. *Arch. Virol.*, **143**: 1847–1851.
Mathews R E F (1991) *Plant Virology*. Third Edition. Academic Press, San Diego, 692.
Mayo M A and Pringle C R (1998) Virus taxonomy – 1997. *J. Gen. Virol.*, **79**: 649–657.
McLain L, Durrani Z, Wisniewski L A, Porta C, Lomonossoff G P and Dimmock N J (1996) Stimulation of neutralizing antibodies to human immunodeficiency virus type 1 in three strains of mice immunized with a 22 amino acid peptide of gp41 expressed on the surface of a plant virus. *Vaccine*, **14**: 799–810.
McLain L, Porta C, Lomonossoff G P, Durrani Z and Dimmock N J (1995) Human immunodeficiency virus type 1 – neutralizing antibodies raised to a glycoprotein 41 peptide expressed on the surface of a plant virus. *AIDS Res. Human Retroviruses*, **11**: 327–334.
Modelska A, Dietzschold B, Sleysh N, Flu Z F, Steplewsky K, Hooper D C, Koprowsky H and Yusibov V (1998) Immunization against rabies with plant derived antigen. *Proc. Natl. Acad. Sci. USA*, **95**: 2481–2485.
Morales F J, Niessen A, Ramirez B and Castano M (1990) Isolation and partial characterization of a geminivirus causing bean dwarf mosaic. *Phytopathology*, **80**: 96–101.

Morinaga T, Ikegami M and Miura K (1990) Physical mapping and molecular cloning of Mung bean yellow mosaic virus DNA. *Intervirology*, **31**: 50–56.

Morinaga T, Ikegami M and Miura K (1993) The nucleotide sequence and genome structure of Mung bean yellow mosaic geminivirus. *Microbiol. Immunol.*, **37**: 471–476.

Nakajima M, Hayakawa T, Nakamura I and Suzuki M (1993) Protection against Cucumber mosaic virus (CMV) strains O and Y and Chrysanthemum mild mottle virus in transgenic tobacco plants expressing CMV-O coat protein. *J. Gen. Virol.*, **74**: 319–322.

Nida D L, Anjos J R, Lomonossoff G P and Ghabrial S A (1992) Expression of Cowpea mosaic virus coat protein precursor in transgenic tobacco plants. *J. Gen. Virol.*, **73**: 157–163.

Noueiry A O, Lucas W J and Gilbertson R L (1994) Two proteins of a plant DNA virus coordinate nuclear and plasmodesmatal transport. *Cell*, **76**: 925–932.

Porta C, Spall V E, Loveland J, Johnson J E, Barker P J and Lomonossoff G P (1994) Development of Cowpea mosaic virus as a high-yielding system for the presentation of foreign peptides. *Virology*, **202**: 949–955.

Powell-Abel P, Nelson R S, De B, Hoffmann N, Rogers S G, Fraley R T and Beachy R N (1986) Delay of disease development in transgenic plants that express the Tobacco mosaic virus coat protein gene. *Science*, **232**: 738–763.

Rojas M R, Noueiry A O, Lucas W J and Gilbertson R L (1998) Bean dwarf mosaic geminivirus movement proteins recognize DNA in a form- and size-specific manner. *Cell*, **95**: 105–113.

Sanford J C and Johnston S A (1985) The concept of parasite-derived resistance: deriving resistance genes from the parasite's own genome. *J. Theort. Biol.*, **113**: 395–405.

Sijen T, Wellink J, Hiriart J B and van Kammen A (1996) RNA-mediated virus resistance: role of repeated transgenes and delineation of targeted regions. *The Plant Cell*, **8**: 2277–2294.

Stauffacher C V, Usha R, Harrington M, Schmidt T, Hosur M V and Johnson J E (1987) The structure of Cowpea mosaic virus at 3.5Å resolution. In: *Crystallography in Molecular Biology* (Eds Moras D, Drenth J, Strandberg B, Suck D and Wilson K), Plenum Press, New York, pp. 293–308.

Swanson M M, Varma A, Muniyappa V and Harrison B D (1992) Comparative epitope profiles of the particle proteins of whitefly-transmitted geminiviruses from nine crop legumes in India. *Ann. Appl. Biol.*, **120**: 425–433.

Timmermans M C P, Das O P and Messing J (1994) Geminiviruses and their uses as extrachromosomal replicons. *Annu. Rev. Plant Physiol. Mol. Biol.*, **45**: 79–112.

Usha R, Rohll J B, Spall V E, Shanks M, Maule A J, Johnson J E and Lomonossoff G P (1993) Expression of an animal virus antigenic site on the surface of a plant virus particle. *Virology*, **197**: 366–374.

Vanitharani R, Karthikeyan A S, Anuradha S and Veluthambi K (1996) Genome homologies among geminiviruses infecting *Vigna*, Cassava, *Acalypha, Croton* and *Vernonia*. *Curr. Sci.*, **70**: 63–69.

Yusibov V M, Modelska A, Steplewski K, Agadjanyan M, Weiner D, Hooper C and Koprowski H (1997) Antigens produced in plants by infection with chimeric plant viruses immunize against rabies virus and HIV-1. *Proc. Natl. Acad. Sci. USA*, **94**: 5784–5788.

MOLECULAR STRATEGIES FOR FUNGAL AND NEMATODE RESISTANCE

N.P. SINGH, S. DATTA AND BANSA SINGH
Indian Institute of Pulses Research, Kanpur – 208 024, India
e-mail: nps@iipr.up.nic.in

Abstract

Fungal pathogens and plant parasitic nematodes are prominent in causing diseases in legume crops. Chemical control of nematodes and fungi are not very effective and also cause product contamination. Molecular biology has made it possible to identify and clone genes associated with pathogenesis related proteins, anti-fungal proteins and compounds, phytoalexins and enzymes controlling oxidative burst, etc. which lead to a number of transgenic plant developments. It has also been possible to overexpress them constitutively in several transgenic crop systems. In a majority of cases it was proved that these anti-fungal proteins are very effective against various fungal pathogens. RFLP (Restriction Fragment Length Polymorphism) has made it possible to generate high-density maps of various commercially important crops including peas. However, progress of molecular mapping and gene tagging is limited due to availability of low degree of polymorphism using conventional DNA markers like RAPD and RFLP. In recent years, development of microsatellite based markers and AFLP has produced adequate polymorphism in a number of grain legumes like chickpea which otherwise was shown to be low polymorphic. Besides, there are also some successful examples of gene tagging using these markers for fungal diseases and nematode resistance in grain legumes.

1. Introduction

Grain legumes, commonly known as pulses, are an important ingredient of vegetarian diet. Besides being very rich in protein, pulses are good source of carbohydrates, minerals and vitamins. They are consumed in several ways, mainly as human food and also as forage crop. Furthermore, grain legumes have special characteristics of biological nitrogen fixation, hence these are very useful component in sustainable agriculture. The productivity of pulses is low, mainly due to their high susceptibility to biotic and abiotic stresses, low response to high agricultural input like fertilizer, irrigation, etc. and high photo-thermal sensitivity. Among biotic stresses, fungal diseases and nematodes cause

P.K. Jaiwal and R.P. Singh (eds.), Improvement Strategies for Leguminosae Biotechnology, 269–277.

very heavy losses to pulse production annually. About one-third of the total food production is lost worldwide due to fungal and other diseases every year.

Conventional methods of plant disease and nematode control using organic pesticides has serious limitations in being ineffective and environmentally unsafe. The most important and practicable approach is to incorporate host resistance genes into improved pulse varieties. Such disease and pest resistant varieties against fungal diseases and nematodes are almost non-existent due to non-availability of suitable donor parents.

In recent years biotechnology and genetic engineering have provided an alternative approach to control disease and nematode pests through genetically engineered crop plants.

2. Molecular tools and techniques

The following tools and techniques of molecular biology and genetic engineering can be exploited for improving grain legumes against fungal diseases and nematodes.

2.1. DNA MOLECULAR MARKER

The DNA molecular marker can be used to detect variation in both coding and non-coding regions of the genome. The main advantages of DNA marker are dominant and co-dominant expression, infinite number, and phenotypical neutrality having no epistatic and developmental effects. Application of molecular markers is thus a new paradigm for a directed improvement of crop plants for disease, insects and nematode resistance through tagging important resistance genes. Various types of markers have been used as explained below:

2.1.1. Restriction Fragment Length Polymorphism (RFLP)

RFLP can be defined as variation in length of DNA fragment produced by digestion of the DNA with restriction endonuclease which is revealed after probing cloned DNA sequences to specific regions of eukaryotic genome.

If RFLP markers tightly linked to resistant genes are available, the plant can be screened at the seedling stage itself for the particular RFLP marker. Some pests/diseases affect the crop at flowering/post-flowering stage and RFLP markers permit to screen for resistance against them in early stage without waiting till the appropriate stage of the crop or infection occurs.

2.1.2. Random Amplified Polymorphic DNA (RAPD)

RAPD is a standard procedure for the identification of polymorphism in plants based on PCR. RAPD markers are based on the use of short synthetic oligonucleotide primers. PCR amplification involves the use of two oligonucleotide primers, repeated cycle of heat denaturation of the DNA, annealing of the primers to their complimentary sequences. The primers flanking the hypervariable regions are designed and used to amplify specific regions. RFLP and RAPD involves the establishment of linkage association between these markers and traits of interest in a segregating population, and both can be used to map the trait of interest.

2.1.3. *Amplified Fragment Length Polymorphism (AFLP)*

AFLP marker is produced when restriction site differences exist between two DNA sources. One of the strengths of AFLP is the large number of markers produced per assay. The complexity (number of fragments) of AFLP fingerprints can be manipulated by increasing or decreasing the number of elective bases and changing base composition. Through AFLP analysis only selected restriction fragments from the digestion of total plant DNA are amplified by the PCR. AFLP analysis is a rapid and efficient method for producing DNA fingerprints and genetic mapping.

2.2. RECOMBINANT DNA TECHNOLOGY

The techniques of inserting genes across species and genetic barriers is termed r-DNA technology because it involves making an incision in DNA of recipient, inserting spliced DNA pieces from a donor biological species and, finally, obtaining a single functional DNA macro molecule. The process of *in vitro* recombination and transfer into the foreign host involves mainly four steps:

(i) Gene identification and isolation
(ii) Splicing of foreign DNA into the vector
(iii) Insertion and maintenance of the recombinant vector in a host
(iv) Selection of desired clone.

Recombinant-DNA technique has been used to generate transgenic plants against insect, pest, fungal, bacterial and viral diseases to identify disease resistance genes and to generate improved strains of biocontrol agents.

Management of the important biotic stresses such as nematodes and fungal pathogens has been done mainly through deployment of resistant varieties bred by conventional means and by chemical pesticides. Toxic pesticides and fungicides though able to control the diseases are also adding pollutants to the environment day by day and are contributing to the degradation of biomass. Hence, there is an immense need for effective alternative methods not only to protect the crops, but also to ensure a safe and clean environment. Genetic engineering holds potential in this respect as it can allow genes that carry instructions for a particular feature to be isolated and moved from one organism to another, resulting in genetically modified organisms or transgenics. Thus, efforts have been made to make it possible to identify and clone genes associated with pathogenesis-related proteins, antifungal proteins and compounds, phytoalexins, enzymes controlling oxidative burst, etc. This gives a chance to develop resistance against diseases through genes encoding chitinases, glucanases, PR-la, stilbene synthase, glucose oxidase, lysozymes, ribosome inactivating proteins (RIPs) constitutively over-expressed in some model systems as well as crop plants under the control of different promoters. Combinatorial expression of chitinases with other compounds render enhanced protection against various fungal pathogens. There are still a host of novel genes that are being cloned and will be available for evaluation in the near future. Expression of these genes singly or in association holds great promise for engineering fungal and nematodal resistance in crop plants.

Development of molecular markers (tag) for resistance gene(s) could lead to quicker assessment of susceptibility, allow screening of seedlings and reduce the need for maintaining virulent isolates of different fungal and nematodal diseases. Isozyme electrophoresis can provide relatively quick and easy access to useful and reliable molecular markers. However, it revealed insufficient polymorphism in grain legumes. A potentially much larger number of markers can be resolved by means of restriction fragment length polymorphism (RFLP). This technology has been used to establish high-density gene maps for an increasing number of crop species but low level of polymorphism has also been identified using this technique in grain legumes. Random amplified polymorphic DNA (RAPD) represents another method for identifying additional genetic variation within populations and there is virtually no limit to the numbers of primers. This technology is also much faster than RFLP and employed widely for mapping genomes and tagging resistance genes. Besides, a new class of molecular markers, the microsatellite markers, has been shown to be more useful and highly polymorphic in comparison with other molecular markers. The ISSR marker procedure is also very simple, quick, reliable and can process many samples compared to RFLP. The amplification conditions with ISSR markers are very stringent, which allows high reproducibility as compared to RAPD markers. It can also provide information for the development of sequence tagged microsatellite site (STMS) linked to fungal diseases such as Fusarium wilt resistance. Finding STMS at a desired locus is very expensive and time consuming. The microsatellite primed PCR can speed up the identification of such markers. Development of molecular markers closely linked to gene resistance to various fungal diseases and nematodes and marker assisted selection (MAS) in combination will make it possible to screen for disease resistance without the need to inoculate the segregating plant material. Therefore, tagging genes and pyramiding them into a single genetic background would be helpful and can play a significant role in developing resistance in crop plants.

3. Molecular strategies for fungal and nematode management in legumes

In their natural habitat plants employ a variety of defense mechanisms to protect themselves against pathogenic fungi and nematodes. Activation of resistance genes upon infection allow plants to become more resistant to attack. Thus, the major biochemical changes are the biosynthesis and accumulation of phytoalexins and secondary metabolites which are toxic to parasites (Darvill and Albersheim, 1984). Plants also accumulate a novel class of protein termed pathogenesis related proteins (PR proteins) in response to pathogen attack (Ven Loon, 1985). Knowledge of a molecular basis of plant-nematode interaction is the key for success of non-chemical nematode control strategies. The major bottleneck in understanding the interaction has been the obligate parasitic nature of nematodes. With the advent of recent molecular techniques, like PCR, cDNA cloning, ELISA, etc., it has been possible to elucidate the molecular basis of plant-nematode interaction in detail. There is no single mechanism to operate either in induction or the progression of hypersensitive cell death. The hypersensitive cell death elicited by infection includes a decrease in passive membrane potential, an aggregation of cytoplasm around the invading pathogen, loss of electrolytes from cell and loss of plasmolyzing ability of cells. Identification and sequencing of genes involved in hyper-sensitive cell death and host-parasite interaction will provide valuable information into molecular basis of phyto-nematode infection.

These induced defenses by which plants protect themselves from nematode attack also involve a number of means. The resistance appears to depend on the ability of the host to recognize pathogen rapidly, in order to limit the spread of pathogen. Many plant genes are up-regulated during a resistance reaction including genes encoding proteins such as glucanase, chitinase and others commonly referred to as PR-proteins. Modern molecular strategies of plant disease control aim to explore all these avenues with the ultimate objective to design crops resistant to a wide array of pathogenic fungi and nematodes.

The important molecular approaches to manipulate resistance to fungal pathogens and nematodes are

(i) Genetic engineering of plants with disease resistance genes.
(ii) Marker aided selection for pyramiding resistance.

The identification of resistance genes and exploitation of chitinase and glucanase are the most widely emphasized areas at present. Besides this, use of molecular marker in selection of resistance lines is also getting attention. This chapter summarizes the molecular strategies to control phyto-parasitic fungi and nematodes in legumes.

3.1. TRANSGENIC APPROACH

Resistance genes involved in race-cultivar-specific interactions are well known from conventional breeding programmes. To isolate such resistance genes, various approaches are being persued, including map based cloning and transposon tagging (Bennetzen and Jona, 1992). After resistance genes have been identified and isolated, they can be transferred to pulse crops of interest to provide resistance to specific races of pathogen. However, the resulting transgenic plants will remain susceptible to other races of pathogen. This in no way undermines the importance of over-expression of anti-fungal and anti-nematode compounds as a feasible approach for enhancement of disease resistance in legumes.

The best example of this approach is development of transgenic tobacco plants that show enhanced resistance against the fungus *Rhizoctonia solani*, which is brought about by constitutive expression of genes encoding protein shown to have *in vitro* anti-fungal activity (Broglie *et al.*, 1991; Logemann *et al.*, 1992). The anti-fungal proteins most frequently described are probably chitinases and β 1–3 glucanases. Chitinases catalyze the hydrolysis of chitin and glucanases hydrolyze β 1–3 glucan. Through breakdown of these two important cell components, these enzymes inhibit fungal growth. Among the four classes of hydrolases, the class I hydrolases are localized in plant vacuoles and show anti-fungal properties. Class I glucanases in combination with chitinases show a very strong growth inhibition on many parasitic fungi.

Plant RIPs (Ribosomal Inactivating Protein) inhibit protein synthesis in target cells by a specific modification of 28S rRNA. The beauty of these plant RIPs are that they don't inhibit protein synthesis machinery of the plants, specifically inhibiting fungal ribosomes. A strong synergy is observed when RIPs are combined with chitinases or glucanases. Thus, RIPs provide another class of inhibiting protein with high degree of specificity.

For colonization in plant tissues, pathogens produce endo polygalacturonases which break down plant cell homogalacturonans. Inhibition of their hydrolysis by the plant's own inhibitor proteins provides resistance to plants against fungal pathogens. The gene encoding the polygalacturonase inhibitor has been cloned and characterized and is being used in fungal resistance programmes (Toubart *et al.*, 1992).

In addition to the above mentioned anti-fungal protein, there is a whole array of basic, small molecules, including proteins exhibiting properties of fungal growth inhibition. These include

(i) Salt stress induced protein, osmotin
(ii) Seed lectin from legume species
(iii) Protease inhibitors from legume grains
(iv) Small anti-microbial components, thionins.

Most of these proteins were originally described as anti-microbial compounds, but they are potent inhibitors of phytoparasitic fungi as well.

The genetic engineering approach has also been able to generate a durable and efficient resistance that controls a broad range of plant parasitic nematodes. Artificial defense systems can be constructed by introducing effector genes into the host plant that have a nematicidal effect. Such transgenes can encode nematicidal

(i) Toxic compounds that are readily ingested
(ii) Compounds that bind to signal molecules
(iii) Enzymes that interact with the nematode
(iv) Substances that cause breakdown of specific feeding structure.

The effective use of anti-nematode genes to generate artificial resistance relies on the specificities for the nematode, phytotoxicity to the host. Nematodes use cysteine proteinases for metabolism and digestion of dietary proteins. Thus, the employment of proteinase inhibitors from plants will inhibit the proteinase, disturbing metabolic balance. To improve the efficacy of inhibitors, site directed mutagenesis may be helpful. Oryzacystatin I gene from rice has been mutagenized and subsequently introduced and expressed under the control of CaMV-35S promoter into tobacco and *Arabidopsis thaliana*. The expression of cystatin in the transgenic plants prevented female nematodes from developing properly, causing reduced size and fecundity (Urwin *et al.*, 1998).

Lectins are another class of putative anti-nematode proteins with high degree of specificity. These sugar binding proteins when bound to the carbohydrate moieties on intestinal gut cells of nematodes cause perforations disturbing nutrient uptake and salt balance. The gene encoding the snowdrop lectin, GNA (*Galanthus nivalis* agglutinin) has been introduced into potato plants and these transgenics express GNA at a very high level. At this level, GNA caused up to 80% reduction in *G. pallida* population (Burrows *et al.*, 1997, 1998).

The degradation of structural components of the nematode catalyzed by enzymes like chitinase or collagenases is an additional objective for emerging nematode resistance. Nematode egg shell contains chitin and thus forms a possible target for chitinases (Mercer *et al.*, 1992). However, to date no attempt has been made to develop transgenic plants over expressing chitinases for this purpose. Human collagenase DNAs have been expressed in transgenic tobacco plant for breaking down the collagen containing cuticle of the nematode. For successful application of this approach, two main obstacles have to be overcome. First, there is evidence that the structure of the collagen from nematodes is different from that of vertebrates, questioning the efficacy of this enzyme, and second, the collagens are not present on the outermost surface of the nematode, and are therefore not exposed to the environment.

The endotoxins from *Bacillus thuringiensis* (Bt) is a potent toxic substance to nematodes. A nematotoxic version of this protein (CryIAb) has been expressed in tomato plants. The transgenic plants showed up to 50% reduction in *Meloidogyne* infestation.

Monoclonal anti-bodies (Mab) bind very specifically to a particular protein. Thus if Mab could be raised against nematode protein, they can be successfully used for inhibition of nematode growth. This concept has been tested by expressing monoclonal anti-bodies specific to stylet of *M. incognita* in tobacco plants (Baum *et al.*, 1996).

Another strategy for engineering an artificial resistance system is to promote the breakdown of specific feeding structures. The most realistic approach is the introduction of genes encoding toxic compounds in combination with proper promoter and regulatory elements to direct their expression exclusively in the cells of the feeding structures. A two component system consisting of a plant-toxic gene under the control of a nematode responsive promoter and a second detoxifying gene under the control of a constitutive promoter have been developed. Thus "leaky" expression of cytotoxic gene can be neutralized by antagonistic gene (Oul *et al.*, 1997).

After isolation of first resistance genes against fungi and nematodes, transformation of crop plants for artificial resistance progressed at a very fast pace. With numbers of plants coming out of the controlled environments of the laboratory, it is being felt that perhaps single genes would not be able to give a durable protection. For wide spectrum resistance, pyramiding several resistance genes into crop species must be accomplished.

3.2. MOLECULAR MARKER APPROACH

Selection and fixation of desirable genotypes occupy a central role in any plant breeding program aimed to improve performance of crops. Selecting resistance genes particularly is troublesome because of their polygenic nature. In recent years numbers of techniques involving molecular markers have made it possible to map both polygenic and monogenic types of resistance. Molecular marker based on restriction fragment length polymorphism (RFLP), amplified fragment length polymorphism (AFLP), random amplified polymorphic DNA (RAPD), simple sequence repeats (SSR) and micro satellites have enabled fine mapping of resistance genes into chromosomes (Table 1). Biochemical

Table 1. Linkage of DNA marker with disease/pest resistance in legume plants

Crop	Disease/Pest resistance	Nature of marker	Reference
Chickpea	Fusarium wilt	RAPD	Mayer *et al.*, 1998; Jung *et al.*, 1993
Common bean	Common blight	RFLP	Nodari *et al.*, 1992
Lentil	Fusarium vascular wilt	RAPD	Chare *et al.*, 1991
Mungbean	Powdery mildew	RAPD	Young and Danesh, 1991

markers based on isozymes have been used for identification of powdery mildew resistance locus in pea. In chickpea Fusarium wilt resistance gene has been mapped and cloned using RAPD markers (Mayer *et al.*, 1997). Rust resistance gene has been isolated in common bean. RFLP markers have been useful in identifying resistance gene against *Phytophthora* in Soybean.

Identification and cloning of natural nematode resistance genes has been a longstanding goal in legume improvement programmes. Resistance genes for the soybean cyst nematode *H. glycines* have been mapped with molecular markers (Webb *et al.*, 1995).

A major partial resistance locus (rhg 1) is located on chromosome 6 of soybean. This locus accounts for up to 50% of the genetic variation. Resistance to the southern root knot nematode is inherited in a quantitative manner. Two major quantitative trait loci (QTL) for resistance to this nematode have been mapped with RFLP markers (Taumolonis *et al.*, 1997). In peanut, the root knot nematode *M. arecarta* causes major problems. Two dominant genes, *Mae* and *Mag*, which reduce nematode egg number or galling have been linked to RAPD, RFLP and SCAR loci (Garcia, 1996). A marker based selection programme has been started for introgression of these genes into cultivars.

4. Conclusions and future prospects

Increasing knowledge of the molecular basis of pathogenesis and resistance has provided opportunity for the scientists to develop new strategies to pathogenic fungi and nematodes. Genetic engineering of resistance for pests and pathogens has resulted in development of several transgenics engineered with natural and artificial resistance genes. These plants will significantly broaden the genetic variability available to plant breeders for genetic enhancement of crop plants. However, such informations are very meager in the case of grain legumes. The presence of multiple genes, each conferring different resistance mechanism in a single plant, will offer the great opportunity to breed varieties that possess durable resistance. Besides, marker technology will also be of great help to select precisely against these adversities in a more refined and accurate manner.

In the past, major advances have been made in developing disease resistant cultivars by application of Mendelian genetics and conventional breeding methods. Recent advances in biotechnology particularly in protoplast culture, gene cloning and transformation enabled scientists to introduce novel genes for resistance into crop plants. Nucleic acid probes and DNA markers will be increasingly used for evaluation of germplasm and for increasing the selection efficiency in breeding programmes. The resistant cultivars thus developed will be the backbone of pest management programmes in the future.

References

Ballvora A, Hesselbach J, Niewohner J, Leister D, Salamini F and Gebhardt C (1995) Marker enrichment and high resolution map of the segment of potato chromosome VII harbouring the nematode resistance gene *GroL*. *Mol. Gen. Genet.*, **249**: 82–90.

Baum T J, Hiatt A, Parrot L H and Hussey R (1996) Expression in tobacco of a functional monocloned antibody specific to stylet secretors of the root knot nematode. *Mol. Plant Microbe Interact.*, **9**: 382–387.

Broglie K, Chet I, Holliday M, Cressman R, Biddle Ph, Knowton S, Mauvais C J and Broglie R (1991) Transgenic plants with enhanced resistance to fungal pathogen *Rhizoctonia solani. Science*, **254**: 1194–1197.

Burrows P R, Barker A D P, Newel C A and Hamilton W D O (1998) Plant desired enzyme inhibitors and lectins for resistance against plant parasitic nematodes in transgenic crops. *Pestic. Sci.*, **52**: 176–183.

Burrows P R, Waele D de and De Waele D (1997) Engineering resistance against plant parasitic nematodes using anti-nematode genes. In: *Cellular and Molecular Aspects of Plant-Nematode Interaction* (Eds Fenoll C, Grundless and Ohl S), Kluwer Academic Publishers, Dordrecht, The Netherlands, pp. 217–236.

Cai D, Kleine M, Kiffe S, Harloff H J, Sandal N N, Marcker K A, Klein-Lankhorst R M, Salentijm E M J, Lange W, Stiekema W J, Wyss U, Grundler F M W and Jung C (1997) Positional cloning of a gene for nematode resistance in sugar beet. *Science*, **275**: 832–834.

Chare C D, Ortega V M and Vallejos C E (1991) DNA restriction fragment length polymorphism correlate with isozyme diversity in *Phaseolus vulgaris. Theort. Appl. Genet.*, **81**: 806–811.

Darvill A G and Albersheim P (1984). Phytoalexins and their exciters – a defense against microbial infection in plants. *Annu. Rev. Plant Physiol.*, **35**: 243–275.

Garcia G M, Stalker H T, Shroeder E and Kochert G (1996) Identification of RAPD, SCAR and RFLP markers tightly linked to nematode resistance genes introgressed for *Arachis cardenasii* into *Arachis hypogea. Genome*, **39**: 836–845.

Jung G, Slorch P W, Cogne D P, Nicnhuis J and Arnaud-Santoma E (1993) Mapping common blight resistance loci in *Phaseolus vulgaris* L. using random amplified polymorphic DNA – *Xanthomonas campestirs* PV *Phaseoli* disease resistance, quantitative trait loci mapping in kidney bean transgenic plant (Abstract). *Hort. Sci.*, **28**: 456.

Kretischmer J M, Chalmess K J, Mannings S, Karakousis A, Barr A R, Islam A K M R, Logue S J, Choe Y W, Barker S J, Lance R C M and Langridge P (1997) RFLP mapping of the Hal cereal cyst nematode resistance gene in barley. *Theort. Appl. Genet.*, **94**: 1060–1064.

Logemann J, Jach G, Tommerup H, Mundy J and Schell J (1992) Expression of a barley ribosome inactivating protein leads to increased fungal protection in transgenic *Arabidopsis* expressing a modified plant cystalin. *Plant. J.*, **12**: 453–461.

Mayer M S, Tulu A, Somon C J, Kumar J S, Kraft J M, Kaiser W J and Muehlaver F J (1997) Development of a DNA marker *Fursarium* wilt resistance in chickpea. *Crop Sci.*, **37**: 1625–1629.

Mercer C F, Greenwood D R and Grant J L (1992) Effect of plant and microbiol chitirases on the eggs and juveniles of *Meloidogyne hapla. Nemotalogica*, **38**: 277–286.

Nodari R O, Koinage E M K, Kelly J D and Gupta P (1992) Towards an integrated linkage map of common bean. Development of Genomic DNA probes and levels of restriction fragment length polymorphism. *Theort. Appl. Genet.*, **84**: 186–192.

Oul S A, Vander Lee F M and Sijmons P C (1997) Anti-feeding structure approaches to nematode resistance. In: *Cellular Aspects of Plant-Nematode Interactions* (Eds Fenoll C, Grundler F and Ohl S), Kluwer Academic Publishers, Dordrecht, The Netherlands, pp. 250–261.

Taumolonis J P, Luzzi B M, Hussey R S, Parrott W A and Boerma H R (1997) RFLP mapping of resistance to southern root knot nematode in soybean. *Crop Sci.*, **37**: 1903–1990.

Toubart P, Desiderio A, Salvi G, Cervone F, Daroda L, De Lorenzo G, Bergmann C, Darvill A G and Alberscheim P (1992) Cloning and characterization of the encoding endopolygalacturonase inhibiting protein (PGIP) of *Phaseolus vulgaris* L. *Plant J.*, **2**: 367–373.

Urwin P E, McPherson M J and Atkinson H J (1997) Enhanced transgenic resistance to plant nematodes by dual protease inhibitor constructs. *Planta*, **204**: 472–479.

Ven Loon L C (1985) Pathogenesis related proteins. *Plant Mol. Biol.*, **4**: 111–116.

Webb D M, Baltazar B M, Rao Arelli A P, Schupp J, Clayton K, Keim P and Beavis W D (1995) Genetic mapping of soybean cyst Nematode race-3 resistance loci in the soybean plant. *Theort. Appl. Genet.*, **91**: 574–581.

Young N D and Danesh D (1991) RFLP analysis of complex interaction between mungbean (*V. radiata*) and powdery mildew (*E. Polygoni* D.O.). *Phytopathology*, **81**: 1028.

PART IV
PRODUCT QUALITY IMPROVEMENT

IMPROVEMENT OF METHIONINE-DEFICIENT LEGUMES THROUGH GENETIC ENGINEERING

LUCILIA HELENA MARCELLINO AND EUGEN SILVANO GANDER
EMBRAPA – Recursos Genéticos e Biotecnologia Parque Estação Biológica PqEB. Final W5 Norte 70770-900 Brasília-DF- Brazil
e-mail: lucilia@cenargen.embrapa.br

Abstract

A review of research aiming at the improvement of methionine-deficiency in legumes is presented. We discuss three possible strategies to reach this objective and present data on the legume crops where recombinant DNA technology has been applied to introduce and express genes coding for methionine-rich proteins. Some considerations concerning the possibility of introducing allergenic proteins into host plants are discussed.

1. Introduction

Jonathan Swift put it in a nutshell: in "Gulliver's Travels", the king of Brobdingnag declares that "… whoever could make two ears of corn, or two blades of grass to grow upon a spot of ground where only one grew before; would deserve better of mankind, and do more essential service to his country, than the whole race of politicians put together" (J. Swift, 1726).

These lines were written in 1726, when the world's population was under 1 billion; they are more true now, when actually close to 7 billion people have to be sustained. It is quite obvious: the necessary sustainment is mostly, if not completely, based on agriculture.

Within the agricultural commodities used for feeding humans and animals, legumes, in particular the grain legumes, i.e. black beans, mung beans, chickpeas, common beans and soybeans, occupy an important place, and in particular in developing countries.

According to the FAO the average daily protein intake of an adult should be around 0.35 g/kg of body weight while a 2 year old needs an average of 2.4 g/kg of body weight. In developed countries these needs are, to a large part, satisfied through the intake of animal proteins, while in developing countries vegetable proteins are a major component of the diet. Animal proteins from meat, milk and eggs are, from a nutritional point of view, more adequate for human nutrition since their amino acid composition, in particular

P.K. Jaiwal and R.P. Singh (eds.), Improvement Strategies for Leguminosae Biotechnology, 281–290.

their content of essential amino acids, is more balanced. Indeed, the major drawback of vegetable proteins is the fact that they are poor in the essential sulfur amino acids methionine and cysteine as well as tryptophan. This is illustrated in Table 1 (Aykroyd and Doughty, 1964) where the amino acid content of some of the more important grain legumes, in comparison with a hypothetical "ideal" protein, is shown.

So far there is no indication that this low content in sulfur amino acids and tryptophan of grain legumes can be overcome by classical breeding methods. The advent of recombinant DNA technology in the beginning of the 80s, however, offered a promising alternative based on the isolation of high-methionine storage protein genes and their introduction and expression in suitable, methionine-deprived and economically interesting host plants.

*Table 1. Amino acid content of legumes (mg/gN)**

Legume species Dry seeds	Isoleucine	Leucine	Lysine	Phenylalanine	Tyrosine	Sulfur (total)	Methionine	Cystin	Threonine	Tryptophan	Valine
FAO-hypothetical ideal protein	270	306	270	180	180	270	144	126	180	90	270
Arachis hypogaea	260	380	220	320	220	150	60	90	170	70	310
Cajanus cajan	380	490	450	540	210	160	70	90	240	30	330
Canavalia ensiformis	280	570	370	390	240	180	110	80	310	–	330
Glycine max	340	480	400	310	200	200	80	110	250	90	330
Lens esculenta	330	440	380	280	170	100	50	50	220	50	340
Mucuna utilis	300	480	390	300	320	130	80	60	250	–	240
Parkia biglobosa	300	500	440	380	310	150	100	50	210	–	–
Phaseolus acutifolius	280	480	410	330	200	150	60	90	250	–	360
P. angularis	280	490	440	340	210	180	110	70	240	–	340
P. aureus	350	560	430	300	100	110	70	40	200	50	370
P. lunatus	360	520	420	370	160	190	100	90	300	60	390
P. mungo	270	490	460	410	210	140	90	60	230	–	370
P. vulgaris	360	540	460	350	240	120	60	60	270	60	380
Pisum sativum	350	520	460	320	250	160	80	80	240	70	350
Vicia faba	390	540	350	260	170	70	30	40	200	60	410
Vigna unguiculata	260	450	410	340	210	230	120	110	220	–	340

*Aykroyd and Doughty (1964).

During the last decade several excellent reviews on the subject of genetic engineering of the sulfur amino acid content of grain legumes have been published (e.g. Altenbach and Simpson, 1990; Tabe *et al.*, 1993; Müntz *et al.*, 1998). Therefore, here we first give an overview of the different approaches used to achieve this goal, the problems encountered, the state of the art of research and then discuss one of the major problems in food-orientated biotechnology, i.e. the possible introduction of allergenic components into food.

2. Biotechnological approaches for protein quality and quantity improvement

Recombinant DNA Technology offers many approaches that may result in a change of the protein profile and/or amino acid composition of model plants or socio-economically important crops. This review is limited to the discussion of three of them: the delivery and expression of suitable transgenes into hosts, the site-directed mutagenesis and subsequent re-introduction of the mutated endogenous or heterologous genes and the engineering of the pathways involved in the biosynthesis of essential amino acids.

2.1. INTRODUCTION OF HETEROLOGOUS HIGH-METHIONINE GENES

Historically the first method to be employed to alter amino acid composition was also the most obvious: introduction of heterologous genes coding for high methionine proteins. So far not many such genes have been isolated: according to Altenbach and Simpson (1990) and Müntz *et al.* (1998) those presently available code for two 10 and 15 kDa maize zeins (22% and 11% methionine, respectively), a 10 kDa rice prolamin (20% methionine), a 10 kDa sunflower albumin containing 15% methionine (Molvig *et al.*, 1997) and a 18% methionine 2S albumin from Brazil nut. Another gene, from wild maize, coding for a 18 kDa zein is quite promising since it contains 25% methionine (Swarup *et al.*, 1995). Furthermore, a gene coding for a nutritionally well balanced seed albumin from *Amaranthus hypochondriacus* has become available (Raina and Datta, 1992), and one can hope that in the near future still more genes will be at hand. Recently, for instance, the primary structure of a methionine-rich albumin from *Cannabis sativa* has been reported (Odani and Odani, 1998).

Supposedly the best (albeit, recently, most controversial) candidate for recombinant DNA technology applied to legumes was one gene of the 2S albumin multigene family from Brazil nut (*Bertholletia excelsa*) coding for an albumin containing 18% methionine and 8% cysteine. Since the discovery and description of this protein by Rotenberg and Iachan (1975) and Youle and Huang (1981), several groups have successfully isolated and characterized physically-chemically as well as biologically the 2S albumin proteins (Ampe *et al.*, 1986). cDNA, genomic sequences and *in vivo* biosynthesis of these proteins have also been investigated extensively (Altenbach *et al.*, 1987; De Castro *et al.*, 1987; Gander *et al.*, 1991). This, true to the dogma that in order to do biotechnology one has to know the product of the gene of interest and the physical-chemical and biological properties of the gene itself.

Most of these groups attempted to introduce this gene into different plant species in order to achieve improvement of their nutritional value. In 1989, Altenbach *et al.* reported for the first time the enhancement of the methionine content of seed proteins after introducing and expressing Brazil nut (BN) 2S albumin cDNA sequences in transgenic tobacco. In this study, the levels of methionine in the seed proteins increased 30%. This strongly indicated that it is possible to improve the nutritional value by means of recombinant DNA technology. Tobacco, however, is a model plant and soon or in parallel several other attempts using similar strategies with different nutritionally and economically important crops were reported by other groups. Table 2 summarizes these efforts.

Table 2. Plant species transformed with transgenes coding for methionine-rich protein

Crop	Transgene	Promoter	Method of transformation	Methionine increase	Authors
N. tabacum	15 kDa zein	ß-phaseolin	*A. tumefaciens*	ND	Hoffman *et al.*, 1987
N. tabacum	BN2S	Phaseolin	*A. tumefaciens*	30%	Altenbach *et al.*, 1989
N. tabacum, A. thaliana, B. napus	BN 2S/AT2S1	AT2S1	*A. tumefaciens*	0.1–0.2%	De Clercq *et al.*, 1990
B. napus	BN2S	Lectin	Electroporation	ND	Guerche *et al.*, 1990
B. napus	BN2S	Phaseolin	*A. tumefaciens*	33%	Altenbach *et al.*, 1992
S. tuberosum	BN2S	CaMV 35S	*A. tumefaciens*	ND	Tu *et al.*, 1994
N. tabacum, V. narbonensis, V. faba and *P. sativum*	Synthesized BN2S gene	CaMV 35S	*A. tumefaciens, A. rhizogenes*	ND	Saalbach *et al.*, 1994
N. tabacum, V. narbonensis	BN2S	CaMV 35S and LeB4	*A. tumefaciens*	3 X [wt]	Saalbach *et al.*, 1995a
V. narbonensis	BN2S	CaMV 35S and LeB4	*A. tumefaciens*	3 X [wt]	Saalbach *et al.*, 1995b
L. angustifolius	SSA	Vicilin	*A. tumefaciens*	94%	Molvig *et al.*, 1997
S. tuberosum	BN2S, modified	CaMV 35S	*A. tumefaciens*	ND	Tu *et al.*, 1998
P. vulgaris	BN2S	CaMV 35S	Particle bombardment	23%	Aragão *et al.*, 1999
S. guianensis	BN2S	*Ats1A*	Particle bombardment, electroporation	ND	Quecini, 1999

BN2S = Brazil nut 2S albumin; LeB4 = legumin B4 from *Vicia faba*; CaMV 35S = cauliflower mosaic vírus 35S; SSA = sunflower seed albumin; ND = not determined; [wt] = wild type concentration.

It is evident from this table that, with the exception of one group (Movig *et al.*, 1997), all others used the Brazil nut 2S albumin gene sequences as transgenes. It is also evident that the concept of introducing and expressing a transgene in order to overcome nutritional deficiencies is correct. However, there exists a puzzling variability with respect to the actual increase of the methionine content: the results vary from 0.1% in *A. thaliana* transformed with BN2S (De Clercq *et al.*, 1990) up to 94% in *L. angustifolius* transformed with SSA (Molvig *et al.*, 1997). It is difficult if not impossible to discuss these discrepancies since not only different promoters were used but, in addition, different host plant species were transformed.

2.2. INTRODUCTION OF *IN VITRO* ENGINEERED TRANSGENES

The *in vitro* modification of available genes, prior to the introduction and expression in host plants, is another possibility, useful for the correction of amino acid deficiencies. This was first shown with the tobacco model by Hoffman *et al.* (1988). These authors increased the methionine codons of ß-phaseolin from three to nine by inserting an oligonucleotide of 45 bases and transforming *N. tabacum* via *A. tumefaciens*. The modified sequence was expressed and processed in the host, however, no accumulation of the transgene's product in the protein bodies could be shown using immunocytochemical techniques. The authors speculate that the modified protein was degraded somewhere between the Golgi complex and the protein bodies.

It is worthwhile to mention here another attempt, although its goal was not the improvement of methionine deficiency but the demonstration that storage proteins can be used as protein factories for the synthesis of pharmaceutically important compounds. Vanderckhove *et al.* (1989) used a 2S albumin gene from *A. thaliana* as a carrier for the sequences coding for the neuropeptide Leu-enkephalin. The chimeric gene was introduced into *A. thaliana* and *Brassica napus*. Leu-enkephaline was recovered from seeds of the transgenic plants in very satisfactory amounts, i.e. 200 nmol/gram of *Arabidopsis* seeds and 50 nmol/gram of *Brassica* seeds. Besides showing the usefulness of storage proteins in plant molecular pharming, this article also demonstrates the high biological plasticity of 2S seed storage albumin genes.

To a certain degree this is also reflected in the work of De Clercq *et al.* (1990) who modified *Arabidopsis* 2S sequences. All these modifications were done in the *Arabidopsis* 2S large subunit hypervariable region, between cysteins # 6 and 7, and consisted of synthetic oligonucleotides. The modified 2S proteins contained 7, 10 and 14 methionine residues, respectively. The modified sequences were introduced into *A. thaliana, B. napus* and *N. tabacum* and the authors determined that between 100 to 200 nmol of modified 2S proteins are present per gram of *Arabidopsis* seed.

It is possible, at least in this hypervariable region, to introduce foreign sequences without affecting transcription or protein processing. Based on this notion and aware of the fact that although 2S from Brazil nut is methionine-rich, it is deficient in tryptophan, another essential aminoacid, we applied site-directed mutagenesis in order to compensate for this deficiency. In a first construct, 5 tryptophan codons were introduced into the hypervariable region of the BN2S cDNA sequence (Marcellino *et al.*, 1996). The modified transgene, under control of the 35S CaMV promoter, was introduced in *N. tabacum* via *A. tumefaciens*. Western analyses have shown that the modified protein was expressed and correctly processed in tobacco seeds.

In addition, two other modifications were introduced into the 2S albumin cDNA. Based on a theoretical model of the tertiary structure, two modifications were planned so that the tryptophan residue would occupy an internal (leucine 44 substitution) or an external position (arginine 80 substitution) in the 2S molecule (Fig. 1). These last two mutants could be used for further structural analyses of the tertiary conformation of the 2S protein by fluorescence quenching techniques. We could show that all the mutants were correctly transcribed, expressed in tobacco seeds and were accumulated in protein bodies, as expected for storage proteins in general. Semi-quantitative analyses indicated that between 0.15 and 0.3% of salt-extractable seed proteins are mutated 2S proteins. This result was somewhere between the values reported by Guerche *et al.* (1990) and Altenbach *et al.* (1992) for transgenic *B. napus*.

Another work that takes advantage of the Brazil nut gene's hypervariable region as an acceptor for additional codons, albeit not for the engineering of legumes but of potatoes, is described by Tu *et al.* (1998). The four mutations constructed by these authors and introduced into the variable region of the Brazil nut 2S large subunit increased the methionine content from an original 18% to 21, 23, 24 and 25% in the coding region. These mutants were put under the control of the CaMV 35S promoter and introduced into *S. tuberosum* via *Agrobacterium*-mediated transformation. All mutant genes were expressed and the proteins correctly processed. The authors emphasize, however, that no increase of the methionine concentration in the transformed plants was obtained and they state that in order to do so, a ten-fold increase of the expression level would be necessary.

The most obvious message here is doubtless the fact that the 2S albumins and in particular the Brazil nut 2S sequences, which, *a priori* are already rich in methionine, can be modified in their variable region, and can be expressed successfully in host plants. These modifications do not seem to influence transcription efficiency nor do they interfere with the correct processing and storage of the modified proteins.

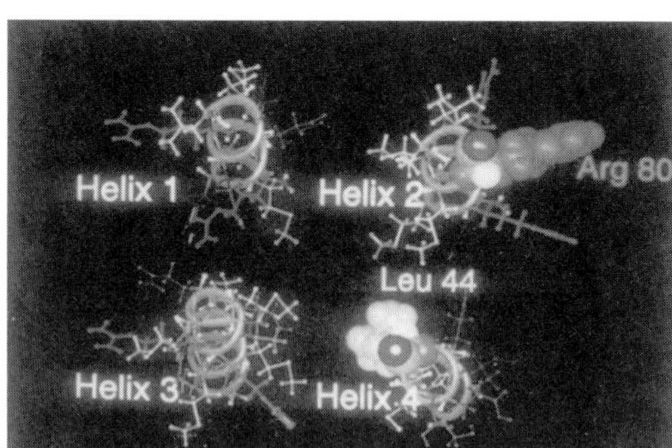

Figure 1. Theoretical structure of the 2S albumin. The photo shows four of the five helices presumably present in Brazil nut 2S albumin. Note that Leu^{44} faces the interior of the molecule while Arg^{80} faces outwards – both amino acids are represented as ball model.

2.3. ENGINEERING OF THE BIOSYNTHESIS PATHWAYS OF METHIONINE

The biosynthesis of methionine is dependent on the citrate-cycle and, consequently, on aspartate as the point of origin. This, in turn, means that oxaloacetate and therefore the citric acid cycle stand, invariably, at the root of methionine biosynthesis. If the assumption is made that in methionine-deficient crops the limiting factor is methionine as a free amino acid, attempts to interfere with methionine's biosynthesis pathway would be justified. The idea would then be to engineer one of the key enzymes that might regulate methionine accumulation through a feed-back mechanism. This type of approach is indeed valid since high-lysin soybeans have been obtained with this strategy (Falco *et al.*, 1995). In addition the biosynthesis of lysine in bacteria, algae, ferns as well as in higher plants also starts from aspartic acid. The perspective, however, to interfere with some of the 9 steps involved in the citrate cycle and/or the 7 steps that are necessary to synthesize methionine from aspartic acid seems to us, for the time being at least, an approach much too complex to be of any practical value in the near future. This, of course, says nothing about the scientific insights and merits that could result from investigations along these lines.

3. Conclusions and further considerations

The examples presented so far of recombinant DNA technologies applied towards the compensation of methionine deficiency of model plants or of species of agro-commercial and/or nutritional importance leave no doubt that genetic engineering of seed protein composition is possible and a reality. However, with the exception of the case of the *Lupinus* transformed with the sunflower SSA gene, no animal feeding tests have been performed in order to evaluate the actual nutritional effect of the transgenics. One reason might very well be that in most of the cases, the fraction of proteins derived from transgene transcription is too small to provoke a nutritionally significant change of the methionine concentration. This indicates that one of the more important topics of future research most probably has to focus on the understanding of promoter and enhancer elements and their interactions with regulatory proteins as well as the identification of strong seed promoters or enhancer-like elements that could be used to construct more effective seed promoters. Along these lines one might mention the work of Vincentz *et al.* (1997) who identified, in the promoter of the Brazil nut 2S albumin gene, boxes containing binding sites for the maize trans-acting factor opaque-2 and showed that this regulatory protein can transactivate promoter constructs consisting of the duplicated boxes and of a 35S minimal promoter in tobacco leaves.

Another important point emerges from the work of Hoffman *et al.* (1988) who showed that the products of modified phaseolin transgenes are not stable in tobacco seeds. The reasons for this instability are not clear, however, it could be a problem of misdirectioning to a wrong cellular compartment. Consequently another important aspect arises: the directioning of the transgene's product to an appropriate cell compartment. This requires the inclusion, in transgenes, of the adequate signal sequences. The work of Tabe *et al.* (1993) illustrates this fact: these authors included, at the 3′ end of the vicilin gene under the control of the 35S promoter, sequences representing a signal for the retention of proteins in the endoplasmic reticulum. They showed that tobacco leaves from plants transformed with this construct accumulated 100 times more vicilin than leaves from plants transformed with a similar construct that did not contain the signal sequence.

Last but not least, it is important to realize that within the programs that regulate the ordinate expression of genes and gene batteries during development and during an organism's lifetime, there must also exist modules responsible for correct structural dimensions. Consequently it is not necessarily obvious that a quantitative increase of proteins in , say, seeds, can simply be achieved through the expression of one or several transgenes. One possibility to cope with this notion would be to knock-out an undesired gene concomitantly with the introduction of a desirable transgene. Evidence for the correctness of this line of reasoning comes from the work of Kohno-Murase *et al.* (1994, 1995) who introduced and expressed either an anti-sense napin or anti-sense cruciferin construct in *Brassica*. In the first case, production of cruciferin was increased whereas cruciferin knockouts produced more napin than wild type plants.

3.1. FOOD FROM TRANSGENIC CROPS – THE ALLERGEN PROBLEM

As mentioned before, when improvement of methionine content is envisioned, the Brazil nut 2S gene was, and still is, one of the most obvious transgenes but, on the other hand, it has become also one of the most controversial.

In 1996 Nordlee *et al.* published a study, supported by Pioneer Hi-Bred International Inc., showing that sera from individuals with the documented allergy to Brazil nuts contain IgE that binds to the Brazil nut 2S albumin present in protein extracts from transgenic soybean. This result confirms the findings of Arshad *et al.* (1991) that Brazil nut proteins are allergenic and establishes that the allergenic principle present in the Brazil nut 2S albumin can be transferred to host soy plants via recombinant DNA technology. These results have been widely used as an argument against transgenic food and biotechnology. The fact is that numerous non-manipulated native proteins used for human consumption, e.g. proteins from peanuts (not only Brazil nuts!), crustaceans, cow milk, legumes, cereals, etc. can be allergenic for a certain segment of the population. Now this, of course, is common knowledge and nobody would dare, for this reason, to take a stance against, say, crustaceae. So why take a stance against the use of Brazil nut 2S albumin in manipulation of food crops? All it takes in order to get into the clear with this admittedly serious aspect are regulations stipulating:

1. a label indicating if the food or crop contains a known allergen and
2. the execution of routine-tests, long since a common practice in the cosmetic industry, to verify if proteins of unknown allergenic properties are to be introduced in the nutritional line.

Tests of this kind should or could include the demonstration that, after processing and/or cooking, the allergenic properties are still maintained, a test not done, to the best of our knowledge, in the case of the soybeans transformed with the BN2S gene. Furthermore, with the introduction of transgenes into food crops being a reality, it is essential to fund more programs that are aimed at the understanding of the molecular basis of allergenicity and at the identification of useful genes in organisms of less antigenic notoriety.

In summary, we are convinced that biotechnology applied to the qualitative improvement of food for human and animal consumption is a valid alternative to the "classical" plant breeding methods. The successful introduction and expression of genes coding for high-methionine storage proteins in crops of nutritional importance illustrate this notion.

However, we are also aware that, parallel to the rapid increase of available genes with desirable traits and the fact that, by now, virtually any crop can be transformed, it is essential to establish risk-assessment procedures and biological tests as foolproof as possible and to inform the consumer clearly on the components of a given food.

Acknowledgements

We would like to thank Drs. I. Miranda dos Santos, F. J. L. Aragão and L. A. B. de Castro for critical reading of the manuscript.

References

Altenbach S B and Simpson R B (1990) Manipulation of methionine-rich protein genes in seeds. *TIBTECH*, **8**: 156–160.

Altenbach S B, Pearson K W, Leung F W and Sun S S M (1987) Cloning and sequence analysis of a cDNA encoding a Brazil nut protein exceptionally rich in methionine. *Plant Mol. Biol.*, **8**: 239–250.

Altenbach S B, Pearson K W, Meeker G, Staraci L C and Sun S M (1989) Enhancement of the methionine content of seed proteins by the expression of a chimeric gene encoding a methionine-rich protein in transgenic plants. *Plant Mol. Biol.*, **13**: 513–522.

Altenbach S S, Kuo Ch-Ch, Staraci L C, Pearson K W, Wainwright C W, Georgescu A and Townsend G (1992) Accumulation of a Brazil nut albumin in seeds of transgenic canola results in enhanced levels of seed protein methionine. *Plant Mol. Biol.*, **18**: 235–245.

Ampe C, Van Damme J, De Castro L A B, Sampaio M J, Van Montagu M and Vandekerckhove J (1986) The amino-acid sequence of the 2S sulphur-rich proteins from seeds of Brazil nut (*Bertholletia excelsa* H.B.K.). *Eur. J. Biochem.*, **159**: 597–604.

Aragão F J L, Barros L M G, De Souza M, Grossi De Sá M F, De Almeida E R P, Gander E S and Rech E L (1999) Expression of a methionine-rich storage albumin gene from Brazil nut (*Bertholletia excelsa* H.B.K.) in transgenic bean plants (*Phaseolus vulgaris* L.). *Genet. Mol. Biol.*, **22**: 445–449.

Aragão F J L, Grossi de Sá M F, De Almeida E R P, Gander E S and Rech E L (1992) Particle bombardment-mediated transient expression of a Brazil nut methionine-rich albumin in bean (*Phaseolus vulgaris* L.). *Plant Mol. Biol.*, **20**: 357–359.

Aykroyd W R Y and Doughty J (1964) Las leguminosas en la nutrición humana. Organización de las Naciónes Unidas para la Agricultura y la Alimentacíon. Roma.

De Castro L A B, Lacerda Z, Aramayo R A, Sampaio M J A M and Gander E S (1987) Evidence for a precursor molecule of Brazil nut 2S seed proteins from biosynthesis and cDNA synthesis. *Mol. Gen. Genet.*, **206**: 338–343.

De Clercq A, Vanderwiele M, Van Damme J, Guerche P, Van Montagu M, Vandekerckhove J and Krebbers E (1990) Stable accumulation of modified 2S albumin seed storage proteins with higher methionine contents in transgenic plants. *Plant Physiol.*, **94**: 970–979.

Falco S C, Guida T, Locke M, Mauvais J, Sanders C, Ward R T and Webber P (1995) Transgenic canola and soybean seeds with increased lysine. *Bio/Technology*, **13**: 577–582.

Gander E S, Holmstroem K O, De Paiva G R, De Castro L A B, Carneiro M and Grossi de Sá M F (1991) Isolation, characterization and expression of a gene coding for a 2S albumin from *Bertholletia excelsa* (Brazil nut). *Plant Mol. Biol.*, **16**: 437–448.

Guerche P, De Almeida E R P, Schwarzstein M A, Gander E, Krebbers E and Pelletier G (1990) Expression of the 2S albumin from *Bertholletia excelsa* in *Brasica napus*. *Mol. Gen. Genetics.*, **210**: 510–520.

Hoffman L A, Donaldson D D and Herman E M (1988). A modified storage protein is synthesized, processed and degraded in seeds of transgenic plants. *Plant Mol. Biol.*, **11**: 717–729.

Hoffman L M, Donaldson D D, Bookland R, Rashka K and Herman E M (1987) Synthesis and protein body deposition of maize 15 kd zein in transgenic tobacco seeds. *EMBO J.*, **6**: 3213–3221.

Kohno-Murase J, Murase M, Ichikawa H and Imamura J (1994) Effects of an antisense napin gene on seed storage compounds in transgenic *Brassica napus* seeds. *Plant Mol. Biol.*, **26**: 1115–1124.

Kohno-Murase J, Murase M, Ichikawa H and Imamura J (1995) Improvement in the quality of seed storage protein by transformation of *Brassica napus* with an antisense gene for cruciferin. *Theort. Appl. Genet.*, **9**: 672–631.

Molvig L, Tabe L M, Eggum B, Moore A E, Craig S, Spencer D and Higgins T J V (1997) Enhanced methionine levels and increased nutritive value of seeds of transgenic lupins (*Lupinus angustifolius* L.) expressing a sunflower seed albumin gene. *Proc. Natl. Acad. Sci. USA*, **94**: 8393–8398.

Müntz K, Christov V, Saalbach G, Saalbach I, Waddell D, Pickardt T, Schieder O and Wuestenhagen T (1998) Genetic engineering for high methionine grain legumes. *Nahrung*, **42**: 125–127.

Nordlee J A, Taylor S L, Townsend J A, Thomas L A and Bush R K (1996) Identification of a Brazil-nut allergen in transgenic soybeans. *N. Engl. J. Med.*, **334**: 688–692.

Odani S and Odani S (1998) Isolation and primary structure of a methionine- and cysteine-rich seed protein of *Cannabis sativa. Biosci. Biotech. Biochem.*, **62**: 650–654.

Quecini V M (1999) Transferência direta de genes para plantas de *Stylosanthes guianensis* (Aubl.) Sw. Ph.D. thesis, University of São Paulo, School of Agriculture "Luiz de Queiroz", Piracicaba-SP.

Raina A and Datta A (1992) Molecular cloning of a gene encoding a seed-specific protein with nutritionally balanced amino acid composition from *Amaranthus. Proc. Natl. Acad. Sci. USA*, **89**: 11774–11778.

Rotenberg B E and Iachan A (1975) Estudo da proteína da Castanha do Pará. *Informativo do Instituto Naciónal de Tecnologia*, **8**: 22–24.

Saalbach I, Pickardt T, Machemehl F, Saalbach G, Shieder O and Müntz K (1994) A chimeric gene encoding the methionine-rich 2S albumin of the Brazil nut (*Bertholletia excelsa* H.B.K.) is stably expressed and inherited in transgenic grain legumes. *Mol. Gen. Genet.*, **242**: 226–236.

Saalbach I, Pickardt T, Waddell D R, Hillmer S, Schieder O and Müntz K (1995) The sulphur-rich Brazil nut 2S albumin is specifically formed in transgenic seeds of the grain legume *Vicia narbonensis. Euphytica*, **85**: 181–192.

Saalbach I, Waddell D R, Pickardt T, Schieder O and Muntz K (1995) Stable Expression of sulphur-rich 2S albumin gene in transgenic *Vicia narbonensis* increases the methionine content of seeds. *J. Plant Physiol.*, **145**: 674–681.

Swarup S, Timmermans M C, Chauduri S and Messing J (1995) Determinants of the high-methionine trait in wild and exotic germplasm may have escaped selection during early cultivation of maize. *Plant J.*, **8**: 359–368.

Swift Jonathan (1993) Gulliver's Travels. Wordsworth Classics, Wordsworth Editions Limited, St. Albans, Hertfordshire, UK.

Tabe L M, Higgins C M, McNabb W C and Higgins T J V (1993) Genetic engineering of grain and pasture legumes for improved nutritive value. *Genetica*, **90**: 181–200.

Tu H M, Godfrey L W and Sun S S M (1998) Expression of the Brazil nut methionine-rich protein and mutants with increased methionine in transgenic potato. *Plant Mol. Biol.*, **37**: 829–838.

Tu H M, Godfrey L W and Sun S M (1994) Expression of the Brazil nut methionine-rich protein in transgenic potato plants. In: *The Molecular and Cellular Biology of the Potato* (Eds Belknap W R, Vayda M E and Park W D), CAB International, Wallingford, pp. 209–220.

Vandekerckhove J, Van Damme J, Van Lijsebettens M, Botterman J, De Block M, Vandewiele M, De Clercq A, Leemans J, Van Montagu M and Krebbers E (1989) Enkephalins produced in transgenic plants using modified 2S storage proteins. *Bio/Technology*, **7**: 929–932.

Vincentz M, Leite A, Neshich G, Vriend G, Mattar C, Barros L, Weinberg D, De Almeida E R, Paes de Carvalho M, Aragão F and Gander E S (1997) ACGT and vicilin core sequences in a promoter domain required for seed-specific expression of a 2S storage protein gene are recognized by the opaque-2 regulatory protein. *Plant Mol. Biol.*, **34**: 879–889.

Youle R and Huang A H C (1981) Occurrence of low molecular weight and high cysteine containing albumin storage protein in oil seeds of diverse species. *Am. J. Bot.*, **68**: 44–48.

DEVELOPMENT OF REGENERATION PROTOCOLS TO EXPLOIT SOMACLONAL VARIATIONS IN *LATHYRUS SATIVUS* FOR DEVELOPING TOXIN FREE CULTIVAR

S.L. MEHTA AND I.M. SANTHA
Division of Biochemistry, Indian Agricultural Research Institute, New Delhi – 110 012, India
e-mail: shantiji@nda.vsnl.net.in

Abstract

Lathyrus sativus is an important grain legume characterized to withstand drought and waterlogging conditions. Due to occurrence of a neurotoxin, β-N-oxalyl-diamino propionic acid, its prolonged consumption can cause lower limb paralysis. This chapter reviews *in vitro* regeneration protocols and biotechnological approaches for removal of antinutrient compounds from *Lathyrus* and improvement in its productivity and yield. Somaclones with high yield and low toxin content (less than 0.1%) have been developed by *in vitro* techniques.

1. Introduction

Lathyrus sativus is an important grain legume characterized by its ability to withstand extreme conditions of drought and waterlogging. However, its cultivation is banned in India since its prolonged consumption is known to cause lower limb paralysis. The causative agent for this has been identified as β-N-oxalyl-diamino propionic acid (ODAP), or it is also called β-N-oxalyl amino alanine (BOAA) (Murti *et al.*, 1964; Rao *et al.*, 1964). Despite efforts by plant breeders using conventional technologies it has not been possible to reduce substantially or to eliminate ODAP. Scientists have developed P-24 cultivar with low ODAP content of ~0.3% as against 0.6–1.2% in commercially grown cultivars at our institution. However, the major problem was the lack of stability of low ODAP strains such as P-24 at different locations. ODAP is a small molecule.

P.K. Jaiwal and R.P. Singh (eds.), Improvement Strategies for Leguminosae Biotechnology, 291–299.

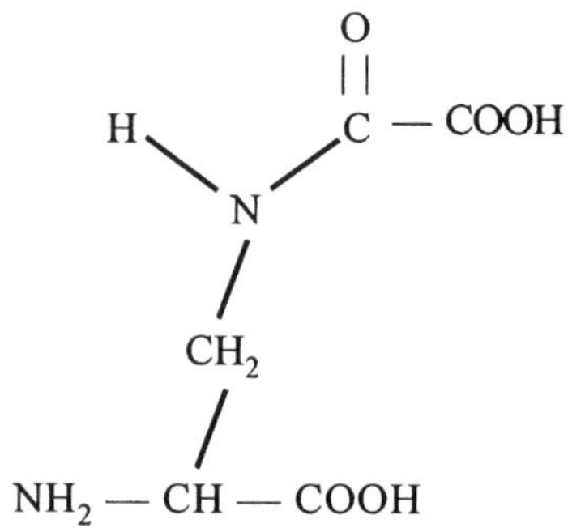

Molecular structure of ODAP

It is synthesized by terminal two enzymes

$$\text{Oxalate} + \text{CoA} + \text{ATP} \xrightleftharpoons[\text{Mg}^{2+}]{\text{Oxalyl CoA Synthetase}} \text{Oxalyl CoA} + \text{AMP} + \text{PPi}$$

$$\text{Oxalyl CoA} + \text{DAP} \xrightarrow[\text{ODAP synthase}]{} \text{ODAP} + \text{CoA}$$

Therefore, the following approaches using biotechnological tools could be utilized for reduction/removal of toxin from *Lathyrus*

1. Somaclonal variation
2. Antisense cloning for any of the enzyme involved in biosynthesis of ODAP
3. Isolation, characterization and cloning of an enzyme degrading ODAP.

However, a prerequisite for applying recombinant DNA techniques is the availability of good regeneration protocols. In our laboratory it has been possible to develop regeneration protocols for *L. sativus* from leaf, root and internode explants. This made it possible to exploit somaclonal variation to develop toxin free cultivars. In this report we have summarized the work relating to the development of somaclones with extremely low ODAP and having high yield potential.

2. Regeneration of *L. sativus*

Grain legumes are generally recalcitrant to shoot and root regeneration. For many legumes even now the protocols are not available for successful regeneration. Many researchers attempted to regenerate *L. sativus*. Mukhopadhyay and Bhojwani (1978), and Gharyal and Maheshwari (1983) got shoot formation in calli from shoot apices and leaf explants but rooting was not obtained. Sinha *et al.* (1983) obtained callus development and shoot regeneration from stem explants but they also could not provide evidence for rooting and development of plants bearing viable seeds. Roy *et al.* (1991, 1992, 1993) from our laboratory reported the successful regeneration of *L. sativus* cultivar P-24 using leaf, internode and root explants. Regeneration was obtained by controlling the phytohormone concentrations as well as the basal media. From leaf discs, callus induction (96%) and shoot regeneration (46%) was obtained on B_5 (Gamborg *et al.*, 1968) medium supplemented with 2.0 mg/L naphthaleneacetic acid (NAA) and 0.5 mg/L benzylaminopurine (BAP). The shoots were rooted on half-strength MS (Murashige and Skoog, 1962) medium containing 0.01 mg/L indole butyric acid. All other concentrations of growth regulators and strength of media failed to show any regeneration. In this way by carefully controlling the media and growth regulators about 300 *in vitro* regenerated plants were taken to the field, of which 102 plants survived and produced viable seeds. The chemical analysis of single plant progeny showed tremendous variation not only in morphological traits but also in the content of ODAP. ODAP content varied as low as 0.03% to as high as 0.89% (Figs 1–3). Analysis of single seeds in some lines showed near absence of ODAP. The seeds from these lines were further advanced. In R_2 plant progeny the ODAP content varied from 0.02 to 0.8%. The frequency distribution of seed and leaf ODAP in R_2 generation of *in vitro* regenerated plants from leaf explants of *L. sativus* as presented (Table 1) show that, in leaf, ODAP varied from 0.016 to 0.257%. Toxin content in about 45.5% of the plants was lower than that of parent cultivar P-24. Seven plants' progeny had ODAP content lower than 0.05%. Plants which had lower ODAP in leaves also had low ODAP content in seeds. The data presented in Table 2 show the characteristics of low toxin R_2 progeny derived from leaf explant. One of the progeny 15–8 had the least content of ODAP and single seed analysis showed near absence of ODAP in some seeds (Fig. 4). The seed yield of some of the individual 15–8 plant progeny was substantially higher (1–5-fold) than the parent P-24.

Development of *in vitro* regeneration protocols to raise complete plants from callus cultures from different explants *viz* leaf, internode and root opened up the possibility of utilization of somaclonal variation for reducing ODAP content in *L. sativus* seeds. Somaclonal variants have been reported in many crops and this has helped in improvement

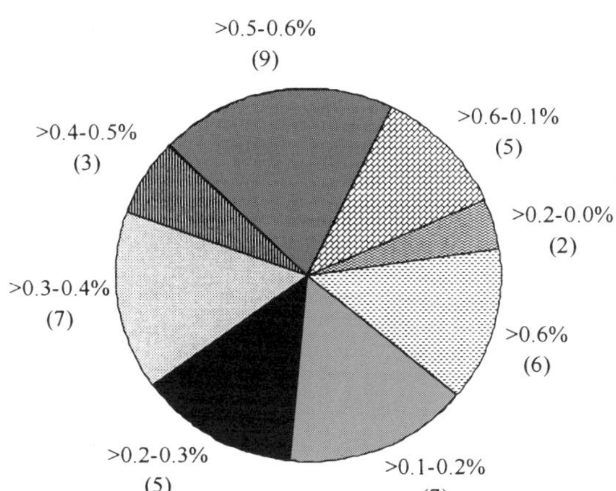

Figure 1. Frequency distribution of ODAP in R_1 plants regenerated from leaf (values in % represent ODAP content and in parentheses no. of plants).

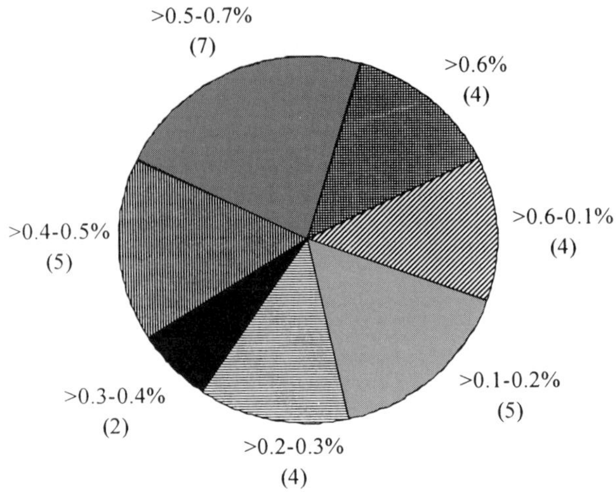

Figure 2. Frequency distribution of ODAP in R_1 plants regenerated from roots (values in % represent ODAP content and in parentheses no. of plants).

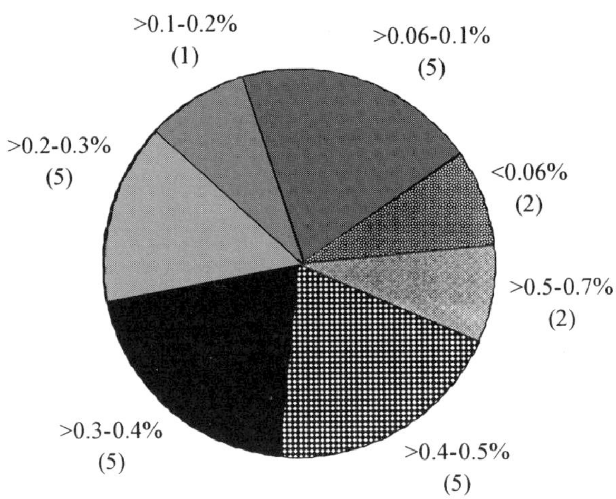

Figure 3. Frequency distribution of ODAP in R_1 plants regenerated from internode (values in % represent ODAP content and in parentheses no. of plants).

of crops. *Lathyrus* somaclones developed in our lab showed variation not only in ODAP content but in several phenotypic characters like leaf size (Fig. 5a), flower color (Fig. 5b), pod marking, seed color (Fig. 5c) and yield (Mehta *et al.*, 1994). Somaclone Bio RO2, Bio R-31, Bio R-33, Bio L-12, Bio L-54, Bio L-56, Bio I-18, Bio I-22, Bio I-30 (L, I, R designate the explant leaf, internode and root respectively followed by numeral which denotes the serial numbers of R_0 plants) had significantly lower ODAP content in seeds (0.03–0.07%) as compared to the parent cultivar which has 0.3–0.4% ODAP. Leaf length

Table 1. Frequency distribution of leaf and seed toxin in R_2 progeny of in vitro *regenerated plants from leaf explant of* Lathyrus sativus

	Seed toxin		Leaf toxin	
Toxin limits (%)	Frequency	Percent	Frequency	Percent
0.000–0.050	7	5.8	11	9.1
0.051–0.100	9	7.4	17	14.0
0.101–0.150	4	3.3	12	9.9
0.151–0.200	6	5.0	24	19.8
0.201–0.250	9	7.4	22	18.2
0.251–0.300	5	4.1	16	13.2
0.301–0.350	12	9.9	8	6.6
0.351–0.400	11	9.1	6	5.0
0.401–0.450	13	10.7	4	3.3
0.451–0.500	19	15.7	0	0.0
0.501–0.550	7	5.8	1	0.8
<0.551	19	15.7	0	0.0
Parent P-24	Seed toxin = 0.366		Leaf toxin = 0.257	

Table 2. Characteristics of low toxin in R_2 progeny of Lathyrus sativus *derived from leaf explant*

Progeny	Seed wt. per plant	Seed toxin	Leaf toxin	Seed wt./ 100 grains	Seed proteins (%)
15.8	32.0	0.016	0.039	6.6	27.9
16.6	13.0	0.018	0.063	7.6	34.4
15.4	26.7	0.019	0.036	6.0	34.4
16.7	6.9	0.020	0.054	5.4	29.9
16.4	18.5	0.023	0.312	6.1	27.9
17.4	25.3	0.044	0.039	7.1	28.0
15.1	2.5	0.049	0.056	4.4	33.2
19.5	15.9	0.052	0.070	5.6	28.2
21.3	9.7	0.054	0.070	6.6	29.5
21.5	8.7	0.062	0.000	5.8	27.9
20.6	22.8	0.063	0.031	7.0	28.0
17.6	1.6	0.070	0.063	6.0	29.5
19.6	33.4	0.078	0.070	7.5	27.6
12.1	3.6	0.082	0.390	4.4	25.4
17.3	2.1	0.083	0.085	4.2	29.1
10.8	40.2	0.085	0.187	6.7	29.9
12.4	2.4	0.105	0.040	4.4	31.7
15.2	23.0	0.109	0.164	5.0	29.7
14.4	14.3	0.125	0.156	6.0	31.9
16.1	41.6	0.125	0.125	5.3	31.3
P-24	33.6	0.366	0.257	7.8	29.8

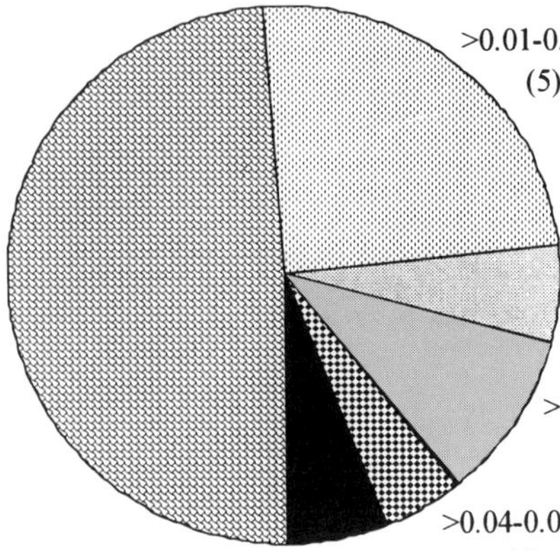

Figure 4. Seed toxin in individual plant seeds of somaclones from 15 to 8 in R_3 generation (values in % represent ODAP content and in parentheses no. of plants).

of many somaclones significantly increased over the parent cultivar, the increase being up to 1.5 times with Bio I-13 and Bio I-30. The leaf breadth was also increased up to a maximum of 3 times in case of Bio L-12. Flower colour among the somaclones showed variation. Bio R-27, Bio L-08 and Bio I-05 had white flowers; 15-8-1 had red and Bio-L-19 had pink flowers while all others along with parent P-24 had blue flower colour. The seed coat colour also varied from white (Bio L-08, Bio R-27, Bio I-05) to red (Bio L-19) to brown (Bio L-56) to grey (parent and all other somaclones). Somaclonal variants with low ODAP and having high yield were advanced to further generation and were found to be stable over generations. Many of them showed yield potential substantially higher than the parent cultivar P-24 (Table 3). Some of the somaclones having low ODAP and high yielding potential were grown at different locations under various field trial programmes and have fared well (Table 4). Based on the data from these trials one of the somaclone Bio L-212 (Ratan) was released for cultivation in central and North Eastern Plains Zones. In all the trials, all our somaclones had substantially lower ODAP content as compared to other entries and also yielded better (Table 4).

The phenotypic variations observed in *Lathyrus* somaclones were found to be stable over generations, therefore it provides opportunity of using new variability in plant breeding programmes to develop high yielding low toxin varieties.

3. Somatic embryogenesis

Somatic embryogenesis and organogenesis were successful with *L. sativus*. For inducing somatic embryogenesis immature leaf segments and internodal segments of cultivar P-24 were cultured on MS medium supplemented with NAA alone or in combination with kinetin for 8–10 days followed by 4 weeks on growth regulator free medium. Direct embryo

Figure 5. Phenotypic variations observed in somaclones of Lathyrus sativus *(a) variation in leaf size and morphology; (b) variation in flower color (i) Bio L-08, (ii) Parent P-24, (iii) Bio L-19; (c) variation in seed coat color.*

formation was observed on the explants without intervening callus phase, only on medium with 2, 4-dichlorophenoxy acetic acid. The embryos formed were bottle shaped to spherical. Supplementation of MS media with either glutamine, asparagine or abscisic acid resulted in the shape differences of the embryo, in the former case they were bottle shaped while in the latter two cases they were spherical in nature. Even though the individual embryos on transfer to hormone free medium germinated (98%) to form roots with slight elongation of hypocotyl, only 2% embryos germinated to form shoots (Barna and Mehta, 1995).

Table 3. ODAP and yield in different somaclones of Lathyrus sativus

	ODAP (%)				
Somaclones	(1994)	(1995)	(1996)	(1997)	Yield (q/ha)
Bio 158	0.044	0.056	0.070	0.038	19.42
Bio 164	0.034	0.031	0.089	0.053	18.75
Bio L203	0.063	0.065	0.081	0.034	22.12
Bio L207	0.050	0.044	0.095	0.069	19.46
Bio L208	0.028	0.047	0.125	0.056	33.65
Bio L212	0.037	0.038	0.087	0.050	23.40
Bio L254	0.046	0.062	0.095	0.050	20.87
Bio L256	0.065	0.050	0.065	0.047	20.00
Bio L257	0.063	0.059	0.105	0.065	23.00
Bio R202	0.034	0.056	0.117	0.072	27.12
Bio R215	0.041	0.100	0.150	0.150	25.65
Bio R224	0.046	0.056	0.097	0.047	20.75
Bio R229	0.063	0.068	0.111	0.063	23.61
Bio R231	0.050	0.046	0.128	0.072	27.51
Bio R233	0.069	0.075	0.132	0.132	28.12
Bio I218	0.059	0.056	0.103	0.041	20.00
Bio I222	0.031	0.040	0.076	0.076	31.65
Bio I230	0.044	0.050	0.130	0.076	22.71
P24 (Check)	0.321	0.320	0.592	0.365	13.00

Table 4. Yield of Lathyrus sativus *somaclones under special CVT (1995–96)*

	Yield (kg ha^{-1})		
Somaclones	NEPZ	CZ	NWPZ
Bio L208	853	767	1016
Bio L212	1350 (I)	1182 (II)	1031
Bio R202	1076	1107	1141
Bio R231	986	1164 (III)	1141
P24 (Check)	742	951	1367

CVT, Central varietal trial; NEPZ, North East plane zone; CZ, Central zone; NWPZ, North West plane zone.
Values given in parentheses indicate rank.

So far there are no reports on development of transgenic *L. sativus*. There is further scope to develop proper transformation protocols which can be used to transform *L. sativus* plants with ODAP degrading gene isolated from a microbe (Sukanya *et al.*, 1993) to develop toxin free transgenic plants.

Development of the low ODAP lines of *L. sativus* and also release of one of them should help in lifting the ban imposed on *Lathyrus*. Toxicological studies are going on at Central Toxicological Research Institute to evaluate the released variety in comparison with its parent P-24 for toxicological effects on feeding them. In the eventuality of lifting the ban, it will still be needed to have much extension work to popularize this pulse crop. Being a bee pollinated crop extra care will be needed to maintain genetic purity for neurotoxin levels by discouraging the growing of high toxin varieties side by side as well as admixture.

4. Conclusions and future prospects

Thus in the present study it has been possible to develop successful regeneration protocols for *L. sativus* using leaf, root and internode as explants. A large number of somaclonal variants with phenotypic variations have been developed some of which will be useful for future breeding programmes. Some of the somaclones developed have very low toxin content (<0.1%) combined with high yielding characteristics. One of the low toxin somaclone Bio L212 (Ratan) has been released for cultivation by Govt. of India, but the ban imposed on the sale of this grain legume should be lifted in order to fully exploit this crop.

Development of regeneration protocols has also opened the way for attempting genetic transformation of *L. sativus*. This is relevant in the context that we have also isolated and characterized seed toxin (ODAP) degrading gene from a soil microbe (Sukanya *et al.*, 1993) which can be introduced into *L. sativus* cultivars for developing transgenic plants devoid of toxin.

References

Barna K S and Mehta S L (1995) Genetic transformation and somatic embryogenesis in *Lathyrus sativus*. *J. Plant Biochem. Biotech.*, **4**: 67–71.

Gharyal P K and Maheshwari S C (1983) Genetic and physiological influences on differentiation in tissue cultures of legume *Lathyrus sativus*. *Theort. Appl. Genet.*, **66**: 123–126.

Mehta S L, Ali K and Barna K S (1994) Somaclonal variation in food legume – *Lathyrus sativus*. *J. Plant Biochem. Biotech.*, **3**: 73–77.

Mukhopadhyay A and Bhojwani S S (1978) Shoot-bud differentiation in tissue cultures of leguminous plants Z. *Pflanzen Physiol.*, **88**: 263–268.

Murti V V S, Seshadri T R and Venkatasubramanian T A (1964) Neurotoxic compounds of the seeds of *Lathyrus sativus*. *Phytochemistry*, **3**: 73.

Rao S L N, Malathi K and Sarma P S (1964) Isolation and characterisation of β-N-oxalyl L-diamino propionic acid, a neurotoxin from seeds of *Lathyrus sativus*. *Biochemistry*, **3**: 432.

Roy P K, Ali K, Gupta N, Barat G K and Mehta S L (1993) β-N-oxalyl L-diamino propionic acid in somaclones derived from internode explants of *Lathyrus sativus*. *J. Plant Biochem. Biotech.*, **2**: 9–13.

Roy P K, Barat G K and Mehta S L (1992) *In vitro* plant regeneration from callus derived from root explants of *Lathyrus sativus*. *Plant Cell Tiss. Org. Cult.*, **29**: 135–138.

Roy P K, Singh B, Mehta S L, Barat G K, Gupta N, Kirti P B and Chopra V L (1991) Plant regeneration from leaf disc of *Lathyrus sativus*. *Indian J. Exp. Biol.*, **29**: 327–330.

Sinha R R, Das K and Sen S K (1983) Plant regeneration from stem derived callus of the seed legume *Lathyrus sativus* L. *Plant Cell Tiss. Org. Cult.*, **2**: 67–76.

Sukanya R, Santha I M and Mehta S L (1993) Cloning and characterisation of a gene responsible for the *Lathyrus sativus* neurotoxin degradation. *J. Plant Biochem. Biotech.*, **4**: 33–36.

PART V
BIODIVERSITY

ORGANISATION AND GENETIC MAPPING OF THE CHICKPEA GENOME

PETER WINTER[1], CHRISTINA STAGINNUS[1], PRAKASH C. SHARMA[2] AND GÜNTER KAHL[1]
1 Plant Molecular Biology, Biocenter, University of Frankfurt/Main, Germany
2 Department of Agricultural Botany, Ch. Charan Singh University, Meerut, India
e-mail: kahl@em.uni-frankfurt.de

Abstract

The improvement of important food legumes such as chickpea by modern biotechnology has been lagging behind achievements in e.g. model plants or cereals. However, substantial advancements have been made in many of these crops in recent years. Using chickpea as an e.g. this article summarises the progress made in our understanding of genome structure and evolution, genetic mapping, mapping of resistance genes for important pathogens, synteny between related species and genera, and plant transformation and regeneration. It is expected, that transfer of knowledge from model plants and advanced crops together with high-throughput technologies will catalyze the analysis of entire transcriptomes and proteomes which will foster the development also in the neglected crops and substantially add to their agricultural value.

1. Introduction

The Leguminosae, comprising about 18,000 species and 650 genera (Pohill and Raven, 1981), is the most important family of the Dicotyledonae, and with an annual world market value of around two billion US $ is second only to cereals in world food production. Legumes are often the only source of high-quality protein for the rural poor in developing countries, and are used as fodder, forages and green manure, are important sources of oils, fatty acids, timber, medicine, tannins and gums, and are also employed for soil recovery and improvement. Many legumes are hardy crops that grow on poor soils under harsh conditions, where cereal production is either not possible or uneconomic (Duke, 1981).

Legumes fix atmospheric nitrogen via symbiosis with bacteria of the genus *Rhizobium* (see Caetano-Anollés and Gresshoff, 1991; Long, 1996). Though this property

P.K. Jaiwal and R.P. Singh (eds.), Improvement Strategies for Leguminosae Biotechnology, 303–351.

Table 1. Some genomic parameters of the important legumes (Bennett et al., 1998)

Species	Genome size (1C) pg	Genome size (1C) Mbp*	Chromosome number (2n)
Chickpea (*Cicer arietinum* L.)	0.95	916	16
Common bean (*Phaseolus vulgaris* L.)	0.60	579	22
Cowpea (*Vigna unguiculata* L.)	0.60	579	22
Faba bean (*Vicia faba* L.)	13.50	13,027	12
Lentil (*Lens culinaris* L.)	4.2	4,053	14
Pea (*Pisum sativum* L)	4.65	4,487	14
Soybean (*Glycine max* L.)	1.13	1,090	40
Model legumes			
Lotus japonicus	0.47	453	12
Medicago truncatula	0.47	453	16
Model plant			
Arabidopsis thaliana	0.17	168	10

*1pg of genomic DNA = 9.65×10^8 bp (Bennett and Smith, 1976).

has attracted the attention of researchers for a long time, the complex mechanisms of host-bacteria interactions are still not fully understood.

With the exception of soybean, which belongs to the four most important crops world-wide, the majority of legumes are minor players on the world market, because many of them are only produced for local consumption. Consequently, most breeding and research efforts have focused on soybean, for which an almost complete set of biotechnological tools and extended genomic maps are available. Since soybean has a large and complex genome (Table 1), and plant transformation and regeneration is difficult, the model legumes *Lotus japonicus* (Handberg and Stougaard, 1992; Jiang and Gresshoff, 1997) and *Medicago truncatula* (Cook *et al.*, 1995, 1997) have attracted much attention, especially since their genome sizes compare well with the model plant *Arabidopsis thaliana* (see Table 1). They are also easy to transform (Oger *et al.*, 1996; Martirani *et al.*, 1999). Moreover, many mutants for both legumes are at hand to decipher the complex interaction(s) between the plant and e.g. bacteria (Penmetsa *et al.*, 1997; Schauser *et al.*, 1998).

Genome research in other important legumes is lagging far behind. Important pulses such as chickpea and lentil, belonging to the founder crops of modern agriculture (Zohary and Hopf, 1993; Lev-Yadun *et al.*, 2000), are probably of monophyletic origin (Zohary, 1999). Therefore, they possess highly invariable genomes which hampered their analysis for a long time. However, the advent of DNA marker technology allowed progress also in the genomics of these crops, and the achievements made in genomics of legumes other than soybean and the model legumes are exemplified with chickpea (*Cicer arietinum*) here.

1.1. CHICKPEA: THE THIRD MOST IMPORTANT FOOD LEGUME WORLD-WIDE

Chickpea (*Cicer arietinum* L.) is mostly grown on the Indian sub-continent, West Asia and North Africa (WANA). The species is often divided into two major groups that differ in size, shape and color of the seeds (Auckland and van der Maesen, 1980): large, round seeds of white or pale cream color are referred to as kabuli type (from Kabul, Afghanistan), the desi type (Hindi: local) produces smaller, angular seeds with sharp edges, that like flowers, stems and sometimes leaves are pigmented. Kabuli chickpea normally has non-pigmented flowers, and is primarily grown in the Mediterranean, the Americas and Australia, whereas desi type chickpeas are preferred on the Indian subcontinent and in East Africa. Though separated for centuries, no hybridisation barriers exist between the two groups (Muehlbauer and Singh, 1987).

Despite considerable international investment in conventional breeding, productivity of the chickpea crop has not yet been improved significantly. Productivity has been stagnant over many decades, though prices in the most important producer country India rose more than twice from 1988 to 1995 (FAOSTAT 2000). Main constraint for increasing yield on the Indian subcontinent is the plant's susceptibility to *Fusarium* wilt, a vascular disease caused by *Fusarium oxysporum* f. sp. *ciceri* (Srivastava *et al.*, 1984; Jiminéz-Diaz *et al.*, 1993), whereas in WANA countries a foliar disease, *Ascochyta* blight, caused by the Ascomycete *Ascochyta rabiei* (Saxena and Singh, 1987; Singh *et al.*, 1992; Kaiser, 1997) is most devastating. Consequently, chickpea breeding aims at developing high yielding cultivars combining long-lasting resistances against *Fusarium* wilt and *Ascochyta* blight with tolerance to abiotic stresses such as drought and cold.

Modern tools like molecular markers, dense genetic maps, and transformation and regeneration techniques significantly increase the speed of crop improvement as proven for tomato or rice (Tanksley and McCouch, 1997). We will summarise the progress made in recent years in our understanding of the general structure of the chickpea genome, the development of molecular markers and marker-based genetic maps, and transformation and regeneration of the plant.

1.2. THE GENUS *CICER*

The genus *Cicer* in the subfamily *Papilionaceae*, tribe *Cicereae* Alef of the *Leguminosae*, involves nine annual and 33 perennial species (van der Maesen, 1987). Whereas the cultivated species *C. arietinum* L. is distributed across many parts of the world, the eight wild annual species are endemic to an area reaching from Turkey to Central Asia and to parts of Egypt, Ethiopia, and Sudan (Robertson *et al.*, 1997). Chickpea's wild relatives have attracted much attention as they are potential sources of a wide range of agronomically favourable traits for the improvement of the gene pool of the cultigen (Muehlbauer *et al.*, 1994; Singh and Ocampo, 1997; Singh *et al.*, 1998).

Evolutionary and genetic relationships between the eight annual species of the section *Monocicer* are well described. *C. chorassanicum*, the ninth known annual species, has been classified into section *Chamaecicer*. Annual species are subdivided into four groups on the basis of crossability relationships (Ladizinsky and Adler, 1976b), karyotype analysis (Ocampo *et al.*, 1992), isozyme and rRNA polymorphisms (Patil *et al.*, 1995; Labdi *et al.*, 1996) and seed storage protein characteristics (Ahmad and Slinkard, 1992). The first group contains the cultigen, its presumed ancestor *C. reticulatum* Ladiz., and *C. echinospermum* P.H. Davis. Both wild species produce fertile hybrids with the cultigen, though fertility barriers exist between chickpea and *C. echinospermum* (Singh and

Ocampo, 1993). The second group comprises *C. bijugum* K.H. Rech, *C. pinnatifidum* Jaub and Spp., *C. judaicum* Boiss and *C. yamashitae* Kitamura. Interspecific hybrids between the members of this crossability group and chickpea may be obtained through *in vitro* culture (Badami *et al.*, 1997), though hybridisation was successful also without these techniques (Verma *et al.*, 1990; Singh *et al.*, 1994). The remaining two species, *C. cuneatum* Hoechst ex. Rich and *C. chorassanicum* (Bge.) M. Pop. can neither be crossed with each other nor with any other species and, therefore, make up the third and fourth group, respectively (Kazan and Muehlbauer, 1991). However, *in vitro* techniques may improve interspecific hybridisation even between very distantly related taxa, since Singh and Singh (1989) recovered rare interspecific hybrids between *C. cuneatum* and chickpea via tissue culture.

Prior to the advent of DNA marker techniques, the relationships between annual *Cicer* species have been inferred from various morphological, crossability, and biochemical data (Ladizinsky and Adler, 1976a, b; Sharma and Gupta, 1988; Kazan and Muehlbauer, 1991; Ahmad and Slinkard, 1992; Ahmad *et al.*, 1992; Tayyar and Waines, 1996; Robertson *et al.*, 1997). DNA marker techniques (Ahmad, 1999; Choumane *et al.*, 2000) in general support these studies, though RAPD marker analysis (115 characters) places *C. yamashitae* and *C. chorassanicum* more closely to members of the first crossability group rather than the other members of the second group (Ahmad, 1999), thereby contradicting isozyme (Kazan and Muehlbauer, 1991) and STMS marker data (Choumane *et al.*, 2000). Though the relative position of *C. chorassanicum* within the section *Monocicer* is disputed, isozyme as well as DNA marker data all justify its placement into this section and reject its former classification into the section *Chamaecicer* (Kazan and Muehlbauer, 1991; Ahmad, 1999; Staginnus *et al.*, 1999; Choumane *et al.*, 2000).

The opposite is true for *C. cuneatum*, which should better be placed into another section of the genus on the basis of isozyme patterns and other features (Ladizinsky and Adler, 1976a, b; Kazan and Muehlbauer, 1991; Labdi *et al.*, 1996; Tayyar and Waines, 1996; Ahmad, 1999; Choumane *et al.*, 2000). The strongest argument for a reclassification of this species, however, is the absence of the two satellite repeats which are major structural components of the genomes of all other species of the section *Monocicer* (Staginnus *et al.*, 1999).

Of the perennial species, only *C. anatolicum* Alef. and *C. songaricum* were studied in more detail by karyotype and isozyme analysis (Ahmad, 1989; Kazan and Muehlbauer, 1991; Tayyar and Waines, 1996) and DNA markers (Choumane *et al.*, 2000). Surprisingly, isozyme analysis led to contradictory results. Kazan and Muehlbauer (1991) found a closer relationship of *C. anatolicum* to the first crossability group as compared to the other annual species. These results were challenged by Tayyar and Waines (1996) who reported a distant relationship of *C. songaricum* to the species of the first crossability group. Further, Kazan and Muehlbauer (1991) advocated a classification of *C. anatolicum*, first grouped into section *Polycicer*, to the section *Monocicer*. Our own molecular studies support this classification (Staginnus *et al.*, 1999; Choumane *et al.*, 2000). However, more research is needed to prove that this perennial is in fact the ancestor of the first crossability group, as suggested by Kazan and Muehlbauer (1991).

2. A closer look into chickpea chromosomes

Availability of detailed cytogenetic information of a crop and its related species is important e.g. alien gene transfer, molecular mapping and systematics. Here, the available information on the large-scale structure of the chickpea genome including recent *in-situ* hybridisation data are presented.

2.1. CHROMOSOME NUMBER

The somatic chromosome number in chickpea and all the nine annual *Cicer* species has been determined as 2n = 16 (x = 8), although some reports suggested 2n = 14 (see Sharma and Gupta, 1982; Singh *et al.*, 1984; Bahl, 1987). It has rightly been argued that plants with a chromosome count of 2n = 14 may either be rare in nature, or the observation an artifact (one chromosome pair in the complement is very small). Considering differences in basic chromosome number, interphase nuclear structure, and scanning electron microscopy of seed testa, the placement of the genus *Cicer* in the tribe *Vicieae* has been often questioned in the past. Therefore, the genus has now been placed in the monogeneric tribe *Cicereae* (see Ahmad, 2000).

2.2. KARYOMORPHOLOGY

Dombrovsky-Sludsky (1927) studied somatic mitosis in *C. arietinum* first. Since then karyotypes in the cultigen and the annual wild species using root tip chromosomes have been worked out (Sharma and Gupta, 1982, 1986; Ahmad, 2000), revealing small variations in chromosome size, position of centromeres, number of satellited chromosomes and number of chromosomes with attached nucleolus. These variations may partly be real, but most are probably artifacts. For example, the observed differences in chromosome length may mainly result from the various pre-treatment agents as e.g. p-dichlorobenzene, 8-hydroxyquinoline, colchicine or cold water. Also, a bias may be introduced by the investigator, selecting cells at different chromosome condensation stages as prophase proceeds to metaphase. One could minimise the variation by scoring data from many cells (Ahmad, 2000).

The consensus haploid chickpea karyotype emerged from different cytological approaches: it comprises (i) one comparatively long, satellited and submetacentric chromosome (>3.5 μm), (ii) six medium sized (1.6–2.75 μm) metacentric to submetacentric chromosomes, and (iii) one very short (<1.5 μm) metacentric chromosome (Sharma and Gupta, 1986; Ahmad, 2000).

Chromosome morphology was also studied at pachytene, when chromosomes are in a more extended state. Chromosome lengths and centromere positions for different chromosomes varied in somatic and pachytene cells, according to the differential condensation of the chromosomes and even arms of the same chromosome at the two stages. Meiosis, however, is normal in chickpea, with eight bivalents at metaphase I, although sometimes precocious disjunction of one or two pairs of chromosomes was observed (Mercy *et al.*, 1974; Sharma and Gupta, 1986). Chiasmata frequency per cell in annual *Cicer* species ranged from 11.5 in *C. reticulatum* to 15.5 in *C. bijugum*.

With the aim of understanding phylogenetic relationships within the genus *Cicer*, karyotypes of wild annual *Cicer* species have also been studied (Sharma and Gupta, 1986; Kabir and Singh, 1990; Ohri and Pal, 1991; Ocampo *et al*., 1992; Ahmad, 2000). Although the relative and absolute chromosome lengths vary in the different reports, the karyotypes could be grouped into two categories according to Stebbins' (1971) classification which is based on chromosome asymmetry. Generally, in cultivated species more asymmetric karyotypes are observed indicating more recent evolution of chromosomes. Also, in the genus *Cicer*, the annual species of chickpea's first crossability group *C. arietinum*, *C. reticulatum* and *C. echinospermum* display a karyotype of the more asymmetric and more evolved 1B category (Fig. 1), while the more basic A type is characteristic for *C. bijugum*, *C. judaicum*, *C. pinnatifidum*, *C. cuneatum*, *C. anatolicum* and *C. yamashitae*. Karyotype asymmetry in chickpea and its closest relatives is associated with somewhat longer chromosomes and higher 2C DNA values than in the other annual species. Additionally, size variation of the different chromosomes in these species is more pronounced than in the other annuals, indicating substantial rearrangement of chromatin (Ahmad, 2000). Interestingly, the amount of heterochromatin did not increase with genome size in *Cicer* (Tayyar *et al*., 1994; Galasso *et al*., 1996), suggesting rearrangements both in heterochromatin and euchromatin during evolution in this genus, possibly very much as occurred in *Lathyrus*, another leguminous genus (Narayan, 1982). Since the karyotype of *C. reticulatum* is most closely related to that of the cultivated *C. arietinum* (Ladizinsky and Adler, 1976a, b; van der Maesen, 1987), *C. reticulatum* is most probably the progenitor of present-day chickpea. Also, *C. arietinum*, *C. reticulatum* and *C. echinospermum* display quite similar karyotypes (Singh, 1988; Ohri and Pal, 1991; Galasso *et al*., 1996). Moreover, chromosome banding and *in-situ* hybridisation techniques

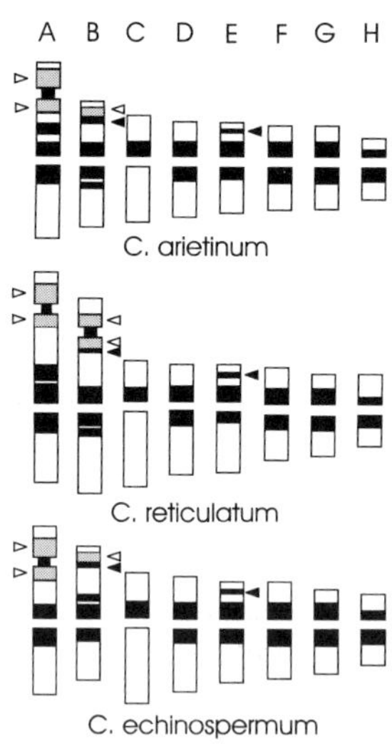

Figure 1. Schematic karyotype of the three annual species of the first crossability group of Cicer *as revealed by* in-situ *hybridisation and different staining techniques (according to Galasso et al., 1996, modified). The position of the 18S-5.8S-28S rDNA clusters is indicated by open arrowheads, 5S rDNA blocks by filled arrowheads.*

demonstrated C-banded karyotypes for *C. arietinum* (Galasso and Pignone, 1992), *C. reticulatum* and *C. echinospermum* (Tayyar *et al.*, 1994). A distinctive class of heterochromatin located at the NOR region of chromosome A as well as the telomeres of the non-satellited chromosome B was detected, whose chemical and structural properties made it a marker for phylogenetic relationships within the genus *Cicer*. The chromosomal distribution of different classes of heterochromatin (as visualised by staining with fluorochromes chromomycin A3, DAPI and Hoechst 33258), and the localisation of the 18S, 5.8S, 25S and 5S rDNA sequences and their expression (Abbo *et al.*, 1994; Galasso *et al.*, 1996) suggested an evolutionary pathway, in which *C. arietinum* and *C. echinospermum* are derived from the ancestral species *C. reticulatum*, losing one pair of satellites. Subsequently, *C. echinospermum* differentiated further by chromosomal rearrangement(s), that decrease the chance to obtain fertile offspring from crosses with the cultigen (Galasso *et al.*, 1996).

3. Much of the genome is reiterated DNA

Repetitive DNA is a ubiquitous and major component of the genomes of higher plants, and at least partly responsible for genome size and complexity. Repetitive sequences vary in length between 1 and 10,000 bp and can be reiterated many hundreds or even several thousand times. Within the genome of a particular species the position of the different repetitive elements varies from being widely dispersed to residing at only a few, well-defined chromosomal sites. Moreover, even between closely related species the absolute amount, dispersion pattern and sequences of repetitive elements differ extensively, whereas genes and their arrangement along the chromosomes are often syntenic within, and in some cases, beyond taxonomic groups. Well defined functions are known only for a few repeat families such as telomeric or centromeric sequences, or 18S-5.8S-25S and 5S rRNA gene clusters encoding ribosomal structural RNA components.

Two main types of repeats can be distinguished by their different overall structure: tandemly repeated and dispersed elements. **Tandem arrays** consist of continuous stretches of repetitive units and comprise rDNA gene clusters, micro- and minisatellites and satellite DNA. The latter may account for several percent of a plant genome, whereas rRNA genes and mini- and microsatellites are only of medium or low abundance. Usually, satellite DNA can be found at few, characteristic loci (Schmidt and Heslop-Harrison, 1998). **Dispersed elements** consist of repeated units that are interspersed with single copy sequences or other repetitive elements. Most of these are transposable elements such as transposons and retrotransposons, or their remnants (Flavell, 1986). Retroelements represent up to 50% of the genomic DNA in some plants (SanMiguel *et al.*, 1996) and are usually dispersed along chromosome arms, a consequence of their transposition via RNA intermediates.

As compared to genes, other types of selective constraints act on repetitive DNA, resulting in sequence and structural differences even among closely related species. Therefore, repetitive sequences are important tools for evolutionary and taxonomic studies. Since genes are embedded in stretches of repetitive DNA with variable length and composition (SanMiguel *et al.*, 1996; Schmidt and Heslop-Harrison, 1998), repetitive sequences should be of basic interest for mapping of a genome or map-based cloning of genes.

3.1. MAJOR REPEATS IN THE CHICKPEA GENOME: *CaSat1*, *CaSat2* AND *CaRep*

The chickpea genome harbours at least three major repetitive elements. Two, *CaSat1* and *CaSat2*, are satellite repeat families, the third, *CaRep1*, is a member of the retrotransposon-like repeat family (Staginnus *et al.*, 1999).

Both **satellite families** display the typical organisation of satellite repeats with long tandem arrays of head-to-tail oriented repetitive units. This arrangement is characterised by conserved restriction sites in the array of repeat units which cause the characteristic, prominent ladder patterns on autoradiograms from Southern blots of restricted DNA after hybridisation to cloned repeat sequences. With its 162–168bp long repeat units, the *CaSat1* family matches the preferred unit length of plant satellite DNA, which is suspected to correspond to the size of a nucleosomal unit (Vershinin and Heslop-Harrison, 1998). However, repeats of the *CaSat2* family are only 100bp in length (Staginnus *et al.*, 1999), and like several tandem repeats from other plants, do not fit into this model. Both satellite sequences are AT-rich, but differ in the level of homology between members of the same family. Whereas *CaSat2* elements are highly conserved, *CaSat1* sequences display more variety reflecting different stages of homogenisation.

The satellite sequences represent a considerable part of the DAPI – stainable, pericentromeric heterochromatin as shown by fluorescence *in-situ* hybridisation (FISH) on metaphase chromosomes (Staginnus *et al.*, 1999). In metaphase nuclei, the eight chromosome pairs (A–H) are visible, which exhibit large regions of bright DAPI fluorescence around centromeres. The largest chromosome A carries a secondary constriction corresponding to the NOR region (Galasso and Pignone, 1992). The rRNA gene clusters are located on chromosome pairs A and B (18S-5.8S-25S rRNA genes) and on B and E (5S rRNA genes; Galasso *et al.*, 1996).

Double-target probing with rDNA sequences and *CaSat1* probes locate the majority of these satellite sequences to clusters in the vicinity of the 18S-5.8S-25S rRNA gene blocks on chromosomes A and B – close to the secondary constriction on A, and within the pericentric heterochromatic block on B. Minor hybridisation signals are visible in the distal regions of all other chromosomes (Staginnus *et al.*, 1999). The independent occurrence of tandemly arranged or dispersed sequences and subrepeats of the rDNA intergenic spacer region (IGR) has been reported for several legumes (Maggini *et al.*, 1991; Unfried *et al.*, 1991; Falquet *et al.*, 1997). Thus, a relationship between *CaSat1* sequences and rDNA spacer subrepeats would not be unexpected. The occurrence of minor *CaSat1* elements on other parts of all chromosomes may be explained by unequal exchange between non-homologous chromosomes and sister chromatids as well as transposition during the evolution of this sequence family as reported for other plant species (Ohtsubo and Ohtsubo, 1994; Kamm *et al.*, 1995; Schmidt and Heslop-Harrison, 1996a).

CaSat2 elements are dominant components of the DAPI-positive pericentric heterochromatin of all chromosome pairs, eventually including the centromeric regions. On chromosomes A and B the clusters reside in close vicinity, but clearly separated from the major *CaSat1* sites. The high intensity of hybridisation signals in metaphase and interphase nuclei suggests that *CaSat2* represents the most abundant sequence family in the chickpea genome.

The third highly abundant family of repetitive elements, *CaRep*, comprises three sets of sequences with homology to different parts of the Ty3-gypsy type retrotransposon *del1* from *Lilium henryi* (Smyth *et al.*, 1989), probably representing the remnants of different segments of a similar retrotransposon in the chickpea genome (Staginnus *et al.*, 1999).

The majority of the *CaRep* family members is clustered in the pericentric heterochromatin on all chromosomes. The uniform hybridisation signals along the DAPI-positive heterochromatic blocks are only interrupted at the centromeric regions – probably consisting of *CaSat2* sequences. *CaSat1* elements on chromosomes A and B do not interfere with *CaRep* signals, but reside in the distal areas of the heterochromatin block, whereas *CaRep* elements are located adjacent to them in the proximal parts of the block. *CaRep* repeats are also absent from the secondary constriction harbouring the 18S-5.8S-25S rRNA gene clusters on chromosome A. The exclusion of **retrotransposons** from telomeres, centromeres, NORs and from regions occupied by prominent tandem repeats is known from other plants (Schmidt and Heslop-Harrison, 1996a; Brandes *et al.*, 1997; Heslop-Harrison *et al.*, 1997). However, *CaRep* elements are not restricted to heterochromatin, but also reside in the most likely gene-rich distal euchromatin of chromosome arms (Staginnus *et al.*, 1999). In this respect, *CaRep* repeats are similar to those from other plants, where retrotransposon-derived sequences intermingle with plant genes (SanMiguel *et al.*, 1996).

3.2. REPETITIVE ELEMENTS SUPPORT THE TAXONOMY OF THE GENUS *CICER*

All three families of highly repetitive elements seem to be specific for the genus *Cicer*. We assume that *CaSat1* and *CaSat2* tandem repeats were already present in high copy numbers in the putative progenitor of the annual *Cicer* species, since monomer units and their tandem organisation are conserved (Staginnus *et al.*, 1999). However, the extensive divergence or even absence of the two satellite repeats in *C. cuneatum* suggest a separate origin of this annual species (also supported by isozyme polymorphism, crossability data, morphological features, and its endemism: Ladizinsky and Adler, 1976a, b; Kazan and Muehlbauer, 1991; Labdi *et al.*, 1996). The classification of *C. cuneatum* into the section *Monocicer* is therefore doubtful (Tayyar and Waines, 1996). The only annual species of the section *Chamaecicer*, *C. chorassanicum*, on the other hand, contains *CaSat1* and *CaSat2* elements of similar structure and abundance as the other members of that section.

Also the perennial *C. anatolicum* harbours these characteristic sequences. Because of the similarity of its repeats (Staginnus *et al.*, 1999), to those of the members of the first crossability group (*C. arietinum*, *C. reticulatum* and *C. echinospermum*), *C. anatolicum* could be an ancestor of this group (Kazan and Muehlbauer, 1991). Using the *CaSat* elements as probes, it will probably be possible to clarify the relationships of the other perennials to the cultigen.

CaRep sequences are conserved within the genus *Cicer*, whereas they are either absent or greatly diverged in other legumes. Specific structures of these gypsy-like elements, though conserved within each crossability group, vary among the species belonging to the first and the second crossability group of the annual *Cicer* species and *C. cuneatum*. Therefore, we assume that these retrotransposon-like sequences were independently amplified and translocated during the evolution of the annual species of the genus *Cicer*.

3.3. LESS ABUNDANT REPEAT FAMILIES

In addition to the three major repeat families, at least four non-tandemly arranged, dispersed repeat element families of middle to low abundance exist in the chickpea genome. All display homologies to transposable elements of various classes as e.g. *Ty-copia*-like elements, non-LTR retrotransposons and *En-Spm*-like transposons of other plants (Table 2). None of these repeat families is restricted to the genus *Cicer*. The majority of these elements is located in the distal parts of the AT-rich heterochromatic blocks or in the euchromatin of several or all chromosomes, but they are excluded from rRNA gene clusters and the highly repetitive satellite sequences.

3.4. MICROSATELLITES AND THEIR DISTRIBUTION IN THE CHICKPEA GENOME

The chickpea genome – like that of all other higher eucaryotes – additionally contains a set of repetitive elements, so-called **microsatellites**, that attracted much attention in recent years, since their variability makes them ideal markers for identification of individuals and genetic mapping.

Microsatellites (MS: Litt and Luty, 1989), also called simple sequence repeats (SSRs), or short tandem repeats (STRs), consist of short, reiterated sequence motifs of about 1 to 5 nucleotides such as $(A)_n$, $(AT)_n$, $(GA)_n$, $(CTT)_n$, $(TAGG)_n$ or $(TTCGG)_n$, to name only a few. Like satellite DNA, MSs are organised in more or less perfect tandem

Table 2. Repetitive sequence families in the chickpea genome (Staginnus et al., *1999)*

Repetitive element	Type	Abundance
CaSat1	Satellite DNA	high
CaSat2	Satellite DNA	high
CaRep	*Ty3-gypsy*-like LTR retroelement	high
CaTy	*Ty1-copia*-like LTR retroelement	middle
CaDis	Retroelement	middle
CaLin	LINE-like non-LTR retrotransposon	low-middle
CaEn/Spm	*En-Spm*-like transposable element	low-middle
Microsatellite	Mono-, di-, tri-, tetra- and pentanucleotide repeats	high

arrays of few to hundreds or even thousands of repeat units. A key feature of MS repeats is a strong tendency to change their overall length by slipped-strand mispairing and other less well understood processes (for review see McMurray, 1995; Sia *et al.*, 1997), leading to variable numbers of tandem repeats (VNTR: Nakamura *et al.*, 1987) and resulting in simple sequence length polymorphisms (SSLPs: Cho *et al.*, 2000). SSRs are abundant and usually more or less evenly dispersed throughout the plant genome, but less frequent in plants as compared to vertebrates (Lagercrantz *et al.*, 1993). High levels of polymorphism, abundance and ubiquitous occurrence recommend SSRs as exceptionally useful molecular markers (see reviews by Weising *et al.*, 1995a, 1998; Powell *et al.*, 1996; Gupta and Varshney, 2000). To test any usefulness for genome mapping, microsatellite distribution in the chickpea genome was probed (Gortner *et al.*, 1998).

First, it should be stressed, that probably different types of microsatellites exist, depending on their genomic organisation and localisation. Some – most probably clustered – motifs can be detected by *in-situ* hybridisation. For example, in *Beta vulgaris* motif-specific hybridisation patterns with the enrichment of some motifs at centromeric or intercalary positions were found (Schmidt and Heslop-Harrison, 1996b). Others give rise to a comparatively small number of strong, often polymorphic bands in restriction fragment length polymorphism (RFLP) fingerprints on Southern blots of restricted genomic DNA after hybridisation with the respective microsatellite motif. These probes certainly recognise SSR-like target sequences. However, the molecular basis of the observed RFLP-type polymorphisms is not yet perfectly clear. While sequenced SSRs generally span about 20 to 500 bp, polymorphic RFLP fingerprint bands are often much larger (up to 10 kb), and their size may vary by several kilobases. Target loci of RFLP fingerprint experiments are most probably part of larger repetitive structures. Thus, the observed polymorphisms are most likely caused by the cumulative action of VNTR mutations, restriction site variations, and possibly transposition events (see e.g. Zischler *et al.*, 1992; Broun and Tanksley, 1993; Kashi *et al.*, 1994; Vosman and Arens, 1997). Genetic mapping based on fingerprinting showed a clustering of several of these motifs in pea (Dirlewanger *et al.*, 1994) and tomato, where large MSs appear to be associated with centromeric regions (Arens *et al.*, 1995; Broun and Tanksley, 1996; Areshchenkova and Ganal, 1999).

The third class is the target of many microsatellite-based marker techniques, because these elements are more evenly distributed, the repeats are short and can be completely sequenced in one stretch together with its surrounding, mostly unique DNA. The distribution pattern of the different motifs in the chickpea genome varies considerably (Gortner *et al.*, 1998). $(CA)_n$ repeats, for example, are clustered in centromeric regions, where they are part of a centromeric satellite repeat (e.g. in sugar beet). Also, the (GATA) motif is preferentially clustered in centromeric areas. $(A)_n$, $(AAC)_n$ and $(TA)_n$ repeats, on the other hand, hybridise predominantly to euchromatin, whereas hybridisation signals in the vicinity of centromeres are weak (Gortner *et al.*, 1998).

4. Molecular markers for genetic mapping of the chickpea genome

Chickpea is probably of monophyletic origin and arose by selection from its ancestor *C. reticulatum* (Ladizinsky and Adler, 1976a; Zohary and Hopf, 1993; Zohary, 1999).

Genetic bottlenecks imposed during domestication and breeding practices narrowed the genetic base of crops as compared to wild founder species. This holds especially for obligatory self-pollinating species as chickpea, resulting in highly invariable genomes (Tanksley and McCouch, 1997). Therefore, the many variations for qualitative and quantitative traits in chickpea, which in many ways is similar to related crops such as pea and lentil, is astonishing. However, they are probably not a result of large-scale intraspecific genetic variation, but instead caused by mutations in single genes. Variations were described for growth, podding, flower and seed colour (Kumar *et al.*, 2000a, b), leaf form and colour, disease resistance and many quantitatively inherited traits (Muehlbauer and Singh, 1987). Only recently a major gene for flowering time was reported (Or *et al.*, 1999, Kumar and van Rheenen, 2000).

Linkages between several of these physiological and morphological characters led to a preliminary genetic map in chickpea (Muehlbauer and Singh, 1987). Subsequently, mapping efforts focussed on biochemical (isozymes, Gaur and Slinkard, 1990a, b; Kazan *et al.*, 1993) and DNA markers (Simon and Muehlbauer, 1997). However, mapping was considerably hampered by the small number of available isozyme polymorphisms (around 30, Kazan and Muehlbauer, 1991) and the monotony of the chickpea genome, which also precluded the large-scale application of particular types of DNA markers such as RFLPs (Udupa *et al.*, 1993) so useful in other crops. To circumvent the problem of low genetic variability within the cultivated species, all published molecular marker-based maps (see Table 4) are derived from interspecies crosses between chickpea and *C. reticulatum* or *C. echinospermum*, respectively.

4.1. SOME REPETITIVE ELEMENTS ARE USEFUL: MICROSATELLITE-BASED MARKERS

Notwithstanding the overall monotony of the chickpea genome, Weising *et al.* (1989, 1991, 1992) found considerable variation at the intra- and interspecific level by RFLP analysis with SSR-specific oligonucleotides, called "**oligonucleotide fingerprinting**" or "multilocus RFLP fingerprinting" (Ali *et al.*, 1986). It relies on hybridisation of end-labelled oligonucleotide probes complementary to SSR motifs (e.g. $[GATA]_4$) to restricted and electrophoretically separated genomic DNA. Hybridisation generates highly informative, multiple banding patterns, so-called multilocus RFLP fingerprints, that often allow the differentiation of individuals.

In one of the most comprehensive oligonucleotide fingerprinting studies in plants, Sharma *et al.* (1995a) analysed four chickpea accessions with 14 different restriction enzymes and 38 di-, tri- and tetra-nucleotide probes. Of these, 35 probes yielded detectable hybridisation signals. The abundance and polymorphism of the target sequences varied considerably, and the number of bands strongly depended on the probe, less the enzyme. No obvious correlation existed between abundance, fingerprint quality, and sequence characteristics of a particular motif. On the basis of hybridisation patterns, the different probes fell into 3 categories. The first category comprised probes yielding distinct, polymorphic banding patterns of probe-specific complexity, whereas the other two categories yielded either distinct, monomorphic banding patterns (category 2), or smeary, diffuse bands superimposed on a high in-lane background, or very weak signals only (category 3). The category 1 probes could be further subdivided according to the level of detected variability. Thus, some probes like $[GATA]_4$ or $[CAA]_5$ revealed considerable inter- as well as intra-accessional polymorphisms. Others, such as $[GCGT]_4$, differentiated between accessions, but produced monomorphic patterns within an accession. The complete set of 20 category 1 probes detected 78 screenable *TaqI*-polymorphisms between two accessions (Sharma *et al.*, 1995a).

Fingerprint polymorphisms can hardly be applied to genetic mapping. One serious problem concerns mutation rate. Considerable numbers of nonparental fingerprint bands indicative of extremely high mutation frequencies (change in fragment size from one generation to the next) were observed in RFLP fingerprints (Hüttel, 1996). A second problem for mapping is the tendency of long SSR arrays (that are presumably detected by RFLP fingerprinting) to cluster, unlike the much shorter evenly distributed SSRs usually analysed by PCR (see below). However, the variability, abundance and random distribution of certain SSR motifs in the chickpea genome led to the exploitation of this rich source of genetic variability by other techniques. In most of these methods, SSR-complementary oligonucleotides serve as PCR primers, either alone or in combination with arbitrary or specific primers, to amplify distinct regions of genomic DNA (see Gupta and Varshney, 2000). Particularly microsatellite-primed (MP)-PCR (Meyer *et al.*, 1993; Gupta *et al.*, 1994), anchored microsatellite-primed (AMP)-PCR (Zietkiewicz *et al.*, 1994), random amplified microsatellite polymorphism (RAMP: Wu *et al.*, 1994), selective amplification of microsatellite polymorphic loci (SAMPL: Vogel and Scolnik, 1997), random amplified microsatellite polymorphism (RAMPO: Richardson *et al.*, 1995), and retrotransposon-microsatellite amplified polymorphisms (REMAP: Kalendar *et al.*, 1999) were tested. Further, methods not directly targeting at VNTR-type polymorphisms, as e.g. DNA amplification fingerprinting (DAF: Caetano-Anolles *et al.*, 1991) and amplified fragment length polymorphism (AFLP: Vos *et al.*, 1995) were also studied.

The suitability of these methods for the detection of polymorphisms between chickpea accessions and *Cicer* species varied considerably. MP-PCR, for example, where SSR-derived primers anneal to the 3′-ends of microsatellite loci allowing the amplification of the inter-simple sequence repeat (ISSR) regions, detects mainly polymorphisms independent of microsatellite length variation, but caused by small insertions/deletions in intervening sequences. A variety of di-, tri-, and tetra-nucleotide repeat-containing oligonucleotides as PCR primers generated distinct, polymorphic banding patterns (Sharma *et al.*, 1995b; Weising *et al.*, 1995b). However, dinucleotide repeats as well as AT-rich trinucleotide repeats often produced a smear, probably a consequence of the large copy number of these motifs in plants. The reproducibility of MP-PCR and its sensitivity to reaction conditions was similar to RAPD analysis, since the majority of MP-PCR bands originated from mismatch annealing of primers (Weising *et al.*, 1995b).

Nevertheless, the pattern complexity and informativeness of MP-PCR bands were considerably lower than patterns derived from AMP-PCR, that applies MP-PCR-primers anchored at the 5′ or 3′ end by additional bases not included in the microsatellite. While MP-PCR usually resulted in species-specific bands, AMP-PCR also detected intraspecific polymorphisms (Hüttel, 1996). Consequently, both MP-PCR and AMP-PCR were used for genome mapping in chickpea (Ratnaparkhe *et al.*, 1998a; Winter *et al.*, 2000). However, MP-PCR is only efficient for mapping in inter-species crosses, whereas AMP-PCR can also be used in intra-species crosses of the cultigen.

RAMP (i.e. the combination of random 10mers with SSR-primers) detected only few intraspecific polymorphisms, though a series of RAPD primers were tested in combination with several SSR-primers. In its present form, RAMP cannot be recommended for genomes with a low level of polymorphism such as chickpea.

The RAMPO technique starts with DNA amplification using random 10mer primers as the RAPD technique (Williams *et al.*, 1990). The amplification products are then hybridised to different radiolabelled SSR motifs, resulting in reproducible, probe-dependent fingerprints, which are often completely different from the ethidium bromide stained bands, and are polymorphic at an intraspecific level (Ramser *et al.*, 1997). RAMPO bands are probably derived from the many minor RAPD or MP-PCR reaction products below the ethidium bromide detection level. Due to the high abundance and ubiquitous presence of SSRs in eukaryotic genomes, sensitive hybridisation with microsatellite motifs visualises a subset of such minor amplification products. RAMPO may be used for mapping in chickpea. However, since the cloning of polymorphic RAMPO bands and their conversion into locus-specific markers is difficult, this technique has been superseded by better methods.

As a non-microsatellite based method, DAF is very similar to the more popular RAPD technique. It applies random primers of various, but short length for the amplification of genomic DNA segments. However, only 10 to 15 ng DNA are used in a PCR reaction as

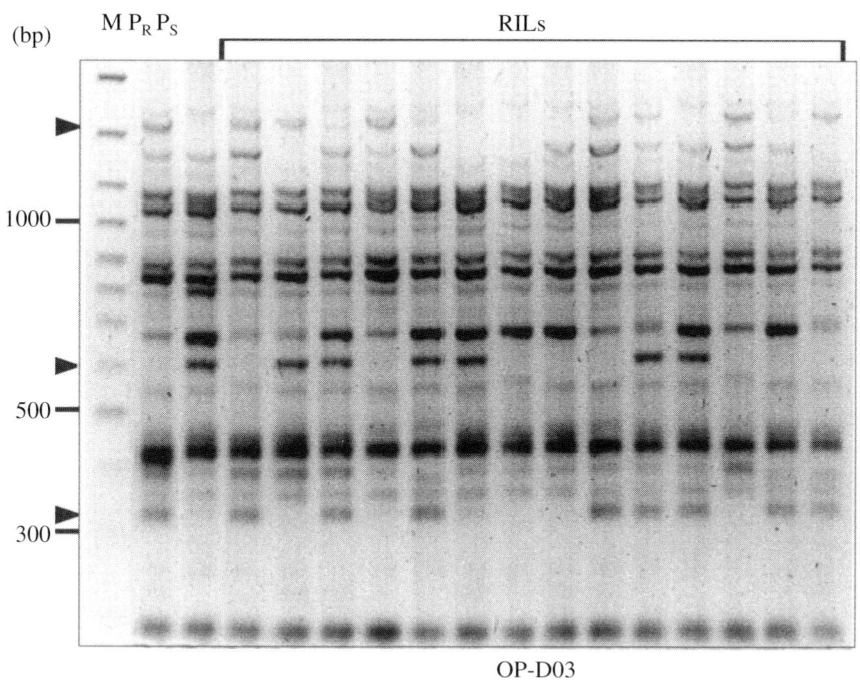

Figure 2. Segregation of DAF marker bands (indicated by arrowheads) in RILs from a cross between chickpea accession ILC4958 (P_R resistant to Fusarium *races 4 and 5) and* C. reticulatum *accession P.I.489777 (P_S susceptible) with the random 10mer primer OP-D03. The molecular weights of marker (M) fragments is given in basepairs.*

compared to the 50 to 100 ng required for RAPD, and up to 10 times higher primer concentrations are employed. Therefore, it is much more reproducible. DAF reveals a few polymorphisms in chickpea, though the amount of intra-species variation detected is much lower than inter-species polymorphisms (Fig. 2). Unexpectedly, 15mer primers were more efficient in generating bands than 8 mers, that detected polymorphisms only when AT-rich sequences were used.

The other non-microsatellite-based technique, AFLP, detects an RFLP-type of polymorphism and was more efficient than RFLP only for its high multiplexing ratio. Even in an inter-species cross, only 70 AFLP markers could be mapped (Winter *et al.*, 2000). Intra-species polymorphism was much lower than inter-species variation, so that inspite of the complex banding patterns, only one or two polymorphisms per reaction were found. Given the technical skills required for AFLP analysis, this technique can only be recommended for mapping in inter-species crosses.

SAMPL, on the other hand, combines the controllable multiplexing ratio of the AFLP technique with the higher level of SSR polymorphisms by utilising AFLP-type primers together with AMP-PCR primers. Therefore, it generates large number of polymorphic bands. Especially the combination of *Eco*RI primers of low selectivity (i.e. with no selective bases at the 3′-end) and primers directed against compound SSRs results in complex, highly polymorphic patterns, that are easy to interpret because of their clarity. Though SAMPL is technically demanding, it is the dominant, SSR-based marker technique of choice for chickpea, speeding up genome mapping tremendously.

4.2. LOCUS-SPECIFIC AMPLIFICATION OF MICROSATELLITES: STMS MARKERS

All described methods provide dominant markers, that can only rarely be transferred from one population to another, and do not reliably detect heterozygotes. However, highly polymorphic, co-dominant markers can be generated by locus-specific amplification of SSRs with specifically designed primers directed towards their flanking sequences (Litt and Luty, 1989; Smeets *et al.*, 1989; Tautz, 1989; Weber and May, 1989). The resulting amplification products often exhibit considerable length variation among different individuals or populations of the same species, mostly due to the variable number of tandem repeats within the SSR. These sequence-tagged microsatellite site (STMS: Beckmann and Soller, 1990) markers are the markers of choice for nearly every organism. They are single-locus, co-dominant, easy-to-use and reliable markers with high polymorphic information content possessing the potential for automated, non-radioactive detection.

In spite of the many advantages of STMS, some limitations have also to be considered: high costs for cloning, sequencing and primer synthesis. Further, standard protocols still use radioisotopes and sequencing gels to detect the amplified SSRs. Finally, the efficiency of primer generation suffers from a number of problems including redundancy of clones and occurrence of artificial chimeras.

In order to determine the most suited microsatellite motifs for the generation of STMS markers for chickpea, Hüttel *et al.* (1999) screened 13,000 colonies from two small insert libraries (250 to 400, and 400 to 600 bp, respectively) with a set of microsatellite

Table 3. Characterisation of microsatellite-containing clones from size-selected, randomly cloned chickpea DNA (Hüttel et al., *1999)*

Probe	Clones identified	Repeat motif	Number sequences	Number repeat units	Number perfect repeats
A/T-pool	27	GAA	1	9	1
		CAA	3	4–7	3
		CAT	1	5	1
G/C-pool	11	CAG	2	4–5	2
$[GA]_8$	26	GA	13	9–32	10
$[GT]_8$	18	GT	5	4–42	3
[TAA]	39	TAA	13	5–54	10
		AT	5	9–20	3

motifs (Table 3). These identified a total of 121 positive colonies, of which 51 were sequenced. [TAA] repeats were most abundant in chickpea. Since TAA arrays were often quite long and mostly perfect, this repeat type together with GA and GAA motifs offer themselves for the generation of large numbers of STMS markers in chickpea (Hüttel *et al.*, 1999).

This pilot study led Winter *et al.* (1999) to screen a much larger genomic library (280,000 colonies), representing about 18% of the chickpea genome, for clones containing $[GA]_n$, $[GAA]_n$ and $[TAA]_n$-motifs, the most abundant SSRs. A total of 389 positive colonies were sequenced, of which 75% contained perfect repeats. Interrupted, compound and interrupted-compound repeats were present from 6 to 9% only. Again, fragments isolated by the [TAA] probe contained the longest repeats with unit numbers from 9 to 131. For 218 loci, primers could be designed and used for the detection of length polymorphisms among eight *Cicer* genotypes, including six relevant chickpea breeding cultivars and *C. reticulatum* and *C. echinospermum*. A total of 174 primer pairs gave interpretable banding patterns, 137 (79%) of which revealed at least two alleles on native polyacrylamide gels.

5. Genome mapping in chickpea

The availability of DNA marker maps for the genomes of many crops facilitated marker-assisted selection, positional cloning of resistance genes, and mapping of quantitative trait loci (QTLs: Paterson *et al.*, 1988; Young, 1996) of agronomic interest (review in Winter and Kahl, 1995). In spite of all progress, chickpea is still awaiting a DNA marker map of sufficient density for use in marker-assisted selection and cloning of important genes. Formerly reported linkages between physiological and morphological characters were at the dawn of a genetic chickpea map (review in Muehlbauer and Singh, 1987). However, expression of these marker types often depend on ever-changing environmental influences, and thus are not suitable for routine use (Winter and Kahl, 1995).

Subsequent mapping efforts by Gaur and Slinkard (1990a, b) focused on biochemical markers such as isozymes. As in former studies (Oram *et al.*, 1987; Tuwafe *et al.*, 1988; Kazan and Muehlbauer, 1989), the monotony of the chickpea genome was reflected by monomorphic isozyme patterns. For example, 31 isozyme loci were invariant between chickpea accessions, but 28 were polymorphic within *C. reticulatum*. Only between-species polymorphisms were abundant enough to allow their mapping, forcing to work on F_2 populations from inter-species crosses rather than crosses between chickpea accessions (Gaur and Slinkard, 1990a). In a first attempt to generate an isozyme map for chickpea, Gaur and Slinkard (1990a) studied the inheritance of 16 isozyme loci in a large number of accessions and wild *Cicer* species. Of a total of 16 loci, 12 could be assigned to 4 linkage groups, and several of these linkages were conserved between the *Cicereae* (*Cicer*) and *Vicieae* (*Pisum* and *Lens*). In a follow-up study, the number of isozyme loci were increased by 14, and 3 morphological traits could also be mapped, resulting in a map of 7 linkage groups, made up of a total of 29 markers (Gaur and Slinkard, 1990b). The largest group contained 9 markers, whereas the two smallest groups were made up of 2 loci each. The map spanned around 200 cM (Gaur and Slinkard, 1990b).

Subsequent mapping by Kazan *et al.* (1993), again applying F_2 populations from inter-specific crosses, detected linkages of morphological and isozyme markers in chickpea, that had previously been reported from pea, indicating at least partial co-linearity between the two genomes. Nevertheless, not all linkages reported from pea could be confirmed in chickpea. In the resulting map, 5 morphological and 23 isozyme markers formed 8 linkage groups, covering a total of 257 cM.

With the work of Simon and Muehlbauer (1997), DNA marker technology entered the chickpea mapping stage. These authors used the populations of Gaur and Slinkard (1990a, b) and Kazan *et al.* (1993) for the mapping of morphological, isozyme, RFLP and RAPD markers. Random cDNAs and genomic sequences from lentil served as RFLP probes. Whereas the cDNAs performed reasonably well, the random genomic sequences hybridised only poorly to chickpea DNA. As already true for isozymes, the DNA markers detected hardly any polymorphism between chickpea accessions. For example, with 26 RFLP probes only monomorphic bands were found between the 4 chickpea parental lines, whereas 10 probes detected polymorphisms between the parents of the inter-species crosses. The level of polymorphism between chickpea accessions revealed by RAPD analysis was as low as RFLP and isozyme variation. However, a reasonable degree of polymorphism between the parents of the inter-species crosses was detected. Astonishingly, the RAPD banding patterns of different *C. reticulatum* individuals (accession P.I.489777) used as parental lines in all three inter-species crosses were polymorphic, indicating a high degree of intra-accession variation within this line. Nevertheless, the invariability of the bands derived from chickpea allowed to integrate the results from the three mapping populations into a single map, because the marker order was almost similar in the three crosses. Differences between the different maps mainly concerned the genetic distances between the markers. The resulting integrated map of 550 cM consisted of 9 morphological, 27 isozyme, 10 RFLP and 45 RAPD markers organised in 10 linkage groups. At several loci the segregation of parental alleles deviated slightly from the expected Mendelian ratio. Five regions of marker co-linearity between the genetic maps of chickpea and pea were detected.

5.1. LANDMARK MAPPING WITH STMS AND EXTENDED MAPS

The type of markers and also populations used in these mapping procedures precluded the generation of extended genetic maps of the chickpea genome for a long time. Co-dominant markers as isozyme and RFLPs detected only low levels of polymorphism and are nearly useless in intra-species crosses. Therefore, their application in normal breeding programs, that mostly rely on intra-species hybridisation, is prohibitive. RAPD markers were only slightly polymorphic, and, as dominant markers, cannot differentiate between homo- and heterozygous allelic states. All the more, populations were either in the F_2 or F_3 generation and still segregated for most traits. Therefore, the introduction of phenotypic characters into maps and the efficient use of dominant markers were hardly possible.

Most of these drawbacks could be overcome with STMS markers and a population of recombinant inbred lines (Winter *et al.*, 1999). Recombinant inbred lines (RILs) are generated by single seed descent from the F_2 to the F_6 generation and beyond, until their genomes can be considered virtually homozygous (Burr *et al.*, 1988; Lister and Dean, 1993). From that point on, the RILs can be increased to provide sufficient seed for evaluation of many traits in different environments. RILs also have the advantage that, contrary to F_2 populations, dominant and co-dominant markers have similar information content. Therefore, the integration of economic dominant-markers such as RAPDs, DAFs, AFLPs and ISSRs into a co-dominant STMS framework is possible.

Recently, more than 200 STMS markers were developed for chickpea (Hüttel *et al.*, 1999; Winter *et al.*, 1999), which are not only applicable in inter-species crosses, but – due

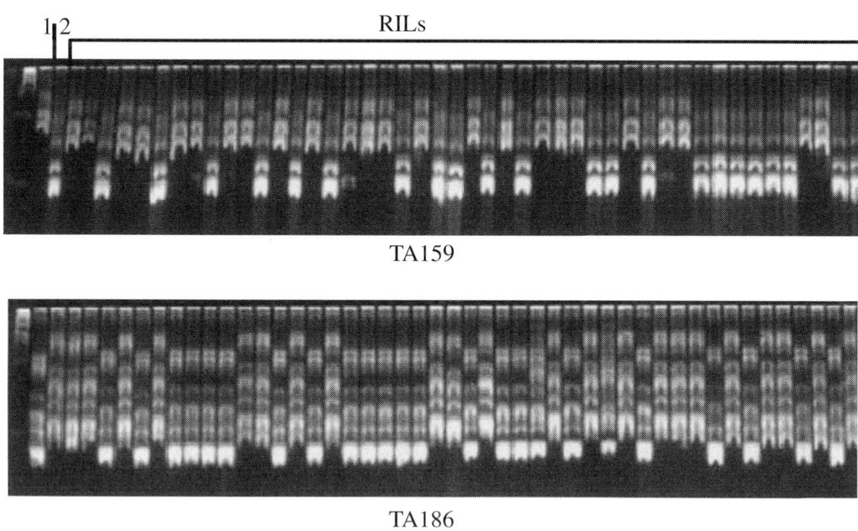

Figure 3. Segregation of STMS markers TA159 and TA186 in RILs from a cross between chickpea accession ILC4958 (1) and C. reticulatum *accession P.I.489777 (2). Note that all bands in an individual lane segregate similarly and are therefore derived from the same locus.*

to the variability of microsatellites – detect also polymorphisms between closely related chickpea accessions normally used by breeders, and are therefore ideal tools for marker-assisted selection. The suitability of STMS markers for genome mapping in chickpea was proven by Winter *et al.* (1999), who studied the segregation of 120 STMS in a population of 90 RILs from an inter-species cross between *C. reticulatum* accession P.I.489777 and

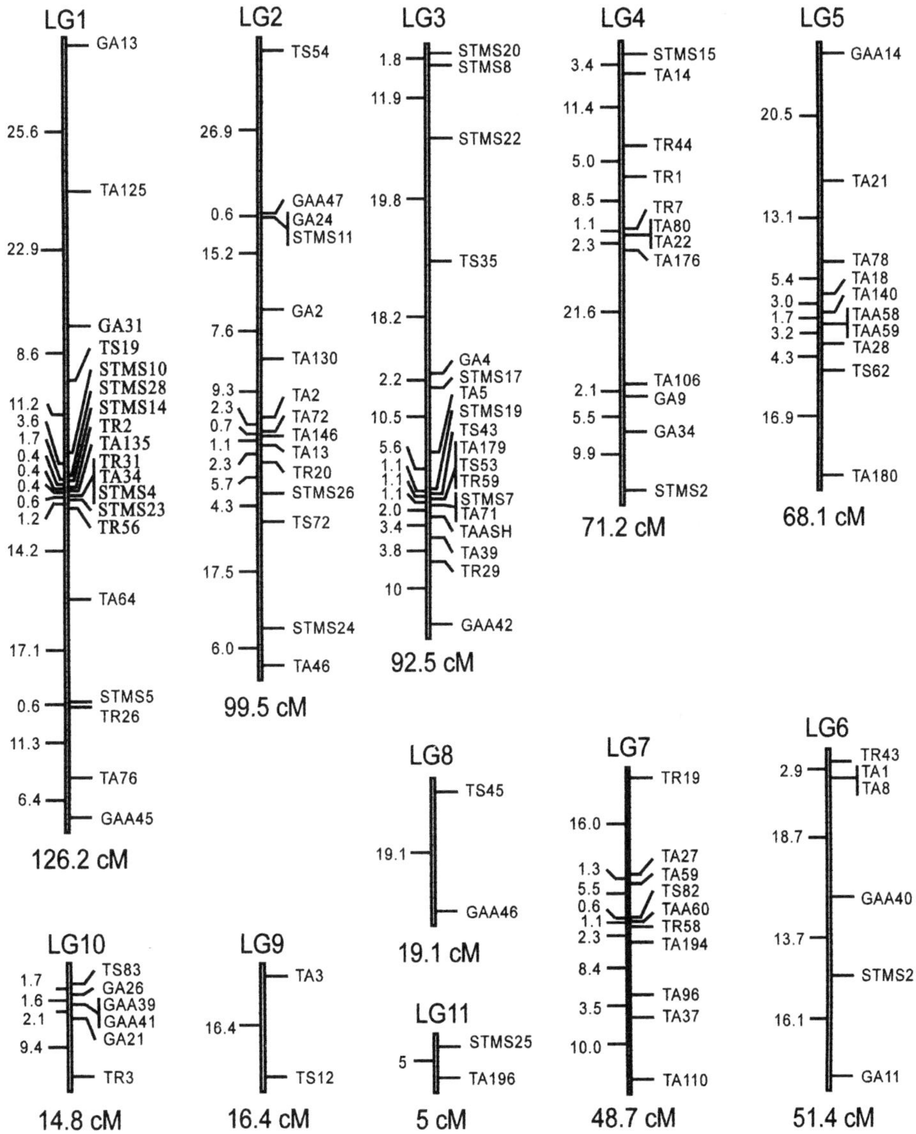

Figure 4. STMS marker map of the chickpea genome (adopted from Winter et al., *1999).*

the cultivar ICC4958. The resulting first co-dominant DNA marker map contained 112 markers in 11 linkage groups covering 613 cM. Clustering as well as random distribution of loci was observed. Segregation of 46 markers (39%) deviated significantly ($P > 0.05$) from the expected 1 : 1 ratio, and the majority of these loci (73%) was located in 3 distinct genomic regions. This STMS marker map represented the first landmark map of the chickpea genome (Winter *et al.*, 1999).

Building on this skeleton map, we used 130 RILs from the above cross to extend the map of Winter *et al.* (1999). Since the parental line ICC4958 is resistant to races 0, 4, and 5 of *Fusarium oxysporum*, and *Cicer reticulatum* accession P.I.489777 is susceptible, the segregation of the underlying resistance loci could also be followed. Beside the resistance loci, 351 other markers were mapped. At a LOD-score of 4.0, 303 markers covered 2077.9 cM in 8 large and 8 small linkage groups at an average distance of 6.8 cM between markers. Fifty-one markers (14.4%) were unlinked. The large linkage groups probably represented the 8 chickpea chromosomes. Clustering of markers in central regions of linkage groups was observed. Markers of the same class except ISSR and RAPD markers tended to generate subclusters. Resistance genes against *Fusarium* wilt races 4 and 5 mapped to the same linkage group together with several types of markers, including STMS and an ASAP marker, previously shown to be linked to *Fusarium* wilt race 1 (Ratnaparkhe *et al.*, 1998a). Clustering of wilt resistance genes around this locus is therefore possible. Resistance against *Fusarium* race 0 could not be placed on any linkage group. Significant deviation from the expected 1 : 1 segregation ratio was observed for 136 markers (38.4%, $P < 0.05$), and segregation was biased towards the wild progenitor in 68% of the cases. Segregation distortion was similar for all marker types except ISSRs, that showed only 28.5% aberrant segregation. The most extended genetic map of chickpea currently available serves for marker-assisted selection and map-based cloning of *Fusarium* wilt resistance genes and other agronomically important genes in future.

6. Mapping of resistance loci: towards map-based cloning of resistance genes

Many factors prevent increases in chickpea yield such as e.g. inadequate amounts of fertilisers and pesticides, and some traditional methods of land preparation and seeding. For example, in the Mediterranean basin and adjacent WANA countries, chickpea yields could be increased for 50 to 100% by switching from the traditional spring to winter sowing. This would extend the vegetative and reproductive growth, which is additionally fuelled by winter rains in these areas. However, in spite of the potential advantages of changing sowing habits, farmers stick to their traditional sowing practices to avert risks associated with environmental uncertainties. The greatest risk in this respect is foliar blight disease caused by *Ascochyta rabiei* (Pass) Labr., that spreads rapidly under favourable winter conditions (>150 mm rainfall, around 20°C), and can reduce yield up to 100% (Nene and Reddy, 1987; Jiminéz-Diaz *et al.*, 1993; Acikgoz *et al.*, 1994). The best control of this disease would be to use elite cultivars combining cold tolerance with *Ascochyta* resistance (Nene and Sheila, 1992; Saxena, 1992; Singh *et al.*, 1992). Therefore, a large part (>15,000 accessions) of the world chickpea germplasm collection was screened for blight resistance, and 15 resistant lines were identified (Singh *et al.*, 1992) for the breeding of resistant elite cultivars. However, conventional breeding can hardly solve the blight problem, because the pathogen diverges extremely fast. For example, 6 races of the fungus have already been identified in Syria and Lebanon (Reddy and Kabbabeh, 1985), and another 12 pathotypes have been reported from northern India (Singh, 1985). Gowen (1983) found an isolate that was lethal to formerly resistant cultivars, indicating the emergence of new pathotypes. The situation becomes even more complicated, because the fungus can also occur in its perfect stage *Didymella* (formerly *Mycosphaerella*) *rabiei* (Kov.) V. Arx (Kovachevsky, 1936; Kaiser, 1973; review in Kaiser, 1992). Marker-assisted pyramiding of resistance genes (Robinson, 1976) and/or transformation with complementary resistance genes are presently the only effective techniques for durable resistance.

Whereas *Ascochyta rabiei* is the major constraint to yield increases in the WANA countries, chickpea productivity on the Indian subcontinent and other parts of the world is mainly restricted by *Fusarium* wilt caused by the soil-borne *Fusarium oxysporum* Schlecht. Emend. Snyd. and Hans f.sp. *ciceri* (Padwick) Snyd. and Hans. From 10 to 90% yield losses are a constant feature (Srivastava *et al.*, 1984; Jiminéz-Diaz *et al.*, 1993). The pathogen invades all tissues of the host. It is hard to control by normal crop rotation, because it survives for more than 6 years in the soil and invades lentils, pigeon peas, and peas as symptomless carriers. Like *A. rabiei*, *F. oxysporum* also appears in different races. Four races have been reported from India (races 1, 2, 3 and 4: Haware and Nene, 1982), three races from Spain (races 0, 1 and 5: Cabrera de la Colina *et al.*, 1985), and race 6 from California, USA (Phillips, 1988). However, most isolates from Mexico and USA probably belong to race 1 (Nene and Sheila, 1992). Like *Ascochyta* blight, *Fusarium* wilt would also be best controlled by resistant cultivars.

In addition to these major fungal pathogens, more than 70 other pathogens attack chickpea around the world, including several viruses, bacteria, fungi, nematodes and insects (reviews by Nene *et al.*, 1984; Nene and Reddy, 1987; Nene and Sheila, 1992).

6.1. GENETICS OF *ASCOCHYTA* BLIGHT RESISTANCE

The genetics of *Ascochyta* blight resistance has extensively been studied, because the disease is of great agronomic importance. However, the emerging picture is confusing (reviews in Singh *et al.*, 1992; Nene and Sheila, 1992; Singh and Reddy, 1993): depending on the fungal isolate and the cultivar, either one dominant (Hafiz and Ashraf, 1953; Vir *et al.*, 1975; Eser, 1976), one recessive and one dominant (Singh and Reddy, 1983; Tewari and Pandey, 1986), or one recessive (Tewari and Pandey, 1986) resistance gene was reported. Also, two complementary recessive (Kusmenoglu, 1990), or two complementary dominant genes (Dey and Singh, 1993) were detected. To complicate the picture even more, other genes may modify the expression of resistance (Muehlbauer and Singh, 1987). At present it is not clear, whether the reported genes represent the same or different loci, because allelic tests were not performed.

Another reason for the apparent confusion is the use of different methods for assessment of disease severity. Moreover, inheritance of resistance was studied in F_2 or backcross populations grown for one single year at only one location. Also, in most cases natural inocula from local sources were used in field tests. As has been shown by oligonucleotide fingerprinting of isolates from a single chickpea field in Tunisia, several different genotypes of the fungus occur even in the same lesion (Morjane *et al.*, 1994). These could comprise different pathotypes or races, which would confuse the results. Since the severity and spread of the disease is highly dependent on environmental conditions (which may change from year to year), testing for disease severity under controlled glasshouse or growth chamber conditions (Singh *et al.*, 1992) would very much help to get reliable and meaningful data.

Only recently, Tekeoglu *et al.* (2001) provided nearly optimal conditions to elucidate the nature of *Ascochyta* blight resistance. Three different RIL populations derived from one inter-specific cross between *C. arietinum* FLIP 84-92C (resistant parent) and *C. reticulatum* P.I.599072 (susceptible parent), and two intra-species crosses of the cultigen were tested in the field for three subsequent years using natural inoculum. The reaction of the lines towards the fungus was scored on the 1 to 9 scale for disease severity developed by ICARDA (Singh *et al.*, 1992), and the area under the disease progress curve (AUDPC: Campbell and Madden, 1990) estimated, which gave reproducible results. This study defined two to three major recessive and complementing resistance genes together with several modifiers. However, the third gene could only be detected in one year (Tekeoglu *et al.*, 2000). The two major genes together accounted for 50.3 and 45.0%, respectively, of the estimated phenotypic variation in two subsequent years as shown by QTL mapping (Santra *et al.*, 2001). Absence of either one or two major genes led to susceptibility, whereas the modifiers determined the degree of resistance (Tekeoglu *et al.*, 2001).

Resistances against Ascochytoses caused by *Ascochyta lentis* in lentil and *A. pisi* in pea, respectively, are also governed by polygenes (Paterson *et al.* 1988) or quantitative trait loci (QTLs, reviews by Young, 1996; Geiger and Heun, 1989). However, resistance in chickpea more closely resembles resistance in pea, where 3 genes confer 71% of the variance (Dirlewanger *et al.*, 1994). In lentil, a single locus accounted for almost 90% of the phenotypic variation between resistant and susceptible individuals of an F_3 population (Ford *et al.*, 1999).

Though now identified, the nature of the resistance genes is still elusive. In fact, Höhl *et al.* (1990) identified a hypersensitive response (HR: Lamb *et al.*, 1989; Dixon and Lamb, 1990) in resistant but not in susceptible chickpea plants after infection with *Ascochyta* spores, which was accompanied by a rapid oxidative burst (Otte and Barz, 1996) and other fast defence reactions (Barz *et al.*, 1993). These observations suggest that the *Ascochyta* resistance genes of chickpea belong to a specific class of genes encoding receptor-like protein kinases sharing homologies, and serving as receptors for pathogen-derived avirulence factors. Whether similar genes confer resistances against *A. rabiei* in chickpea will be revealed via map-based cloning. The work of Tekeoglu *et al.* (2001) will certainly provide a basis for such experiments.

6.2. MAPPING OF *ASCOCHYTA* BLIGHT RESISTANCE GENES

A major prerequisite for the mapping of resistance genes is a large population of individuals reliably rated for resistance against the pathogen, and genetically diverse enough for the efficient application of DNA and other markers for linkage analysis. The intra-species population of RILs, rated by Tekeoglu *et al.* (2000) and described above, satisfies these requirements. The population comprising 142 F5:8 RILs, is large enough, and also

Table 4. Linkage maps of the chickpea genome: a summary

Reference	Populations	Markers	Linkage groups	Map size (cM)
Gaur and Slinkard (1990a, b)	F_2 populations from inter- and intra-species crosses between five accessions of *C. arietinum*, six accessions of *C. reticulatum* and one accession of *C. echinospermum*	3 morphological 26 isozymes Total: 29	7	~200
Kazan *et al.* (1993)	Eight F_2 populations from inter-species crosses between two *C. arietinum* varieties (desi and kabuli) and *C. reticulatum* and *C. echinospermum*	5 morphological 23 isozymes Total: 28	8	257
Simon and Muehlbauer (1997)	Three F_2 and one F_3 populations from inter-species crosses between *C. arietinum* × *C. reticulatum*	9 morphological 27 isozymes 10 RFLP 45 RAPD Total: 91	10	550
Winter *et al.* (1999)	Interspecific cross of *C. arietinum* × *C. reticulatum* (ICC4958 × P.I.489777) with 131 RILs	120 STMS	11	613
Winter *et al.* (2000)	Same as in Winter *et al.* (1999)	118 STMS 96 DAF 70 FLP 37 ISSR 17 RAPD 8 isozymes 3 cDNAs 3 *Fusarium* resistance loci Total: 354	16	2078
Santra *et al.* (2001)	Inter-species cross of *C. arietinum* × *C. reticulatum* (FLIP 84-92C × P.I.599072) with 142 RILs	111 RAPD 21 ISSR 11 isozymes 1 morphological 3 QTLs for *Ascochyta* blight resistance Total: 146	9	981.6

genetically diverse, because it descended from a cross between the cultigen and an accession of *C. reticulatum*. This population was exploited by Santra *et al.* (2001) for the generation of a genetic map of 116 markers covering 981.6 cM in 9 linkage groups and for mapping of *A. rabiei* resistance loci. Interval mapping revealed two QTLs, QTL-1 and QTL-2, with LOD scores of 17.2 and 7.3 located on linkage group 6 and 1 of the maps of Gaur and Slinkard (1990b) and Kazan *et al.* (1993), respectively. Individually, the two loci accounted for 42.5% and 19.9% of the resistance against *A. rabiei*. Two RAPD markers (UBC733b and UBC181a) flanked QTL-1 and were 10.9 cM apart from each other, whereas QTL-2 was flanked by an ISSR and an isozyme marker 5.9 cM apart. The two major QTLs were detected using resistance scoring data from two subsequent years. However, a third QTL-3 could only be detected using the data from one year. This QTL was located on linkage group 4 at a LOD score of 3.04 – slightly above the threshold of acceptance – and was flanked by a RAPD and an ISSR marker at a distance of 11.7 cM (Santra *et al.*, 2001).

Though the map established by the Pullman group (Santra *et al.*, 2001; Tekeoglu *et al.*, 2001) provides a firm basis from which attempts to clone the resistance genes can start, the density of markers is still too low for map-based cloning. Further, the positions of the resistance genes on the most extended genomic map of chickpea was still unknown.

However, Rakshit *et al.* (in preparation) used 94 RILs from the cross of Santra *et al.* (2001) to identify more, and more closely linked, markers for the major resistance locus QTL-1, applying DAF in combination with bulked segregant analysis (BSA: Michelmore *et al.*, 1991). Of 312 random 10 mers, three produced five polymorphic bands between the bulks. Two of these markers, OPS06-1 and OPS03-1 were linked at LOD-scores above 5 to markers UBC733B and UBC181a, flanking the major *Ascochyta* resistance locus. OPS06-1 was located at distances of 4.4 cM from marker UBC733B, and 9.5 cM from UBC181a, between the two loci, and is probably the most closely linked marker for the major *A. rabiei* resistance locus available to date. OPS03-1 mapped about 25 cM away from UBC181a on one flank of the resistance locus, whereas a third DAF marker, OPK06-5, was located on the other flank at the distal end of the linkage group around 30 cM away from UBC733b.

For the localisation of QTL-1 on the map of Winter *et al.* (2000), OPS03-1 was transferred to the population on which this map is based. It mapped to linkage group 4, which identifies this linkage group as linkage group 6 of Gaur and Slinkard (1990a, b) and Kazan *et al.* (1993). STMS markers located on linkage group 4 were transferred to the population segregating for *Ascochyta* resistance. Three of these markers were closely linked to the major resistance locus. Twelve out of 14 STMS markers could be used in both populations. The order of STMS markers was essentially similar in both populations, but map distances between them differed between both populations. The availability of STMS markers for the major *A. rabiei* resistance locus will elucidate the complex resistance against different fungal races.

6.3. GENETICS OF *FUSARIUM* WILT RESISTANCE

Any research into the genetics of *Fusarium* resistance must take into account that different races of the fungus occur in different parts of a country (see above). So far only resistance to race 1, prevalent in India, has been studied in detail.

Earlier field studies on segregating populations suggested a single recessive gene to be sufficient for resistance (Haware and Nene, 1982; Sindhu *et al.*, 1983). However, in one of these crosses segregation of resistance deviated from the expected ratio, which makes more than one gene necessary to obtain resistant offspring (Kumar and Haware, 1982). In subsequent studies, Upadhyaya *et al.* (1983a, b) observed differences in the time-course of wilting in two susceptible cultivars, which was governed by a single recessive gene. In these studies, the early wilting phenotype was partially dominant over

Table 5. Genetic stocks for resistance to Fusarium *race 1, their geno- and phenotypes (van Rheenen* et al.*, 1992)*

Chickpea line	Genotype	Phenotype
WR 315	$h_1h_1\ h_2h_2\ h_3h_3$	resistant
CPS 1	$h_1h_1\ h_2h_2\ h_3h_3$	resistant
H 208	$H_1H_1\ H_2H_2\ H_3H_3$	late wilting
C 104	$H_1H_1\ h_2h_2\ h_3h_3$	late wilting
K 850	$h_1h_1\ H_2H_2\ h_3h_3$	late wilting
JG 62	$H_1H_1\ H_2H_2\ h_3h_3$	early wilting

the late wilting character. After crossing two late wilting lines, Singh *et al.* (1987) observed a resistant individual in the offspring for which two different genes for late wilting were postulated, complementing each other. The resistant individual was thought to be homozygous for both of them. More crosses between early and late wilters and resistant lines detected at least three resistance loci, H_1, H_2 and H_3. Either two of the partially recessive loci H_1 and H_2, respectively, or the dominant locus H_3 confer the late wilting phenotype, whereas a combination of three loci is necessary for complete resistance (Upadhyaya *et al.*, 1983b; Singh *et al.*, 1987; van Rheenen *et al.*, 1992). The genetic repertoire for wilt resistance and carriers of the respective genes are listed in Table 5.

6.4. MAPPING OF *FUSARIUM* WILT RESISTANCE GENES

In a first attempt to tag *Fusarium* resistance loci with DNA markers, an intra-species cross between resistant accession WR315 and late wilting line C104 – differing in only the H_1 resistance locus for *Fusarium* race 1 – was used to map this locus with RAPD markers (Mayer *et al.*, 1997). This study identified two markers (UBC170 and CS27) linked to the resistance locus at 7% recombination. About 6% recombination was observed between the markers themselves. Since RAPD banding patterns and especially that of the UBC170 marker were ambiguous (a second faint band co-migrated with the linked marker band $UBC170_{550}$), allele-specific associate primer pairs (ASAPs: Weeden *et al.*, 1992; Yu *et al.*, 1995) were generated by sequencing the marker band and designing primers against the ends of the fragment. Interestingly, the $UBC170_{550}$ fragment contained a $(CA)_{14}$ microsatellite repeat at one end. However, only the ASAP generated from CS27 worked well, whereas the $UBC170_{550}$ ASAP produced similar bands from resistant and susceptible lines. Linkage relationships to other markers suggested, that the two markers were located on the same side of the resistance locus rather than flanking it (Mayer *et al.*, 1997). These linkages also allowed to localise the resistance locus on linkage group 6 of the map of Simon and Muehlbauer (1997). Consequently, this linkage group is the same as linkage group 2 of the map of Winter *et al.* (2000).

Applying 5′ and 3′ anchored AMP-PCR markers to 131 F_6-derived F_7 RILs from an intra-species cross between a cultivated chickpea line resistant to races 4, 5 and 0 (ICC4958) and the susceptible *C. reticulatum* accession P.I.489777, Ratnaparkhe *et al.*

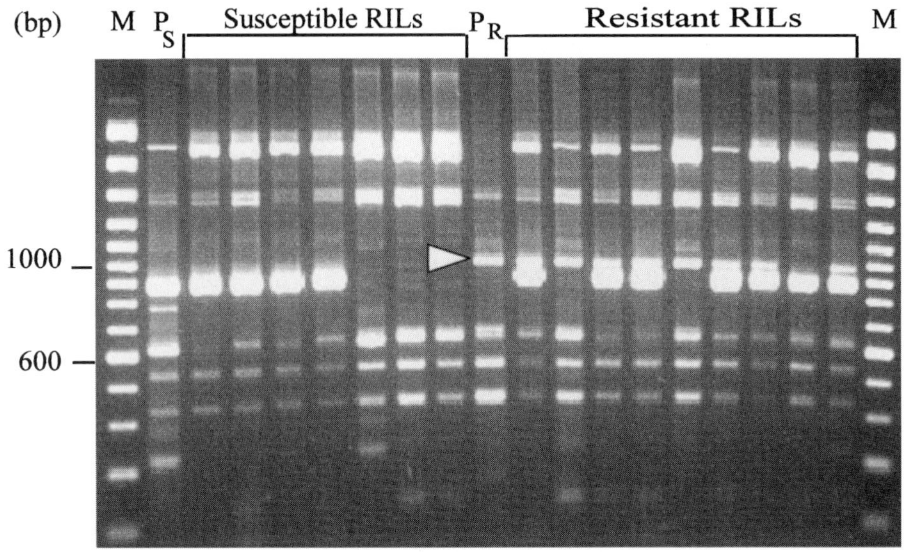

Figure 5. Segregation of DAF marker R260-09-1 in RILs, susceptible or resistant to Fusarium *races 4 and 5, respectively, from a cross between chickpea accession ILC4958 (PR, resistant to* Fusarium *races 4 and 5) and* C. reticulatum *accession P.I.489777 (PS, susceptible). The marker, though closely linked to the locus for resistance to* Fusarium *race 4, nevertheless shows some recombination as indicated by the lack of the marker band at 1000 bp in the resistant RIL second from right.*

(1998a) demonstrated linkage between marker $UBC855_{550}$ (derived from the 3′ anchored AMP-PCR primer $(AC)_8YT$) and the *Fusarium* race 4 resistance locus at a distance of 5.2 cM. This marker was coupled at 0.6 cM to marker $CS27_{700}$ of Mayer *et al.* (1997) (closely linked to the race 1 resistance gene), indicating a clustering of resistance genes for races 1 and 4. Since both linked markers contained $(AC)_n$ microsatellites, Ratnaparkhe *et al.* (1998b) performed an extended study with $(AC)_n$-containing primers for AMP-PCR. Indeed, primer UBC825 [sequence $(AC)_8T$] produced a band ($UBC825_{1200}$), which was linked to race 4 resistance locus at a distance of 5.0 cM and thus was closer to the resistance locus than the previous markers. Another $(AC)_n$-containing primer (UBC857, sequence $(AC)_8YG$) produced marker $UBC857_{800}$, linked to $UBC855_{500}$ at a distance of 5.5 cM and also to the race 4 resistance locus. Primer UBC860RG, being almost complementary to UBC857, produced marker $UBC860_{600}$, which was located on the other side of the resistance locus, but did not co-segregate with the UBC857-derived marker.

In this study, the resistance locus for race 5 mapped to the same linkage group as race 1 and 4 resistance genes, 41 cM away from the race 4 resistance locus. Such co-localisation was expected, since Tullu (1996) already demonstrated close linkage between resistance genes for races 1, 2, 4 and 5 in a narrow cross between chickpea accessions WR315 and C104. Hence, also the resistance gene for race 2 occurs in close vicinity to the genes for races 1, 4 and 5, though this linkage needs to be confirmed with molecular markers (see Fig. 6).

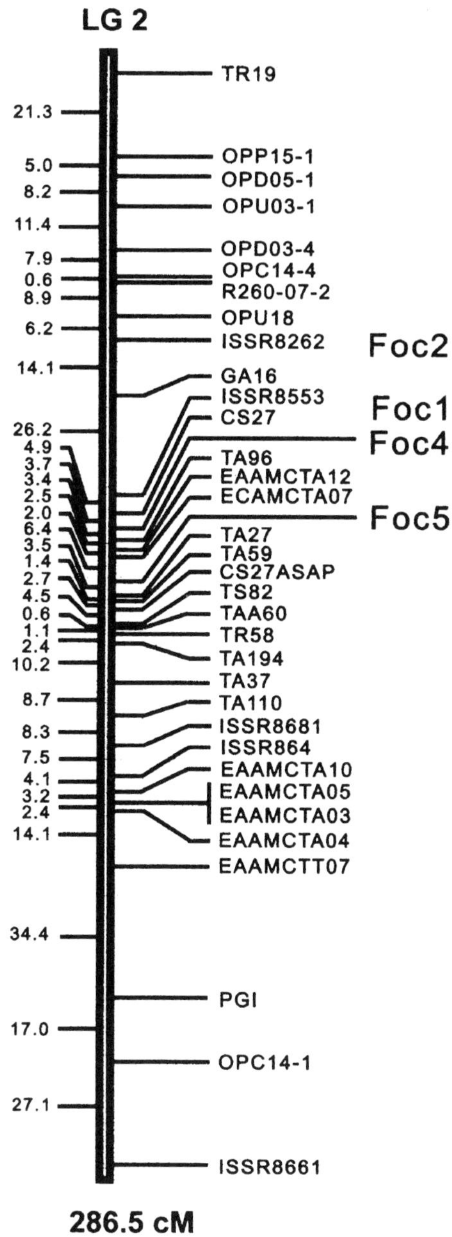

Figure 6. Genetic map around the Fusarium *resistance gene cluster on linkage group 2 of the map of Winter* et al. *(2000). The positions of resistances against races 1 and 2 of the fungus originate from Tullu (1966).*

The marker density in and around the *Fusarium* resistance gene cluster is too low to allow the map-based cloning of resistance genes a situation, similar to the *Ascochyta* resistance genes. Therefore, a set of 432 different arbitrary 10mers was tested with DAF in combination with BSA using 94 RILs from the inter-species cross ICC4958 × *C. reticulatum*. After several screening rounds 19 markers were located on the same linkage group as the resistance genes. The most closely linked markers were R-260-9-1 and OP-U17-1, flanking the Foc-4 locus on either side at map distances of 2.0 and 4.1 cM, respectively (Fig. 5). Chi-square analysis revealed, that 10 out of 19 loci deviated significantly ($P = 0.05$) from the expected Mendelian segregation ratio of 1 : 1. Clustering of distorted loci was most pronounced in the neighbourhood of Foc-4, which also showed distorted segregation. The generation of single copy markers for marker-assisted selection and map-based cloning of the Foc-4 gene started with the sequencing of marker bands of the five most tightly linked loci. Five out of seven clones showed significant homology to already identified genes, some of them related to putative resistance factors. For six of the cloned DAF amplicons ASAP primer pairs were designed. Of these, 4 generated a single band monomorphic between the parental lines. The other two primer pairs, OP-M20-1/2 and OP-P15-3/2, generated polymorphic bands between the parents and resistant and susceptible bulks.

7. Marker-assisted breeding: towards application of marker technology

All published genetic maps of the chickpea genome are based on inter-species crosses between the cultigen and the wild relative *C. reticulatum*. Also, mapping of resistance loci for chickpea's most important fungal pathogens relied on such crosses. These restrictions were dictated by the low level of genetic polymorphism between chickpea accessions. However, chickpea breeding normally applies intra-specific crosses for crop improvement. The present marker technologies can deal with the monotony of the chickpea genome and support routine breeding work with simple, fast and cheap allelic tests. Especially the STMS marker technology is best suited since (i) STMS loci are sufficiently variable to detect differences even between closely related chickpea accessions and (ii) the technique has the potential for high-throughput screening and automation.

However, the successful application of marker technologies requires:

(1) a dense intra-species map linking traits of agronomic importance such as yield and flowering time (Or *et al.*, 1999; Kumar and van Rheenen, 2000), tolerances for the most important stresses such as cold, drought and also pest resistances to highly polymorphic, co-dominant markers in sufficiently close vicinity to allow marker-assisted selection also in offspring from intra-species crosses. Such an intra-species map of chickpea will be available soon (J. Kumar, ICRISAT, personal communication).
(2) tailoring of STMS markers and reactions in combination with high-throughput screening techniques to speed up and simplify the application of marker technology. Successful tailoring of STMS markers for multiplex PCR and electrophoresis has been reported for soybean, where up to 8 STMS markers can be amplified in a single PCR reaction and separated in a single lane of a sequencing gel. More than 70 of the tailored

markers cover the whole genome (Narvel *et al.*, 2000). However, this advancement built on the availability of more than 600 STMS markers for soybean (Cregan *et al.*, 1999). In chickpea, the number of STMS markers is still too low for such applications.

7.1. TRANSFERABILITY OF STMS MARKERS BETWEEN LEGUME SPECIES

The major limitation to a broader application of STMS technology is the species-specificity of the markers. Unless sufficient database information is available, microsatellites have to be cloned and sequenced for every species under study, which is laborious and cumbersome. The situation would be greatly improved, if primer binding sites and microsatellite loci were conserved among different taxa. The need for marker-assisted exploitation of the primary and secondary gene pool of crops for quality improvement has triggered research aiming at transfer of STMS markers from one species to another. In general, the extent of STMS marker transferability between species depends on the evolutionary rate of the microsatellite-flanking sequences as well as on the microsatellites themselves. However, the speed of evolution is highly variable, and depends on both the particular locus and the taxon. Systematic surveys of STMS marker transferability in large plant families revealed that the majority of markers work well within the genus of origin and closely related taxa, but less so in less related species and even worse or not at all in species from other genera (Gupta and Varshney, 2000). Hence, MSs and their flanking regions appear to be less conserved in plants than in animals, and the transferability of STMS markers across genus borders is quite limited. An improvement of transferability of STMS marker to divergent plant taxa may be to tap cDNA and EST libraries for marker generation, because SSRs and their flanking sequences, if present in exons, are likely to be more conserved. Recent studies with STMS markers from EST libraries from grape (Scott *et al.*, 2000) and pea suggest that this may indeed be the case.

We wanted to explore whether and to what extent the 200 STMS from chickpea could be transferred to species within the genus *Cicer*. Therefore, the DNAs of five accessions each of the eight annual *Cicer* species and one accession of *C. anatolicum* were amplified with primers derived from 90 SSR-flanking sequences from chickpea. In general, the primers successfully amplified sequences in the related species, which suggested good conservation of most SSR-flanking sequences among chickpea's relatives. Size and number of amplification products varied considerably between and within species. The extent of primer site conservation in the various *Cicer* species was in accordance with their known phylogenetic relationship(s) to chickpea, and ranged from 92.2% in *C. reticulatum*, chickpea's closest relative and potential ancestor, to 50% in *C. cuneatum*. A phylogenetic tree, based on presence or absence (null-alleles) of amplification products revealed a close relationship between cultivated chickpea and its crossability group on one, and the perennial *C. anatolicum* on the other hand. Both groups were more closely related to each other than to other annual species of the genus.

Null-alleles, generated by mutations in SSR-flanking sequences, may be more reliable for phylogenetic studies across and even within species than the actual size variation of an SSR locus, because allele size homoplasy – i.e. loci of similar size do not share the same sequence – has often been encountered. Size homoplasy is not only a problem in nuclear SSRs, but has also been detected in chloroplast MSs of wild perennial relatives of soybean (Doyle *et al.*, 1998). Sequence comparisons of highly divergent amplicons from the genomes of chickpea and its relatives revealed that variations are either caused by large differences in the number of SSR repeat units (as with *C. reticulatum*), or by the amplification of a locus unrelated to the one amplified from chickpea DNA (other crossability groups). Sequence information altogether suggests that STMS markers derived from chickpea may efficiently and reliably be used for synteny studies and map-based exploitation of chickpea's primary gene pool (i.e. chickpea's crossability group including *C. anatolicum*). However, care should be taken when applying chickpea-derived STMS markers to more distant species, because in many cases loci will not be syntenic.

8. Regeneration and transformation of chickpea: a step towards improved chickpea in the field

Conventional breeding of grain legumes did not meet its objectives very successfully. *In vitro* techniques, however, offered novel avenues for crop improvement, and the two main routes were the exploitation of somaclonal variation and the production of transgenic plants. Progress in grain legume tissue culture generally, and genetic transformation in particular, has earlier been reviewed (Tabe *et al.*, 1993; Christou, 1994). However, since that time, remarkable advancements were made. Yet for several important grain legumes including chickpea, routine and reproducible high-frequency regeneration protocols are still not available (Singh *et al.*, 1998).

For chickpea, several regeneration regimes were developed, exploiting different regeneration events and explants. These can arbitrarily be grouped under (i) direct regeneration from explants with or without preexisting meristems, (ii) regeneration via a callus phase, (iii) regeneration through direct somatic embryogenesis or somatic embryogenesis after a callus phase, and (iv) regeneration from wounded explants (Singh *et al.*, 1998). Kartha *et al.* (1981) regenerated multiple shoots from short meristems cultured on MS medium with an optimal N^6-benzylaminopurine (BAP) concentration of 0.02–0.22 mg/L. Malik and Saxena (1992) used 0.22–11 mg/L thidiazuron (TDZ) for multiple shoot development from cotyledonary nodes and adjacent areas of whole seed explants. When immature chickpea cotyledons were cultured on B_5 medium supplemented with zeatin (1–10 mg/L) and indole-3-acetic acid (IAA, 0.04–0.4 mg/L), cotyledon-like structures (CLS) emerged from their proximal ends, some of which developed into shoots. Singh *et al.* (1998) reported similar results on CLS formation, although efforts to maintain CLS tissues in a regenerating state failed. Shoot regeneration was also possible from chickpea calli derived from different meristematic and non-meristematic explants. For instance, calli derived from shoot tips (Sharma *et al.*, 1979; Rao and Chopra, 1989a; Surya-Prakash *et al.*, 1992), sliced shoot tips (Singh *et al.*, 1982), immature embryonic axes, immature leaflets (Barna and Wakhlu, 1994) and cotyledons (Rao and Chopra, 1989a; Surya-Prakash *et al.*, 1992) were induced for shoot regeneration. Invariably, shoots regenerated from calli of explants with pre-existing meristems better than from differentiated tissue. As with most of the legumes, the employed cytokinin and auxin concentrations ranged between 0.5–3 mg/L and 0.1–1.0 mg/L, respectively. Rooting was generally induced by placing regenerated shoots on MS medium supplemented with 1 mg/L naphthalene acetic acid (NAA).

Somaclonal variation in callus cultures has also been exploited for chickpea improvement. Cell lines were selected for resistance to (i) abiotic stresses (Pandey and Ganapathy, 1984) and insecticides (endosulfan: Saxena and Beg, 1988; Seetaram *et al.*, 1991), or (ii) biotic stress by *Fusarium* culture filtrate (Singh *et al.*, 1998). However, only Rao and Naidu (1989) could regenerate plants after such *in vitro* selection.

Somatic embryogenesis (SE), either directly from the explant or after a variable callus phase was reported for a number of legumes including chickpea. Murthy *et al.* (1996) observed direct SE in kabuli chickpea seed explants in the presence of 10 mM L-proline along with 2.2 mg/L TDZ. Ramana *et al.* (1996) also reported direct SE in the cultivar Annigeri from cultured immature cotyledons on B_5 medium supplemented with 2,4,5-trichlorophenoxy acetic acid (2,4,5-T) or 2,4-dichlorophenoxy acetic acid (2,4-D) in combination with BAP or kinetin (KN). Direct SE is promising for transformation experiments and clonal propagation. Moreover, absence of a callus phase minimises the risk of undesirable genetic variation and also shorten the time to develop transformed plants. SE after a callus phase has been reported from a variety of explants from immature cotyledons, embryonic axes (Sagare *et al.*, 1993; Eapen and George, 1994), leaflets (Rao and Chopra, 1989b; Barna and Wakhlu, 1993; Kumar *et al.*, 1994, 1995a, b; Singh *et al.*, 1998) and young seedlings (Murthy *et al.*, 1996). In these studies, somatic embryos were induced on MS medium supplemented with various combinations/concentrations of growth hormones (e.g. 2,4,5-T, 2,4-D, picloram, NAA, indole acetic acid (IAA), BAP and putrescine). The somatic embryos were matured on basal MS medium or MS supplemented with 2,4,5-T, $AgNO_3$, 2,4-D, indole butyric acid (IBA) + BAP, and IAA + BAP. Germination of somatic embryos required various concentrations and combinations of zeatin, gibberellic acid (GA_3), abscisic acid (ABA), NAA, BAP, KN, B_5 vitamins added to MS or B_5 basal medium. The explant genotype also had a strong influence on SE induction and its further development (Barna and Wakhlu, 1993; Sagare *et al.*, 1993). Although SE induction frequencies are high (57–100% of explants cultured), only low numbers of embryos really matured. Moreover, somatic embryos from immature cotyledons and embryonic explants matured less frequently (3–10%) than those from leaflets (up to 45%). Dessication of fully developed somatic embryos improves their conversion (Barna and Wakhlu, 1993).

Regeneration from wound sites is necessary for *Agrobacterium* mediated-gene transfer. Fontana *et al.* (1993) observed direct adventitious shoot regeneration from the cut ends of decapitated embryonic axis cultured on 1 mg/L kinetin. Mostly single, occasionally multiple shoots were obtained from about 40 to 50% of the explants. Singh *et al.* (1998) recorded multiple shoot regeneration in MS + zeatin (1–2 mg/L) or TDZ (0.1–2 mg/L) only, though various combinations of BAP, kinetin and TDZ and auxins were tested. The addition of zeatin to the medium induced more frequent multiple shoot regeneration in comparison to TDZ. Recently, Polisetty *et al.* (1997), Kar *et al.* (1996) and Krishnamurthy *et al.* (2000) developed protocols for multiple shoot regeneration from decapitated embryonic axis for the design of *Agrobacterium*-mediated gene transfer systems for chickpea.

8.1. TRANSFORMATION SYSTEMS

In recent years, genetic engineering allowed introduction of new traits (as e.g. resistance to viruses, bacteria, fungi, nematodes, insects and herbicides) and improvement of the nutritive and commercial value (to name only a few) of agronomically important plants producing a combination of desirable features, which is difficult or impossible to achieve through conventional breeding. The genetic engineering of a desired trait into plants still involves the following steps:

(i) the identification of a desirable trait,
(ii) isolation of the gene(s) specifying the trait,
(iii) modification of the gene for expression at an appropriate level in the target tissue (with e.g. regulatable promotors),
(iv) generation of transgenic plants,
(v) analysis and optimization of transgene expression, and
(vi) release of a transgenic plant as commercial variety after appropriate field trials and other regulatory tests.

Although wild *Cicer* species possess many desirable traits, the incompatibility between them and the cultigen has deterred chickpea improvement through conventional plant breeding. Therefore, the introduction of specific genes into chickpea through genetic engineering is mandatory for chickpea improvement. One of the necessary pre-requisites for the production of transgenics is an efficient, reproducible and genotype-independent, routine transformation system. Only a few reports, however, are available on the transformation of chickpea. Progress in this area has been limited as compared to other important grain legumes like pea, lupin, soybean and peanut, and biolistic methods

Table 6. Regeneration of chickpea: an overview

Regeneration event	Explant	References
Shoot regeneration via callus phase	Immature cotyledon leaflets, shoot tip	Sharma *et al.*, 1979; Singh *et al.*, 1982; Rao and Chopra, 1989a; Surya-Prakash *et al.*, 1992; Barna and Wakhlu, 1994
Soot regeneration via cotyledon-like structures	Cotyledon	Shri and Davis, 1992
Direct shoot regeneration	Shoot tip, embryonic axis; different combinations of embryonic axis and cotyledons; seed	Kartha *et al.*, 1981; Malik and Saxena, 1992; Surya Prakash *et al.*, 1992; Murthy *et al.*, 1996; Polisetty *et al.*, 1996; Polisetty *et al.*, 1997
Direct somatic embryogenesis	Seed, immature cotyledonary segments	Murthy *et al.*, 1996; Ramana *et al.*, 1996
Somatic embryogenesis via callus phase	Cotyledon, embryonic axis; immature leaflets	Sagare *et al.*, 1993; Barna and Wakhlu, 1993; Kumar *et al.*, 1994, 1995a
Shoot regeneration from wound site	Embryo axis devoid of either shoot meristem, or root and shoot meristems	Fontana *et al.*, 1993; Kar *et al.*, 1996; Krishnamurthy *et al.*, 2000

have never been attempted. At present, all transformation protocols for chickpea involve *Agrobacterium*-mediated gene transformation only.

Islam *et al*. (1994) obtained transformed calli from leaf and stem explants with wild strains of *Agrobacterium*. Hairy roots were induced with *Agrobacterium rhizogenes* (Siefkes-Boer *et al*., 1995). Others employed immature cotyledons, embryo axis devoid of apical (shoot) meristem (Fontana *et al*., 1993), and shoot and root meristems (Kar *et al*., 1996; Krishnamurthy *et al*., 2000) as explants. The use of injured embryo axis explants leads to multiple shoot formation without an intervening callus phase. Moreover, somaclonal variation can influence the phenotype of the regenerated plants. Chickpea embryo axis deprived of the apical dome are preferred, since the cells around the wound reconstitute the meristematic apex under a cytokinin stimulus without callus formation. They consequently are the targets for a successful *Agrobacterium*-mediated transformation. Fontana *et al*. (1993) and Kar *et al*. (1996) used gene constructs harbouring the neomycin-phosphotransferase (*nptII*) gene as plant selectable marker and the β-glucuronidase (GUS) gene (*uidA*) as reporter. The regenerated transformants could therefore be selected on kanamycin (50–100 mg/L). Krishnamurthy *et al*. (2000) used PPT (phosphinothricin) as an additional selectable marker. The integration of the transgenes was confirmed through histochemical and fluorogenic GUS assay, neomycin phosphotransferase II enzyme assay, and Southern hybridization using *uidA*/*nptII* probes. Although successful regeneration of transgenic chickpeas was achieved, no attempt was made to follow the inheritance of transgenes in subsequent generations. Hamblin *et al*. (1998) reviewed the development and present status of transgenic pulses in Australia and reported the existence of transgenic chickpeas in their T_5 generation. These transgenics harboured herbicide resistances (*bar*/PAT, bromoxynil), fungal disease resistances (chitinase, osmotin), insect resistances (α-amylase inhibitor, protease inhibitor), and seed quality factors (high methionine protein), although details have not yet been published. Very recently, Krishnamurthy *et al*. (2000) confirmed the inheritance of the *nptII* gene in four T_1 plantlets through PCR amplification. However, none of these plantlets was GUS positive.

The efficiency of a transformation system is prerequisite to routine production of transgenic plants. Although very high transformation frequencies were reported for transformed calli (Srinivasan *et al*., 1991) and somatic embryos (Ramana *et al*., 1996), plant regeneration in other experiments was accompanied by transformation efficiencies of only 4% (Fontana *et al*., 1993), or less than 0.4% (Hamblin *et al*., 1998; Krishnamurthy *et al*., 2000).

All these experimental details and contradictions do not obscure the fact, that presently no efficient and reliable transformation protocol is at hand. Given the rapid retrieval of genes from sequence data banks or genome-wide expression profiling, it may soon turn out, that all is prepared for a successful improvement of whatever quality of chickpea – except the once-thought trivial but apparently tricky transformation and recalcitrant regeneration process.

9. Legume genomics: a perspective

As witnessed in this article, there has been substantial progress made in legume genomics over the past decade. Marker techniques were adopted or newly designed for the major legumes. The diverse molecular markers (spanning from dominant RAPD-, DAF-, and microsatellite fingerprint markers to co-dominant RFLP or simple sequence repeat markers) were used to detect genetic diversity in lines and populations, and to establish genetic maps for nearly all the more agronomically important legumes. It comes as no surprise that soybean genomics is presently most advanced, given the number of laboratories working on this crop and the support being granted for soybean research. In spite of all these advances, legumes as a whole are still under-researched, the research generally underfinanced, and the genomics of these important crops necessarily underdeveloped, especially if compared with other crops such as rice, maize, or potato, or even the so-called model plant *Arabidopsis thaliana*. However, it is expected that legume research will intensify in the near future and probably catch up with research on other crop plants. The major areas on which progress is programmed, will be portrayed shortly.

9.1. MODEL LEGUMES: *MEDICAGO TRUNCATULA* AND *LOTUS JAPONICUS*

The size, complexity and high repeated sequence content of the genomes of most crop legumes (e.g. pea genome: 5×10^9 bp) forces us to develop more simple and more easy-to-treat systems: model legumes. And there is no doubt that more research will be focussing on the genomics of such models rather than the agronomically relevant legumes in the near future. However, the reward will come in the form of genes and sequences also of use for the crop legumes. Two model legumes are now at hand: *Medicago truncatula* and *Lotus japonicus*. Both plants have a small genome and a rapid life cycle. Extensive linkage maps have already been prepared for both species. The map for the *M. truncatula* genome covers the eight linkage groups and includes morphological, enzyme and DNA markers with an average marker spacing of less than 3.5 cM. For *L. japonicus*, a map based on a wide cross between *L. japonicus* (Gifu) and *L. filicaulis* covers roughly 400 cM and harbours more than 450 RFLP and AFLP markers. Markers closely linked to the supernodulation locus sym16 at a distance of 0.2 cM will allow the map-based cloning of the underlying gene(s). And this is only one out of several examples.

Aside of the saturated linkage maps and BAC libraries of both genomes, there is an increasing collection of expressed sequence tags (ESTs) available in public databases (e.g. the expressed portion of the *L. japonicus* genome is now covered by more than 10,000 non-redundant entries). About 40% of these ESTs are unique to *L. japonicus*, representing an enormous source for the detection of novel and species-specific genes. *M. truncatula* is equally well represented: close to 15,000 ESTs from roots, nodulated roots and arbuscular mycorrhizal roots, about 17,000 ESTs derived from pathogen and symbiont-exposed *M. truncatula* plants as well as 27,000 ESTs from various tissues of the plant have been collected. Global gene expression patterns in both plants will be assessed for a variety of treatments using microarrays and serial analysis of gene expression (SAGE). The genome of *L. japonicus* will probably be the first legume genome to be sequenced. Also, several successful approaches for the stable transformation of these model legumes exist, and some of them are based on relatively easy-to-perform in planta infiltrations with *Agrobacterium tumefaciens* cells (Trieu *et al.*, 2000). Last but not least, a wealth of different transposon insertional mutants are available for both plants, a rich source for many fundamental experiments. For example, mutants that either lost or gained a specific trait, as e.g. root nodule formation upon interaction with *Rhizobia*, allow the straightforward isolation of the responsible genes (see e.g. Wopereis *et al.*, 2000). Also, promoter tagging comes to the fore: out of 200 promoter-tagged *L. japonicus* lines one individual was isolated, that showed increased root-specific expression of a calcium-binding protein upon contact with *Mesorhizobium loti*, demonstrating the potential of this technique (Webb *et al.*, 2000).

All these developments mirror the successful approach to the *Arabidopsis thaliana* genome (Rounsley *et al.*, 2000). The wealth of new data expected from research into the model legumes will certainly catalyze many areas of legume genomics and biology.

10. Comparative genomics

The continuous development of all kinds of molecular markers has added to our knowledge of plant genome structure, at the least to establish linkage maps of a series of plants. Such maps developed for different species have been compared using common sets of DNA markers, which revealed an unexpected degree of conservation between the genomes of several monocot (Moore *et al.*, 1995; Tikhonov *et al.*, 1999) and dicot species (Cavell *et al.*, 1998; Lagercrantz, 1998), allowing us to identify collinear chromosome regions and orthologous loci. Even bridging the monocot-dicot gap by synteny is possible (Paterson *et al.*, 1996). Therefore, comparative mapping and sequence analysis of collinear regions will tell us about the extent of macrosynteny maintained at the inter- and intragenic levels. Naturally, closely related species show extensive macro- and also microsynteny. For example, a high gene sequence conservation can be found between *Arabidopsis* and *Brassica* species (Cavell *et al.*, 1998; O'Neill and Bancroft, 2000) as well as in the Gramineae (Moore *et al.*, 1995), so that markers designed for *Arabidopsis*

Table 7. Attempted transformations in chickpea: a summary

Reference	Explant	Selectable marker	Reporter gene	Transformation efficiency (%)	Progress status
Srinivasan *et al.*, 1991	Leaf, stem	*nptII*	–	77 ± 9	Callus
Fontana *et al.*, 1993	Embryo axis (without vegetative apex)	*nptII*	*uidA*	4	T_0 plants
Ramana *et al.*, 1996	Immature cotyledonary segments	*nptII*	*uidA*	30	Somatic embryos
Kar *et al.*, 1996	Embryo axis (without root and shoot meristems)	*nptII*	*uidA*	–	T_0 plants
Hamblin *et al.*, 1998	–	*bar/pat* Bromoxynil	*uidA*	0.1–0.03	T_5 plants
Krishnamurthy *et al.*, 2000	Embryonic axis (without root and shoot meristems)	*nptII*	*uidA*	0.0–1.4 (average <0.04)	T_1 plants

can be exploited to anchor homologous regions in *Brassica* genomes. Also, large-scale synteny exists among genomes of solanaceous plants (as e.g. potato and tomato; Bonierbale *et al.*, 1988; Tanksley *et al.*, 1992) or grasses (Moore, 2000; Keller and Feuillet, 2000). However, large-scale duplications of chromosomes and selective losses of genes frequently leave a patchwork of syntenic regions so characteristic for differential evolution (as e.g. the *Arabidopsis* and tomato genomes; Ku *et al.*, 2000). In most cases, however, large-scale microlinearity has been conserved between plant genomes (Bennetzen, 2000).

10.1. LEGUME COMPARATIVE GENOMICS IS LAGGING FAR BEHIND

Comparative linkage analysis with RFLP markers in mungbean (*Vigna radiata*), cowpea (*V. unguiculata*), and common bean (*Phaseolus vulgaris*), all from the subtribe *Phaseolinae*, and soybean (*Glycine max*), subtribe *Glycininae* detected extensive linkage conservation between the genomes of the *Vigna* species and *Phaseolus*, and also considerable collinearity among the genomes of these species and soybean. Linkage groups from chickpea have been related to those from pea (Gaur and Slinkard, 1990a, b; Kazan *et al.*, 1993; Weeden *et al.*, 1993; Simon and Muehlbauer, 1997) and lentil (Muehlbauer *et al.*, 1989; Weeden *et al.*, 1992), using morphological, isozyme and RFLP markers from lentil developed by Havey and Muehlbauer (1989). Efforts to exploit these markers for mapping in chickpea suffered from the low level of isozyme and DNA marker polymorphism in chickpea, and the little homology of lentil RFLPs to restriction fragments of chickpea (excluding many lentil RFLPs from mapping in chickpea). Consequently, conserved arrangement of DNA markers in pea and chickpea could only be detected for short segments of the chickpea map (Simon and Muehlbauer, 1997). Whereas Weeden *et al.* (1992) detected eight syntenic regions between lentil and pea maps, Simon and Muehlbauer (1997) found only five conserved groups between chickpea and the other two legumes. Most pronounced synteny occurred at linkage group 7 of the chickpea map, with two clusters of loci syntenic to pea linkage group 7 (Weeden *et al.*, 1993) and lentil linkage group 5 (Weeden *et al.*, 1992). Reduced synteny between chickpea and pea/lentil reflects their taxonomic relationships: pea and lentil ($2n = 14$) belong to the tribe *Vicieae*, chickpea ($2n = 16$) to the monogeneric tribe *Cicereae*.

However, there are other sources of synteny. TBLASTX conceptual translations of DNA sequences from *Arabidopsis thaliana* and soybean revealed substantial colinearity between linkage groups of soybean and chromosomes of *Arabidopsis*. For example, loci from three soybean linkage groups had homologs on *Arabidopsis* chromosome I (arabI), and especially soybean linkage group 2 (soyA2) shares substantial synteny with arabI. Only one translocation and one inversion in the evolution of soyA2 basically were sufficient to arrange the map orders in both plants as they appear today. And this in spite of the fact, that Cruciferae and legumes separated about 92 million years ago (Grant *et al.*, 2000).

For all of these reasons, comparative genomics already gained momentum, and the model genome data (as e.g. *Arabidopsis thaliana*, more so *Lotus japonicus* and *Medicago truncatula*) can surely be exploited for genomics in all legumes: a vast source for syntenic sequences, markers and genes.

11. Transcriptomics: EST collections and high-throughput transcript mapping

The past decade has seen the development of a plethora of genome sequencing, genome-wide transcript mapping and proteome techniques, that altogether has catalyzed the advances in genomics, transcriptomics and proteomics, at least in medical and animal sciences. However, some of these techniques have more recently been adopted for plant science, especially model plants (basically *Arabidopsis thaliana*, but also rice, and most recently, model legumes). These techniques span from the use of locked nucleic acids (LNAs), peptide nucleic acids (PNAs), or nanoparticle probes for nanoparticle-based DNA detection, to fiber-optic DNA arrays, quantitative DNA fiber mapping (QDFM), massively parallel signature sequencing (MPSS), cDNA and oligonucleotide arrays (in their macro- or microformat), serial analysis of gene expression (SAGE) and its many variants (as e.g. microSAGE or SAGElite), total gene expression analysis (TOGA), polymer-based biosensors, microfluidics and nanodroplet techniques, protein chips (living chips with dotted cells or colonies, and chemical chips with synthetic peptides) and translation state array analyses (TSAAs). These are only a few examples and do by no means reflect the repertoire of modern genomics, which is driven by ever-increasing throughput and miniaturization. The success of this technical revolution can be measured by the output of new data and gaining of new insights into almost every facet of life.

It is mandatory to tap this source of techniques, together with the experiences of the scientists who use them, for genomics of legumes. First and rather low-scale trials with cDNA microarrays, gene macroarrays, oligonucleotide arrays, and genome-wide transcript profiling methods such as SAGE already began and have shown the great potential of these novel approaches for some legumes (e.g. chickpea). What can be expected from this development? Certainly we will soon know better what genes are expressed under various environmental or experimental conditions, and how strongly they are expressed, and how they are regulated. Of course, at the same time, we can retrieve the corresponding genes from the expression arrays. This can be done genome-wide, i.e. including all genes. It goes without explanation that the most important genes can then be transferred into any background of choice. Also, the proteome will attract much more attention than in the past. Protein chips already appeared, though the technology is not advanced. But they will eventually allow to construct protein-protein interaction maps for the whole cell, or screen for a multitude of ligands interacting with one of a few peptide domains. Again, it is clear that the experiments will fuel the design of novel and more effective proteins (e.g. receptors reacting more effectively with their ligands, or possessing changed affinities).

It is certainly beyond the scope of this review to introduce all developments in a compact package. However, it is intended to simply portray few of the possibilities to suggest more research into the genomics and transcriptomics of legumes.

12. A deeper look into pathogenesis

Race-specific resistance in plants is frequently based on a gene-for-gene relationship, in which a resistance gene-encoded R protein acts as a receptor of either the corresponding avirulence (Avr) protein of the pathogen or a molecule produced by the Avr protein. Upon recognition of an attacking or invading pathogen, the plant initiates a characteristic sequence of defense reactions, including a rapid and localised cell death (hypersensitive response; Hammond-Kosack and Jones, 1997).

Among the four presently known R gene classes, the largest one is the nucleotide binding site (NBS)- leucine-rich repeat (LRR) class. The members of this group have been identified in monocots and dicots and confer resistance to a whole variety of phytopathogens such as viruses, bacteria, ascomycetes, oomycetes, nematodes and aphids (Rossi *et al.*, 1998; Vos *et al.*, 1998). No other R gene class confers such a broad resistance to such a diverse set of pathogens. Therefore, much future research will certainly be focussed on this category. Moreover, more than 30 *avr* genes have now been cloned from *Pseudomonas syringae* and *Xanthomonas* (Alfano and Collmer, 1996; Hutcheson, 1998). Many corresponding Avr proteins seem to be transported into the host cell (requiring the *hrp* gene cluster, encoding components of the bacterial type III secretion system; Gopalan *et al.*, 1996). For the tomato R gene *Pto* and the *Arabidopsis thaliana* gene *RPS2*, a complex formation with the cognate *avr* gene products AvrPto of *P. syringae* and AvrRpt2, respectively, has been shown (Scofield *et al.*, 1996; Tang *et al.*, 1996; Leister and Katagiri, 2000), so that the first steps in signal recognition are no more hypothetic. However, there is definitely a gap in our knowledge of how the signal(s) is transmitted to the genome where it induces the transcription of a set of defense genes. The intracellular signaling of stress challenges as e.g. fungal attack is mediated by mitogen-activated protein kinase (MAPK) cascades, where an inactive dephosphorylated MAPK is activated by phosphorylation of both threonine and tyrosine residues in the sequence TxY, catalyzed by an upstream MAPK kinase (MAPKK). An inactive dephosphorylated MAPKK is in turn activated by phosphorylation of both serine and serine/threonine residues in the sequence SxxxS/T, catalyzed by an immediate upstream MAPKK kinase (MAPKKK). The signal, that is only needed transiently, is abolished by protein phosphatases as e.g. serine/threonine phosphatases (PPases such as PP1, PP2A, and PP2C) and protein tyrosine phosphatases (PTPases). The genes for most of the kinases and phosphatases are activated in response to stress, but the fast kinase action is independent of gene activation (see Suzuki and Shinshi, 1995; Mizoguchi *et al.*, 1997; Luan, 1998; Xu *et al.*, 1998; Desikan *et al.*, 1999).

Though research into signaling in plants is accelerating, a huge gap remains in comparison to signal transduction in animals. However, a deeper understanding of plant resistance to various environmental stresses presupposes a detailed knowledge of the molecular events leading to the response. The design of plants with an improved and durable resistance via gene transfer would benefit from the deciphering of the chain leading from signal perception to signal response, be it an engineered faster response or stronger reaction of the plant.

13. Outlook

The Leguminosae as the third largest angiosperm family engulf many agronomically important crop species. To name a few, soybean, chickpea, bean, pea, peanut and lentil are major protein sources for millions of people, soybean and peanut provide valuable edible oils, many legumes are a rich source for secondary metabolites, legumes are used as forage crops, and, in combination with *rhizobia*, play a central role in the global nitrogen cycle. In view of their importance for the developing, but also the developed world the legumes should receive more attention in the future, both by funding agencies as well as the breeder and molecular biologist.

Acknowledgements

Research of the authors was supported by grants from the following sources: Deutsche Forschungsgemeinschaft (grant Ka332/17.1-3), International Atomic Energy Agency, Breeding Section (grant CRP 302-02-GFR-8148), Bundesminsterium fuer Technische Zusammenarbeit (BMZ grant 89.7860-01.130), UNESCO (Paris, France), Fazit-Stiftung (Frankfurt, Germany) and Deutscher Akademischer Austauschdienst (DAAD). G. Kahl appreciates the hospitality of the Iwate Biotechnology Research Center (Director: Dr. Hitoshi Enei) at Kitakami, Japan and STA support.

References

Abbo S, Miller T E, Reader S M, Dunford R P and King I P (1994) Detection of ribosomal DNA sites in lentil and chickpea by fluoresence *in situ* hybridization. *Genome*, **37**: 713–716.

Acikgoz N, Karaca M, Er C and Meyveci K (1994) Chickpea and lentil production in Turkey. In: *Expanding the Production and Use of Cool Season Food Legumes* (Eds Muehlbauer F J, Kaiser W J), Kluwer Academic Publishers, Dordrecht, The Netherlands.

Ahmad F (1989) The chromosomal architecture of *Cicer anatolicum* Alef., a wild perennial relative of chickpea. *Cytologia*, **54**: 753–757.

Ahmad F (1999) Random amplified polymorphic DNA (RAPD) analysis reveals genetic relationships among the annual *Cicer* species. *Theort. Appl. Genet.*, **98**: 657–663.

Ahmad F (2000) A comparative study of chromosome morphology among the nine annual species of *Cicer* L. *Cytobios*, **101**: 37–53.

Ahmad F and Hymowitz T (1993) The fine structure of chickpea (*Cicer arietinum* L.) chromosomes as revealed by pachytene analysis. *Theort. Appl. Genet.*, **86**: 637–641.

Ahmad F and Slinkard A E (1992) Genetic relationships in the genus *Cicer* L. as revealed by polyacrylamide gel electrophoresis of seed storage proteins. *Theort. Appl. Genet.*, **84**: 688–692.

Ahmad F, Gaur P M and Slinkard A E (1992) Isozyme polymorphism and phylogenetic interpretations in the genus *Cicer* L. *Theort. Appl. Genet.*, **83**: 620–627.

Alfano J R and Collmer A (1996) Bacterial pathogens in plants: Life up against the wall. *Plant Cell*, **8**: 1683–1698.

Ali S, Müller C R and Epplen J T (1986) DNA fingerprinting by oligonucleotide probes specific for simple repeats. *Hum. Genet.*, **74**: 239–243.

Arens P, Odinot P, Van Heusden A W, Lindhout P and Vosman B (1995) GATA- and GACA-repeats are not evenly distributed throughout the tomato genome. *Genome*, **38**: 84–90.

Areshchenkova T and Ganal M W (1999) Long tomato microsatellites are predominantly associated with centromeric regions. *Genome*, **42**: 536–544.

Auckland A K and van der Maesen L J G (1980) Chickpea. In: *Hybridization of Crop Plants* (Eds Fehr W R, Hadley H H), American Society of Agronomy: Crop Science Society of America, pp. 249–259.

Badami P S, Mallikarjuma N and Moss J P (1997) Interspecific hybridization between *Cicer arietinum* and *Cicer pinnatifidum*. *Plant Breed.*, **116**: 393–395.

Bahl P N (1987) Cytology of chickpea. In: *The Chickpea* (Eds Saxena M C and Singh K B), CAB International, Wallingford, UK, pp. 83–98.

Barna K S and Wakhlu A K (1993) Somatic embryogenesis and plant regeneration from callus cultures of chickpea (*Cicer arietinum* L.). *Plant Cell Rep.*, **12**: 521–524.

Barna K S and Wakhlu A K (1994) Whole plant regeneration of *Cicer arietinum* from callus cultures via organogenesis. *Plant Cell Rep.*, **13**: 510–513.

Barz W, Bless W, Gunia W, Höhl B, Mackenbrock U, Meyer D, Tenhaken R and Vogelsang R (1993) Mechanism of resistance to fungal pathogens in cool-season food legumes. In: *Breeding for Stress Tolerance in Cool-Season Food Legumes* (Eds Singh K B and Saxena M C), John Wiley and Sons, New York, pp. 193–210.

Beckmann J S and Soller M (1990) Towards a unified approach to genetic mapping of eucaryotes based on sequenced tagged microsatellite sites. *Bio/Technology*, **8**: 930–932.

Bennett M D and Smith J B (1976) Nuclear DNA amounts in angiosperms. *Philos. Trans. Roy. Soc. London*, **B**: 227–274.

Bennett M D, Leitch I J and Leitch I J (1998) Angiosperm DNA C-value database: http://www.rbgkew.org.uk/cval/database1.html

Bennetzen J L (2000) Comparative sequence analysis of plant nuclear genomes: Microlinearity and its many exceptions. *Plant Cell*, **12**: 1021–1029.

Bonierbale M W, Plaisted R L and Tanksley S D (1988) RFLP maps based on a common set of clones reveal modes of chromosome evolution in potato and tomato. *Genetics*, **120**: 1095–1103.

Brandes A, Heslop-Harrison J S, Kamm A, Kubis S, Doudrick R L and Schmidt T (1997) Comparative analysis of the chromosomal and genomic organization of Ty1-copia-like retrotransposons in pteridophytes, gymnosperms and angiosperms. *Plant Mol. Biol.*, **33**: 11–21.

Broun P and Tanksley S D (1993) Characterization of tomato clones with sequence similarity to human minisatellites 33.6 and 33.15. *Plant Mol. Biol.*, **23**: 231–242.

Broun P and Tanksley S D (1996) Characterization and genetic mapping of simple repeat sequences in the tomato genome. *Mol. Gen. Genet.*, **250**: 39–49.

Burr B, Burr F A, Thompson K H, Albertsen M C and Stuber C W (1988) Gene mapping with recombinant inbreds in maize. *Genetics*, **118**: 519–526.

Cabrera de la Colina J, Trapero-Casas A and Jiminéz-Diaz R M (1985) Races of *Fusarium oxysporum* f.sp. *ciceri* in Andalucia, southern Spain. *Int. Chickpea Newslett.*, **13**: 24–26.

Caetano-Anollés G and Gresshoff P M (1991) Plant genetic control of nodulation in legumes. *Annu. Rev. Microbiol.*, **45**: 345–382.

Caetano-Anollés G, Bassam B J and Gresshoff P M (1991) DNA amplification fingerprinting: a strategy for genome analysis. *Plant Mol. Biol. Rep.*, **9**: 292–305.

Campbell C and Madden L V (1990) Temporal analysis of epidemics: Description and comparison of disease progress curves. In: *Introduction to Plant Disease Epidemiology*. John Wiley and Sons Inc, USA, pp. 161–202.

Cavell A C, Lydiate D C, Parkin I A P, Dean C and Trick M (1998) Collinearity between a 30 centimorgan segment of *Arabidopsis thaliana* chromosome 4 and duplicated regions within the *Brassica napus* genome. *Genome*, **41**: 62–69.

Cho Y G, Ishii T, Temnykh S, Chen X, Lipovich L, McCouch S R, Park W D, Ayres N and Cartinhour S (2000) Diversity of microsatellites derived from genomic libraries and GeneBank sequences in rice (*Oryza sativa* L.). *Theort. Appl. Genet.*, **100**: 713–722.

Choumane W, Winter P, Weigand F and Kahl G (2000) Conservation and variability of sequence-tagged microsatellite sites from chickpea (*Cicer arietinum* L.) within the genus *Cicer*. *Theort. Appl. Genet.*, **101**: 269–278.

Christou P (1994) The biotechnology of crop legumes. *Euphytica*, **74**: 165–185.

Cook D, Dreyer D, Bonnet D, Howell M, Nony E, and VandenBosch K (1995) Transient induction of a peroxidase gene in *Medicago truncatula* precedes infection by *Rhizobium meliloti*. *Plant Cell*, **7**: 43–55.

Cook D, VandenBosch K, de Bruijn F and Huguet T (1997) Model legumes get the nod. *Plant Cell*, **9**: 275–281.

Cregan P B, Jarvik T, Bush A L, Shoemaker R C, Lark K G, Kahler A I, Kaya N, VanThoai T T, Lohnes D G, Chung J and Specht J (1999) An integrated genetic linkage map of the soybean genome. *Crop Sci.*, **39**: 211–217.

Desikan R, Clarke A, Atherfold P, Hancock J T and Neill S J (1999) Harpin induces mitogen-activated protein kinase activity during defense responses in *Arabidopsis thaliana* suspension cultures. *Planta*, **210**: 97–103.

Dey S K and Singh G (1993) Resistance to *Ascochyta* blight-Genetic basis. *Euphytica*, **68**: 147–153.

Dirlewanger E, Isaac P G, Ranade S, Belajouza M, Cousin R and de Vienne D (1994) Restriction fragment length polymorphism analysis of loci associated with disease resistance genes and developmental trials in *Pisum sativum* L. *Theort. Appl. Genet.*, **88**: 17–27.

Dixon R A and Lamb C J (1990) Molecular communication in interactions between plants and microbial infection in plants. *Annu. Rev. Plant Physiol. Plant Mol. Biol.*, **41**: 339–367.

Dombrovsky-Sludsky L (1927) La cinese somatique de *Cicer arietinum. Zhurnal Russk Botaniki Obshch*, **12**: 163–171.

Doyle J J, Morgante M, Tingey S V and Powell W (1998) Size homoplasy in chloroplast microsatellites of wild perennial relatives of soybean (*Glycine* subgenus *Glycine*). *Mol. Biol. Evol.*, **15**: 215–218.

Duke J A (1981) *Handbook of Legumes of World Economic Importance*. Plenum Press, New York.

Eapen S and George L (1994) Somatic embryogenesis in *Cicer arietinum* L.: Influence of genotype and auxins. *Biol. Plant.*, **36**: 343–349.

Eser D (1976) Heritability of some important plant characters, their relationship with plant yield and inheritance of *Ascochyta* blight resistance in chickpea (*Cicer arietinum* L.), Ankara Universitesi, Ziraat Fakultesi Yayinlari 620, 40.

Falquet J, Creusot F and Dron M (1997) Molecular analysis of rDNA unit and characterization of a satellite DNA homologous to IGS subrepeats. *Plant Physiol. Biochem.*, **35**: 611–622.

Flavell R B (1986) Repetitive DNA and chromosome evolution in plants. *Philos. Trans. Roy. Soc. London*, B **312**: 227–242.

Fontana G S, Santini L, Caretto S, Frugis G and Mariotti D (1993) Genetic transformation in grain legume *Cicer arietinum* L. (chickpea). *Plant Cell Rep.*, **12**:194–198.

Ford R, Pang E C K and Taylor P W J (1999) Genetics of resistance to ascochyta blight (*Ascochyta lentis*) of lentil and the identification of closely linked RAPD markers. *Theort. Appl. Genet.*, **98**: 93–98.

Galasso I and Pignone D (1992) Characterization of chickpea chromosomes by banding techniques. *Genet. Res. Crop. Evol.*, **39**: 115–119.

Galasso I, Frediani M, Maggiani M, Cremonini R and Pignone D (1996) Chromatin characterization by banding techniques, *in situ* hybridization and nuclear DNA content in *Cicer* L. (Leguminosae). *Genome*, **39**: 258–265.

Gaur P M and Slinkard A E (1990a) Inheritance and linkage of isozyme coding genes in chickpea. *J. Hered.*, **81**: 455–461.

Gaur P M and Slinkard A E (1990b) Genetic control and linkage relations of additional isozyme markers in chickpea. *Theort. Appl. Genet.*, **80**: 648–656.

Geiger H H and Heun M (1989) Genetics of quantitative resistance to fungal diseases. *Annu. Rev. Phytopathol.*, **27**: 318–341.

Gopalan S, Bauer D W, Alfano J R, Loniello A O, He S Y and Collmer A (1996) Expression of the *Pseudomonas syringae* avirulence protein AvrB in plant cells alleviates its dependence on the hypersensitive response and pathogenicity (Hrp) secretion system in eliciting genotype-specific hypersensitive cell death. *Plant Cell*, **8**: 1095–1105.

Gortner G, Nenno M, Weising K, Zink D, Nagl W and Kahl G (1998) Chromosomal localization and distribution of simple sequence repeats and the *Arabidopsis*-type telomere sequence in the genome of *Cicer arietinum* L. *Chrom. Res.*, **6**: 97–104.

Gowen S R (1983) Pathogenic variation in *Ascochyta rabiei. Proc. 10th Intl. Cong. Plant Protection*, Brighton, UK, The British Plant Protection Council, Croydon, UK, Vol. 2: 831.

Grant D, Cregan P and Shoemaker R C (2000) Genome organization in dicots: Genome duplications in *Arabidopsis* and synteny between soybean and *Arabidopsis*. *Proc. Natl. Acad. Sci. USA*, **97**: 4168–4173.

Gupta M, Chyi Y S, Romero-Severson J and Owen J L (1994) Amplification of DNA markers from evolutionary diverse genomes using single primers of simple sequence repeats. *Theor. Appl. Genet.*, **89**: 998–1006.

Gupta P K and Varshney R K (2000) The development and use of microsatellite markers for genetic analysis and plant breeding with emphasis on bread wheat. *Euphytica*, **113**: 163–185.

Hafiz A and Ashraf M (1953) Studies on the inheritance of resistance to *Mycosphaerella* blight in gram. *Phytophathology*, **43**: 580–581.

Hammond-Kosack K E and Jones J J (1997) Plant disease resistance genes. *Annu. Rev. Plant Physiol Plant Mol. Biol.*, **48**: 575–607.

Handberg K and Stougaard J (1992) *Lotus japonicus*, an autogamous, diploid legume species for classical and molecular genetics. *Plant J.*, **2**: 487–496.

Haware M P and Nene Y L (1982) Races of *Fusarium oxysporum* f.sp. *ciceri. Plant Dis.*, **66**: 809–810.

Heslop-Harrison J S, Brandes A, Taketa S, Schmidt T, Vershinin A V, Alkhimova E G, Kamm A, Doudrick R L, Schwarzacher T, Katsiotis A, Kubis S, Kumar A, Pearce S R, Flavell A J and Harrison G E (1997) The chromosomal distribution of Ty1-copia group retrotransposable elements in higher plants and their implications for genome evolution. *Genetica*, **100**: 197–204.

Höhl B, Pfautsch M and Barz W (1990) Histology of disease development in resistant and susceptible cultivars of chickpea (*Cicer arietinum* L.) inoculated with spores of *Ascochyta rabiei*. *J. Phytopathol.*, **129**: 31–45.

Hutcheson S W (1998) Current concepts of active defense in plants. *Annu. Rev. Phytopathol.*, **36**: 59–90.

Hüttel B (1996) Mikrosatelliten als molekulare Marker in der Kichererbse (*Cicer arietinum* L.), Ph.D. Thesis, University of Frankfurt, Germany.

Hüttel B, Winter P, Weising K, Choumane W, Weigand F and Kahl G (1999) Sequence-tagged microsatellite site markers for chickpea (*Cicer arietinum* L.). *Genome*, **42**: 210–217.

Islam R, Malik T, Husnain T and Riazuddin S (1994) Strain and cultivar specificity in the *Agrobacterium*-chickpea interaction. *Plant Cell Rep.*, **16**: 32–37.

Jiang Q and Gresshoff P M (1997) Classical and molecular genetics of the model legume *Lotus japonicus*. *Mol. Plant Microbe Interact.*, **10**: 559–568.

Jiminéz-Diaz R M, Crino P, Halila M H, Masconi C and Trapero-Casas A T (1993) Screening for resistance to fusarium wilt and ascochyta blight in chickpea. In: *Breeding for Stress Tolerance in Cool-Season Food Legumes* (Eds Singh K B and Saxena M C), John Wiley and Sons, New York, pp. 77–95.

Kabir G and Singh R M (1990) Karyotype analysis of six species of *Cicer* L. *Bangladesh J. Bot.*, **19**: 175–181.

Kaiser W J (1973) Factors affecting growth, sporulation, pathogenicity and survival of *Ascochyta rabiei*. *Mycologia*, **65**: 444–457.

Kaiser W J (1992) Epidemiology of *Ascochyta rabiei*. In: *Breeding for Stress Tolerance in Cool-Season Food Legumes* (Eds Singh K B and Saxena M C), John Wiley and Sons, New York, pp. 117–134.

Kaiser W J (1997) Inter- and intranational spread of *Ascochyta* pathogens of chickpea, faba bean, and lentil. *Can. J. Plant Path.*, **19**: 25–224.

Kalendar R, Grob T, Regina M, Suoniemi A and Schulman A (1999) IRAP and REMAP: two new retrotransposon-based DNA fingerprinting techniques. *Theort. Appl. Genet.*, **98**: 705–711.

Kamm A, Galasso I, Schmidt T and Heslop-Harrison J S (1995) Analysis of a repetitive DNA family from *Arabidopsis arenosa* and relationships between *Arabidopsis* species. *Plant Mol. Biol.*, **27**: 853–862.

Kar S, Johnson T M, Nayak P and Sen S K (1996) Efficient transgenic plant regeneration through *Agrobacterium*-mediated transformation of chickpea (*Cicer arietinum* L.). *Plant Cell Rep.*, **16**: 32–37.

Kartha K K, Pahl K, Leung N L and Mroginski L A (1981) Plant regeneration from meristems of grain legumes: Soybean, cowpea, groundnut, chickpea and bean. *Can. J. Bot.*, **59**: 1674–1679.

Kashi Y, Nave A, Darvasi A, Gruenbaum Y, Soller M and Beckmann J S (1994) How is it that microsatellites and random oligonucleotides uncover DNA fingerprint patterns? *Mamm. Genome*, **5**: 525–530.

Kazan K and Muehlbauer F J (1989) Linkage of isozyme loci and morphological characters in chickpea. *Am. Soc. Agron. Abst.*, Madison, WI, 89.

Kazan K and Muehlbauer F J (1991) Allozyme variation and phylogeny in annual species of *Cicer* (Leguminosae). *Plant Syst. Evol.*, **175**: 11–21.

Kazan K, Muehlbauer F J, Weeden N F and Ladizinsky G (1993) Inheritance and linkage relationships of morphological and isozyme loci in chickpea (*Cicer arietinum* L.). *Theort. Appl. Genet.*, **86**: 417–426.

Keller B and Feuillet C (2000) Colinearity and gene density in grass genomes. *Trends Plant Sci.*, **5**: 246–251.

Kovachevsky I C (1936) The blight of chickpea (*Cicer arietinum*) *Mycosphaerella* n.sp. (In Russian, English summary), Ministry of Agriculture and National Domains, Sofia, Bulgaria, p. 80.

Krishnamurthy K V, Suhasini K, Sagare A P, Meixner M, de Kathen A, Pickardt T and Schieder O (2000) *Agrobacterium*-mediated transformation of chickpea (*Cicer arietinum* L.) embryo axes. *Plant Cell Rep.*, **19**: 235–240.

Ku H M, Vision T, Liu J P and Tanksley S D (2000) Comparing sequence segments of the tomato and *Arabidopsis* genomes: Large-scale duplication followed by selective gene loss creates a network of synteny. *Proc. Natl. Acad. Sci. USA*, **97**: 9121–9126.

Kumar J and Haware M P (1982) Inheritance of resistance to *Fusarium* wilt in chickpea. *Phytopathology*, **72**: 1035–1036.

Kumar J and van Rheenen H A (2000) A major gene for time to flowering in chickpea. *J. Hered.*, **91**: 67–68.

Kumar P A, Bisaria S and Sharma R P (1995b) Comparative shoot regeneration in different genotypes of chickpea (*Cicer arietinum* L.). *Indian J. Exp. Biol.*, **33**: 77–78.

Kumar J, Srivastava R K and Ganesh M (2000a) Penetrance and expressivity of the gene for double podding in chickpea. *J. Hered.*, **91**: 234–236.

Kumar J, Vijayalakshmi N V S and Rao T N (2000b) Inheritance of flower color in chickpea. *J. Hered.*, **91**: 416–417.

Kumar V D, Kirti P B, Sachan J K S and Chopra V L (1994) Plant regeneration via somatic regeneration in chickpea (*Cicer arietinum* L.). *Plant Cell Rep.*, **13**: 468–472.

Kumar V D, Kirti P B, Sachan J K S and Chopra V L (1995a) Picloram induced somatic embryogenesis in chickpea (*Cicer arietinum* L.). *Plant Sci.*, **109**: 207–213.

Kusmenoglu I (1990) *Ascochyta* blight of chickpea: Inheritance and relationship to seed size, morphological traits and isozyme variation. M.S. thesis, Washington State University, Pullman, WA, USA: 81.

Labdi M, Robertson L D, Singh K B and Charrier A (1996) Genetic diversity and phylogenetic relationships among the annual *Cicer* species as revealed by isozyme polymorphisms. *Euphytica*, **88**: 181–188.

Ladizinsky G and Adler A (1976a) The origin of chickpea (*Cicer arietinum* L.). *Euphytica*, **25**: 211–217.

Ladizinsky G and Adler A (1976b) Genetic relationships among the annual species of *Cicer* L. *Theort. Appl. Genet.*, **48**: 197–203.

Lagercrantz U (1998) Comparative mapping between *Arabidopsis thaliana* and *Brassica nigra* indicates that *Brassica* genomes have evolved through extensive genome replication accompanied by chromosome fusions and frequent rearrangements. *Genetics*, **150**: 1217–1228.

Lagercrantz U, Ellegren H and Andersson L (1993) The abundance of various polymorphic microsatellite motifs differs between plants and vertebrates. *Nucleic Acids Res.*, **21**: 1111–1115.

Lamb C J, Lawton M A, Drou M and Dixon R A (1989) Signals and transduction mechanism for activation of plant defences against microbial attack. *Cell*, **56**: 215–224.

Leister R T and Katagiri F (2000) A resistance gene product of the nucleotide – binding site – leucine rich repeats class can form a complex with bacterial avirulence proteins *in vivo*. *Plant J.*, **22**: 345–354.

Lev-Yadun S, Gopher A and Abbo S (2000) The cradle of agriculture. *Science*, **288**: 1602–1603.

Lister C and Dean C (1993) Recombinant inbred lines for mapping RFLP and phenotypic markers in *Arabidopsis thaliana*. *Plant J.*, **4**: 745–750.

Litt M and Luty J A (1989) A hypervariable microsatellite revealed by *in vitro* amplification of a dinucleotide repeat within the cardiac muscle actin gene. *Am. J. Hum. Genet.*, **44**: 397–401.

Long S (1996) *Rhizobium* symbiosis: Nod factors in perspective. *Plant Cell*, **8**: 1885–1898.

Luan S (1998) Protein phosphatases and signaling cascades in higher plants. *Trends Plant Sci.*, **3**: 271–275.

Maggini F, Cremonini R, Zolfino C, Tucci G F, D'Ovidio R, Delre V, DePace C, Scarascia Mugnozza G T and Cionini P G (1991) Structure and chromosomal localization of DNA sequences related to ribosomal subrepeats in *Vicia faba*. *Chromosoma*, **100**: 229–234.

Malik K A and Saxena P K (1992) Thidiazuron induces high-frequency shoot regeneration in intact seedling of pea (*Pisum sativum*), chickpea (*Cicer arietinum*) and lentil (*Lens culinaris*). *Aust. J. Plant Physiol.*, **19**: 731–740.

Martirani L, Stiller J, Mirabella R, Alfano F, Lamberti A, Radutoiu S E, Iaccarino M and Gresshoff P M (1999) T-DNA tagging of nodulation and root-related genes in *Lotus japonicus*. Expression patterns and potential for promotor-trapping and insertional mutagenesis. *Mol. Plant Microbe Interact.*, **12**: 275–284.

Mayer M S, Tullu A, Simon C J, Kumar J, Kaiser W J, Kraft J M and Muehlbauer F J (1997) Development of a DNA marker for *Fusarium* wilt resistance in chickpea. *Crop Sci.*, **37**: 1625–1629.

McMurray C T (1995) Mechanisms of DNA expansion. *Chromosoma*, **104**: 2–13.

Mercy S T, Kakar S N and Chowdhury J B (1974) Cytological studies in three species of the genus *Cicer*. *Cytologia*, **39**: 383–390.

Meyer W, Mitchell T G, Freedman E Z and Vilgalys R (1993) Hybridization probes for conventional DNA fingerprinting used as single primers in the polymerase chain reaction to distinguish strains of *Cryptococcus neoformans*. *J. Clin. Microbiol.*, **31**: 2274–2280.

Michelmore R W, Paran I and Kesseli R V (1991) Identification of markers linked to disease resistance genes by bulked segregant analysis: A rapid method to detect markers in specific genomic regions using segregating populations. *Proc. Natl. Acad. Sci. USA*, **88**: 9828–9832.

Mizoguchi T, Ichimura K and Shinozaki K (1997) Environmental stress response in plants: the role of mitogen-activated protein kinases. *TIBTECH*, **15**: 15–19.

Moore G (2000) Cereal chromosome structure, evolution and painting. *Annu. Rev. Plant Physiol. Plant Mol. Biol.*, **51**: 195–222.

Moore G, Foote T, Helentjaris T, Devos K, Kurata N and Gale M (1995). Was there a single ancestral cereal chromosome? *Trends Genet.*, **11**: 81–82.

Morjane H, Geistlinger J, Harrabi M, Weising K and Kahl G (1994) Oligonucleotide fingerprinting detects genetic diversity among *Ascochyta rabiei* isolates from a single chickpea field in Tunisia. *Curr. Genetics*, **26**: 191–197.

Muehlbauer F J and Singh K B (1987) Genetics of chickpea. In: *The Chickpea* (Eds Saxena M C and Singh K B), CAB International, Wallingford, Oxon, pp. 99–125.

Muehlbauer F J, Kaiser W J and Simon C J (1994) Potential for wild species in cool season food legume breeding. *Euphytica*, **73**: 109–114.

Muehlbauer F J, Weeden N F and Hoffman D L (1989) Inheritance and linkage relationships of several isozyme loci in lentil (*Lens* Miller). *J. Hered.*, **80**: 298–303.

Murthy B N S, Victor J, Singh R P, Fletcher R A and Saxena P K (1996) *In vitro* regeneration of chickpea (*Cicer arietinum* L.): Stimulation of direct organogenesis and somatic embryogenesis by thidiazuron. *Plant Growth Reg.*, **19**: 233–240.

Nakamura Y, Leppert M, O'Connel P, Wolff R, Holm T, Culver M, Martin C, Fujimoto E, Hoff M, Kumlin E and White R (1987) Variable number of tandem repeat (VNTR) markers for human gene mapping. *Science*, **235**: 516–522.

Narayan R K J (1982) Discontinuous DNA variation in the evolution of plant species: the genus *Lathyrus*. *Evolution*, **36**: 877–891.

Narvel J M, Chu W C, Fehr W R, Cregan P B and Shoemaker R C (2000) Development of multiple sets of simple sequence repeat DNA markers covering the soybean genome. *Mol. Breed.*, **6**: 175–183.

Nene Y L and Reddy M V (1987) Chickpea diseases and their control. In: *The Chickpea* (Eds Saxena M C and Singh K B), CAB International, Wallingford, Oxon, UK.

Nene Y L and Sheila V K (1992) Important disease problems of kabuli chickpea. In: *Disease Resistance Breeding in Chickpea* (Eds Singh K B and Saxena M C), ICARDA, Aleppo, Syria, pp. 11–22.

Nene Y L, Sheila V K and Sharma S B (1984) A world list of chickpea (*Cicer arietinum* L.) and pigeonpea (*Cajanus cajan* (L.)) pathogens. *ICRISAT Pulse Pathol. Rep.*, **32**, 19.

Ocampo B, Venora G, Errico A, Singh K B and Saccardo F (1992) Karyotype analysis in the genus *Cicer*. *J. Genet. Breed.*, **46**: 229–240.

Oger P, Petit A and Dessaux Y (1996) A simple technique for direct transformation and regeneration of the diploid legume species *Lotus japonicus*. *Plant Sci.*, **116**: 159–168.

Ohri D and Pal M (1991) The origin of chickpea (*Cicer arietinum* L.): Karyotype and nuclear DNA content. *Heredity*, **66**: 367–372.

Ohtsubo H and Ohtsubo E (1994) Involvement of transposition in dispersion of tandem repeat sequences (TrsA) in rice genomes. *Mol. Gen. Genet.*, **245**: 449–455.

O'Neill C M and Bancroft I (2000) Comparative physical mapping of segments of the genome of *Brassica oleracea* var. *alboglabra* that are homologous to sequence regions of chromosomes 4 and 5 of *Arabidopsis thaliana*. *Plant J.*, **23**: 233–243.

Or E, Hovav R and Abbo S (1999) A major gene for flowering time in chickpea. *Crop Sci.*, **39**: 315–322.

Oram R N, Shaikh M A Q, Zaman K M S and Brown A H D (1987) Isozyme similarity and genetic difference in morphology between Hyprosola, a high yielding, high protein mutant of chickpea (*Cicer arietinum* L.) and its parental cultivar. *Environ. Exp. Bot.*, **27**: 455–462.

Otte O and Barz W (1996) The elicitor-induced oxidative burst in cultured chickpea cells drives the rapid insolubilization of two cell wall structural proteins. *Planta*, **200**: 238–246.

Pandey R and Ganapathy P S (1984) Isolation of sodium chloride-tolerant callus lines of *Cicer arietinum* L. cv BG-203. *Plant Cell Rep.*, **10**: 35–38.

Paterson A, Lander E, Hewitt J D, Peterson S, Lincoln S E and Tanksley S D (1988) Resolution of quantitative traits into Mendelian factors by using a complete linkage map of restriction fragment length polymorphisms. *Nature*, **335**: 721–726.

Paterson A H, Lan T H and Reischmann K P (1996) Toward a unified genetic map of higher plants, transcending the monocot-dicot divergence. *Nature Genet.*, **14**: 380–382.

Patil P B, Vrinten P L, Scoles G J and Slinkard A E (1995) Variation in the ribosomal RNA units of the genus *Lens* and *Cicer*. *Euphytica*, **83**: 33–42.

Penmetsa R and and Cook D (1997) A legume ethylene-insensitive mutant hyperinfected by its rhizobial symbiont. *Science*, **275**: 527–530.

Phillips J C (1988) A distinct race of chickpea wilt in California. *Int. Chickpea Newslett.*, **18**: 19–21.

Pohill R M and Raven P H (eds) (1981) *Advances in Legume Systematics*. Royal Botanical Garden, Kew.

Polisetty R, Patil P, Deveshwar J J, Khetarpal S and Chandra R (1996) Rooting and establishment of *in vitro* grown shoot tip explants of chickpea (*Cicer arietinum* L.). *Indian J. Exp. Biol.*, **34**: 806–809.

Polisetty R, Paul V, Deveshwar J J, Khetarpal S, Suresh K and Chandra R (1997) Multiple shoot induction by benzyladenine and complete plant regeneration from seed explants of chickpea (*Cicer arietinum* L.). *Plant Cell Rep.*, **16**: 565–571.

Powell W, Machray G C and Provan J (1996) Polymorphism revealed by simple sequence repeats. *Trends Plant Sci.*, **1**: 215–222.

Rakshit S, Winter P, Tekeoglu M, Pfaff T, Juarez Muñoz J, Benko-Iseppon A M, Huettel B, Muehlbauer F J and Kahl G. A major locus for *Ascochyta* blight resistance is located on linkage group 4 of the genome map of chickpea (in preparation).

Ramana R V, Venu C, Jayasree T and Sadanadam A (1996) Direct somatic embryogenesis and transformation in *Cicer arietinum* L. *Indian J. Exp. Biol.*, **34**: 716–718.

Ramser J, Weising K, Chikaleke V and Kahl G (1997) Increased informativeness of RAPD analysis by detection of microsatellite motifs. *BioTechniques*, **23**: 285–290.

Rao B G and Chopra V L (1989a) Morphogenesis in callus cultures of chickpea. *Int. Chickpea Newslett.*, **21**: 7–11.

Rao B G and Chopra V L (1989b) Regeneration in chickpea (*Cicer arietinum* L.) via somatic embryogenesis. *J. Plant Physiol.*, **134**: 637–638.

Rao S and Naidu M M (1989) A tissue culture derived pesticide tolerant line of chickpea. *Proc. Indian Acad. Sci. (Plant Sci)*, **99**: 523–527.

Ratnaparkhe M B, Santra D K, Tullu A and Muehlbauer F J (1998a) Inheritance of inter-simple-sequence polymorphisms and linkage with a fusarium wilt resistance gene in chickpea. *Theort. Appl. Genet.*, **96**: 348–353.

Ratnaparkhe M B, Tekeoglu M and Muehlbauer F J (1998b) Inter-simple-sequence-repeat (ISSR) polymorphisms are useful for finding markers associated with disease resistance gene clusters. *Theort. Appl. Genet.*, **97**: 515–519.

Reddy M V and Kabbabeh S (1985) Pathogenic variability in *Ascochyta rabiei* (Pass.) Lab. in Syria and Lebanon. *Phytophatologia Mediterranea*, **24**: 265–266.

Richardson T, Cato S, Ramser J, Kahl G and Weising K (1995) Hybridization of microsatellites to RAPD: a new source of polymorphic markers. *Nucleic Acids Res.*, **23**: 3798–3799.

Robertson L D, Ocampo B and Singh K B (1997) Morphological variation in wild annual *Cicer* species in comparison to the cultigen. *Euphytica*, **95**: 309–319.

Robinson R A (1976) *Plant Pathosystems*. Springer-Verlag, Berlin.

Rossi M, Goggin F L, Milligan S B, Kaloshian I, Ullman D E and Williamson V M (1998) The nematode resistance gene Mi of tomato confers resistance against the potato aphid. *Proc. Natl. Acad. Sci. USA*, **95**: 9750–9754.

Rounsley S, *et al.* (2000) Analysis of the genome sequence of the flowering plant *Arabidopsis thaliana*. *Nature*, **408**: 796–815.

Sagare A P, Suhasini K and Krishnamurthy K V (1993) Plant regeneration via somatic embryogenesis in chickpea (*Cicer arietinum* L.). *Plant Cell Rep.*, **12**: 652–655.

SanMiguel P, Tikhonov A, Jin Y K, Motchoulskaia N, Zakhrov D, Melake-Berhan A, Springer P S, Edwards K J, Lee M, Avramova Z and Bennetzen J L (1996) Nested retrotransposons in the intergenic regions of the maize genome. *Science*, **274**: 765–768.

Santra D K, Tekeoglu M, Kaiser W J and Muehlbauer F J (2001) Identification and mapping of QTLs conferring resistance to *Ascochyta* blight. *Crop. Sci.* (in press).

Saxena M C (1992) Current status and prospects of Kabuli chickpea production. In: *Disease Resistance Breeding in Chickpea* (Eds Singh K B and Saxena M C), ICARDA, Aleppo, Syria, pp. 1–10.

Saxena M C and Singh K B (eds) (1987) *The Chickpea*. CAB International, Wallingford, UK.

Saxena R P and Beg M U (1988) Selection of endosulfan-resistant gram cell line. *Plant Cell Tiss. Org. Cult.*, **12**: 273–277.

Schauser L, Handberg K, Sandal N, Stiller J, Thykjaer T, Pajuelo E, Nielsen A and Stougaard J (1998) Symbiotic mutants deficient in nodule establishment identified after T-DNA transformation of *Lotus japonicus*. *Mol. Gen. Genet.*, **259**: 414–423.

Schmidt T and Heslop-Harrison J S (1996a) High-resolution mapping of repetitive DNA by *in-situ* hybridization: Molecular and chromosomal features of prominent dispersed and discretely localized DNA families from the wild beet species *Beta procumbens*. *Plant Mol. Biol.*, **30**: 1099–1114.

Schmidt T and Heslop-Harrison J S (1996b) The physical and genomic organization of microsatellites in sugar beet. *Proc. Natl. Acad. Sci. USA*, **93**: 8761–8765.

Schmidt T and Heslop-Harrison J S (1998) Genomes, genes and junk: the large-scale organization of plant chromosomes. *Trends Plant Sci.*, **3**: 195–199.

Scofield S R, Tobias C M, Rathjen J P, Chang J H, Lavelle D T, Michelmore R W and Staskawicz B J (1996) Molecular basis of gene-for-gene specificity in bacterial speck disease of tomato. *Science*, **274**: 2063–2065.

Scott K D, Eggler P, Seaton G, Rosetto M, Ablett E M, Lee L S and Henry R J (2000) Analysis of SSRs derived from grape ESTs. *Theort. Appl. Genet.*, **100**: 723–726.

Seetaram A, Ramulu C A and Rao D (1991) Selection and isolation of pesticide resistant cotyledonary derived calli clones in *C. arietinum*. *In Vitro*, **27**: 113.

Sharma D R, Kumari R and Chowdhury J B (1979) Plant regeneration in *Cicer* species through tissue culture. *Indian J. Exp. Biol.*, **17**: 607–609.

Sharma P C and Gupta P K (1982) Karyotypes in some pulse crops. *Nucleus*, **25**: 181–185.

Sharma P C and Gupta P K (1984) Pachytene chromosome morphology in *Cicer bijugum* rech. In: *Perspectives in Cytology and Genetics* (Eds Manna G K and Sinha U), **4**: 485–489.

Sharma P C and Gupta P K (1986) Cytogenetics of legume genera *Cicer* L. and *Lens* L. In: *Genetics and Crop Improvement* (Eds Gupta P K and Bahl J R), Rastogi and Company, Meerut, India, pp. 321–339.

Sharma P C, Huttel B, Winter P, Kahl G, Gardner R C and Weising K (1995b) The potential of microsatellites for hybridization- and PCR-based fingerprinting of chickpea (*Cicer arietinum* L.) and related species. *Electrophoresis*, **16**: 1755–1761.

Sharma P C, Winter P, Bunger T, Huttel B, Weigand F, Weising K and Kahl G (1995a) Abundance and polymorphism of di-, tri- and tetranucleotide tandem repeats in chickpea (*Cicer arietinum* L.). *Theort. Appl. Genet.*, **90**: 90–96.

Shri P V and Davis T M (1992) Zeatin-induced shoot regeneration from immature chickpea (*Cicer arietinum* L.) cotyledons. *Plant Cell Tiss. Org. Cult.*, **28**: 45–51.

Sia E A, Jink-Robertson S and Petes T D (1997) Genetic control of microsatellite instability. *Mutation Res.*, **383**: 61–70.

Siefkes-Boer H J, Noonan M J, Bullock D W and Conner A J (1995) Hairy root transformation system in large-seeded grain legumes. *Israel J. Plant Sci.*, **43**: 1–5.

Simon C J and Muehlbauer F J (1997) Construction of a chickpea linkage map and its comparison with maps of pea and lentil. *J. Hered.*, **88**: 115–119.

Sindhu J S, Singh K P and Slinkard A E (1983) Inheritance to *Fusarium* wilt in chickpea. *J. Hered.*, **74**: 68.

Singh B D, Tiwari K N, Srivastava K, Singh R and Srivastava D K (1998) Regeneration systems in chickpea and other legumes and their possible applications. In: *Genetics and Biotechnology in Crop Improvement* (Eds Gupta P K, Singh S P, Balyan H S, Sharma P C and Ramesh B), Rastogi Publications, Meerut, India, pp. 341–369.

Singh G (1985) Biological specialization and other related studies on parasitisation by *Ascochyta rabiei* on gram. In: *Proc. Workshop on Rabi Pulses*, AICPIP, SKUAT, Srinagar, India.

Singh H, Kumar J, Smithson J B and Haware M P (1987) Complementation between genes for resistance to race 1 of *Fusarium oxysporum* f.sp. *ciceri* in chickpea. *Plant Pathol.*, **36**: 539–543.

Singh K B, Malhotra R S, Halila M H, Knights E J and Verma M M (1994) Current status and future strategy in breeding chickpea for resistance to biotic and abiotic stresses. *Euphytica*, **73**: 137–149.

Singh K B, Malhotra R S and Muehlbauer F J (1984) An annotated bibliography of chickpea genetics and breeding. ICRISAT/ICARDA joint publication. ICARDA, PO Box 5466, Aleppo, Syria.

Singh K B and Ocampo B (1993) Interspecific hybridization in annual *Cicer* species. *J. Genet. Breed.*, **47**: 199–204.

Singh K B and Ocampo B (1997) Exploitation of wild *Cicer* species for yield improvement in chickpea. *Theort. Appl. Genet.*, **95**: 418–423.

Singh K B, Ocampo B and Robertson L D (1998) Diversity for abiotic and biotic stress resistance in the wild annual *Cicer* species. *Genet. Res. Crop Evol.*, **45**: 9–17.

Singh K B and Reddy M V (1983) Inheritance of resistance to *Ascochyta* blight in chickpea. *Crop Sci.*, **23**: pp. 9–10.

Singh K B and Reddy M V (1993) Sources of resistance to *Ascochyta* blight in wild *Cicer* species. *Neth. J. Plant Path.*, **99**: 163–167.

Singh K B, Reddy M V and Haware M P (1992) Breeding for resistance to ascochyta blight in chickpea. In: *Disease Resistance in Chickpea* (Eds Singh K B and Saxena M C), ICARDA, Aleppo, Syria, pp. 23–54.

Singh O (1988) Induced Mutations and Cytogenetic Studies in Chickpea (*Cicer arietinum* L.). Ph.D. thesis, Meerut University, Meerut, India.

Singh R P and Singh B D (1989) Recovery of rare interspecific hybrids of gram *Cicer arietinum* × *C. cuneatum* L. through tissue culture. *Curr. Sci.*, **58**: 874–876.

Singh R P, Singh B D, Jaiswal H K, Singh R M and Singh R B (1982) Organogenesis in callus cultures of chickpea. *Indian J. Agric. Sci.*, **52**: 86–90.

Smeets H J M, Brunner H G, Ropers H H and Wieringa B (1989) Use of variable simple sequence motifs as genetic markers: application to study of myotonic dystrophy. *Hum. Genet.*, **83**: 245–251.

Smyth D R, Kalitsis P, Joseph J L and Sentry J W (1989) Plant retrotransposon from *Lilium henryi* is related to Ty3 of yeast and the gypsy group of *Drosophila*. *Proc. Natl. Acad. Sci. USA*, **86**: 5015–5019.

Srinivasan, Mohapatra T and Sharma R P (1991) *Agrobacterium*-mediated genetic transformation of chickpea (*Cicer arietinum* L.). *Indian J. Exp. Biol.*, **29**: 758–761.

Srivastava S K, Singh S N and Khare M N (1984) Assessment of yield losses in some promising gram cultivars due to fusarium wilt. *Indian J. Plant Prot.*, **12**: 125–126.

Staginnus C, Winter P, Desel C, Schmidt T and Kahl G (1999) Molecular structure and chromosomal localization of major repetitive DNA families in the chickpea (*Cicer arietinum* L.) genome. *Plant Mol. Biol.*, **39**: 1037–1050.

Stebbins G L (1971) *Chromosome Evolution in Higher Plants*. Edward Arnold Ltd., London, p. 219.

Stougaard J (2000) Regulators and regulation of legume root nodule development. *Plant Physiol.*, **124**: 531–540.

Surya-Prakash, Chowdhury J B, Jain P K and Chowdhury V K (1992) Factors affecting plant regeneration in chickpea (*Cicer arietinum* L.). *Indian J. Exp. Biol.*, **30**: 1149–1153.

Suzuki K and Shinshi H (1995) Transient activation and tyrosine phosphorylation of a protein kinase in tobacco cells treated with a fungal elicitor. *Plant Cell*, **7**: 639–647.

Tabe L M, Higgins C M, McNabb W C and Higgins T J V (1993) Genetic engineering of grain and pasteur legumes for improved nutritive value. *Genetica*, **90**: 181–200.

Tang X, Frederick R D, Zhou J, Halterman D A, Jia Y and Martin G B (1996) Initiation of plant disease resistance by physical interaction of AvrPto and Pto kinase. *Science*, **274**: 2060–2063.

Tanksley S D, Bernatzky R, Lapitan N L and Prince J P (1988) Conservation of gene repertoire but not gene order in pepper and tomato. *Proc. Natl. Acad. Sci. USA*, **85**: 6419–6423.

Tanksley S D, Ganal M W, Price J P, de Vincente M C, Bonierbale M W, Broun P, Fulton T M, Giovannoni H, Grandillo S, Martin G B, Messeguer R, Miller J C, Miller L, Patterson A H, Pineda O, Röder M, Wing R A, Wu W and Young N D (1992) High density molecular linkage maps of tomato and potato genomes. *Genetics*, **132**: 1141–1160.

Tanksley S D and McCouch S R (1997) Seed banks and molecular maps: Unlocking genetic potential from the wild. *Science*, **277**: 1063–1066.

Tautz D (1989) Hypervariability of simple sequences as a general source for polymorphic DNA markers. *Nucleic Acids Res.*, **17**: 6463–6471.

Tayyar R I, Lukaszewski A J and Waines J G (1994) Chromosome banding patterns in the annual species of *Cicer*. *Genome*, **37**: 656–663.

Tayyar R I and Waines J G (1996) Genetic relationships among annual species of *Cicer* (Fabaceae) using isozyme variation. *Theort. Appl. Genet.*, **92**: 245–254.

Tekeoglu M, Santra D K, Kaiser W J and Muehlbauer F J (2001) *Ascochyta* blight resistance inheritance in three chickpea recombinant inbred lines. *Crop Sci.* (in press).

Tewari S K and Pandey M P (1986) Genetics of resistance to *Ascochyta* blight in chickpea (*Cicer arietinum* L.). *Euphytica*, **35**: 211–215.

Tikhonov A P, SanMiguel P J, Kakajima Y, Gorenstein N M, Bennetzen J L and Avramova Z (1999) Colinearity and its exception in orthologous *adh* regions of maize and sorghum. *Proc. Natl. Acad. Sci. USA*, **96**: 7409–7414.

Trieu A T, Burleigh S H, Kardailsky I V, Maldonado-Mendoza I E, Versaw W K, Blaylock L A, Shin H, Chiou T, Katagi H, Dewbre G R, Weigel D and Harrison M J (2000) Transformation of *Medicago truncatula* via infiltration of seedlings or flowering plants with *Agrobacterium*. *Plant J.*, **22**: 531–541.

Tullu A (1996) Genetics of fusarium wilt resistance in chickpea. Ph.D thesis, Crop and Soil Sciences Department, Washington State University, Pullman, USA.

Tullu A, Muehlbauer F J, Simon C J, Mayer M S, Kumar J, Kaiser W J and Kraft J M (1998) Inheritance and linkage of a gene for resistance to race 4 of fusarium wilt and RAPD markers in chickpea. *Euphytica*, **102**: 227–232.

Tuwafe S, Kahler A L, Boe A and Ferguson M (1988) Inheritance and geographical distribution of allozyme polymorphisms in chickpea (*Cicer arietinum* L.).

Udupa S M, Sharma A, Sharma A P and Pai R A (1993) Narrow genetic variability in *Cicer arietinum* L. as revealed by RFLP analysis. *J. Plant Biochem. Biotechnol.*, **2**: 83–86.

Unfried K, Schiebel K and Hemleben V (1991) Subrepeats of rDNA intergenic spacer present as prominent independent satellite DNA in *Vigna radiata* but not in *Vigna angularis*. *Gene*, **99**: 63–68.

Upadhyaya H D, Haware M P, Kumar J and Smithson J B (1983a) Resistance to wilt in chickpea. I. Inheritance of late-wilting in response to race 1. *Euphytica*, **32**: 447–452.

Upadhyaya H D, Smithson J B, Kumar J and Haware M P (1983b) Resistance to wilt in chickpea. II. Further evidence for two genes for resistance to race 1. *Euphytica*, **32**: 749–755.

van der Maesen L J G (1987) Origin, history and taxonomy of chickpea. In: *The Chickpea* (Eds Saxena M C and Singh K B), CAB International Publications, Wallingford, UK, pp. 11–34.

van Rheenen H A, Reddy M V, Kumar J and Haware M P (1992) Breeding for resistance to soil-borne diseases in chickpea. In: *Disease Resistance in Chickpea* (Eds Singh K B and Saxena M C), ICARDA, Aleppo, Syria, pp. 55–70.

Verma M M, Sandhu J S, Brar H S and Brar J S (1990) Crossability studies in different species of *Cicer* (L.). *Crop Improv.*, **17**: 179–181.

Vershinin A V and Heslop-Harrison J S (1998) Comparative analysis of the nucleosomal structure in rye, wheat and their relatives. *Plant Mol. Biol.*, **36**: 149–161.

Vir S and Grewal J S (1974) Physiologic specialization in *Ascochyta rabiei*, the causal organism of gram blight. *Indian Phytopathol.*, **27**: 355–360.

Vogel J M and Scolnik P A (1997) Direct amplification from microsatellites: detection of simple-sequence repeat-based polymorphism without cloning. In: *DNA Markers: Protocols, Aplications and Overviews* (Eds Caetano-Anollés G and Gresshoff P M), Wiley-Lis, John Wiley and Sons Inc., New York, pp. 133–150.

Vos P, Hogers R, Bleeker M, Reijans M, van de Lee T, Hoernes M, Frijters A, Pot J, Peleman J, Kuiper M and Zabeau M (1995) AFLP: A new technique for DNA fingerprinting. *Nucleic Acids Res.*, **23**: 4407–4414.

Vos P *et al.* (1998) The tomato Mi-1 gene confers resistance to both root-knot nematodes and potato aphids. *Nature Biotechnol.*, **16**:1365–1369.

Vosman B and Arens P (1997) Molecular characterization of GATA/GACA microsatellite repeats in tomato. *Genome*, **40**: 25–33.

Webb K J, Skot L, Nicholson M N, Jorgensen B and Mizen S (2000) *Mesorhizobium loti* increases root-specific expression of a calcium-binding protein homologue identified by promoter tagging in *Lotus japonicus*. *Mol. Plant Microbe Interact.*, **13**: 606–616.

Weber J L and May P E (1989) Abundant class of human DNA polymorphisms which can be typed using the polymerase chain reaction. *Am. J. Hum. Genet.*, **44**: 388–396.

Weeden N F, Ambrose M and Swiecicki W (1993) *Pisum sativum*, pea. In: *Genetic Maps*. 6th ed. (Ed O'Brian S J), Cold Spring Harbor Laboratory Press, Cold Spring Harbor, New York, pp. 624–634.

Weeden N F, Muehlbauer F J and Ladizinsky G (1992) Extensive conservation of linkage relationships between pea and lentil genetic maps. *J. Hered.*, **83**: 123–129.

Weising K, Atkinson R G and Gardner R C (1995b) Genomic fingerprinting by microsatellite-primed PCR: a critical evaluation. *PCR Meth. Appl.*, **4**: 249–255.

Weising K, Kaemmer D, Weigand F, Epplen J T and Kahl G (1992) Oligonucleotide fingerprinting reveals probe-dependent levels of informativeness in chickpea (*Cicer arietinum* L.). *Genome*, **35**: 436–442.

Weising K, Nybom H, Wolff K and Meyer W (1995a) *DNA Fingerprinting in Plants and Fungi*. CRC Press, Boca Raton, FL.

Weising K, Ramser J, Kaemmer D, Kahl G and Epplen J T (1991) Oligonucleotide fingerprinting in plants and fungi. In: *DNA Fingerprinting: Approaches and applications* (Eds Burke T, Dolf G, Jeffreys A J and Wolff R), Birkhäuser, Basel, pp. 313–329.

Weising K, Weigand F, Driesel A J, Kahl G, Zischler H and Epplen J T (1989) Polymorphic simple GATA/GACA repeats in plant genomes. *Nucleic Acids Res.*, **17**: 10128.

Weising K, Winter P, Hüttel B and Kahl G (1998) Microsatellite markers for molecular breeding. *J. Crop Prod.*, **1**: 113–143.
Williams J G K, Kubelik A R, Livak K J, Rafalski J A and Tingey S V (1990) DNA polymorphisms amplified by arbitrary primers are useful as genetic markers. *Nucleic Acids Res.*, **18**: 6531–6535.
Winter P, Benko-Iseppon A M, Hüttel B, Ratnaparkhe M, Tullu A, Sonnante G, Pfaff T, Tekeoglu M, Santra D, Sant V J, Rajesh P N, Kahl G and Muehlbauer F J (2000) A linkage map of the chickpea (*Cicer arietinum* L.) genome based on recombinant inbred lines from a *C. arietinum* × *C. reticulatum* cross: Localization of resistance genes for Fusarium races 4 and 5. *Theort. Appl. Genet.*, **101**:1155–1163.
Winter P and Kahl G (1995) Molecular marker technologies for crop improvement. *World J. Microbiol. Biotechnol.*, **11**: 449–460.
Winter P, Pfaff T, Udupa S M, Hüttel B, Sharma P C, Sahi S, Arreguin-Espinoza R, Weigand F, Muehlbauer F J and Kahl G (1999) Characterization and mapping of sequence-tagged microsatellite sites in the chickpea (*Cicer arietinum* L.) genome. *Mol. Gen. Genet.*, **262**: 90–101.
Wopereis J, Pajuela E, Dazzo F B, Jiang Q, Gresshoff P M, de Bruijn F J, Stougaard J and Szczyglowski K (2000) Short root mutant of *Lotus japonicus* with a dramatically altered symbiotic phenotype. *Plant J.*, **23**: 97–114.
Wu K, Jones R, Danneberger L and Scolnik P A (1994) Detection of microsatellite polymorphisms without cloning. *Nucleic Acids Res.*, **22**: 3257–3258.
Xu Q, Fu H H, Gupta R and Luan S (1998) Molecular characterization of a tyrosine-specific protein phosphatase encoded by a stress-responsive gene in *Arabidopsis*. *Plant Cell*, **10**: 849–857.
Young N D (1996) QTL mapping and quantitative disease resistance in plants. *Annu. Rev. Phytopathol.*, **34**: 479–501.
Zischler H, Kammerbauer C, Studer R, Grzeschik K H and Epplen J T (1992) Dissecting (CAC_5/GTG_5) multilocus fingerprints from man into individual locus-specific, hypervariable components. *Genomics*, **13**: 983–990.
Zohary D (1999) Monophyletic vs. polyphyletic origin of the crops on which agriculture was founded in the near east. *Genet. Res. Crop Evol.*, **46**: 133–142.
Zohary D and Hopf M (1993) *Domestication of Plants in the Old World*. Clarendon Press, Oxford, pp. 1–271.

BIODIVERSITY IN INDIAN UNDEREXPLOITED/TRIBAL PULSES

K. JANARDHANAN[1], V. VADIVEL[2] AND M. PUGALENTHI[1]
1 Seed Technology Laboratory, Department of Botany, Bharathiar University, Coimbatore – 641 046 (T.N.), India
2 Present address : Indian Cardamom Research Institute, Spices Board, Myladumpara, Kailasanadu – 685 553, Kerala, India
e-mail: karnam-janardhan@usa.net

Abstract

Pulses are an important source of proteins to the vegetarian and socio-economically weaker sections of the population. With rapid increase in population coupled with stagnation in pulses production at around 12 to 14 million tonnes, the per capita availability of protein has decreased causing Protein-Energy-Malnutrition (PEM) especially among children below the age of five years in India. This has necessitated exploration of alternate sources of protein to bridge the gap for protein requirement of the various sections of vegetarian population. In this context, alternate sources like untraditional legumes (underexploited/tribal pulses) assume significance. Therefore it is time to understand their biodiversity with a view to exploit such genetic resources for additional supply of plant protein, so that they can augment the availability of protein from conventional pulse crops in India. Around 40 legumes belonging to the genera, *Abrus*, *Acacia*, *Alysicarpus*, *Atylosia*, *Bauhinia*, *Canavalia*, *Cassia*, *Dolichos*, *Entada*, *Erythrina*, *Indigofera*, *Lens*, *Mucuna*, *Parkia*, *Parkinsonia*, *Phaseolus*, *Pongamia*, *Prosopis*, *Sesbania*, *Tamarindus*, *Vicia*, *Vigna* and *Xylia* are known to be consumed by as many as 550 tribal communities in India who account for 67.76 million of total Indian population. Besides there are another 10 underexploited/little known legumes, which have not been exploited to the extent they deserve. The importance of Indigenous Knowledge (IK) including the traditional medicinal properties of these wild legumes have been enumerated. The authors' team of workers have gathered 192 germplasm (accessions) of wild legumes including 45 tribal pulses from different agroclaimatic regions in India. Marked genetic variation among the germplasm/varieties of several tribal pulses at the time of collection and evaluation for agrobotanical characters have been recorded. The genetic erosion of germplasm of tribal pulses, such as *Entada scandense*, *Mucuna monosperma*, *M. gigantea*, *Atylosia goensis*, *Bauhinia vahlii*, *Xylia xylocarpa*, *Parkia roxburghii*, *Vigna radiata* var. *sublobata*, *V. mungo* var. *silvestris* and *V. grandis*, has been noticed. Several germplasm conservation strategies and their utility in this context have been stressed. Chemical investigations on

P.K. Jaiwal and R.P. Singh (eds.), Improvement Strategies for Leguminosae Biotechnology, 353–405.

levels of crude protein and crude lipid also have revealed the existence of genetic diversity not only among different varieties but also among different germplasm of the same species. Several germplasm/varieties/species such as *Atylosia goensis*, *Canavalia ensiformis*, *C. gladiata*, *Erythrina indica*, *Indigofera linifolia*, *Lens esculenta*, *Mucuna* spp., *Psophocarpus* spp., *Pongamia pinnata*, *Phaseolus lunatus*, *Sesbania bispinosa*, *Vicia faba*, *Vigna trilobata* and *Xylia xylocarpa* possess 20–80% (w/w) more crude protein compared to the conventional pulses. Similarly several germplasm/varieties/species like *Alysicarpus rugosus, Bauhinia* spp., *Cassia* spp., *Entada scandense, Erythrina indica, Mucuna* spp., *Parkia roxburghii, Pongamia pinnata, Prosopis chilensis, Psophocarpus* spp., *Tamarindus indica* and *Xylia xylocarpa* contain high levels of crude lipid (10–30%, w/w). Human nutrition requires a balanced source of essential amino acids. Generally seed proteins of several legumes are deficient in sulfur – containing amino acids, cysteine and methionine. Exceptionally the seed proteins of *Alysicarpus rugosus*, *Indigofera linifolia* and *Tamarindus indica* contain sulfur containing amino acids more than FAO/WHO (1991) requirement pattern. Similarly seed proteins of certain varieties/germplasm of tribal pulses like *Abrus precatorius*, *Acacia leucophloea*, *A. nilotica*, *Bauhinia malabarica*, *Canavalia gladiata*, *Cassia laevigata*, *Entada scandens*, *Indigofera linifolia*, *Mucuna pruriens* var. *pruriens*, *M. pruriens* var. *utilis*, *Parkia roxburghii*, *Phaseolus lunatus*, *Prosopis chilensis*, *Parkinsonia aculeata*, *Pongamia pinnata*, *Vigna aconitifolia*, *V. capensis*, *V. sesquipedalis*, *V. sinensis*, *V. sublobata*, *V. umbellata*, *V. vexillata* and *Xylia xylocarpa* contain tryptophan more than FAO/WHO (1991) requirement pattern. The germplasm of untraditional pulses such as *Canavalia ensiformis*, *C. gladiata*, *Mucuna pruriens* var. *pruriens*, *M. pruriens* var. *utilis* and *Psophocarpus tetragonolobus* have been advocated for large scale cultivation and consumption as alternate and additional sources of plant protein after toxicological evaluation and technological processing.

1. Introduction

1.1. BIODIVERSITY CONCEPT

Biodiversity is the variety of the world's living species, including their genetic diversity and the communities and ecosystems that they form (Sinha, 1996). Biodiversity is dynamic: the genetic composition of species changes over time in response to natural and human induced selection pressures. The occurrence and relative abundance of species in ecological communities changes as a result of ecological and physical factors.

Biologists have identified India as one of the top twelve megadiversity countries of the world; harbouring an estimated 500,000 out of some 10 to 30 million species of living organisms (Gadgil, 1996). According to Khan and Al-Ajmi (1999) out of an estimated total number of species on earth at between 13 and 14 million, only 1.75 million have been fully described.

India being one of the biodiversity centres has contributed to world agriculture at least 167 plant species (Khoshoo, 1991; Zeven and de Wet, 1982). Within the overall mega Indian Centre of Diversity as recognized by Vavilov, there are at least nine subcentres of diversity, where wild relatives of cultivated plants still occur (Arora and Pandey, 1996). In

India, Eastern Himalayas and Western Ghats have been recognized as two of the world's 18 Biodiversity Hot Spots. India ranks tenth among the plant rich countries of the world and sixth among the centres of diversity (Gadgil, 1996). It is estimated that India accounts for 8% of the global biodiversity existing in only 2.4% of land area of the world.

The International Convention on Biological Diversity (CBD) initiated at the Rio Earth Summit in 1992 for the first time recognizes the sovereign right of all countries over their biodiversity resources. CBD commits countries to prepare inventories of their biodiversity resources, to monitor their fate, to organize adequate information systems for such resources and to take steps to conserve them. CBD also realizes the role of indigenous communities in conservation and sustainable use of biodiversity resources and enjoins member countries to share benefits of utilization of biodiversity resources with the indigenous communities (Gadgil, 1996).

The task of inventorying and conserving biodiversity is immense, for valuable elements like genetic resources of a particular category of plants are not restricted to a few pockets in a country. Plant Genetic Resources (PGR) (germplasm) of several tribal/underexploited pulses have been gathered mostly from natural stands in Western Ghats, Eastern Ghats, Deccan plateau and northeast India; such a task cannot be carried out in a routine fashion since ecological systems are highly variable in space and time and must be investigated and managed in ways sensitive to this variation. So it is necessary to organize an effective network covering the entire country.

As part of national programs on agricultural research and development, recently in India, a series of meetings, dialogues and workshops were conducted on issues related to plant genetic resources (PGR)/genetic resources for food and agriculture (GRFA). The more prominent were a two dialogue series by the Indian Society of Plant Genetic Resources (ISPGR) on national policy options and PGR management, held in December 1993 and December 1998, and a national workshop on concern for management, conservation and use of agricultural biological diversity (commonly referred to as agrobiodiversity), held in October 1997. These dialogues/workshops were attended by a broad range of interested people such as scientists, experts, members of governmental and non-governmental organizations, representatives of agricultural universities and others. It was unanimously resolved that the importance of agrobiodiversity must be recognized as a distinct national issue of prime concern within the broader area of biodiversity (Paroda *et al.*, 1999).

Plant Genetic Resources for Food and Agriculture (PGRFA) provide the biological basis for world food security and directly or indirectly support the livelihoods of every person on the earth (FAO, 1996). The genetic diversity within traditional crop species is now under threat, because modern and intensive forms of agriculture – which produce much of the world's food supply – are based on genetically uniform crops. Genetic diversity within the edible plant species is being eroded by the destruction of the habitats of wild related plant species (Khan and Al-Ajmi, 1999). Scientists agree that, at current rates of tropical forest loss and disturbance, roughly 5–10% of all tropical forest species will be lost per decade during the next quarter century (Sinha, 1996). According to Switzerland based International Union for Conservation of Nature and Natural Resources (IUCN), every year about 1,200 plant species at global level and as many as 16 plant species in northeast India alone become extinct. IUCN also has listed about 25,000 plant species as endangered which may become extinct in the near future.

India has dense forest cover of only about 12%, which is much less than the goal of having 33% cover. There is also a need to shift the focus from merely forest protection to managing forests by defining and incorporating the role of communities into the policy guidelines. As India's forests are extremely biodiverse, the focus must be on the protection of pristine forests. But there is no information about the extent of such forests and what is happening to them.

Conservation will succeed only if causes of environmental degradation and species loss are addressed. The complex threats to biological diversity call for a wide range of responses from many private and public sectors and these responses must be adjusted to suit local conditions.

1.2. REASONS FOR WIDE PREVALENCE OF PROTEIN-ENERGY-MALNUTRITION (PEM) AND ITS IMPLICATIONS

There is an ever widening gap between food supplies and population growth, particularly in the Third World countries. This food shortage is serious when per capita protein intake is considered (Amubode and Fetuga, 1984). The protein occupies an important and unique place in man's diet. Although various classes of compounds can fulfil our energy needs, the protein requirement is basic and irreplaceable (Casey and Wrigley, 1982). Proteins are required for maintenance (replacing the wear and tear in tissue) in adults, for growth in infants and children, for foetal development in pregnancy and milk output during lactation. The relative requirements of proteins of the latter groups are higher than in adults (Narasinga Rao *et al.*, 1989). Seed products are our most important dietary sources of both energy and protein. Cereals and legumes supply about two-thirds of man's protein intake (Pimentel *et al.*, 1975).

Inadequate availability and consumption of protein food coupled with both population explosion and urbanization result in Protein-Energy-Malnutrition (PEM) or Protein-Calorie-Malnutrition (PCM), in Third World countries including India. The PEM problem will remain, if efforts are not made towards finding other available cheaper sources of protein (Prakash and Misra, 1988; Waterlow, 1994). About 80% of the protein consumed by humans in developing countries is supplied by plants (Singh and Singh, 1992). Legumes are some of the low priced sources of protein-rich foods that seem to be important in alleviating PEM (Aykroyd *et al.*, 1982). The conventional method of improvement has, however, not been so successful in meeting the vegetable protein demand of ever increasing population due to their narrow genetic variability and sexual incompatibility. India has attained food security, but not nutrition security as the production of protein rich pulses do not keep pace with the population growth and consequently its per capita availability has declined from 60.7 g in 1951 to 33.2 g in 1998 (only provisional). Nonetheless, FAO has recommended a per capita consumption of 80 g per day (Satya Sundaram, 2000).

There is a huge shortfall in the demand and supply of pulses because pulse crops have not experienced the impact of Green Evolution. Ironically, India, despite being the largest producer, is also the biggest importer of pulses. The latest estimate of production of pulses for 1996–97 is 14.8 million tonnes as against its requirements of 18 million tonnes. About 0.5 million tonnes of pulses have been imported and thus a net shortage of 2.7 million tonnes existed by the terminal year (1996–97) of the VIII Plan period (Ali, 1997).

In 1995, it has been estimated that currently more than 200 million under five-year-old children are malnourished in terms of being underweight. Currently over two-thirds (79%) of the world's malnourished children live in Asia (especially southern Asia), whilst 17% are in Africa and 3% in Latin America (WHO, 1995).

Malnutrition is widespread in India. Based on the studies undertaken by the National Nutrition Monitory Bureau (NNMB), the major nutritional deficiency diseases identified in India are PEM, vitamin A deficiency, iron deficiency, anemia and endemic goitre. PEM is by far the most widespread nutritional problem affecting mainly children in the preschool age (1–5 years), particularly among the poor socio-economic groups of the population. Malnutrition affects the physical and mental well-being of individuals (Devadas, 1993).

Using the WHO Global Database on Child Growth, the worldwide magnitude of PEM has been reviewed by de Onies *et al.* (1993). The International Conference on Nutrition (ICN) held in December, 1992, identifies the root causes of malnutrition as poverty, deprivation, social inequality, lack of education and growth of population out of proportion to the natural resource endowment (WHO, 1993; Ramalingaswami, 1995).

In the past, investment in pulse research has been very low as compared to rice and wheat, primarily due to national priority for attaining self-sufficiency in food security. Nutritional security, especially in the context of widespread protein malnutrition in growing children, is now being recognized and prioritized. During VII Plan period, research investment on pulse crops was Rs. 101.2 million, which has been raised to Rs. 301.9 million for the VIII Plan. The nodal research Organization on pulses, the Directorate of Pulses Research was strengthened and upgraded as the Indian Institute of Pulses Research (IIPR) in 1993 and the erstwhile national network, All-India Co-ordinated Pulses Improvement Project was trifurcated and independent co-ordinated research projects on chickpea, pigeonpea and other pulses were created in 1995. The proposed financial outlay for the IIPR and three co-ordinated projects for IX Plan is over Rs. 1,000 million (Ali, 1997).

The search for novel high quality but cheap sources of protein and energy has continued to be a major concern of governments and bodies charged with the responsibility for food and nutrition in many parts of the developing world, including India (Balogun and Fetuga, 1986). Projections based on current trends indicate a gap between human population and protein supply. Hence, conscious efforts resulted in the development of alternative sources of protein such as single cell proteins, algal proteins; insect proteins and leaf protein concentrates (see Rajaram and Janardhanan, 1992a). At present sincere and concerted efforts are being made to identify and evaluate underexploited legume food sources as alternate protein crops for the future (Egbe and Akinyele, 1990). A perusal of the available literature reveals that at present concerted and systematic efforts are being made for nutritional evaluation of several Indian underexploited/tribal pulses (Janardhanan and Lakshmanan, 1985; Babu *et al.*, 1988; Ravindran and Ravindran, 1988; Rajaram and Janardhanan, 1990, 1991a, 1992a, b; Siddhuraju *et al.*, 1992a, b, 1994, 1995a, b, 1996a, b; Vijayakumari *et al.*, 1993a, b, 1994, 1995; Rajyalakshmi and Geervani, 1994; Mohan and Janardhanan, 1993a, b, 1994a, b, 1995a, b; Vadivel and Janardhanan, 1999, 2001).

1.3. TRADITIONALLY CULTIVATED PULSE CROPS IN INDIA

A wide range of legume crops has an important, even vital, place in many agricultures, ranging from arid to wet, and cold temperate to tropical, sometimes as essential staples, but otherwise providing welcome dietary diversity. The current worldwide interest in legume crops, however, has been stimulated by the high protein content and favourable nutritional, storage and processing qualities of their seeds. The legumes have an amino acid balance nutritionally complementary to that of cereal proteins for man and monogastric animals (Thomson, 1975).

The following six legumes are extensively cultivated as grain crops in the world, viz. chickpea (*Cicer arietinum* L.), cowpea (*Vigna unguiculata* (L.) Walp.), dry bean (*Phaseolus vulgaris* L.), peanut (*Arachis hypogaea* L.), soybean (*Glycine max* (L.) Merrill.) and pigeonpea (*Cajanus cajan* Mill sp.) (Roberts, 1970).

In India as many as 17 legume crops are cultivated (Shiv Raj, 1978). They are *Cicer arietinum* L. (chickpea or Bengal gram), *Cajanus cajan* Mill sp. (pigeonpea or red gram), *Pisum sativum* L. (peas), *Glycine max* (L.) Merrill. (soybean), *Phaseolus aureus* Roxb. (mungbean or green gram), *P. mungo* L. (urdbean or black gram), *Vigna unguiculata* (L.) Walp. (cowpea), *Phaseolus vulgaris* L. (french beans), *Lens culinaris* Medic (masur), *Vicia faba* L. (vetches), *Phaseolus aconitifolius* Jacq. (moth bean), *P. calcaratus* Roxb. (rice bean), *P. lunatus* L. (lima bean), *Lathyrus sativa* L. (khesari), *Cyamopsis tetragonolobus* L. (cluster beans), *Dolichos lablab* L. (field bean) and *Arachis hypogaea* L. (groundnut). Of these green gram (*Vigna radiata*), black gram (*Vigna mungo*), cowpea (*Vigna unguiculata*) and pigeonpea (*Cajanus cajan*) (Udayasekhara Rao and Deosthale, 1982) and chickpea (*Cicer arietinum*) (Jambunathan and Singh,1980) are the commonly cultivated legume crops. In addition, rice bean (*V. umbellata*) and moth bean (*V. aconitifolia*) are also cultivated as pulse and forage crops (personal communication from Dr. Z. Abraham, Sr. Scientist, NBPGR Regional station, Trichur, Kerala).

In India, pulses were brought within the ambit of Technology Mission only in 1991. Various Micro missions operate under this to deal with different aspects of pulses program. Micro mission-I is operated by the ICAR in collaboration with Department of Agriculture and Co-operations (DAC), Government of India. It envisages evolving profitable crop production and protection and Technologies for various regions and crop growing situations. Five pulse crops, viz. chickpea, pigeonpea, urdbean, mungbean and lentil, have been covered under this mission (Ali, 1997).

1.4. IMPORTANCE OF INDIGENOUS KNOWLEDGE (IK)

According to Jain (1995) an interesting area of work in ethnobotany is the study of selection brought about by the tribals in certain economic plants, and the conservation of germplasm through patronage of land races.

Many tribals and other orthodox cultivators have not adopted all new or improved varieties of crops and have continued to grow the traditional (even less productive) land races or wild relatives of commonly cultivated crops, thereby maintaining their genetic material or germplasm. Specific character of hardiness, disease resistance and adaptability to special conditions like water-logging, extreme drought or cold, etc. in the land races have been continuously utilized by the plant breeders and on some occasions helped averting widespread famines (Arora, 1981).

A revitalized investigation of all plant resources available worldwide, including little known plant species, is the need of the hour, to feed, clothe and house, an ever increasing population (Vietmeyar, 1986). Presently the importance of ethnobotanical researches mainly in ethnomedicine and the group or regional ethnobotany is keenly felt and it represents one of the best avenues for searching new economic plants for food, medicine, etc. as well as for a gene-pool source for development of agricultural and medicinal crops (Shah, 1995). The tribal sects of India live on a vast treasure of germplasm, which has hitherto remained untapped. Some of the underutilized legumes/tribal pulses may fit well into subsistence agriculture as alternate protein sources (Rajaram and Janardhanan, 1993b).

A well-documented information is available on genetic resources of wild/less known cultivated plants including pulses in India (Bhandari, 1974; Jain, 1981, 1991, 1996; Arora, 1981, 1985a; Paroda and Bhag Mal, 1989; Janardhanan, 1990; Janardhanan and Rajaram, 1993; Bhag Mal and Joshi, 1991; Sing and Premanath, 1992; Rajaram and Janardhanan, 1993b; Siddhuraju *et al.*, 1993; Arora and Pandey, 1996; Negi, 1996; Vadivel and Janardhanan, 1997; Vadivel *et al.*, 1994, 1998; Anonymous, 1996).

The Indian subcontinent is ethnically, floristically and agriculturally very diverse. A wide variety of land races, primitive cultivars, wild semi-domesticated/domesticated and gathered diversity of potential use is held by native communities including tribals who are the custodians of such priced resources (Arora, 1995, 1996).

Tribal people through their hereditary traditional knowledge infer what to eat and what not to eat. In fact food habits of various tribes developed on the basis of experiences and survival through successive generations. However, further study of the food value of the roots and tubers and other plant materials (including seeds) eaten by tribal people may be undertaken in co-operation with Nutrition Research Laboratories, with a view to advise the tribal people on better use of such plant materials, which have already helped these people to have extraordinary vigour and stamina to sustain the hardships in such harsh and inhospitable environment (Gupta, 1995).

Ethnobotanical surveys and field work in tribal areas and amongst primitive human societies and cultures is an important and critical area of research, because the habitat and the environment where the tribals and primitive people learnt useful folklore and carried out on-site experimentation with plants for millennia are fast disappearing. This work is in progress, both in the New world and Old world at several centres of research. An All India Co-ordinated Research Project on Ethnobiology has recently been formulated with project centres at Lucknow, Calcutta, Coimbatore, Jammu, Srinagar (Garhwal), Port Blair, etc. An experienced plant taxonomist, an anthropologist, a linguist or folklorist could form an excellent team for identification, collection, documentation and inventorization of plants used by ethnic races and cultures, namely for food, medicine, housing, etc.

With the advent of greater awareness on Biodiversity Conservation/Intellectual Property Rights/Plant Breeders' Rights, etc., concern on national sovereignty and role of farmers has gained momentum (Swaminathan, 1995). The various tribal sects live in perfect harmony with nature. It is believed that their knowledge about the utility of various plants is passed on orally from one generation to another. The keepers of much of this knowledge are often the elders. An increasing and rapid cultural and mass rural-urban migration among the tribes is witnessed in the latter half of the 20th century, a danger that indigenous knowledge may not be passed on to younger generations and could be lost forever exists. In view of these developments, the United Nations declared the year 1993 as "International year of Indigenous People" based on the recommendation of the Rio Earth Summit (Shankar, 1995).

The Convention on Biological Diversity (CBD) contains important provisions regarding indigenous and other local communities, including the need for countries to respect the knowledge and practice of their people that are relevant to bio-diversity, to seek their permission for the wide use of such knowledge, and to share equability and benefits arising from such use. The Fourth Conference of Parties of the Convention on Biological Diversity held in Bratislava, Slovakia, from May 4 to 15, 1998 highlighted the importance of traditional community knowledge and practices for the conservation and sustainable use of the ecosystems (Kothari, 1998). In national action plan on agrobiodiversity in India, the role of Ethnic communities, particularly of women, in conserving traditional varieties in fragile ecosystems has been stressed. While implementing farm conservation strategies, the Ethnic people, it is advocated, be rewarded (Paroda *et al.*, 1999).

In the 1980s in India 28 wild legumes were known to be consumed as pulses by different tribal sects (Gunjatkar and Vartak, 1982; Jain, 1981; personal communication from Dr. B.D. Sharma, Scientist, SE, Central National Herbarium, Howrah, India). As many as 19 species of the genus, *Cassia*, which are economically and medically useful to mankind and their distribution in the Indian sub-continent are documented (Singh and Premanath, 1992). An account of 9 species of *Mucuna*, found in the Indian sub-continent and Burma is offered with keys and distribution maps (Wilmot-Dear, 1987). Subsequently our team of researchers during the course of two major funded projects, viz. "Exploitation of pulses of tribal utility" sponsored by ICAR, New Delhi (1987–90) and "South Indian tribal pulses: Germplasm resource management and biochemical studies" sponsored by the Ministry of Environment and Forests, New Delhi (1992–96), by undertaking extensive explorative field trips have collected and documented as many as 45 species of little-known/tribal pulses including their wild related crops. Their vernacular names, distribution and mode of consumption (only for edible legumes) are tabulated in detail (Table 1). Out of the 45 documented legumes only 39 wild species are known to be consumed by different tribal sects by employing different modes of consumption as pulses (Janardhanan, 1990; Janardhanan and Rajaram, 1993; Rajaram and Janardhanan, 1993b; Siddhuraju *et al.*, 1993; Mohan, 1992; Vadivel and Janardhanan, 1999; Vadivel, 1998). The tribals of Shahdol district of Rewa division (M.P.) are reported to utilize the seeds of *Vicia sativa* Linn. as a pulse during famine and acute food shortage (Khan *et al.*, 2000).

The majority of bio-assets and traditional medicinal remedies in India such as Ayurveda, Siddha and Unani are a multiplicity of indigenous plants, most of them unique to this country. Wherever appropriate, geographical indications have to be invoked to protect plants and germplasm from biopiracy. To ensure effectiveness, the country needs to document and catalogue its bio-assets including germplasm resources in all details and make databases available to all interested international organizations/search agencies. In this connection, southern countries rich in biodiversity should develop common strategies to protect their common interest and heritage from the economically rich northern countries of the world.

Table 1. Distribution and mode of consumption of underexploited/tribal pulses of India

Species and vernacular name	Distribution in India	Mode of consumption
Abrus precatorius L. Tamil – Kuntumani Tel – Guruginja chettu Mal – Kunni Hindi – Gunchi	Distributed in all parts of tropical zone and Andaman Islands	During extreme famine boiled seeds are cooked and eaten by Onges of Andamans & Katkharis of Pune District, Maharashtra
Acacia catechu Willd. Hindi – Khair	Distributed in Himalayas from Punjab to Sikkim ascending from 1,300 to 1,700 m	Seeds are cooked and eaten by tribals of Madhya Pradesh MSL; Bihar and Madhya Pradesh
Acacia leucophloea (Roxb.) Willd Tamil – Velvelam Tel – Tellatuma Hindi – Nimbar	Deccan and Carnatic dry deciduous forests	Mature seeds are cooked and eaten by the tribal people living in the hilly region of Pune district, Maharashtra
Acacia nilotica L. Del. Tamil – Karuvelam	Cultivated and naturalised in dry localities especially on old tank beds and black cotton soils throughout India	Pods and seeds are cooked and eaten by tribal people living in Western Rajasthan
Alysicarpus rugosus (Willd) DC	Distributed in the hills of Deccan, Mysore, Coimbatore, Western Ghats, Anamalais, Pulneys and Travancore hills up to 1,200 m MSL	Seeds are cooked and eaten by the tribes, Ghonds and Khatkharis
Atylosia goensis Dalz	Distributed in the hills of Western Ghats, The Nilgiris and Travancore up to 1,200 m MSL	Roasted seeds are eaten by Dravidian group of tribes
Bauhinia malabarica Tamil – Atthy Tel – Puli shinta Hindi – Amlosa	Distributed in Eastern and Western Ghats, Ramanathapuram and Tirunelveli districts	Tender shoots and seeds are cooked and eaten by the tribal sects, Ghonds and Khatkharis
Bauhinia purpurea L. Tamil – Mandari Tel – Kanchanam Hindi – Khainwal	Cultivated in almost all parts of the country; but wild in the forests of Deccan plateau and Western Himalayas and Khasi hills	Young pods and seeds are cooked and eaten by Khatkharis and Ghonds
Bauhinia racemosa Lamk. Tamil – Archi, Atti Tel – Kanchini Hindi – Gurial	Distributed in Western Ghats, Maharashtra, Uttar Pradesh, Madhya Pradesh, Punjab and Tamil Nadu	Young pods and seeds are cooked and eaten by Khatkharis and Ghonds
**Bauhinia tomentosa* L. Tamil – Kanchini Tel – Kanchini Hindi – Kanchnar	Circars and Carnatic in dry forests from Chilk Lake to Tirunelveli	–

Table 1. Continued.

Species and vernacular name	Distribution in India	Mode of consumption
Bauhinia vahlii W. & A. Tel–Addi Hindi–Maljami	Distributed at the foot of central and eastern Himalayas, Bihar, Madhya Pradesh, Gorakpur, Dehradun, Saharanpur forests, Rohilkhand, Assam Chot Nagpur, Western and Southern India	The roasted seeds are eaten by the tribes Chenchus, Mundari and the various tribes of Araku valley in Visakhapatnam district in Andhra Pradesh like Kondakapulu and Baagethelu
Canavalia ensiformis Dc Tamil – Segappu thambatti Tel – Thammakai chettu	Everywhere in tropics but rarely cultivated in South India	The cooked seeds and young pods are eaten by The Kurumbas, Irulas, Malayali, Mudugu and other Dravidian tribes
Canavalia gladiata Jacq. Dc. Tamil - Koli Avari Tel - Thamma, Thamba	Cultivated in Northern and Peninsular India, wild forms are distributed in Eastern and Western Ghats and towns	Originally a tribal pulse, now the tender buds and seeds are cooked with spices and consumed by civilised people in villages
Cassia floribunda Cav.	Widespread in tropics	Mature seeds are cooked and eaten by tribal people of Kolli hills, Namakkal district, Tamil Nadu
**Cussia hirsuta* L.	Deccan and Carnatic and hills of Mysore up to 1,300 m MLS	–
Cassia laevigata Willd	Nilgiri and Pulney hills at 1,300–2,600 m MSL introduced from tropical America and now run wild	Yound pods and seeds are cooked and eaten by Kurumbas
Cassia obtusifolia L. Tamil–Tagerai Tel–Tantipu	Widespread in waste lands throughout India	The roasted seed are eaten by Ghonds, Katkharis and Dravidian group of tribes
**Cassia occidentalis* L. Tamil – Payaverai Tel – Kasintha, Eddhu Kommaku	Distributed in Deccan and Camatic regions in waste lands	–
Dolichos lablab var. *vulgaris* (L.) Savi.	Distributed in subtropical and tropical regions	Cooked seeds are eaten by Kadars of Valparai hills in Coimbatore district, Tamil Nadu
Entada scandens Benth (= *Entada phaseoloides* (L.) Merri.) Tamil – Vattavalli, Chillu	Distributed in sub-Himalayan tracts, Sikkim, Assam, West Bengal, Western and Eastern Ghats and Andaman and Nicobar islands	Boiled seeds are eaten by the Karib tribes of Assam and Oceanic group of tribes such as Onges and Great

Table 1. Continued

Species and vernacular name	Distribution in India	Mode of consumption
Tel – Gilatige Mal – Malamanchadi Hindi – Gila		Andamanese
Indigofera linifolia (L.F.) Retz.	Distributed in Deccan and Carnatic Western Ghats, Andhra Pradesh and Tamil Nadu	Seeds are cooked along with other cereals and eaten by the tribe Khatkharis
**Mucuna atropurpurea* DC.	Deccan, Hills of Mysore, Coimbatore 1,000 m MSL, Western Ghats, Hills of Pulney, Sirumalai and Tirunelveli	–
Mucuna gigantea (Willd) DC.	Coastal areas in Tamil Nadu, Kerala and Andamans	Boiled seeds are eaten by the Oceanic group of tribes such as Onges, Great Andamanese and Sompens
**Mucuna hirsuta* W & A	Western Ghats, the Nilgiris and Pulney Hills	–
Mucuna monosperma DC. Ex. Wight Tamil – Thellukkhai Mal – Kakkavalli	Tamil Nadu, Kerala, Andhra Pradesh, Karnataka, Bihar, West Bengal, Assam, Chittagong, Pegu, Hills of Western Peninsula and Andamans	Boiled seeds are eaten by tribes of north-eastern India, the Oceanic group of tribes, the Onges, Great Andamanese and Sompens
Mucuna pruriens var. *pruriens* (L.) DC. Tamil – Poonaykali Tel – Pilladugu, Doolagondi Mal – Naikumi Hindi – Kiwach	Tamil Nadu, Kerala, Andhra Pradesh, Dehradun, Siwalikrange, sub Himalayan tracts, Bundel khand, Merwara, Punjab Plains, Madhya Pradesh, Gujarat, Orissa, Gorakhpur (U.P.) and Lucknow (U.P.)	Mature seeds, seeds from unripe pods and young pods are soaked and boiled/roasted and eaten as such or mixed with salt by the north-east Indian tribes such as Khasi, Naga, Kuki
Mucuna pruriens var. *pruriens* (L.) DC.		Jaintia, Chakma and Mizo; tribes of north-western part of Madhya Pradesh like Abujh-Maria, Muria, Gond and Halba and tribes of South India like Mundari, Dravidian, Kani, Kader and Muthuvan, Savera, Jatapu, Gadebe and Kondadora
Mucuna pruriens var. *utilis* Wall ex Wight Tamil – Perumpayaru Mal – Kongna	Mostly in Western Ghats in wild condition. Recently cultivated by certain tribes in Tirunelveli district in Tamil Nadu in their vicinity	Kanikkars, a hill tribe of Kerala consume the seeds after repeated boiling

Table 1. Continued

Species and vernacular name	Distribution in India	Mode of consumption
**Purkia biglandulosa* W & A	Distributed in north-eastern India	–
Parkia roxburghii G. Don	Mostly in north-eastern region in India	Yound pods and seeds are cooked and eaten by Manipur tribes
Parkinsonia aculeata L. Tel – Simatumma	Almost naturalised and cultivated throughout India	The seeds are cooked and eaten by tribals of Madhya Pradesh
Phaseolus lunatus L. Tamil – Khasikollu, Soleavari Hindi – Zohiya	Cultivated in Uttar Pradesh, Assam, Sikkim, Maharashtra and wild forms are distributed in hilly regions of Tamil Nadu	Cooked seeds are eaten by the tribes, Irulas and Malayalis of Kolli hills and Valparai in Tamil Nadu and rural people
Pongamia pinnata L. Pierre Tamil – Ponga Tel – Kanuga Hindi – Karanj	Cultivated in almost all parts of the country and common on the banks of the rivers and streams and coastal forests	Raw seeds are eaten by Onges of Andaman and Nicobar Islands
Prosopis chilensis (Molina) Stunz Tamil – Vaelimul, Seemakaruvel Tel – Karra Thumma	Distributed as common hedge plant all over the country	The seeds are cooked and eaten by tribals of Madhya Pradesh
Sesbania bispinosa (Jacq) W.F. Wight Tamil – Netti Hindi – Kathsola	Distributed in semi-evergreen forests along the Western Ghats and Bengal	Seeds are cooked and eaten by the tribe, Khatkharis
Tamarindus indica L. Tamil – Pulie Tel – Chinta Hindi – Imli	Cultivated and naturalised in plains of almost all parts of India	Seeds are roasted and soaked in water and eaten by Kurumba, Irulas, Malayali and other Dravidian group of tribals
Vigna aconitifolia (Jacq) Marechal Tamil – Naripayaru Tel – Kuncuma Pesula Bengal – Kheri Hindi – Moth	Cultivated in Panjab, Haryana, Rajasthan, Uttar Pradesh, Bihar, Assam and South India and wild in Orissa	Originally a tribal pulse. Now the seeds are cooked and eaten by rural people
Vigna calcaratus Roxb.	Distributed in Western Ghats, the Nilgiris, Pulney and Travancore hill up to 2,000 m MSL	Mature seeds are cooked and eaten by north-east Indian tribes
Vigna capensis Walp.	Distributed mostly in north-eastern region of India	Seeds are boiled and eaten by north-east Indian tribals

Table 1. Continued

Species and vernacular name	Distribution in India	Mode of consumption
Vigna sesquipedalis (L.) Verdc. (= *V. unguiculata* subsp. *sequipedalis*) Tel – Allasanda	Cultivated in Tamil Nadu, Madhya Pradesh, Bihar, Orissa, Assam and north-eastern India	Seeds and pods are cooked and eaten by tribes of the Dravidian group and also by civilised people in villages and towns
Vigna sinensis (L.) Savi-ex-Hassk	Originally wild in the forests of North India; but now mostly cultivated in upper Gangetic plains	Originally a tribal pulse. Now consumed by civilised rural and urban people after soaking overnight and boiling with spices
Vigna sublobata (Roxb.) Babu et Sharma	Hills of Deccan and Western Ghats up to 1,800 m MSL	Boiled seeds are consumed by rural people
Vigna trilobata (L.) Verdc Tamil – Panipayir Tel – Pilli pesera Malayalam–Ceruvidukol Hindi – Mugani	Distributed in wild form throughout India; cultivated in Rayalaseem and coastal districts of Andhra Pradesh	Seeds are roasted and eaten by the tribal sect, Kurumbas, seeds are cooked and eaten by poor people in villages and towns
Vigna umbellata (Thumb) Ohwi & Ohashi	Found both in cultivated and wild forms from Kerala to Himalayas ascending to an altitude of 1500 m MSL. Frequent in North-eastern India, cultivated in Kitchan gardens in Assam	Seeds are boiled with or without rice and eaten by tribals of India. Recently consumed as a pulse in certain areas of Uttar Pradesh
Vigna vexillata (L.) A. Rich	Distributed in Tamil Nadu, Kerala, Karnataka and Central and Eastern India	Seeds are roasted and eaten by Khatkharis and Ghonds
Xylia xylocarpa (Roxb) Taub Tamil – Irul Tel – Konda tangedu Malayalam–Irumalla Hindi – Jambu	Distributed in Western Ghats, Deccan and Hills from Ganjam to Godavari	Seeds are roasted and eaten by Ghonds and Santal tribals of central India

*Indicates wild related species of underexploited/tribal pulse, but not consumed.

1.5. UNDEREXPLOITED/TRIBAL PULSES

With ever increasing population pressure and fast depletion of natural resources, it has become extremely important to diversify the present day agriculture in order to meet various human needs. The underutilized legumes/wild tribal pulses, have tremendous potential for commercial exploitation but have remained ignored; they offer good scope to meet the ever increasing demands for vegetable protein. Apart from the most common conventional legume crops, chickpea, pigeonpea, mungbean, urdbean and lentils, several less common legume crops belonging to the genera, *Vigna*, *Phaseolus*, *Cyamopsis*, *Psophocarpus*, *Dolichos*, *Macrotyloma*, have not been fully exploited to the extent they deserve.

Certain wild legumes are known to be consumed by different tribal sects in India (Ragunathan, 1978; Jain, 1981; Gunjatkar and Vartak, 1982; Janardhanan, 1990; Janardhanan and Rajaram, 1993; Rajaram and Janardhanan, 1993b; Singhvi and Bhandari, 1992; Mohan, 1992; Siddhuraju *et al.*, 1993; Siddhuraju, 1994; Vijayakumari, 1994; Rajyalakshmi and Geervani, 1994; Sahu, 1996; Radhakrishnan *et al.*, 1996; Vadivel, 1998; Vadivel and Janardhanan, 1999, 2001; Khan *et al.*, 2000).

In India, legumes constitute an important foodstuff and are chief economic sources of protein in the diets of economically weaker sections of population (Kumar *et al.*, 1991). Partly due to scarcity of conventional pulses and partly due to exorbitant cost of animal protein, several novel sources of protein that have hitherto not been employed in human nutrition are being investigated (Yadav and Bharadwaj, 1971; Pant *et al.*, 1974; Bhag Mal, 1992). Some of the wild nuts and seeds used as food in several parts of the world have considerable promise as protein source (Amubode and Fetuga, 1983). Information on seeds of wild species and wild progenitors of cultivated crops is relatively meager. Recently with increasing interest in new food sources, and in improved genetic diversity within domesticated lines, the seeds of wild legumes, including the tribal pulses are receiving more attention (Bewley and Black, 1985; Maikhuri *et al.*, 1991; Rajaram and Janardhanan, 1993b; Siddhuraju *et al.*, 1993; Rajyalakshmi and Geervani, 1994; Vadivel and Janardhanan, 1999, 2000).

Tribals in India are the indigenous, autochthonous people of the land. They are differently named as Adivasis (original settlers), Adim Nivasis (oldest ethnological sector of the population), Aboriginals (indigenous), Vanavasis (forest inhabitants) and several such other names signifying either their ecological, economic or historical or cultural characteristics (Jain, 1987).

There are more than 550 tribal communities under 227 ethnic groups who account for 8.08% or 67.76 million of total Indian population (Census of India, 1991). They inhabit varied geographical and climatic zones of the country covering 5000 villages inside forests (Vadivel, 1998).

2. Tribal demography and geography of India

The Indian subcontinent is divided into three major tribal zones on the basis of tribal demography and geography of India (Azit Raizada, 1984).

(a) *The North and North-Eastern Zone.* It includes the area of eastern Jammu and Kashmir, Himachal Pradesh, the Tarai area of Uttar Pradesh, mountain region of West Bengal, Assam, Meghalaya, Arunachal Pradesh, Nagaland, Manipur, Mizoram and Tripura.

(b) *The Central or Middle Zone.* It includes West Bengal, Bihar, southern Uttar Pradesh, Madhya Pradesh, Orissa, Rajasthan, Gujarat and Maharashtra.

(c) *The Southern Zone.* It includes Andhra Pradesh, Karnataka, Kerala, Tamil Nadu, Lakshadweep and Andaman and Nicobar Islands.

Tribals in India occupy about 18.74% of the total area of the country spread mainly over 21 States and Union Territories. Two-thirds of the total tribal population is concentrated

Table 2. Regions of tribal concentration and main tribal communities

Region/State or Union Territory	Main tribal communities
CENTRAL TRIBAL REGION	
South Central	
Andhra Pradesh, South Madhya Pradesh, South Orissa	Khondadoras, Valmikis, Gonds, Oraon, Bhilkonda, Saravas, Reddies, Koyan, Konda, Kodagu, Pradhan, Khonda
North Central	
Bihar, West Bengal, North Orissa, East Madhya Pradesh	Santhal, Oraon, Munda, Kolha, Bhag Mal Pahariyar, Bhills, Dhodia, Gamit, Waril, Naika Rathwa, Kolidher
WESTERN TRIBAL REGION	
Rajasthan, Gujarat, Maharashtra, Dadar and Nagar Haveli, Goa, Daman and Diu	Bhills, Dhodia, Gamit, Waril, Naika Rathwa, Kolidher
NORTH-EASTERN TRIBAL REGION	
Meghalaya, Nagaland, Arunachal Pradesh, Manipur, Assam, Mizoram, Tripura	Nagal, Khasi, Mizo, Miri, Tripurim, Adi Aka, Apatan, Maripa Mishmi, Garos, Tagin, Naga, Singhpho, Sherdukpen
NORTH-WESTERN TRIBAL REGION	
Western Uttar Pradesh, Himachal Pradesh	Jaunsaries, Laholies, Kinnoras, Tibetians, Battis, Bhotai
SOUTHERN TRIBAL REGION	
Karnataka, Kerala, Tamil Nadu	Malayali, Kurumba, Kani, Muthuvan, Soluga
ISLAND REGION	
Andaman and Nicobar, Lakshadweep	Onges, Sentinelese, Great Andamanese, Jarwas, Nicobarese, Shompen

Source: Anonymous, 1994. Ethnobiology in India, a status report (All India Coordinated Research Project in Ethnobiology), Ministry of Environment and Forests, Government of India, New Delhi, India.

in Madhya Pradesh, Orissa, Bihar, Gujarat and Rajasthan (Arora, 1995). Regions of tribal concentration and main tribal communities are listed in Table 2.

2.1. COLLECTION AND DOCUMENTATION OF GERMPLASM SEED MATERIALS OF UNDEREXPLOITED/TRIBAL PULSES IN INDIA

2.1.1. Definition of germplasm

The hereditary material as a whole is sometimes referred to as the germplasm. In plants, the plastids located in the cytoplasm also constitute part of germplasm (Altenburg, 1976). Adrin *et al.* (1979) has defined the germplasm as the minute and delicately organized material that forms the physical basis for the transmission of inherited qualities.

The germplasm resources of a particular crop species may be divided into three groups. The first group consists of all individuals within a genus or broader cluster, within which sexual compatibility exists, including wild or weedy representatives. The second group is a subset of the first and includes all individuals in a working collection of the breeders. Plant breeding operations are largely limited to this group. The third group is even more restricted and includes the improved cultivars in commercial use and their potential replacement (Sprague, 1980).

2.1.2. Germplasm collection and conservation

Plant Genetic Resources for Food and Agriculture (PGRFA) are the biological basis of world food security and directly or indirectly support the livelihoods of every person on earth. PGRFA comprises the diversity of genetic material contained in traditional varieties and modern cultivars, as well as crop wild relatives and other wild plant species that can be used now or in the future for food and agriculture (Paroda, 1987; Paroda and Arora, 1991; FAO, 1996).

These resources represent both the raw materials used in the production of new cultivars – either through traditional plant breeding or through biotechnology – and are a reservoir of genetic adaptability which act as a buffer against potentially harmful environmental and economic changes (FAO, 1996).

The monoculture practices, environmental degradation and urban development – all have contributed to loss of plant genetic resources (Van Sloten, 1990a, b) and erosion of these resources poses a severe threat to the world's food security in the long term (FAO, 1996). The need to protect and conserve these resources more systematically is getting recognized now more and more and over 100 national programs are presently operating (Van Sloten, 1990b).

The terms "Genetic Resources" started with the technical conference organized by the Food and Agriculture Organization (FAO) in Rome in 1967, wherein scientific principles, methodologies and strategies of exploration, conservation, evaluation and documentation are elaborately discussed for the first time (Arora *et al.*, 1991). A follow up of this resulted in a report on the survey of crop genetic resources (Frankel, 1970) and through subsequent conferences, in two more very timely publications (Frankel and Bennet, 1970; Frankel and Hawkes, 1975). Since then the activities have developed fast in national and international perspectives (Arora *et al.*, 1991). The history of plant genetic resources movement is well documented (Pistoris, 1997). Recently in India the importance of agrobiodiversity has been recognized as a distinct national issue of prime concern within the broader area of biodiversity. Several recommendations on various policy options and Action Plan on conservation of agrobiodiversity have emerged and recorded (Paroda *et al.*, 1999).

2.1.3. Indian national plant genetic resource system

In India, a nuclear plant exploration and collection unit was established in 1946, in the Division of Botany at the Indian Agricultural Research Institute (IARI), New Delhi. Later the Indian Council of Agricultural Research (ICAR) created a separate organization named National Bureau of Plant Genetic Resources (NBPGR) in New Delhi in 1976. The NBPGR has five major Divisions, namely Division of Plant Exploration and Collection, Division of Germplasm Exchange, Division of Plant Quarantine, Division of Germplasm Evaluation, and Division of Germplasm Conservation. The Bureau also has 12 Regional Stations/Basic Centres/Quarantine Stations/Experimental Farms located in different agroclimatic zones throughout India. In addition, three all India coordinated crop improvement projects on Medicinal and Aromatic plants, cluster bean, underutilized/under-exploited plants are also located at the Bureau (Rana and Arora, 1991). Goutam and Kochhar (1999) have reviewed the evolution of plant genetic resources (PGR) activity, plant genetic resources scenario, various dimensions of plant genetic resources activity in India; besides PGR policy area and role of NBPGR in conserving India's plant diversity. According to Paroda *et al.* (1999) a base collection of nearly 1.75 lakh germplasm of various crop species and their relatives is conserved *ex situ* in the National Gene Bank at NBPGR.

2.1.4. Underutilized/little known plant genetic resources (PGR)

Genetic variability of native, underutilized species of food crops, fruits, medicinal, aromatic and other economic plants should be documented on priority (Paroda *et al.*, 1999). This interest is due to their poor representation in germplasm collections, their high value as large stands of genetic variation, their potential value as a source of novel variants for plant breeding and their high value as new source of food, medicine, etc. which meet the need of the increasing population (Frankel, 1975; Harlan, 1976; Janardhanan, 1990; Cohen *et al.*, 1991; Bianchi-Hall *et al.*, 1993; Clements and Cowling, 1994; Laghetti *et al.*, 1994; Vadivel and Janardhanan, 1998). Need-based introduction of useful species may also be encouraged and some selected, hitherto unexploited future potential species studied for adoption. Characterization, evaluation and documentation of Plant Genetic Resources (PGR) should receive a high priority. Relevant important tools and technologies, such as biotechnology, should be employed in future. The National Information Network and Database on germplasm should be strengthened (Paroda *et al.*, 1999).

In view of the aforesaid background, in the past 20 years, the authors and other team members have made sincere attempts to gather germplasm seed materials for 45 species of little known/underexploited/tribal pulses including their wild relatives from their natural stands by undertaking extensive field trips throughout India including Andaman and Nicobar Islands.The total number of germplasm (accessions) of seed materials for all the 45 species account for 192 (Table 3). This exercise has been accomplished mainly during

*Table 3. Accession numbers of germplasm of underexploited/tribal pulses of India with particulars of place of collection, district and State/Union Territory from where collected/ procured**

Accession No. (IC No. if already procured)	Place of collection	District	State/Union Territory
L..N.: *Abrus precatorius* L			
BU/Bot./KJ.1	Perumalmudi	Coimbatore	Tamil Nadu
BU/Bot./KJ.2	Ayyanar Koil forest	Virudhunagar	Tamil Nadu
BU/Bot./KJ.3	Kalakad Wildlife Sanctuary	Tirunelveli	Tamil Nadu
BU/Bot./KJ.4	Mundanthurai Wildlife Sanctuary	Tirunelveli	Tamil Nadu
BU/Bot./KJ.5	Vanavasi	Salem	Tamil Nadu
L.N. *Acacia catechu* Willd			
BU/Bot./KJ.6	Peripheral areas of Morena town	Moren	Madhya Pradesh
L.N. *Acacia leucophloea* (Roxb) Willd			
BU/Bot./KJ.7	Bharathiar University Campus	Coimbatore	Tamil Nadu

Table 3. Continued

Accession No. (IC No. if already procured)	Place of collection	District	State/Union Territory
L.N. *Acacia nilotica* (L.) Del.			
BU/Bot./KJ.8	Bharathiar University Campus	Coimbatore	Tamil Nadu
L.N. *Alysicarpus rugosus* (Willd) Dc.			
BU/Bot./KJ.9	Mallar	Salem	Tamil Nadu
L.N. *Atylosia goensis* Dalz.			
BU/Bot./KJ.10	Gudalur	The Nilgiris	Tamil Nadu
L.N. *Bauhinia malabarica* Roxb.			
BU/Bot./KJ.11	Siruvani Reserve forest	Coimbatore	Tamil Nadu
L.N. *Bauhinia purpurea* L.			
BU/Bot./KJ.12 (IC No. 83198)	Empress garden, Pune	Pune	Maharashtra
BU/Bot./KJ.13	Coimbatore city	Coimbatore	Tamil Nadu
L.N. *Bauhinia recemosa* Lamk.			
BU/Bot./KJ.14	The garden of Allahabad Botanical Survey of India (Central Circle, Allahabad)	Allahabad	Uttar Pradesh
BU/Bot./KJ.15	Ayyanar Koil forest	Virudunagar	Tamil Nadu
BU/Bot./KJ.16	Mundanthurai Wildlife Sanctuary	Tirunelveli	Tamil Nadu
L.N. *Bauhinia tomentosa* L.			
BU/Bot./KJ.17	Mallerpuram	Tuticorin	Tamil Nadu
L.N. *Bauhinia vahlii* W. & A.			
BU/Bot./KJ.18 (IC No. 83211)	Rollapenta (Nallamalais)	Kurnool	Andhra Pradesh
BU/Bot./KJ.19	Araku valley	Visakhapatnam	Andhra Pradesh
L.N. *Canavalia ensiformis* Dc.			
BU/Bot./KJ.20 (IC No. 83201)	Jalagandapuram	Salem	Tamil Nadu
BU/Bot./KJ.21	Mullai Nagar (Coimbatore)	Coimbatore	Tamil Nadu
BU/Bot./KJ.22	Mukkali (Silent valley)	Palkkad	Kerala
BU/Bot./KJ.23	Pathanamthitta	Pathanamthitta	Kerala
BU/Bot./KJ.24	Valacode	Kottayam	Kerala
BU/Bot./KJ.25	Neyar Dam	Thiruvanan-thapuram	Kerala
BU/Bot./KJ.26	Kodacherry	Malapuram	Kerala
BU/Bot./KJ.27	Vazhikkadavu	Malapuram	Kerala
BU/Bot./KJ.28	Mananthavaadi	Wynaad	Kerala
BU/Bot./KJ.29	Naduvil	Cannur	Kerala
BU/Bot./KJ.30	Dasukuppam	Chittoor	Andhra Pradesh
BU/Bot./KJ.31	Gundalpet	Mysore	Karnataka
L.N. *Canavalia gladiata* (Jacq.) Dc.			
BU/Bot./KJ.32	Tamil Nadu Agriculture University, Coimbatore	Coimbatore	Tamil Nadu
BU/Bot./KJ.33	Kalakkad Wildlife Sanctuary	Tirunelveli	Tamil Nadu
BU/Bot./KJ.34	Mundanthurai Wildlife Sanctuary	Tirunelveli	Tamil Nadu
BU/Bot./KJ.35	Maruthamalai hills	Coimbatore	Tamil Nadu
BU/Bot./KJ.36	Thoppur Reserve forest	Dharmapuri	Tamil Nadu

Table 3. Continued

Accession No. (IC No. if already procured)	Place of collection	District	State/Union Territory
BU/Bot./KJ.37	Mukkali (Silent valley)	Palkkad	Kerala
BU/Bot./KJ.38	Alandurai	Coimbatore	Tamil Nadu
BU/Bot./KJ.39	Gopichettipalayam	Erode	Tamil Nadu
BU/Bot./KJ.40	Courtallam	Tirunelveli	Tamil Nadu
BU/Bot./KJ.41	Palamalai	Salem	Tamil Nadu
BU/Bot./KJ.42	Elagiri hills	Vellore	Tamil Nadu
BU/Bot./KJ.43	Kolli hills	Namakkal	Tamil Nadu
BU/Bot./KJ.44	Aliyar	Coimbatore	Tamil Nadu
BU/Bot./KJ.45	Hogenekkal	Dharmapurai	Tamil Nadu
BU/Bot./KJ.46	Yercaud	Salem	Tamil Nadu
BU/Bot./KJ.47	Suruli falls	Madurai	Tamil Nadu
BU/Bot./KJ.48	Paramangalam	Thrissur	Kerala
BU/Bot./KJ.49	Allapuzha	Allapuzha	Kerala
BU/Bot./KJ.50	Kumlza	Pathanamthitta	Kerala
BU/Bot./KJ.51	Goolikadivu	Palkkad	Kerala
BU/Bot./KJ.52	Adoor	Kasargode	Kerala
BU/Bot./KJ.53	Nagarjuna Sagar (Lake area)	Guntur	Andhra Pradesh
BU/Bot./KJ.54	Nagarjuna Sagar (Hill area)	Nalgonda	Andhra Pradesh
BU/Bot./KJ.55	Yethipothala (Water falls)	Guntur	Andhra Pradesh
BU/Bot./KJ.56	Dasukuppam	Chittoor	Andhra Pradesh
BU/Bot./KJ.57	Rajugunta	Chittoor	Andhra Pradesh
BU/Bot./KJ.58	Beerakuppam	Chittoor	Andhra Pradesh
BU/Bot./KJ.59	Lower kodaiyar	Kanyakumari	Tamil Nadu
BU/Bot./KJ.60	Thachanmalai	Kanyakumari	Tamil Nadu
BU/Bot./KJ.61	Myllaru (Upper papanasam)	Tirunelveli	Tamil Nadu
BU/Bot./KJ.62	Sirumalai	Dindigul	Tamil Nadu
BU/Bot./KJ.63	Vanavasi	Salem	Tamil Nadu
BU/Bot./KJ.64	Visakhapatnam	Visakhapatanam	Andhra Pradesh
L.N. *Cassia floribunda* Cav.			
BU/Bot./KJ.65	Kolli hills	Namakkal	Tamil Nadu
BU/Bot./KJ.66	Erattupettai	Kottayam	Kerala
BU/Bot./KJ.67	Kumlza	Pathanamthitta	Kerala
BU/Bot./KJ.68	Gundalpet	Mysore	Karnataka
BU/Bot./KJ.69	Pattanapura	Mysore	Karnataka
BU/Bot./KJ.70	Hunsur	Mysore	Karnataka
BU/Bot./KJ.71	Alalur	Mysore	Karnataka
BU/Bot./KJ.72	Kushalanagar	Kodagu (Medikeri)	Karnataka
BU/Bot./KJ.73	Sidapura	Kodagu (Medikeri)	Karnataka
BU/Bot./KJ.74	Dandeli	Uttarakannada	Karnataka
BU/Bot./KJ.75	Kargal	Uttarakannada	Karnataka
L.N. *Cassia hirsuta* L.			
BU/Bot./KJ.76	Siruvani Reserve Forest	Coimbatore	Tamil Nadu
BU/Bot./KJ.77	Kolli hills	Namakkal	Tamil Nadu
BU/Bot./KJ.78	Thengumarahada	Erode	Tamil Nadu
BU/Bot./KJ.79	Keriparai	Kanyakumari	Tamil Nadu
BU/Bot./KJ.80	Yercaud	Salem	Tamil Nadu

Table 3. Continued

Accession No. (IC No. if already procured)	Place of collection	District	State/Union Territory
BU/Bot./KJ.81	Courtallum	Tirunelveli	Tamil Nadu
BU/Bot./KJ.82	Hogenekkal	Dharmapuri	Tamil Nadu
BU/Bot./KJ.83	Sirumalai	Dindigul	Tamil Nadu
BU/Bot./KJ.84	Idukki	Idukki	Kerala
L.N. *Cassia laevigata* Willd			
BU/Bot./KJ.85	Maruthamalai hills	Coimbatore	Tamil Nadu
L.N. *Cassia obtusifolia* L.			
BU/Bot./KJ.86	Mundanthurai Wildlife Sanctuary	Tirunelveli	Tamil Nadu
BU/Bot./KJ.87	Kolli hills	Namakkal	Tamil Nadu
BU/Bot./KJ.88	Thengumarahada	Erode	Tamil Nadu
BU/Bot./KJ.89	Siruvani Reserve Forest	Coimbatore	Tamil Nadu
BU/Bot./KJ.90	Maruthamali hills	Coimbatore	Tamil Nadu
BU/Bot./KJ.91	Courtallum	Tirunelveli	Tamil Nadu
BU/Bot./KJ.92	Kalakkad	Tirunelveli	Tamil Nadu
BU/Bot./KJ.93	Topslip	Coimbatore	Tamil Nadu
BU/Bot./KJ.94	Keriparai	Kanyakumari	Tamil Nadu
BU/Bot./KJ.95	Chamundi hills	Mysore	Karnataka
BU/Bot./KJ.96	Pattanapura	Mysore	Karnataka
BU/Bot./KJ.97	Bhagamandala	Kodagu	Karnataka
BU/Bot./KJ.98	Sidapura	Kodagu	Karnataka
BU/Bot./KJ.99	Karwar	Uttarakannada	Karnataka
BU/Bot./KJ.100	Dandeli	Uttarakannada	Karnataka
L.N. *Cassia occidentalis* L.			
BU/Bot./KJ.101	Kolli hills	Namakkal	Tamil Nadu
BU/Bot./KJ.102	Manimutharu	Tirunelveli	Tamil Nadu
BU/Bot./KJ.103	Maruthamalai hills	Coimbatore	Tamil Nadu
BU/Bot./KJ.104	Kalakkad	Tirunelveli	Tamil Nadu
BU/Bot./KJ.105	Courtallum	Tirunelveli	Tamil Nadu
BU/Bot./KJ.106	Hogenekkal	Dharmapuri	Tamil Nadu
BU/Bot./KJ.107	Pathanamthitta	Pathanamthitta	Kerala
BU/Bot./KJ.110	Karwar	Uttarakannada	Karnataka
BU/Bot./KJ.111	Araku valley	Visakhapatnam	Andhra Pradesh
L.N. *Dolichos lablab* var. *vulgaris* (L.) Savi.			
BU/Bot./KJ.112	Valparai Reserve Forest	Coimbatore	Tamil Nadu
L.N. *Entada scandens* Benth (= *Entada phaseoloides* (L.) Merr.)			
BU/Bot./KJ.113 (IC No. 83209)	Kolli hills	Namakkal	Tamil Nadu
BU/Bot./KJ.114	Siruvani Reserve Forest	Coimbatore	Tamil Nadu
BU/Bot./KJ.115	Myllaru (Papanasam)	Tirunelveli	Tamil Nadu
L.N. *Indigofera linifolia* L.f. Retz.			
BU/Bot./KJ.116	Bharathiar University Campus	Coimbatore	Tamil Nadu
L.N. *Mucuna atropurpurea* Dc.			
BU/Bot./KJ.117 (IC No. 83197)	Maankarai	Coimbatore	Tamil Nadu
BU/Bot./KJ.118	Aliyar	Coimbatore	Tamil Nadu
BU/Bot./KJ.119	Courtallum	Tirunelveli	Tamil Nadu
BU/Bot./KJ.120	Kolli hills	Namakkal	Tamil Nadu

Table 3. Continued

Accession No. (IC No. if already procured)	Place of collection	District	State/Union Territory
BU/Bot./KJ.121	Marapalam (Connoor)	The Nilgiris	Tamil Nadu
BU/Bot./KJ.122 (IC No. 83196)	Myllaru	Tirunelveli	Tamil Nadu
L.N. *Mucuna gigantea* (Willd) Dc.			
BU/Bot./KJ.123 (IC No. 83295)	Chiriatapu (South Andaman)	Andaman	Andaman & Nicobar Islands
L.N. *Mucuna hirsuta* W & A.			
BU/Bot./KJ.124	Balamaadis (Naduvattam)	The Nilgiris	Tamil Nadu
L.N. *Muncuna monosperma* Dc. ex Wight			
BU/Bot./KJ.125 (IC No. 83196)	Tirunelly Reserve Forest	Wynaad	Kerala
BU/Bot./KJ.126	Siruvani Reserve Forest	Coimbatore	Tamil Nadu
BU/Bot./KJ.127	Aliyar	Coimbatore	Tamil Nadu
BU/Bot./KJ.128	Chittoor	Palkkad	Kerala
BU/Bot./KJ.129	Bogalthode	Mysore	Karnataka
BU/Bot./KJ.130	Borra caves (Araku valley)	Visakhapatnam	Andhra Pradesh
BU/Bot./KJ.131	Kargal	Uttarakannada	Karnataka
BU/Bot./KJ.132	Dandeli	Uttarkannada	Karnataka
L.N. *Mucuna pruriens* var. *pruriens* (L.) Dc.			
BU/Bot./KJ.133 (IC No. 83193)	Begur Reserve Forest	Wynaad	Kerala
BU/Bot./KJ.134 (IC No. 83194)	Mukkali (Silent valley)	Palkkad	Kerala
BU/Bot./KJ.135 (IC No. 83195)	Mothimahal campus	Lucknow	Uttar Pradesh
BU/Bot./KJ.136	Yercaud hills	Salem	Tamil Nadu
BU/Bot./KJ.137	Aliyar	Coimbatore	Tamil Nadu
BU/Bot./KJ.138	Elagiri hills	Vellore	Tamil Nadu
BU/Bot./KJ.139	Thachanmalai (Pecheparai)	Kanyakumari	Tamil Nadu
BU/Bot./KJ.140	Arunooli	Thrissur	Kerala
BU/Bot./KJ.141	Paramangalam	Thrissur	Kerala
BU/Bot./KJ.142	Kursimalai	Kottayam	Kerala
BU/Bot./KJ.143	Alapuzha (Peripheral region)	Alapuzha	Kerala
BU/Bot./KJ.144	Salem (Peripheral region)	Salem	Tamil Nadu
BU/Bot./KJ.145	Karwar	Uttarakannada	Karnataka
BU/Bot./KJ.146	Andhra University Agricultural farm (Waltair)	Visakhapatnam	Andhra Pradesh
L.N. *Mucuna pruriens* var. *utilis* Wall ex Wight			
BU/Bot./KJ.147	Procured from Chennai	Chennai	Tamil Nadu
BU/Bot./KJ.148	Myllaru	Tirunelveli	Tamil Nadu
BU/Bot./KJ.149	Mundanthurai (Wildlife Sanctuary)	Tirunelveli	Tamil Nadu
BU/Bot./KJ.150	Thachanmalai (White-coloured seed coat)	Kanyakumari	Tamil Nadu
BU/Bot./KJ.151	Thachanmalai (Black-coloured seed coat)	Kanyakumari	Tamil Nadu
BU/Bot./KJ.152	Valanad	Thiruvananthapuram	Kerala

Table 3. Continued

Accession No. (IC No. if already procured)	Place of collection	District	State/Union Territory
BU/Bot./KJ.153	Keriparai	Kanyakumari	Tamil Nadu
L.N. *Parkia biglandulosa* W. & A.			
BU/Bot./KJ.154	Procured through Dr. J.K. Maheswari, National Botanical Research Institute (NBRI), Lucknow, collected from Dehra Dun, Uttar Pradesh (1990)	–	–
L.N. *Parkia roxburghii* G. don.			
BU/Bot./KJ.155	Procured from Indian Council of Agricultural Research (ICAR) complex for North-Eastern Hill (NEH) Region, Imphal, Manipur	–	–
L.N. *Parkinsonia aculeata* L.			
BU/Bot./KJ.156	Peripheral region of Morena	Morena	Madhya Pradesh
L.N. *Phaseolus lunatus* L.			
BU/Bot./KJ.157	Procured from Tamil Nadu Agricultural University	Coimbatore	Tamil Nadu
BU/Bot./KJ.158	Valparai (red variety)	Coimbatore	Tamil Nadu
BU/Bot./KJ.159	Kolli hills (Black variety)	Namakkal	Tamil Nadu
BU/Bot./KJ.160	Kolli hills (Brown variety)	Namakkal	Tamil Nadu
BU/Bot./KJ.161	Medikeri	Kodagu	Karnataka
BU/Bot./KJ.162	Karwar	Uttarakannada	Karnataka
BU/Bot./KJ.163	Borra caves (Araku valley)	Visakhapatanam	Andhra Pradesh
L.N. *Pongamia pinnata* L. Pierre			
BU/Bot./KJ.164	Mechery	Salem	Tamil Nadu
BU/Bot./KJ.165	Bharathiar University Campus	Coimbatore	Tamil Nadu
L.N. *Prosopsis chilensis* (Molina) Stunz.			
BU/Bot./KJ.166	Peripheral region of Morena	Morena	Madhya Pradesh
BU/Bot./KJ.167	Manickampalayam	Erode	Tamil Nadu
L.N. *Sesbania bispinosa* (Jacq) W.E. Wight			
BU/Bot./KJ.168	Bharathiar University Campus	Coimbatore	Tamil Nadu
L.N. *Tamarindus indica* L.			
BU/Bot./KJ.169	Pollachi	Coimbatore	Tamil Nadu
BU/Bot./KJ.170	Uthukottai	Thiruvallur	Tamil Nadu
L.N. *Vigna aconitifolia* (Jacq.) Marechal			
BU/Bot./KJ.171	Vanavasi	Salem	Tamil Nadu
BU/Bot./KJ.172	Karwar	Uttarakannada	Karnataka
L.N. *Vigna calcaratus* Roxb.			
BU/Bot./KJ.173 (IC No. 83207)	T.N. Agricultural University	Coimbatore	Tamil Nadu
Vigna capensis Walp			
BU/Bot./KJ.174 (IC No. 89808)	NBPGR Regional Station	Shillong	Meghalaya

Table 3. Continued

Accession No. (IC No. if already procured)	Place of collection	District	State/Union Territory
L.N. *Vigna sesquipedalis* (L.) Verdc (= *V. unguiculata* subsp. *sesquipedalis*)			
BU/Bot./KJ.175 (IC No. 83208)	Tamil Nadu Agricultural University	Coimbatore	Tamil Nadu
L.N. *Vigna sinensis* (L.) Savi. ex. Hassk.			
BU/Bot./KJ.176 (IC No. 83204)	Peripheral region of Allahabad (Maroon-coloured seed coat)	Allahabad	Uttar Pradesh
BU/Bot./KJ.177 (IC No. 83205)	Peripheral region of Allahabad seed coat (pale pink in colour)	Allahabad	Uttar Pradesh
BU/Bot./KJ.178	Borra caves Araku valley (Maroon-coloured seed coat)	Visakhapatnam	Andhra Pradesh
L.N. *Vigna sublobata* (Rox) et Sharma			
BU/Bot./KJ.179 (IC No. 83206)	T.N. Agricultural University	Coimbatore	Tamil Nadu
L.N. *Vigna trilobaia* (L) Verdc.			
BU/Bot./KJ.180 (IC No. 83210)	Bharathiar University Campus	Coimbatore	Tamil Nadu
BU/Bot./KJ.181	Dasukkupam	Chittoor	Andhra Pradesh
L.N. *Vigna umbellata* (Thunb). Ohwi and Ohashi			
BU/Bot./KJ.182 (IC No. 83202)	*V. umbellata* var K1 (procured from Calcutta, West Bengal)	–	–
BU/Bot./KJ.183 (IC No. 83203)	*V. umbellata* var RBL 40 (procured from Tamil Nadu Agricultural University, Coimbatore, Tamil Nadu)	–	–
Procured from ICAR (Regional Station) Shillong, Meghalaya			
BU/Bot./KJ.184 (IC No. 89807)	*V. umbellata* var. RCRB 1–6	–	–
BU/Bot./KJ.185 (IC No. 89809)	*V. umbellata* var. RCRB 1–7	–	–
BU/Bot./KJ.186 (IC No. 89810)	*V. umbellata* var. RCRB 1–8	–	–
BU/Bot./KJ.187 (IC No. 89811)	*V. umbellata* var. RCRB 1–12	–	–
BU/Bot./KJ.188 (IC No. 89812)	*V. umbellata* var. RCRB 1–13	–	–
BU/Bot./KJ.189 (IC No. 89813)	*V. umbellata* var. RCRB 1–15	–	–
L.N. *Vigna vexillata* (L.) A. Rich.			
BU/Bot./KJ.190	Procured from Tamil Nadu Agricultural University, Coimbatore, Tamil Nadu	–	–
L.N. *Xylia xylocarpa* (Roxb.) Taub.			
BU/Bot./KJ.191	Mukkali (Silent valley)	Palkkad	Kerala
BU/Bot./KJ.192	Athirapally	Thrissur	Kerala

* Database created in website: http://www.bharathi.ernet.in
L.N. – Latin name.
IC No. – Indigenous code number procured from NBPGR, New Delhi.

the tenure of two major funded schemes, viz. "Exploitation of pulses of tribal utility" sponsored by ICAR New Delhi (1987–1990) and "South Indian tribal pulses: Germplasm resource management & biochemical studies" sponsored by Ministry of Environment and Forests, Government of India, New Delhi (1992–96).

All the collected 192 germplasm seed materials are compiled and tabulated furnishing place of collection, district and State/Union Territory from where collected/procured (Table 3). As many as 25 accessions of these pulses have already been deposited in NBPGR for long term conservation. After documentation, the Indigenous Code numbers (IC Nos.) were procured.

Legumes in which a rich diversity occurs in India include pigeonpea, chickpea, mungbean, urdbean, horsegram, mothbean, ricebean, clusterbean, tamarind, sword bean and velvet bean (Gautam and Kochhar, 1999). In South India, occurrence of diversity in wild legumes like jack bean, sword bean, velvet bean, itching bean, *Cassias*, *Abrus precatorius*, *Phaseolus lunatus* (limabean) and several cultivated and wild species of *Vigna* is recorded (Arora, 1985b; Rajaram, 1990; Rajaram and Janardhanan, 1993b; Siddhuraju, 1994; Siddhuraju *et al.*, 1993; Anonymous, 1996; Vadivel, 1998; Vadivel and Janardhanan, 1997; Vadivel *et al.*, 1998).

Genus, *Vigna*, comprising prominent grain legumes such as green gram, black gram, mothbean and ricebean are cultivated as pulse and forage crops. Their progenitors are predominantly wild and occur in natural/disturbed habitats. As many as 16 cultivated/wild species from four phytogeographical zones: Western Ghats, Eastern Ghats, north-eastern region and Western Himalayas, have been collected (Arora, 1985b) (Table 4). Further, more diversity occurs in the wild species related to the economically important cultivated taxa in Western Ghats than in other areas where as many as 13 species have been documented (Tables 5 and 6). This region also exhibits a high degree of endemism for wild *Vigna* species (personal communication from Dr. Z. Abraham, Sr. Scientist, NBPGR Regional station, Trichur, Kerala).

The wild types related to black gram, i.e. *V. mungo* var. *silvestris*, is endemic to Western Ghats and its variability is collected from Khandala Ghats and Ratnagiri. These naturally occurring populations twine round the shrubs along forest edges. Two distinct types in green gram, i.e. *V. radiata* var. *sublobata* and *V. radiata* var. *setulosa*, occur in Western Ghats. The *setulosa* type is endemic to Konkan region; while *sublobata* types have much wider distribution. Wild types in ricebean collected in the region belong to two different species, *V. umbellata* and *V. dalzelliana*. *V. dalzelliana* is endemic to Western Ghats and its variability is collected from Khandala Ghats in Maharashtra. The wild *umbellata* types have much wider distribution in the Western Ghats, North East Hill (NEH) region and foot hills of Himalayas being more variable in pod size/seed size/colour. The wild populations in *Vigna aconitifolia* are observed in grass mixed vegetation near Poona in Maharashtra and Kadiyur, Kangeyam in Tamil Nadu. *V. trilobata* types are observed in Anamalai and Coimbatore in Tamil Nadu. *Vigna grandis* (= *V. kandalensis*) is endemic to Khandala (personal communication from Dr. Z. Abraham, Sr. Scientist, NBPGR Regional station, Trichur, Kerala).

Vigna vexillata is a twiner with dark brown pubescence and large purple flowers which is collected from Kasaragod (Kerala). *Vigna wightii* is almost similar to *Vigna vexillata*, has pink flowers and its variability is collected from Wynaad. In *V. vexillata* immature

Table 4. Vigna *species occurring in India*

Species	Status		Remarks: wild taxa
	Cultivated	Wild	
Vigna mungo	+	+	Wild types occur var. *silvestris*
Vigna radiata	+	+	More diversity in wild types – var. *sublobata* and var. *setulosa*
Vigna umbellata	+	+	Wild types occur – var. *gracilis* not much variable, conspecific with the cultigen type
Vigna aconitifolia	+	+	Wild occurring variability conspecific with the cultigen type
Vigna trilobata	+	+	Wild occurring variability conspecific with the cultigen type
Vigna dalzelliana	–	+	Only wild types occur; much akin to *V. umbellata* wild types
Vigna grandis	–	+	Only wild types occur; much akin to *V. radiata/V. mungo* complex
Vigna capensis	–	+	Only wild types occur; exhibit much variability
Vigna pilosa	–	+	Only wild types occur; narrow range of variability
Vigna wightii			Endemic to Western Ghats. It is imperfectly known

Source: Arora, 1985b.

Table 5. Wild relatives of Vigna *species occurring in southern region of India*

Species	Cultivated	Wild	Wild taxa
Vigna mungo	+	+	Wild types occur var. *silvestris*
Vigna radiata	+	+	More diversity in wild types var. *sublobata* and var. *setulosa*
Vigna umbellata	+	+	Wild types occur var. *gracelis*
Vigna aconitifolia	+	+	Wild occurring conspecific with cultigen types
Vigna trilobata	+	+	Wild occurring conspecific with cultigen types
Vigna dalzelliana	–	+	Only wild types occur
Vigna grandis	–	+	Only wild types occur
Vigna minima	–	+	Only wild types occur
Vigna capensis	–	+	Only wild types occur
Vigna vexillata	–	+	Only wild types occur
Vigna wightii	–	+	Only wild types occur
Vigna pilosa	–	+	Only wild types occur
Vigna bourneae	–	+	Only wild types occur

Table 6. Major vegetation types associated with wild Vigna *species in southern region of India*

Vegetation type	Distribution/Phytogeo-graphical zone	Wild *Vigna* species present
Tropical moist evergreen Scrub	Western Ghats	*Vigna radiata* var. *sublobata* *V. radiata* var. *setulosa*, *V. mungo* var. *silvestris*, *V. capensis*, *V. dalzelliana*, *V. umbellata*, *V. minima*, *V. grandis*, *V. bourneae*, *V. trilobata*, *V. vexillata*, *V. pilosa*, *V. wightii*
	Eastern Ghats	*Vigna radiata* var. *sublobata*, *V. umbellata*, *V. aconitifolia*, *V. trilobata*, *V. pilosa*, *V. capensis*
Coastal scrub	Western and Eastern Ghats	*Vigna radiata* var. *sublobata*, *V. trilobata*, *V. bourneae*, *V. umbellata*
Tropical thorn scrub	Deccan plateau	*Vigna trilobata*, *V. aconitifolia*

pods are with nearly black hairs and later become glabrous. However in *V. wightii*, the pods are with pale brown hairs. *V. capensis* is distinct from all other species in its large pinkish pea-like flowers and much longer pods with mottled black, comparatively longer sub kidney shaped seeds. This species is collected from Sholayar in Attapadi (Kerala). *V. pilosa* types are slender climbers. Populations of this species are distributed throughout southern parts of Western Ghats. *V. bourneae*, occurs in more humid parts of southern Western Ghats. The above account in brief describes the areas of diversity for wild *Vigna* species, in Western Ghats. From exploration point of view, about 4 species exhibit sympatric distribution in Western Ghats and they include types of *Vigna radiata* var. *sublobata*, *Vigna umbellata*, *Vigna bourneae* and *Vigna pilosa*. Wild *Vigna* species occur in open habitats, usually in grass mixed vegetation or are seen twining on bushes in scrub vegetation. More diversity occurs in the moist tropical habitats than in the sub drier habitats in the region. The fruiting period for wild *Vigna* species is from September to December. In the Western Ghats and Eastern Ghats ripe pods can be collected in late September to October, while in northern plains/north-western plains, September is ideal for collection. In the north-eastern region, fruiting gets delayed and October–November is the ideal period for collecting ripe pods. In the Western Himalayas, in the foothills, fruiting starts in September and continues up to October in higher elevation. Additionally, in the Western Ghats, it is observed that *silvestries* and *sublobata* types behave as perennials, bearing ripe pods in March also (Arora, 1985b). Besides *V. pilosa* and *V. vexillata* also behave as perennials. In the latter case the tuberous rootedness confers perenniality (personal communication from Dr. Z. Abraham, Sr. Scientist, NBPGR Regional station, Trichur, Kerala).

Apart from our team of workers, NBPGR, different circles of Botanical Survey of India (BSI), National Botanical Research Institute (NBRI) (Lucknow), International Crop Research Institute for Semi-Arid Tropics (ICRISAT) (Patencheru, Hyderabad),

Department of Botany, Delhi University (Delhi), Tropical Botanical Garden and Research Institute (TBGRI, Pallode, Thiruvananthapuram), M.S.Swaminathan Research Foundation (MSSRF, Chennai), etc. have been evincing keen interest in collection, documentation and conservation of germplasm seed materials of underexploited/tribal pulses.

At global level, International Institute of Tropical Agriculture (IITA) Ibadan, Nigeria; Asian Vegetable Research and Development Centre (AVRDC); Republic of China, Taiwan; and Food and Agricultural Organization (FAO), Rome, Italy have been concentrating on germplasm collection and conservation of little known legumes including tribal pulses.

2.1.5. Recorded genetic diversity in tribal pulse germplasm at the time of collection

- Fertility index, number of flowers per cluster/per plant and pod setting percentage are found to be higher in *Alysicarpus rugosus*, *Cassia laevigata*, *Indigofera linifolia* and *Sesbania bispinosa*. "Silent Valley" and "Kolli Hills" germplasm of *Entada scandens* exhibit a low fertility index (Siddhuraju *et al.*, 1993).
- Comparatively, "Silent Valley" germplasm of *Entada scandens* registers the highest value for 100-seed-weight compared to "Kolli Hills" germplasm (Siddhuraju *et al.*, 1993).
- Seed samples of *Phaseolus lunatus* collected from Valparai Hills (red variety) exhibit a higher number of branches/plant, number of leaves/plant, plant height, robust growth, early flowering and high pod setting percentage compared to the other two varieties collected from Kolli Hills (black and brown varieties).
- The 100-seed-weight of *Phaseolus lunatus* (black variety) seems to be higher than the other two varieties (red and brown varieties) (Siddhuraju *et al.*, 1993).
- *Vigna aconitifolia* exhibits early flowering and high fertility index when compared with other two *Vigna* species studied (*V. trilobata* and *V. vexillata*) (Siddhuraju *et al.*, 1993).
- The plant height, number of branches/plant and number of leaves/plant are higher in *V. trilobata* than the other two *Vigna* species (*V. aconitifolia* and *V. vexillata*) (Siddhuraju *et al.*, 1993).
- *Acacia leucophloea*, *Sesbania bispinosa* and *Vigna trilobata* are observed to be highly tolerable to drought conditions (Siddhuraju *et al.*, 1993).
- There is distinct diversity in pod nature between the collected germplasm of *Mucuna pruriens* var. *utilis* (velvet bean) and *M. pruriens* var. *pruriens* (itching bean). The pods of itching bean are characterized by the presence of stinging hairs, whereas the pods of velvet bean are devoid of stinging hairs (Anonymous, 1996).
- In general, all the collected germplasm seeds of velvet bean are larger in size than the germplasm seeds of itching bean (Anonymous, 1996).
- Marked variation in seed coat colour within the collected germplasm of tribal pulses such as *Abrus precatorius*, *Canavalia ensiformis*, *C. gladiata*, *Mucuna pruriens* var. *pruriens*, *M. pruriens* var. *utilis*, *Phaseolus lunatus* and *Vigna sinensis* has been noticed.
- Within the same agroclimatic zone (Vanavasi, Salem district in Eastern Ghats) Siddhuraju (1994) has gathered germplasm of *Abrus precatorius* exhibiting three distinct patterns of seed coat colour viz. the normal pattern of red and black-dotted seed coat, completely white-coloured seed coat and completely black-coloured seed coat.

- Most of the gathered germplasm of *Canavalia ensiformis* exhibit red-coloured seed coat. Exceptionally the germplasm of Pathanamthitta (Kerala), Vazhikkadavu (Kerala) and Mananthavadi (Kerala) exhibit maroon-coloured seed coat (Anonymous, 1996).
- Out of a total collection of 33 germplasm of *C. gladiata*, 28 germplasm exhibit white-coloured seed coat, whereas the germplasm of Dasukuppam (Chittoor district, A.P.); Hogenekkal (Dharmapuri district, T.N.); Allapuzha and Kumlza (Kerala) and Tamil Nadu Agricultural University (Coimbatore, T.N.) show maroon-coloured seed coat (Anonymous, 1996; Vadivel *et al.*, 1998).
- Interestingly the germplasm seed samples of *Mucuna pruriens* var. *utilis* with two distinct coloured seed coats (white and black) have been gathered within the same agro-climatic region, viz. Thachanmali in Kanyakumari district and Mundanthurai Wildlife Sanctuary in Tirunelveli district, Tamil Nadu (Vadivel and Janardhanan, 1997).
- Vadivel and Janardhanan (1999) have gathered *Phaseolus lunatus* white-coloured seed coat germplasm from Karwar (Uttarakannada) and Borra caves (Araku valley, A.P.) and white-coloured seed coat with red patches from Medikeri (Kodagu district, Karnataka).
- Both the germplasm of *Vigna sinensis* gathered from Allahabad (U.P.) exhibit diversity in seed coat colour (pale pink and maroon) (Rajaram and Janardhanan, 1993b).

3. Germplasm resource management of underexploited/tribal pulses

It is increasingly felt that there is an urgent need for exploration, inventorization and conservation of gene pools of wild species because they are needed for various uses in agriculture, horticulture and forestry (Frankel, 1975; Gadgil, 1996; Paroda *et al.*, 1999). Wild germplasm, if found suitable, may be utilized in breeding against pest and pathogen resistance, drought resistance and increasing the yield potential (Harlan, 1976; Khoshoo, 1996).

3.1. EVALUATION OF AGROBOTANICAL CHARACTERS

The agrobotanical characters are the ideal tools for the analysis and evaluation of better germplasm. Keeping this in view, all the collected germplasm seeds of the underexploited/tribal pulses and their wild related species, *Bauhinia malabarica*, *B. purpurea*, *B. racemosa*, *B. tomentosa*, *B. vahlii*, *Canavalia ensiformis*, *C. gladiata*, *Cassia floribunda*, *C. obtusifolia*, *C. occidentalis*, *Entada scandense*, *Mucuna gigantea*, *M. monosperma*, *M. pruriens* var. *pruriens*, *M. pruriens* var. *utilis*, *Phaseolous lunatus*, *Sesbania bispinosa*, *Vigna aconitifolia*, *V. sinensis*, *V. calcaratus*, *V. sesquipedalis*, *V. sublobata*, *V. trilobata*, *V. umbellata* and *V. vexillata* have been evaluated for agrobotanical/agronomical characters listed below in the agroclaimatic zone of Coimbatore.

1. Germination percentage
2. Day of plumule emergence
3. Day of first leaf formation
4. Plant height (cm)
5. Number of leaves
6. Number of branches
7. Leaf area (cm^2)

8. Chlorophyll content
9. Moisture content
10. Day of flower initiation
11. Number of clusters per plant
12. Number of flowers per cluster/plant
13. Fertility index or pod setting percentage
14. Pod length (cm)
15. Number of seeds per pod
16. Seed weight per pod
17. Seed recovery percentage
18. 100-seed-weight (g)
19. Life span of the crop.

The seeds of evaluated germplasm are sown in Randomized Block Design (RBD) in the experimental plots of botanical garden, Bharathiar University, Coimbatore, India. The plants are raised in open sunlight and watered once in a week regularly. The agroclimate of Coimbatore in general and the Bharathiar University Campus in particular come under semi-arid zone and alfisols type soil.

The distinct and salient agrobotanical traits in respect of species/germplasm alone are recorded.

The assessment of agrobotanical characters of the germplasm of different species of the genus, *Bauhinia*, reveals that *B. purpurea* exhibits the following advantageous traits: fast vegetative growth, early flowering and high fertility index (Rajaram, 1990). Early flowering is an important character because it assists in reduction of incidence of numerous pests and diseases. This results in consequent reduction of yield loss and better quality of pods and seeds.

The investigated different germplasm of *Canavalia ensiformis* (jack bean) reveals that Naduvil (Kerala) accession shows several favourable agrobotanical traits (Anonymous, 1996).

All the 33 germplasm of *Canavalia gladiata* (sword bean) assessed for agrobotanical characters reveal significant differences among all the accessions for all the characters investigated (Vadivel *et al.*, 1994, 1998). Of all the germplasm, Maruthamalai (Coimbatore) accession exhibits several favourable traits such as more biomass and better yield per unit area. These results prove the existence of a wide genetic base among sword bean germplasm. This variation could be used in breeding programs to improve yield (Anonymous, 1996).

Among the 14 germplasm of *Cassia obtusifolia* evaluated for agrobotanical characters, Thengumarahada (Erode, Tamil Nadu) germplasm registers the highest germination percentage, plant height, fertility index, 100-seed-weight and early flowering. One of the significant findings on the assessment of agrobotanical characters of the germplasm of *Cassia floribunda* is their failure to flower (Anonymous, 1996; Vadivel and Janardhanan, 1998).

Among all the germplasm of *Mucuna pruriens* var. *pruriens* in the present study, two germplasm have been identified to be superior. Silent valley (Palakkad, Kerala) is identified as the best germplasm based on the traits: more number of fertile flowers produced per cluster, robust vegetative growth, high yielding potential and resistance against the attack of pests and pathogens (Rajaram and Janardhanan, 1993b). In some other aspects, the Salem germplasm exhibits superior traits like more plant height and fertility index, yield based parameters and resistance to the attack of pests and pathogens (Anonymous, 1996).

Though all the collected germplasm seed samples of *Mucuna monosperma* and *M. gigantea* have germinated well and exhibited vegetative growth; they fail to flower. The exact reason for this is not known (Rajaram and Janardhanan, 1993b).

Of all the germplasm of *Mucuna pruriens* var. *utilis* (velvet bean) evaluated, except Mundanthuri (Tirunelveli, Tamil Nadu), all the other germplasm show higher fertility index and 100-seed-weight and register higher values compared to all the investigated germplasm of *M. pruriens* var. *pruriens*. The Keeripari (Kannyakumari) germplasm of *M. pruriens* var. *utilis* registers the highest plant height (Anonymous, 1996; Vadivel and Janardhanan, 1997).

Of all the germplasm of *Vigna calcaratus*, *V. sesquipedalis*, *V. sublobata*, and *V. vexillata* evaluated for agrobotanical characters, *V. sesquipedalis* alone exhibits cent per cent seed germination, maximum biomass production, cent per cent fertility, maximum number of seeds per pod and high value for 100-seed-weight. From these characters, it is inferred that *V. sesquipedalis* appears to be superior to the other species of *Vigna* taken up in the present investigation (Rajaram and Janardhanan, 1992c). The 100-seed-weight of *V. vexillata* seems to be higher and closer to that of *V. radiata* (Ignacimuthu and Babu, 1987). *V. aconitifolia* exhibits high germination percentage, early flowering and high fertility index when compared with *V. trilobata* and *V. vexillata*. Nonetheless, plant height, number of branches/plant and number of leaves/plant are higher in *V. trilobata* than *V. vexillata* (Siddhuraju *et al.*, 1993).

The wild species, *Vigna sublobata*, shows close affinity to its related cultivated species, *V. radiata* (Yohe and Poehlman, 1972). In some other characters like number of clusters per plant and number of days required for flowering it exhibits close affinity with another *Vigna* species, *V. aconitifolia* (Jindal, 1988).

The other wild species, *Vigna glaberesence*, shows robust vegetative growth (Rajaram and Janardhanan, 1992c). The value of its 100-seed-weight is closer to that of *V. mungo* (Ignacimuthu and Babu, 1987). Among the various germplasm of *V. umbellata* (ricebean) evaluated, the variety K1 has been found to be superior in several agrobotanical characters (Rajaram, 1990).

Nonetheless, all the germplasm of *Vigna sinensis* evaluated for agrobotanical traits fail to exhibit normal growth in the Coimbatore agroclimatic region (Rajaram, 1990).

3.2. EVALUATION OF MAJOR FOOD CONSTITUENTS

The nutritive value of legume crops for human diet and animal feeds and how they might be improved is well documented (Bressani, 1973; Bressani and Elias, 1980). The nutrients that must be supplied in the daily food to keep man in good health belong to the group of proteins, fats and carbohydrates – the macronutrients. Man acquires his essential nutrients from foods obtained from the plant and animal world (Subbulakshmi and Chitra, 1996).

Proteins are an essential component of the diet needed for survival of animals and humans. Proteins basic function in nutrition is to supply adequate amounts of needed amino acids (Friedman, 1996). The proteins not only provide the essential amino acids, which can not be synthesised in the tissues of animals and humans but also fix nitrogen for the formation of non-essential amino acids and many other nitrogenous compounds (Cuthbertson, 1989). Fats provide the most concentrated form of energy in our diet. They facilitate the absorption and transport of fat-soluble vitamins and provide the essential polyunsaturated fatty acids (Bamji, 1991). Lipids are a heterogeneous group, which include fatty acids: mono-, di- and tri-acylglycerols; phospholipids; sterols; sterolesters; glycolipids and lipoproteins (Pattee *et al.*, 1982).

The amount of protein present in seeds varies from −10% (in cereals) to −40% (in certain legumes and oil seeds) of the dry weight, forming a major source of dietary protein (Shewry *et al.*, 1995).

Several species/varieties/germplasm of underexploited/tribal pulses such as *Canavalia ensiformis* (Rodrigues and Torne, 1991; Rajaram and Janardhanan, 1992a; Mohan and Janardhanan, 1994c; *Canavalia gladiata* (Rodrigues and Torne, 1991; Mohan and Janardhanan, 1994c; Vadivel, 1998); *Erythrina indica* (Banerji and Dixit, 1998); *Indigofera linifolia* (Siddhuraju *et al.*, 1995c); *Lens esculenta* (Dhindsa *et al.*, 1985); *Mucuna deeringiana* (Banerji and Dixit, 1998); *M. gigantea* (Rajaram and Janardhanan, 1991a); *M. pruriens* var. *pruriens* (Pant *et al.*, 1974; Mary Josephine and Janardhanan, 1992; Siddhuraju *et al.*, 1996b); *M. pruriens* var. *utilis* (Mohan and Janardhanan, 1995b; Vadivel and Janardhanan, 2001); *Phaseolus lunatus* (Rajaram and Janardhanan, 1993a; Vadivel and Janardhanan, 1999); *Pongamia pinnata* (Banerji and Dixit, 1998); *Psophocarpus tetragonolobus* (Banerji, 1985; Misra *et al.*, 1987; Lal and Kumar, 1997; Banerji and Dixit, 1998; Rangalakshmi, 1998); *P. marsupium* (Banerji and Dixit, 1998); *Sesbania bispinosa* (Siddhuraju *et al.*, 1995c; Banerji and Dixit, 1998); *Vicia faba* (Sharma and Sehgal, 1992); *Vigna trilobata* (Vadivel and Janardhanan, 1999) and *Xylia xylocarpa* (Siddhuraju *et al.*, 1995a; Banerji and Dixit, 1998; Vadivel and Janardhanan, 1999) contain crude protein which is 20–80% (w/w) more when compared with the most commonly consumed (traditional) pulses of India (Tables 7 and 8).

Crude lipid contents of most of the germplasm of the wild legumes (underutilized/tribal pulses) range between 10 and 30% (w/w). They are *Alysicarpus rugosus*

Table 7. Crude protein and crude lipid contents of some of the commonly cultivated legume crops

Botanical name	Crude protein (%)	Crude lipid (%)	Source
Cajanus cajan	21.4–25.4	1.3–2.8	Jambunathan and Singh, 1981a and b Oshodi and Ekperigin, 1989 Apata and Ologhobo, 1994
Cicer arietinum	22.0–24.0	4.1–6.69	Jambunathan and Singh, 1980 Attia *et al.*, 1994
Glycine max	34.3–40.7	17.6–18.3	Saxena *et al.*, 1994
Vigna mungo	23.3–23.8	1.25–2.7	Gupta and Wagle, 1978 Rajyalakshmi and Geervani, 1994
Vigna radiata	22.1–23.8	1.12–2.41	Gupta and Wagle, 1978 Sharma *et al.*, 1991

Table 8. Range in levels of crude protein and crude lipid in underexploited/tribal pulses

Species	Crude protein (%)	Crude lipid (%)	Source
Abrus precatorius	16.1–23.0	2.0–9.4	Rajaram and Janardhanan, 1992b; Mohan and Janardhanan, 1995a; Banerjii and Dixit, 1998; Vadivel and Janardhanan, 1999
Acacia catechu	16.3–48.9	2.2–7.3	Yadav and Bharadwaj, 1971; Pant *et al.*, 1974; Chowdhury *et al.*, 1983; Rajaram and Janardhanan, 1991b; Banerji and Dixit, 1998
Acacia leucophloea	25.1–26.5	5.1–8.1	Vijayakumari *et al.*, 1994; Banerji and Dixit, 1998
Acacia nilotica	14.0–23.4	3.5–6.7	Kumaresan *et al.*, 1984; Chowdhury *et al.*, 1983; Siddhuraju *et al.*, 1996a; Banerji and Dixit, 1998
Alysicarpus rugosus	27.1	14.0	Siddhuraju *et al.*, 1992b
Bauhinia malabarica	16.1–27.0	13.7–17.8	Pant *et al.*, 1974; Vijayakumari *et al.*, 1993a
Bauhinia purpurea	25.6–27.2	12.3–14.3	Rajaram and Janardhanan, 1991c; Vijayakumari *et al.*, 1997b
Bauhinia racemosa	16.8–26.9	2.4–9.5	Rajaram and Janardhanan, 1991c; Mohan and Janardhanan, 1994b; Banerji and Dixit, 1998
Bauhinia vahlii	19.3–24.6	18.0–23.3	Rajaram and Janardhanan, 1991c; Vadivel and Janardhanan, 1999
Canavalia ensiformis	29.7–32.2	2.9–8.8	Rodrigues and Torne, 1991; Rajaram and Janardhanan, 1992a; Mohan and Janardhanan, 1994c; Vadivel, 1998
Canavalia gladiata	19.9–33.9	0.8–9.9	Rodrigues and Torne, 1991; Rajaram and Janardhanan, 1992a; Mohan and Janardhanan, 1994c; Vadivel, 1998
Canavalia virosa	17.4	1.6	Rodrigues and Torne, 1991
Cassia floribunda	10.4–24.7	6.3–7.4	Katoch and Bhowmik, 1983; Janardhanan, 1993
Cassia laevigata	20.4–20.8	3.0–5.3	Siddhuraju *et al.*, 1995b; Banerji and Dixit, 1998
Cassia obtusifolia	18.7–23.9	5.2–12.7	Pant *et al.*, 1974; Vijayakumari *et al.*, 1993b; Mohan and Janardhanan, 1994a; Mohan and Janardhanan, 1995a
Cassia tora	18.6–23.9	6.6–10.6	Talpada *et al.*, 1980; Katoch and Bhowmik, 1983; Banerji and Dixit, 1998
Cassia uniflora	27.5	11.7	Vijaya Bhanu *et al.*, 1991

Table 8. Continued

Species	Crude protein (%)	Crude lipid (%)	Source
Dolichos biflorus (= *Macrotyloma uniflorum*)	17.9–25.3	0.7–3.0	Rajyalakshmi and Geervani, 1994; Sudha *et al.*, 1995
Dolichos lablanb var. *vulgaris*	22.1	4.9	Logamadevi, 1991
Entada scandens (= *E. phaseoloides*)	17.4–27.7	3.1–10.8	Chowdhury *et al.*, 1984; Janardhanan and Nalini, 1991; Mohan and Janardhanan, 1993a; Vijayakumari *et al.*, 1993b; Banerji and Dixit, 1998
Erythrina indica	23.1–30.0	11.2–15.7	Pant *et al.*, 1974; Benerji and Dixit, 1998
Indigofera linifolia	32.1	4.7	Siddhuraju *et al.*, 1995c
Lens esculenta	26.3–31.1	3.2–3.9	Dhindsa *et al.*, 1985
Mucuna deeringiana	30.3	4.4	Banerji and Dixit, 1998
Mucuna gigantea	30.6	9.0	Rajaram and Janardhanan, 1991a
Mucuna monosperma	20.1–26.4	4.4–16.7	Arulmozhi and Janardhanan, 1992; Mohan and Janardhanan, 1995b; Benerji and Dixit, 1998
Mucuna puriens var. *pruriens*	27.3–35.0	4.5–9.1	Pant *et al.*, 1974; Mary Josephine and Janardhanan, 1992; Rajyalakshmi and Geervani, 1994; Siddhuraju *et al.*, 1996b; Benerji and Dixit, 1998
Mucuna pruriens var. *utilis*	20.2–29.6	6.3–10.8	Janardhanan and Lakshmanan, 1985; Mohan and Janardhanan, 1995b; Vadivel and Janardhanan, 2001
Parkia roxburghii	21.9	16.6	Mohan and Janardhanan, 1993a
Parkinsonia aculeata	21.3	2.5	Rajaram and Janardhanan, 1991b
Phaseolus lunatus	20.8–32.1	2.2–7.8	Rajaram and Janardhanan, 1993a; Vijayakumari *et al.*, 1993b; Vadivel and Janardhanan, 1999
Pongamia pinnata	30.4	30.0	Banerji and Dixit, 1998
Prosopis chilensis	26.7	12.3	Rajaram and Janardhanan, 1991b
Psophocarpus tetragonolobus	29.8–39.1	14.1–20.0	Banerjee, 1985; Mistra *et al.*, 1987; Lal and Kumar, 1997; Banerji and Dixit, 1998; Rangalakshmi, 1998
Psophocarpus marsupium	42.7	14.0	Banerji and Dixit, 1998
Sesbania bispinosa	29.7–38.8	3.8–6.4	Siddhuraju *et al.*, 1995c; Banerji and Dixit, 1998

Table 8. Continued

Species	Crude protein (%)	Crude lipid (%)	Source
Tamarindus indica	13.0–26.9	5.4–10.9	Ishola *et al.*, 1990; Bhattacharya *et al.*, 1994; Siddhuraju *et al.*, 1995b
Trigonella foenugraecum	25.4	7.9	Udhayasekhara Rao and Sharma, 1987
Uraria picta	23.3	4.5	Pandey and Srivastava, 1991
Vicia faba	28.7–29.2	1.8–2.2	Sharma and Sehgal, 1992
Vigna aconitifolia	22.9–23.9	1.6–3.5	Pawar and Ingle, 1988; Siddhuraju *et al.*, 1994
Vigna calcaratus	22.1	3.5	Rajaram and Janardhanan, 1992c
Vigna capensis	22.4	4.2	Mohan and Janardhanan, 1993b
Vigna sesquipedalis (= *V. unguiculata* subsp. *sesquipedalis*)	24.3	4.3	Rajaram and Janardhanan, 1990
Vigna sinensis	19.3–28.6	2.6–6.4	Rajaram and Janardhanan, 1990; Mohan and Janardhanan, 1993b; Vadivel and Janardhanan, 1999
Vigna sublobata	25.4	1.6	Rajaram and Janardhanan, 1992c
Vigna trilobata	20.3–30.3	5.0–5.5	Siddhuraju *et al.*, 1992a; Vadivel and Janardhanan, 1999
Vigna umbellata	17.5–29.3	1.0–5.4	Singh *et al.*, 1985; Malhotra *et al.*, 1988; Rajaram and Janardhanan, 1990; Mohan and Janardhanan, 1994d; Vadivel and Janardhanan, 1999
Vigna vexillata	24.5	1.6	Siddhuraju *et al.*, 1994
Xylia xylocarpa	29.5–49.7	14.8–23.3	Siddhuraju *et al.*, 1995a; Banerji and Dixit, 1998; Vadivel and Janardhanan, 1999

(Siddhuraju *et al.*, 1992b); *Bauhinia malabarica* (Pant *et al.*, 1974; Vijayakumari *et al.*, 1993a); *B. purpurea* (Rajaram and Janardhanan, 1991c; Vijayakumari *et al.*, 1997b); *B. vahlli* (Rajaram and Janardhanan, 1991c; Vadivel and Janardhanan, 1999); *Cassia obtusifolia* (Mohan and Janardhanan, 1995a); *Cassia tora* (Banerji and Dixit, 1998); *C. uniflora* (Vijaya Bhanu *et al.*, 1991); *Entada scandens* (Mohan and Janardhanan, 1993a); *Erythrina indica* (Pant *et al.*, 1974; Banerji and Dixit, 1998); *Mucuna monosperma* (Mohan and Janardhanan, 1995b; Banerji and Dixit, 1998); *M. pruriens* var. *utilis* (Mohan and Janardhanan, 1995b); *Parkia roxburghii* (Mohan and Janardhanan, 1993a); *Pongamia pinnata* (Banerji and Dixit, 1998); *Prosopis chilensis* (Rajaram and Janardhanan, 1991b); *Psophocarpus tetragonolobus* (Banerji, 1985; Misra *et al.*, 1987; Lal and Kumar, 1997; Banerji and Dixit, 1998; Rangalakshmi, 1998); *P. marsupium* (Banerji and Dixit, 1998); *Tamarindus indica* (Ishola *et al.*, 1990) and *Xylia xylocarpa*

(Siddhuraju *et al.*, 1995a; Banerji and Dixit, 1998; Vadivel and Janardhanan, 1999). On the contrary the crude lipid content of commonly consumed (traditional) pulses range only between 2 and 6.7% (w/w).

The available information shows that the food constituents of the pulses is largely influenced by different environmental factors such as soil type, climate, location, fertilizers, crop rotation and other agricultural practices (Grover and Gupta, 1972; Kapoor *et al.*, 1972; Agrawal and Bhattacharya, 1980; Monti and Grillo, 1983; Singh *et al.*, 1983; Sakla *et al.*, 1988; Mary Josephine and Janardhanan, 1992; Janardhanan and Vadivel, 1994; Mohan and Janardhanan, 1995a).

The protein quality, also known as the nutritional or nutritive value of a food, depends on its amino acids content and on the physiological utilization of specific amino acids after digestion, absorption and minimal obligatory rates of oxidation (Friedman, 1996).

The requirement of dietary protein consists of two components: (1) the requirement for nutritional indispensable amino acids (histidine, isoleucine, leucine, lysine, methionine, phenylalanine, threonine, tryptophan and valine) under all conditions and for conditionally indispensable amino acids (cysteine, tyrosine, taurine, glycine, arginine, glutamine and proline) under specific physiological and pathological conditions and (2) the requirement for nonspecific nitrogen for the synthesis of the nutritionally dispensable amino acids (aspartic acid, asparagine, glutamic acid, alanine and serine) and other physiologically important nitrogen compounds such as nucleic acids, creatinine and porphyrins. The nutritive value of various food protein sources are to a large extent determined by the concentration and availability of the individual indispensable amino acids (Young and Pellett, 1994).

Recently, the nutritive value of proteins from different food sources are reviewed by Friedman (1996) and plant proteins are discussed in relation to their amino acids content, human amino acid requirement and dietary protein quality (Young and Pellett, 1994).

Chemical analysis of some of the cultivated legumes show promising results in respect of their protein content, amino acid pattern and other food nutrients (FAO/WHO/UNU, 1985; FAO/WHO, 1991; ICMR, 1992). However, the nutritive value of legumes in general is considered to be inferior since all the essential amino acids are not present in the required proportion as per the recommendation of FAO/WHO (1991).

In general, seed proteins of legumes are deficient in sulfur containing amino acids, cysteine and methionine, and tryptophan but are rich in lysine when compared with FAO/WHO (1991) recommended pattern. Exceptionally, the proteins of certain germplasm/varieties of underexploited/tribal pulses such as *Alysicarpus rugosus* (Siddhuraju *et al.*, 1992b); *Indigofera linifolia* (Siddhuraju *et al.*, 1995c) and *Tamarindus indica* (Siddhuraju *et al.*, 1995b) contain sulfur containing amino acids more than the FAO/WHO (1991) recommended pattern.

It is noteworthy that the seed proteins of *Dolichos lablab* varieties attracted the attention as potential sources of methionine rich proteins (Laurena *et al.*, 1991). Similarly the seed proteins of *Abrus precatorius* (Rajaram and Janardhanan, 1992b); *Bauhinia purpurea* (Rajaram and Janardhanan, 1991c; Vijayakumari *et al.*, 1997b); *Vigna capensis* (Mohan and Janardhanan, 1993b); *V. sinensis* (Rajaram and Janardhanan, 1990; Mohan and Janardhanan, 1993b); *V. sesquipedalis* (Rajaram and Janardhanan, 1990); *V. umbellata* (Rajaram and Janardhanan, 1990; Mohan and Janardhanan, 1994d); *V. vexillata* (Siddhuraju *et al.*, 1994) and *Xylia xylocarpa* (Siddhuraju *et al.*, 1995a) are rich in methionine. On the contrary the seed proteins of *Acacia leucophloea* (Vijayakumari *et al.*, 1994) and *Phaseolus lunatus* (both Kolli hills and Valparai germplasm) (Vijayakumari *et al.*, 1993b) are found to be rich sources of cysteine.

In general, legume proteins exhibit a wide range of variation in their essential amino acid composition (Salunkhe *et al.*, 1985). Legume seed proteins of different species of the same genus and different varieties/germplasm of the same species are known to exhibit variations in their content of amino acid profiles (Rajaram and Janardhanan, 1990; Laurena *et al.*, 1991; Arulmozhi and Janardhanan, 1992; Mary Josephine and Janardhanan, 1992; Vijayakumari *et al.*, 1993b; Siddhuraju *et al.*, 1994; Mohan and Janardhanan, 1994b, c, 1995a, b).

Certain conventional or commonly consumed pulse seed proteins contain adequate levels of the essential amino acid, tryptophan (Sastry and Murray, 1986). In many developing countries including India, much of the population is still restricted to diets made up almost entirely of cereal grain. The cereal diets of vegetarian and low income populations are traditionally supplemented by legumes as a source of more concentrated protein and one whose amino acid pattern complements that of cereal diets (Scrimshaw and Young, 1977). The cereal proteins are usually rich in sulfur-containing amino acids and tryptophan and deficient in lysine when compared with FAO/WHO (1991) recommended pattern.

The content of tryptophan in legume seed proteins that might complement cereals should, therefore, be considered whenever seed protein sources are evaluated (Sastry and Murray, 1986). The data on various other essential amino acids of underexploited/tribal pulse seed proteins are available (Janardhanan and Lakshmanan, 1985; Rajaram and Janardhanan, 1990, 1991a, b, c, 1992a, b; Mary Josephine and Janardhanan, 1992; Arulmozhi and Janardhanan, 1992; Ravindran and Ravindran, 1988; Mohan and Janardhanan, 1993a, b, 1994b, c, 1995a, b; Siddhuraju *et al.*, 1992a, b, 1994, 1995a, b, c, 1996a, b; Vijayakumari *et al.*, 1993a, b, 1994; Vadivel and Janardhanan, 2001). For the first time Siddhuraju *et al.* (1997) have estimated the tryptophan contents of total proteins and protein fractions, albumins and globulins, of several germplasm seed samples of wild legumes, including underexploited/tribal pulses (Table 9).

Tryptophan contents of total proteins range from 0.58 (*Mucuna atropurpurea*) to 2.03% (*Prosopis chilensis*). Total seed proteins of *Abrus precatorius* "white variety", *Acacia nilotica*, *Prosopis chilensis*, *Vigna aconitifolia* and *V. sinensis* reveal the presence of higher tryptophan contents than that present in whole hen's egg protein. In most of the investigated underexploited/tribal pulses, the albumins seem to be a rich source of tryptophan when compared to globulins (Siddhuraju *et al.*, 1997).

3.3. TRADITIONAL MEDICINAL PROPERTIES

By trial and error, the aboriginal tribals have learnt to employ certain plants/parts/extracts to cure several diseases/ailments. In this context, certain tribal pulses also merit consideration for their therapeutic properties which are getting commercialized.

3.3.1. Abrus precatorius

The decoction prepared from fresh roots is used three times daily in abortion (Maheshwari *et al.*, 1991). Leaves are used as ingredients in cough mixtures. Powdered

Table 9. Tryptophan (%, w/w) content of total proteins and protein fractions (albumins and globulins) of the various wild legumes including underexploited/tribal pulses

Name of the legume	Total protein	Albumins	Globulins
Abrus precatorius L.			
"Black variety"	1.22	0.52	0.71
"Red and black-dotted variety"	1.80	1.07	1.20
"White variety"	1.66	0.96	0.62
Acacia leucophloea (Roxb) Willd	1.21	0.74	1.35
A. nilotica (L.) Del	1.87	2.09	0.82
Alysicarpus rugosus (Willd) DC	0.92	1.25	0.90
Atylosia goensis (Dalz)	0.88	1.93	0.63
Bauhinia malabarica (Roxb)	1.30	1.52	trace
B. purpurea L.	0.78	2.01	trace
B. racemosa Lamk	0.97	0.84	0.79
**B. tomentosa* L.	0.60	0.78	0.34
B. vahlii W & A	0.84	trace	1.09
Canavalia ensiformis DC	0.96	1.94	1.03
C. gladiata (Jacq) DC. "White variety"	1.26	0.87	0.86
C. gladiata (Jacq) DC. "Brown variety"	1.03	0.84	0.76
Cassia laevigata Wild	1.02	0.78	0.64
C. obtusifolia L.	0.82	1.56	1.21
Entada scandens Benth	1.09	1.26	0.82
Indigofera linifolia (L.F.) Retz	1.28	0.91	1.04
**Mucuna atropurpurea* DC	0.58	1.46	0.42
M. gigantea (Willd) DC	1.35	1.25	1.03
M. pruriens (L.) DC	1.35	1.25	1.03
M. utilis Wall ex Wight	1.31	0.97	1.26
**Parkia biglandulosa* W & A	trace	0.82	0.71
P. roxburghii G Don	1.19	0.56	1.02
Phaseolus lunatus L.			
"Brown variety"	1.26	1.92	1.03
"Black variety"	1.08	1.43	1.05
"Red variety"	0.80	1.02	1.17
Prosopis chilensis (Molina) Stunz	2.03	trace	1.36
Parkinsonia aculeata L.	1.04	0.72	trace
Pongamia pinnata (L.) Pierre	1.53	1.03	0.90
Sesbania bispinosa (Jacq) W.F. Wight	0.91	1.11	1.42
Tamarindus indica L.	0.96	0.74	0.37
Vigna aconitifolia (Jacq) Marechal	1.67	1.62	0.43
V. capensis Walp	1.26	0.71	1.02
V. glabrescens Marechal	0.75	0.98	0.52
V. sesquipedalis (L.) Verdc	1.17	0.83	0.49
V. sinensis (L.) Savi. ex Hassk	1.82	1.97	0.73
V. sublobata (Roxb) Babu and Sharma	1.05	0.81	0.76
V. trilobata (L.) Verdc	0.94	trace	1.15
V. umbellata (Thumb) Ohwi and Ohashi (Var. K_1)	1.06	2.11	0.61
V. vexillata (L.) A. Rich	1.13	0.70	1.21
Xylia xylocarpa (Roxb) Taub	1.36	1.22	0.92
FAO/WHO (1990) suggested pattern	1.01	–	–
Whole hen's egg[a]	1.62	–	–
Soybean[a]	1.37	–	–

[a]Reported by Oyenuga and Fetuga (1975).
*Wild legume (not edible).

leaves are employed for treatment of convulsion in children and for conjunctivitis. Juice of leaves is given for sore throat and to purify blood. Seeds are used in small doses as strong purgative and emetic. Seed infusion is used to hasten labour. Seeds are also used by women to prevent conception and seed paste with water and salt is applied on boils to promote suppuration. Powdered seeds are boiled with milk, which is found to be a powerful tonic with aphrodisiac action. Extracts of root are given for chronic coughs, sore throat and rheumatism (Gill and Nyawuame, 1994).

3.3.2. Acacia catechu

Root is made into a paste and applied on the joints for seven days for rheumatism (Tarafder and Chaudhuri, 1991).

3.3.3. Acacia nilotica

The leaf juice is taken along with fennel in blood dysentery (Maheshwari and Singh, 1991). Powdered bark with a little salt is given for acute diarrhoea. Extracts of root and bark are used as taenifuge and gum is used as emollient (Gill and Nyawuame, 1994).

3.3.4. Alysicarpus rugosus

Hot decoction of leaves and roots is given for cough, cold and fever. Leaf juice is used as an antidote for snakebite and also for skin diseases (Gill and Nyawuame, 1994).

3.3.5. Bauhinia purpurea

The stem bark decoction is used three times daily in body pain and fever (Maheshwari *et al.*, 1991). The flowers and buds are ground into a paste, mixed with water and taken for indigestion. A paste from bark is given in cancerous growth in the stomach (Rao, 1991).

3.3.6. Bauhinia vahlii

The root paste mixed with jaggery and ghee is used to cure bone fracture (Maheshwari *et al.*, 1991).

3.3.7. Cassia *species*

Cassia species contain a large number of known and unknown anthroquinones used medically as purgatives. There are also reports that such compounds cause adverse effects on muscular functions in rabbits, rodents, calves and cattle (Chan *et al.*, 1976).

3.3.8. C. obtusifolia

The leaves of this species are useful against tuberculosis and ringworms. The roots are rubbed on a stone with limejuice and applied on ringworms. The grounded seeds with sour buttermilk are used to ease the irritation of itchy eruptions. In Bombay, Assam and Goa, the seeds are used as substitute for coffee. The leaves are gently fried in castor oil and applied to foul ulcers. In China, the pods are used in dysentery and opthalmia treatment (Singh and Premnath, 1992). The seed preparations have been used as a purgative in Japan and China (Crawford *et al.*, 1990).

3.3.9. C. obtusa (= C. obtusifolia L.)

In traditional medicine it is administered to cure fever, sores, diabetes, diarrhoea, skin diseases, helmintaesis and ophthalmia. Recently two new anthroquinons have been isolated and reported (Sekar *et al.*, 1999).

3.3.10. Cassia tora

Tender leaves are eaten to prevent skin diseases (Jain, 1991). The seed powder mixed with tea is used two or three times daily in cough, headache and fever (Maheshwari *et al.*, 1991). Root is made into a paste and along with the powder prepared from the horn of a cow is given orally once daily in high fever, and to a patient who is unable to speak and hear (Tarafder and Chaudhuri, 1991). Leaves and seeds are used as remedy for ringworm and skin diseases (Vasudeva Rao and Shanpru, 1991). The ash of the seed is applied with butter in anal diseases of children. The seeds are soaked in water and taken in spermatorrhoea (Maheshwari and Singh, 1991). The leaves and seeds are useful in skin diseases, chiefly for itch and ringworm. Roots are applied on snakebite. In China, the seeds are employed for all kinds of eye diseases (Singh and Premanath, 1992). Decoction of the leaves is a mild laxative. Leaf and root powders are used for skin diseases like ringworm, scabies and eczema. A paste of the roots made with limejuice is a remedy for ringworm (Gill and Nyawuame, 1994).

3.3.11. Canavalia ensiformis

Seed decoction or powdered seeds are used as an antibiotic and antiseptic (Gill and Nyawuame, 1994).

3.3.12. Mucuna *species*

In general, the seeds of different species of the genus *Mucuna* are extensively utilized in Indian system of medicine. The seeds are a rich source of the bioactive chemical compound, L-DOPA, a valuable drug used in the treatment of Parkinsonism and other muscular disorders.

3.3.13. Mucuna monosperma

Seeds of this species possess expectorant properties and are used in treating cough, asthma and infections of the tongue (Prakash and Misra, 1987). Mucuna monosperma is used to cure acute spasm (Mittre, 1991).

3.3.14. Mucuna pruriens *var.* pruriens

The non-protein amino acid, L-DOPA, is extracted from the seeds to provide commercial drugs for the treatment of Parkinson's disease (Haq, 1983). The seed powder is known to exhibit faster hypothermic (Rajendran *et al.*, 1996) and anti-Parkinsonian activity than synthetic L-DOPA (Rajendran *et al.*, 1996; Hussain and Manyan, 1997). The seed powder is known to stimulate more sexual activity in male albino rats than L-DOPA and also is reported to arouse sexual desire in patients suffering from Parkinson's disease (Anantha Kumar *et al.*, 1994).

Vigorex-SF capsule, an ayurvethic herbo-mineral formulation consisting of the seed powder is found to have adaptogenic effect to improve libido, disturbed due to psychological fear and emotional imbalance and other allied ailments (Shaw and Bera, 1993). Alcohol extracts of leaves and fruit trichomes of this wild legume are found to increase the pain threshold and decrease body temperature. The extract also exhibits anti-inflammatory activity as it is known to inhibit carrageenin – induced edema (Iauk *et al.*, 1993). Seeds of this wild legume are widely used for treating male sexual dysfunction in Unani Medicine (Amin *et al.*, 1996).

The blocking effect of King Cobra venom at the neuro-muscular junction is removed by the aqueous extract of the seeds of this plant species (Aguiyi *et al.*, 1997). Rhinax, a herbal formulation comprising this wild pulse possesses anti-hepatotoxic activity (Dhuley and Naik, 1997). The tribe, Garos of Maghalaya, India consume the seeds for increasing potency and the hairs of the pod are used as vermifuge (Vasudeva Rao and Shanpru, 1991). In Nigeria, powdered hairs on pods are administered with honey for expelling intestinal worms (Gill and Nyawuame, 1994).

3.3.15. Mucuna pruriens *var.* utilis

The beans are employed as a powerful aphrodisiac (Jayaweera, 1981); emetic (Duke, 1981) and are widely used to treat nervous disorders (Jayaweera, 1981; Wijeratne, 1987) and arthritis (Wijeratne, 1987). The beans if applied as a paste on scorpion stings are thought to absorb poison (Jayaweera, 1981). Aqueous extracts of leaves show anti-venom activity (Houghton and Skari, 1994). Further, L-DOPA has been extracted from the seeds to provide commercial drugs for the treatment of Parkinson's disease (Haq, 1983).

3.3.16. Pongamia pinnata (= P. glabra)

The oil is extracted from the seeds and used in skin diseases. The seed powder is used three times daily in piles (Maheshwari *et al.*, 1991). Seed powder is used to cure leucorrhoea (Jain *et al.*, 1991). Seed oil is reported to exhibit medicinal properties such as curative effect on eye diseases, rheumatism, leucoderma and itching wounds (Banerji and Dixit, 1998).

3.3.17. Tamarindus indica

The seed powder is recommended as a valuable remedy in diarrhoea and dysentery (Rao and Srivastava, 1974). Decoction of the leaves is used as a vermifuge. Poultice of the leaves is applied over inflammatory swellings of ankles and joints, sprains and boils. Ash of the bark is mixed with a pinch of salt and given for colic and indigestion. Pulp is an antiscorbutic, digestive, carminative, laxative, febrifuge and refrigerant and also administered to cure disorders of spleen. In small quantities diluted with cold water with sugar and salt added, the macerated pulp forms an agreeable refrigerating drink in febrile diseases. A hot poultice of the pulp is applied over inflammatory swellings and boils (Gill and Nyawuame, 1994).

3.3.18. Uraria picta

Leaves are employed as a snakebite antidote (Gill and Nyawuame, 1994).

3.4. IDENTIFICATION OF ELITE GERMPLASM

i. The germplasm of the tribal pulses/underexploited pulses such as *Acacia leucophloea*, *Prosopis chilensis*, *Parkinsonia aculeata*, *Sesbania bispinosa*, *Vigna trilobata* and *Tamarindus indica* are highly tolerant to drought conditions (Rajaram, 1990; Siddhuraju, 1994; Siddhuraju *et al*., 1993).

ii. Another significant observation is that all evaluated germplasm of *Mucuna pruriens* var. *pruriens* and *M. pruriens* var. *utilis* exhibit luxuriant growth. They can be employed in soil reclamation and improvement of soil fertility (Anonymous, 1996).

iii. *Mucuna pruriens* var. *pruriens* germplasm gathered from Silent valley (Palakkad district, Kerala) and Salem (Salem district, T.N.) exhibit resistance against the attack of pests and pathogens (Rajaram and Janardhanan, 1993b; Anonymous, 1996).

iv. Several evaluated germplasm of *M. pruriens* var. *utilis* exhibit robust vegetative growth, high yielding potential and resistance to pests and pathogens (Anonymous, 1996; Vadivel and Janardhanan, 1997).

v. Assessment for major food components reveal that several species/varieties/germplasm of underexploited/tribal pulses like *Canavalia ensiformis*, *C. gladiata*, *Erythrina indica*, *Indigofera linifolia*, *Lens esculenta*, *Mucuna deeringiana*, *M. gigantea*, *M. pruriens* var. *pruriens*, *Mucuna pruriens* var. *utilis*, black and brown varieties of *Phaseolus lunatus* Kolli hills (Namakkal district) germplasm, *Pongamia pinnata*, *Psophocarpus tetragonolobus*, *P. marsupium*, *Sesbania bispinosa*, *Vicia faba*, *V. trilobata* and *Xylia xylocarpa* possess crude protein content, 20–80% (w/w) more when compared with the most common and conventional pulses, mungbean (green gram), urdbean (black gram), pigeonpea (red gram) and chickpea (Bengal gram).

vi. Most of the investigated accessions of tribal pulses such as *Alysicarpus rugosus*, *Bauhinia malabarica*, *B. purpurea*, *B. vahlii*, *Cassia obtusifolia*, *C. tora*, *C. uniflora*, *Entada scandens*, *Erythrina indica*, *M. pruriens* var. *utilis*, *M. monosperma*, *Parkia roxburghii*, *Pongamia pinnata*, *Prosopis chilensis*, *Psophocarpus tetragonolobus*, *P. marsupium*, *Tamarindus indica* and *Xylia xylocarpa* contain crude lipid content ranging between 10 and 30% (w/w).

vii. Seed proteins of legume seeds are invariably deficient in the sulphur-amino acids, cysteine and methionine (Murray, 1991). Nonetheless, seed proteins of the tribal pulses, *Alysicarpus rugosus*, *Indigofera linifolia* and *Tamarindus indica* contain the aforesaid amino acids more than FAO/WHO (1991) requirement pattern.

viii. Seed proteins of *Abrus precatorius*, *Bauhinia purpurea*, *Vigna capensis*, *V. sesquipedalis*, *V. umbellata*, *V. vexillata* and *Xylia xylocarpa* are found to be rich in methionine.

3.5. GENETIC EROSION AND CONSERVATION STRATEGIES

The conservation of genetic resources is increasingly accepted as a social responsibility. The same is true for the preservation of ecosystems and of the environment within the framework of natural and environmental protection (Frankel, 1975). This becomes especially clear by the signing of the Convention on Biological Diversity on the occasion of the United Nations Conference on Environment and Development (UNCED) conference of 1992 in Rio de Janiero and the political responsibility for the preservation of various components of biological diversity including the genetic resources in many countries (Brockhaus and Oetmann, 1996). In the last two decades, in India, the genetic erosion of several tribal pulses, mainly the woody climbers such as *Entada scandens*, *Mucuna monosperma*, *M. gigantea*, *Atylosia goensis*, *Bauhinia vahlii* tree pulses such as *Xylia xylocarpa* and *Parkia roxburghii* (Mohan, 1992; Rajaram and Janardhanan, 1993b; Siddhuraju *et al.*, 1993) and wild *Vigna* species such as *V. radiata* var. *sublobata*, *V. mungo* var. *silvestris* and *V. grandis* due to overexploitation (Arora, 1985b) has occurred.

National and International efforts to collect and conserve crop plant genetic resources during the last four decades have resulted in a very large number of germplasm accessions of cultivated, obsolete/primitive and wild forms of crop plants being stored *ex situ* at various genetic resources conservation centres around the world (Botanical gardens, Zoological gardens and genebanks) (Damania, 1996). *Ex situ* conservation implies conservation of germplasm outside the natural habitat of the plants concerned. In crops, this is in the form of samples of seeds stored at subzero temperature in airtight containers, tissue culture in glass vials or complete plant in field genebanks where seeds are recalcitrant, cannot be dried and frozen for storing in a genebank, for example, *Entada scandense* (Rajaram and Janardhanan, 1993b); *Pongamia pinnata* (Siddhuraju, 1994). Cohen *et al.* (1991) broadly classified *ex situ* genetic resources conservation and use into four major time eras: 1. the era of plant genetic resource exploration and introduction (1850–1950), 2. the era of conservation (1950–1980), 3. the era of regeneration and new international linkages (1980s) and 4. the era of most efficient utilization (2010 and beyond).

In situ conservation (National Parks, Protected areas, Biosphere Reserves and World Heritage Sites), on the other hand, is the continuing maintenance of a plant population within the ecological community of which it forms a part as well as in the environment to which it is adapted. It is usually applied to wild progenitors of crop plants and forest flora and wild fauna (Damania, 1996; Paroda *et al.*, 1999). *In situ* methods should be provided with backup from *ex situ* collection as a sort of insurance policy against total loss or extinction of germplasm. Thus, *in situ* conservation provides a broad genetic base maintenance of population structures, stability of population numbers and opportunities for future adaptive expansion (Chang, 1994). Threats to destruction of biodiversity in general is due to overexploitation of the biotic and physical environment, habitat fragmentation and subsequent loss, pollution, microclimate change, large-scale but unsustainable agriculture and forestry projects especially in the developing countries (Bajaj and Williams, 1995).

The depletion of crop plant genetic resources especially in their centres of diversity can be associated with the spread of modern agriculture practices, which gained ground mainly after the Second World War. However, agricultural crop diversity (agro-biodiversity) cannot be preserved *in situ* without simultaneously encouraging the traditional culture of the farm community, which fostered it, and protecting its rights (Paroda *et al.*, 1999). In this context, under Integrated Gene Management, community (tribal and rural families) conservation, besides *in situ* and *ex situ* conservation of germplasm resources has been emphasized. Community conservation is further subdivided into *in situ* on farm and *ex situ* on farm. *In situ* on farm community conservation takes into account conservation of landraces and folk varieties; whereas *ex situ* on farm community conservation takes into account the conservation of wild germplasm in Sacred Groves.

Both the species of the genus, *Parkia*, are endemic to north-east India (Mohan, 1992). The germplasm of *Mucuna gigantea*, *Atylosia goensis*, *Bauhinia vahlii*, *Entada scandense*, *Sesbania bispinosa*, *Vigna capensis*, *V. calcaratus*, *V. sublobata*, *V. vexillata*, *V. glabrescens* and *Xylia xylocarpa* are at the verge of extinction (Rajaram and Janardhanan, 1993b; Siddhuraju *et al.*, 1993).

Suggestions for sustainable conservation of tribal pulse genetic resources are listed below (Arora, 1985b; Rajaram and Janardhanan, 1993b; Siddhuraju *et al.*, 1993).

- Extensive field trips should be undertaken to all the remote tribal hamlets throughout India to collect the germplasm seed samples of tribal pulses and their wild related species over a period of 2–3 years as a network involving Universities, Colleges, National Research Institutions, NGO organizations (Gadgil model for Western Ghats, 1996).
- Among the collected seeds the recalcitrant seeds should be separated from the orthodox seeds. The recalcitrant seeds require special attention as mentioned elsewhere for their conservation.
- *In situ*, *ex situ* and community conservation strategies have to be followed urgently for the conservation of germplasm of *Mucuna gigantea*, *Entada scandens*, *Atylosia goensis*, *Bauhinia vahlii*, *Sesbania bispinosa*, *Vigna capensis*, *V. calcaratus*, *V. grandis*, *V. sublobata*, *V. radiata* var. *sublobata*, *V. vexillata*, *V. glabresces*, *V. mungo* var. *silvestris* and *Xylia xylocarpa* which are rare and threatened.
- Among wild species/germplasm of *Vigna*, the situation of *V. grandis* is alarming due to narrow endemism in Western Ghats. Conservation of the germplasm of the aforesaid species has been advocated long back (Arora, 1985b).
- Germplasm of *Atylosia goensis*, *Entada scandens* and *Mucuna gigantea* are advocated for large scale raising in Botanical gardens and replanting in their natural ecosystems.
- Establishment of local genebank centres with cryopreservation facilities has to be given top priority.

4. Conclusions and future prospects

Among the various underexploited/tribal pulses assessed for agronomical traits and nutrient and antinutrient profiles *Canavalia ensiformis*, *C. gladiata* (Canavalia beans), *Mucuna pruriens* var. *pruriens*, and *M. pruriens* var. *utilis* (Mucuna beans) merit a wide use in South Asian countries and other parts of the tropics as food legumes. In addition, *Psophocarpus tetragonolobus* (winged beans) is another promising underexploited food legume in the aforesaid regions of the world. Though winged beans, Mucuna beans and Canavalia beans contain high quantity of both proteins and lipids in comparison to other traditional pulses, nothing is known about the technological properties (milling, cooking and product making qualities) of these pulses. This knowledge is essential for their proper utilization.

An important problem, in legumes, is the presence of antinutritional factors. Ameliorating the nutritional quality of the legume crops involves three possible strategies: (a) expressing a desirable heterologous storage protein, (b) increasing the level of a desirable, but little expressed endogenous protein and (c) suppressing the expression of antinutritional factors. Information on detection and detoxification of the various toxic/antinutrients in the aforesaid beans appear to be meager. It is also suggested that suitable animal feeding experiments have to be undertaken and detoxification techniques have to be formulated for their effective utilization as a future source of food for man and feed for animals. In these lines an International Workshop on "Food and Feed from Mucuna Current Use, Limitation and the Way Forward" has been conducted in April, 2000 in Tegucigalna, Honduras (personal communication from Dr. Marjatta Eilitta, Gainnesville, USA).

Besides conventional plant breeding methods, one can leave the genetic properties of crop plants as they are and add to them well defined new properties through the addition of single genes to the genetic programme of plant cells.

In near future, one can expect improvement in nutritional quality of seed proteins of the aforesaid wild pulses by genetic engineering techniques. It would be possible to identify and clone the genes responsible for high levels of cysteine and methionine in the seed proteins from wild legumes such as *Alysicarpus rugosus, Indigofera linifolia*, and *Tamarindus indica*. Similarly through gene cloning technology it will be possible to produce Genetically Modified Plants (GMP) having properties that will help to stabilize yields and render their cultivation sustainable from an ecological perspective. Presently significant research is geared towards the production of crop plants with a high degree of tolerance or resistance to pests and diseases or various climatic stresses (drought, heat, cold, salinity, etc.) (Schell, 1999).

Acknowledgements

The senior author K.J. is thankful to ICAR, New Delhi and Ministry of Environment and Forests, Government of India, New Delhi for the financial assistance in the form of two Research Projects on Indian tribal pulses. V.V. is grateful to Ministry of Environment and Forests, Government of India, New Delhi for the award of Junior Project Fellowship (JPF). M.P. wishes to thank CSIR, New Delhi for the award of Senior Research Fellowship (SRF).

References

Adrin M S R B, Owen R D and Edger R S (1979) General genetics. In: *Mendelism*. Eurasia Publishing House (P) Ltd., New Delhi, India.

Agrawal P and Bhattacharya L (1980) Proximate composition of seeds of improved varieties of Bengal gram (*Cicer arietinum* Linn.). *Seed Res.*, **8**: 5–14.

Aguiyi J C, Uguru M O, Johnson P B, Obi C L and Adogo G I (1997) Effects of *Mucuna pruriens* seed extract on smooth and skeletal muscle preparations. *Fitoterapia*, **68**: 366–370.

Ali M (1997) Pulses for nutritional food security. *Indian farm.*, **47**: 31–37.

Amin K M Y, Khan M N, Zillur-Rehman S and Khan N A (1996) Sexual function improving effect of *Mucuna pruriens* in sexually normal rats. *Fitoterapia*, **67**: 58–68.

Amubode F O and Fetuga B L (1983) Proximate composition and chemical assay of the methionine, lysine and tryptophan concentrations of some forest tree seeds. *Food Chem.*, **12**: 67–72.

Amubode F O and Fetuga B L (1984) Amino acid composition of seeds of some lesser known tree crops. *Food Chem.*, **13**: 299–307.

Anantha Kumar K V, Srinivasan K K, Shanbhag T and Rao S G (1994) Aphrodisiac activity of seeds of *Mucuna pruriens. Indian Drugs*, **31**: 321–327.

Anonymous (1994) *Ethnobiology in India, A Status Report* (All India coordinated research project in Ethnobiology), Ministry of Environment and Forests, Government of India, New Delhi, India.

Anonymous (1996) Final technical report of the research project, *South Indian Tribal Pulses: Germplasm Resource Management and Biochemical Studies (1992–1996)* submitted to Ministry of Environment and Forests, Government of India, New Delhi, India.

Apata D F and Ologhobo A D (1994) Biochemical evaluation of some Nigerian legume seeds. *Food Chem.*, **49**: 333–338.

Arora R K (1981) Native food plants of northeastern tribals. In: *Glimpses of Indian Ethnobotany* (Ed Jain S K), Oxford and IBH Publishing Co., New Delhi, India, pp. 91–106.

Arora R K (1985a) *Genetic Resources of Less-known Cultivated Food Plants*. NBPGR Sci. Monogr. No. 9, ICAR, New Delhi, India.

Arora R K (1985b) Diversity and collection of wild *Vigna* species in India. FAO/IBPGR *Plant Genet. Resour. Newslett.*, **63**: 26–33.

Arora R K (1995) Ethnobotanical studies on plant genetic resources – National efforts and concern. *Ethnobot.*, **7**: 125–136.

Arora R K (1996) Role of Ethnobotany in the conservation and use of plant genetic resources in India. In: *Ethnobotany in Human Welfare* (Ed Jain S K), Deep Publications, New Delhi, India, pp. 287–296.

Arora R K and Pandey M (1996) *Wild Edible Plants of India*, NBPGR, New Delhi, India.

Arora R K, Paroda R S and Engels J M M (1991) Plant genetic resources activities: International perspectives. In: *Plant Genetic Resources Conservation and Management* (Eds Paroda R S and Arora R K), IBPGR Regional Office for South and Southeast Asia, New Delhi, India, pp. 351–378.

Arulmozhi M and Janardhanan K (1992) The biochemical composition and nutritional potential of the tribal pulse, *Mucuna monosperma* DC. ex Wight. *Plant Foods Hum. Nutr.*, **42**: 45–53.

Attenburg E (1976) Genetics. In: *The Physiological Basis of Heredity* Oxford and IBH Publishing Co., New Delhi, India.

Attia R S, El-Tabey Shehata A M, Aman M E and Hamza M A (1994) Effect of ripening and parching on the quality characteristics, the chemical composition and the nutritive value of chickpea (*Cicer arietinum* L.). *J. Sci. Food Agricult.*, **65**: 385–390.

Aykroyd W R, Doughty J and Walker A (1982) *Legumes in Human Nutrition*. FAO Food and Nutrition paper No. 20, FAO, Rome, Italy, pp. 67–78.

Azit Raizada (1984) Historical backround. In: *The Tribal Development in Madhya Pradesh: A Planing Perspective*. Inter-India Publications, New Delhi, India, pp. 13–20.

Babu C R, Sharma S K, Chatterjee S R and Abrol Y P (1988) Seed protein and amino acid composition of wild *Vigna radiata* var. *sublobata* (Fabaceae) and two cultigens, *V. mungo* and *V. radiata. Econ. Bot.*, **42**: 54–61.

Bajaj M and Williams J T (1995) *Healing Forests – Healing People*. IDRC, New Delhi, p. 62.

Balogun A M and Fetuga B L (1986) Chemical composition of some underexploited leguminous crop seeds in Nigeria. *J. Agricult. Food Chem.*, **34**: 189–192.

Bamji M S (1991) Optimum nutrition. *Curr. Sci.*, **61**: 319–330.

Banerjee A (1985) Nutritional potential of the winged bean. *Indian J. Nutr. Dietet.*, **22**: 49–51.

Banerjee R and Dixit B S (1998) Potential under-exploited minor oilseeds resources for oil-based industries. *Appl. Bot. Abstr.*, **18**: 134–150.

Bewley J D and Black M (1985) Seeds germination, structure and composition. In: *Seeds: Physiology of Development and Germination* (Eds Bewley J D and Black M), Plenum Press, New York, USA, pp. 1–27.

Bhag M (1992) Biodiversity utilization and conservation in underutilized plants: Indian perspective. *Indian J. Plant Genet. Resour.*, **5**: 1–22.

Bhag M and Joshi V (1991) Under-utilized plant resources. In: *Plant Genetic Resources Conservation and Management* (Eds Paroda R S and Arora R K), IBPGR Regional Office for South and Southeast Asia, New Delhi, India, pp. 211–229.

Bhandari M M (1974) Famine foods in the Rajasthan desert. *Econ. Bot.*, **28**: 73–81.

Bhattacharya S, Bal S, Mukherjee R K and Bhattacharya S (1994) Functional and nutritional properties of tamarind (*Tamarindus indica*) kernel protein. *Food Chem.*, **49**: 1–9.
Bianchi-Hall G M, Keys R D, Stalker H T and Murphy J P (1993) Diversity of seed protein patterns of wild peanuts *Arachis* (Fabaceae) species. *Plant Syst. Evol.*, **186**: 1–15.
Bressani R (1973) Legumes in human diets and how they might be improved. In: *Nutritional Improvement of Food Legumes by Breeding* (Ed Milner M), United Nations, New York, USA, pp. 15–42.
Bressani R and Elias L G (1980) Nutritional value of legume crops for humans and animals. In: *Advances in Legume Sciences* (Eds Summerfield R J and Bunting A H), Kew, Richmond, Surrey, UK, Royal Botanic Gardens, pp. 135–155.
Brockhaus R and Oetmann A (1996) Aspects of the documentation of *in situ* conservation measures of genetic resources. A review. *Plant Genet. Resour. Newslett.*, **108**: 1–16.
Casey R and Wrigley C W (1982) The importance of seed proteins in human nutrition. *Qual. Plant-Plant Foods Hum. Nutr.*, **31**: 189–190.
Census of India (1991) Series 1, India, Paper 2 of 1992, Final population totals. Brief analysis of primary census abstract.
Chan S H H, Koo A and Li K M (1976) The involvement of medullary reticular formation in the hypotensive effect of extracts from seeds of *Cassia tora*. *Am. J. Clin. Med.*, **4**: 383–389.
Chang M C J, Bailey J W and Collins J L (1994) Dietary tannins from cowpeas and tea transiently alter apparent calcium absorption but not absorption and utilization of protein in rats. *J. Nutr.*, **124**: 283–288.
Chowdhury A R, Banerji R, Misra G and Nigam S K (1983) Chemical composition of *Acacia* seeds. *J. Am. Oil Chem. Soc.*, **60**: 1893–1894.
Chowdhury A R, Banerji R, Misra G and Nigam S K (1984) Studies on leguminous seeds. *J. Am. Oil Chem. Soc.*, **61**: 1023–1024.
Clements J C and Cowling W A (1994) Patterns of morphological diversity in relation to geographical origins of wild *Lupinus angustifolius* from the Asian region. *Genet. Resour. Crop Evol.*, **41**: 109–122.
Cohen J I, Williams J T, Plucknett D L and Shands H (1991) *Ex situ* conservation of plant genetic resources: Global development and environmental concern. *Science*, **253**: 866–872.
Crawford L, McDonald G M and Friedman M (1990) Composition of sicklepod (*Cassia obtusifolia*) toxic weed seeds. *J. Agricult. Food Chem.*, **38**: 2169–2175.
Cuthbertson W F J (1989) What is a healthy food? *Food Chem.*, **33**: 53–80.
Damania A B (1996) Biodiversity conservation: A review of options complementary to standard *ex situ* methods. A review. *Plant Genet. Resour. Newslett.*, **107**: 1–18.
de Onis M, Monteiro C, Akre J and Clugston G (1993) The worldwide magnitude of protein–energy malnutrition: A overview from the WHO global database on child growth. *Bullet. WHO.*, **71**: 703–712.
Devedas R P (1993) Towards better nutrition in the 21st century. *Everyman's Sci.*, **28**: 40–44.
Dhindsa K S, Sood D R and Chaudhary M S (1985) Nutritional evaluation of some varieties of lentil (*Lens esculenta*). *Indian J. Nutr. Dietet.*, **22**: 186–189.
Dhuley J N and Naik S R (1997) Protective effect of Rhinax, a herbal formulation against CC14-induced liver injury and survival in rats. *J. Ethnopharmoc*, **56**: 159–164.
Duke J A (1981) *Handbook of Legumes of World Economic Importance*. Plenum Press, New York, USA.
Egbe I A and Akinyele I O (1990) Effect of cooking on the antinutritional factors of lima beans (*Phaseolus lunatus*) *Food Chem.*, **35**: 81–87.
FAO (1996) *The State of the World's Plant Genetic Resources for Food and Agriculture*. FAO, Rome, Italy.
FAO/WHO (1991) *Protein Quality Evaluation*. Food and Agricultural Organization of the United Nations, Rome, Italy, 66.
FAO/WHO/UNU (1985) *Energy and Protein Requirements*. Report of a joint FAO/WHO/UNU expert consultation, WHO technical report series, 724. World Health Organization, Geneva, Switzerland.
Frankel O H (Ed) (1970) *Survey of Crop Genetic Resources in Their Centers of Origin*. FAO/IBPGR, Rome, Italy.
Frankel O H (1975) Genetic conservation: Our evolutionary responsibility. *Genetics*, **78**: 53–65.
Frankel O H and Bennett E (Eds) (1970) *Genetic Resources in Plants: Their Exploration and Conservation*. IBH Handbook No. 11, Blackwell Scientific Publishers, Oxford, UK.
Frankel O H and Hawkes J G (Eds) (1975) *Crop Genetic Resources for Today and Tomorrow*. Cambridge University Press, Cambridge, UK.

Friedman M (1996) Nutritional value of proteins from different food sources. A review. *J. Agricult. Food Chem.*, **44**: 6–29.
Gadgil M (1996) Documenting diversity: An experiment. *Curr. Sci.*, **70**: 36–44.
Gautam P L and Kochhar S (1999) Indian national plant genetic resources system (IN-PGRS) In: *Proceeding of the Golden Jubilee National Symposium on Spices, Medicinal and Aromatic Plants – Biodiversity, Conservation and Utilization* (Eds Sasikumar B, Krishnamoorthy B, Rema J, Ravindran P N and Peter K V), Indian Institute of Spices Research, Calicut, Kerala, India, pp. 1–15.
Gill L S and Nyawuame H G K (1994) Leguminosae in Ethnomedicinal practices of Nigeria. *Ethnobotany*, **6**: 51–64.
Grover H L and Gupta V P (1972) Estimation of tryptophan in cereals and pulses. *Indian J. Agricult. Res.*, **6**: 267–272.
Gunjatkar N and Vartak V D (1982) Enumeration of edible legumes from Pune district, Maharashtra state. *J. Econ. Tax. Bot.*, **3**: 1–9.
Gupta S P (1995) Study of plants during ethnological research among the Tribals. In: *A Manual of Ethnobotany*, (Ed S K Jain), Scientific Publishers, Jodhpur, India, pp. 9–18.
Gupta K and Wagle D S (1978) Proximate composition and nutritive value of *Phaseolus mungoreous*, a cross between *Phaseolus mungo* and *Phaseolus aureus*. *J. Food Sci. Technol.*, **15**: 34–35.
Haq N (1983) New food legume crops for the tropics. In: *Better Crops for Food* (Eds Nugent J and Connor M O), Pitman Books, London (Cuba Foundation Symposium, 97), pp. 144–160.
Harlan J R (1976) Genetic resources of wild relatives of crops. *Crop Sci.*, **16**: 329–333.
Houghton P J and Skari K P (1994) The effect of blood clotting of some west African plants used against snakebite. *J. Ethnopharmoc.*, **44**: 99–108.
Hussain G and Manyam B V (1997) *Mucuna pruriens* prove more effective than L-DOPA in Parkinson's disease animal model. *Phytotherapy Res.*, **11**: 419–423.
Iauk L, Galati E M, Kirarainen S, Forestieri A M and Trorato A (1993) Analgesic and antipyretic effects of *Mucuna pruriens*. *Int. J. Pharmacogn.*, **31**: 213–216.
ICMR (1992) *Nutrient Requirement and Recommended Dietary Allowances for Indians*. A report of the expert group of the Indian Council of Medical Research (ICMR) National Institute of Nutrition, Hyderabad, India.
Ignacimuthu S and Babu C R (1987) *Vigna radiata* var. *sublobata* (Fabaceae): Economically useful wild relative of urd and mung bean. *Econ. Bot.*, **41**: 418–422.
Ishola M M, Agabaji E B and Agabaji A S (1990) A chemical study of *Tamarindus indica* (Tsamiya) fruits grown in Nigeria. *J. Sci. Food Agricult.*, **51**: 141–143.
Jain S K (Ed) (1981) *Glimpses of Indian Ethnobotany*. Oxford and IBH Publishing Co., New Delhi, India.
Jain S K (1987) Ethnobotany – its concepts and relevance. *Tenth Botanical Conference, Indian Bot. Soc.*, Department of Botany, University of Patna, Patna, India, pp. 3–12.
Jain S K (1991) Observations on Ethnobotany of the tribals of central India. In: *Contributions to Ethnobotany of India* (Ed Jain S K), Scientific Publishers, Jodhpur, India, pp. 69–73.
Jain S K (1995) Ethnobotany – its scope and various subdisciplines. In: *A Manual of Ethnobotany* (Ed Jain S K), Scientific Publishers, Jodhpur, India, pp. 1–8.
Jain S K (1996) *Ethnobotany in Human Welfare*. Deep Publications, New Delhi, India.
Jain S P, Abraham Z and Shah N C (1991) Herbal remedies among "Ho" tribe in Bihar. In: *Contribution to Ethnobotany of India* (Ed Jain S K), Scientific Publishers, Jodhpur, India, pp. 115–122.
Jambunathan R and Singh U (1980) Studies on Desi and Kabuli chickpea (*Cicer arietinum* L.) cultivars. 1. Chemical composition. In: *Proceedings of the International Workshop on Chickpea Improvement*. ICRISAT, Patancheru, Hyderabad, India, pp. 61–66.
Jambunathan R and Singh U (1981) Grain quality of pigeonpea. In: *Proceedings of the International Workshop on Pigeonpeas*, Vol. 1. ICRISAT, Patancheru, Hyderabad, India, pp. 351–356.
Janardhanan K (1990) Germplasm resources of tribal utility in India. In: *Proceedings of the National Seminar on Advances in Seed Science and Technology* (Eds Shetty H S and Prakash H S), University of Mysore, Mysore, India, pp. 407–409.
Janardhanan K (1993) Chemical composition and nutritional value of the tribal pulse, *Cassia floribunda* Cav. *Adv. Plant Sci.*, **6**: 137–144.
Janardhanan K and Lakshmanan K K (1985) Studies on the pulse, *Mucuna utilis*: Chemical composition and antinutritional factors. *J. Food Sci. Technol.*, **22**: 369–371.

Janardhanan K and Nalini K (1991) Studies on the tribal pulse, *Entada scandens* Benth: Biochemical composition and antinutritional factors. *J. Food Sci. Technol.*, **28**: 249–251.

Janardhanan K and Rajaram N (1993) Tribal pulses and their wild related species as future protein crops. *Seed Res.* (Special Vol.) **2**: 931–942.

Janardhanan K and Vadivel V (1994) Biochemical composition of different germplasm seed materials of South Indian tribal pulse, *Canavalia gladiata*. In: *Proceedings of the National Seminar on Biodiversity: Strategies for Conservation and Future Challenges* (Eds Udaiyan K, Janardhanan K, Manian S and Reddy V R K), Bharathiar University, Coimbatore, India, pp. 93–97.

Jayaweera D M A (1981) *Medicinal Plants Used in Ceylon*. National Science Council of Sri Lanka, Colombo, Sri Lanka.

Jindal S K (1988) Intraspecific divergence in moth bean grain yield and its composition. *Ann. Arid Zone.*, **27**: 267–270.

Kapoor H C, Srivastva V K and Gupta Y P (1972) Estimation of methionine in black gram (*Phaseolus mungo* Roxb.), green gram (*P. aureus* Roxb.) and soybean (*Glycine max* (L.) Merr.) *Indian J. Agricult. Sci.*, **42**: 296–299.

Katoch B S and Bhowmik K B D (1983) Biological evaluation of *Cassia* seeds (*Cassia tora* and *Cassia floribunda*) as poultry feed ingredients. *Indian Vet. J.*, **60**: 390–395.

Khan T I and Al-Ajmi D N (1999) In: *Biodiversity: An Overview* (Eds Khan T I and Al-Ajmi D N), *Global Biodiversity Conservation Measures*. Pointer Publishers, Jaipur, India.

Khan A A, Shukla K M L and Khan I M (2000) Enumeration of wild food plants of ethnobotanical significance in central India. *Adv. Plant Sci.*, **13**: 277–281.

Khoshoo T N (1991) Conservation of biodiversity in biosphere. In: *Indian Geosphere Biosphere Programme. Some aspects* (Eds Khoshoo T N and Sharma M), National Academy of Sciences, Allahabad, India, pp. 178–233.

Khoshoo T N (1996) India needs a national biodiversity conservation board. *Curr. Sci.*, **71**: 506–513.

Kothari A (1998) Seeds of the future. Conference of Parties (Fourth conference) (COP4) *Frontline*, pp. 86–89.

Kumar S, Kumar S, Singh G K, Kumar R, Bhatia N K and Awasthi C P (1991) Variation in quality traits of pigeonpea (*Cajanus cajan* L. Mill sp.) varieties. *J. Food Sci. Technol.*, **28**: 173–174.

Kumaresan A, Mshelia T A and Aliu Y O (1984) Biochemical evaluation of bagaruwa seeds (*Acacia nilotica*) for use as livestock feed. *Anim. Feed Sci. Technol.*, **11**: 45–48.

Laghetti G, Pienaar B J, Braun K and Perrino P (1994) Collecting wild *Vigna* in Natal and Transvaal (South Africa) and Swaziland. *FAO/IPGRI Plant Genet. Resour. Newslett.*, **98**: 21–23.

Lal S and Kumar A (1997) Wingedbean – a multipurpose legume. *Indian Farm.*, **46**: 16–17.

Laurena A C, Rodriguez F M, Sabino N G, Zamora A F and Mendoza E M T (1991) Amino acid composition, relative nutritive value and *in vitro* protein digestibility of several Philippine indigenous legumes. *Plant Foods Hum. Nutr.*, **41**: 59–68.

Logamadei A (1991) Biochemical Studies on the Tribal Pulse, *Dolichos lablab* var. *vulgaris* L. Savi. M.Sc. dissertation, Bharathiar University, Coimbatore, India.

Maheshwari J K, Painuli R M and Dwivedi R P (1991) Notes on Ethnobotany of the Oraon and Korwa tribes of Madhya Pradesh. In: *Contribution to Ethnobotany of India* (Ed Jain S K), Scientific Publishers, Jodhpur, India, pp. 75–90.

Maheshwari J K and Singh H (1991) Ethnobotanical notes from Banda District, Uttar Pradesh. *J. Econ. Bot. Phytochem.*, **2**: 16–20.

Maikhuri R K, Nautiyal M C and Khali M P (1991) Lesser-known crops of food value in Garhwal Himalaya and strategy to conserve them. *FAO/IBPGR Plant Genet. Resour. Newslett.*, **86**: 33–36.

Malhotra S, Malik D and Dhindsa K S (1988) Proximate composition and antinutritional factors in ricebean (*Vigna umbellata*). *Plants Foods Hum. Nutr.*, **38**: 75–81.

Mary Josephine R and Janardhanan K (1992) Studies on chemical composition and antinutritional factors in three germplasm seed materials of the tribal pulse, *Mucuna pruriens* (L.) DC. *Food Chem.*, **43**: 13–18.

Misra P S, Misra G, Prakash D, Tripathi R D, Chaudhary A R and Misra P N (1987) Assay of some nutritional and antinutritional factors in different cultivars of winged bean (*Psophocarpus tetragonolobus* (L.) DC) *Plant Foods Hum. Nutr.*, **36**: 367–371.

Mittre V (1991) Wild plants in Indian folk life – a historical perspective. In: *Contribution to Ethnobotany of India* (Ed Jain S K), Scientific Publishers, Jodhpur, India, pp. 35–39.

Mohan V R (1992) Studies on Tribal Pulses in India. Ph.D. thesis. Bharathiar University, Coimbatore, India.

Mohan V R and Janardhanan K (1993a) Chemical and nutritional evaluation of raw seeds of the tribal pulses *Parkia roxburghii* G. Don. and *Entada phaseoloides* (L.) Merr. *Inter. J. Food Sci. Nutr.*, **44**: 47–53.

Mohan V R and Janardhanan K (1993b) Chemical composition and nutritional evaluation of two little-known species of *Vigna*. *Food Chem.*, **48**: 367–371.

Mohan V R and Janardhanan K (1994a) Studies on Indian tribal pulses. In: *Proceedings of the National Seminar on Biodiversity: Strategies for Conservation and Future Challenges* (Eds Udaiyan K, Janardhanan K, Manian S and Reddy V R K), Bharathiar University, Coimbatore, India, pp. 116–125.

Mohan V R and Janardhanan K (1994b) Chemical and nutritional evaluation of two germplasm of the tribal pulse, *Bauhinia racemosa* Lamk. *Plant Foods Human Nutrn.*, **46**: 367–374.

Mohan V R and Janardhanan K (1994c) The biochemical composition and nutritional assessment of less known pulses of the genus, *Canavalia*. *Inter. J. Food Sci. Nutrn.*, **45**: 255–262.

Mohan V R and Janardhanan K (1994d) Chemical composition and nutritional evaluation of raw seeds of six ricebean varieties. *J. Indian Bot. Soc.*, **73**: 259–263.

Mohan V R and Janardhanan K (1995a) Chemical determination of nutritional and antinutritional properties in tribal pulses. *J. Food Sci. Technol.*, **32**: 465–469.

Mohan V R and Janardhanan K (1995b) Chemical analysis and nutritional assessment of the less known pulses of the genus, *Mucuna*. *Food Chem.*, **52**: 275–280.

Monti L M and Grillo S (1983) Legume seed improvement for protein content and quality. *Qual. Plant- Plant Foods Hum. Nutr.*, **32**: 253–266.

Murray D R (1991) Breeding Plants for the Twenty-first Century (Ed David R Murray) *Advanced Methods in Plant Breeding and Biotechnology*. C.A.B. International, Wallingford, UK, pp. 1–22.

Narasinga Rao B S, Deosthale Y G and Pant K C (1989) *Nutritive Value of Indian foods*. National Institute of Nutrition, Indian Council of Medical Research, Hyderabad, India.

Negi K S (1996) Wild food plants of the Uttar Pradesh hills, India. *FAO/IPGRI Plant Genet. Resour. Newslett.*, **108**: 33–52.

Oshodi A A and Ekperigin M M (1989) Functional properties of pigeonpea (*Cajanus cajan*) flour. *Food Chem.*, **34**: 187–191.

Oyenuga V A and Fetuga B L (1975) Some aspects of the biochemistry and nutritive value of watermelon seed (*Citrullus vulgaris* Schrad) *J. Sci. Food Agric.*, **26**: 843–854.

Pandey V N and Srivastava A K (1991) Yield and *in vitro* nutritional quality of some leguminous seed protein isolates. *Plant Foods Hum. Nutr.*, **41**: 247–251.

Pant R, Rajagopalan Nair C, Singh K S and Koshti G S (1974) Amino acid composition of some wild legumes. *Curr. Sci.*, **43**: 235–239.

Paroda R S (1987) Conservation of plant genetic resources: some basic issues. *Indian J. Genet.*, **47**: 225–237.

Paroda R S and Arora R K (1991) Plant genetic resources: General perspective. In: *Plant Genetic Resources Conservation and Management* (Eds Paroda R S and Arora R K), IBPGR Regional Office for South and Southeast Asia, New Delhi, India, pp. 1–23.

Paroda R S and Bhag Mal (1989) New plant sources for food and industry in India. In: *New Crops for Food and Industry* (Eds Wickens G E, Haq N and Day P), Chapman Hall, London, UK, pp. 135–149.

Paroda R S, Mangala Raj and Gautam P L (1999) *National Action Plan on Agrobiodiversity in India*. National Academy of Agricultural Sciences (NAAS), Indian Council of Agricultural Research (ICAR) and Indian Society of Plant Genetic Resources (ISPGR), New Delhi.

Pattee H E, Salunkhe D K, Sathe S K and Reddy N R (1982) Legume lipids. *CRC Crit. Rev. Food Sci. Nutr.*, **17**: 97.

Pawar V D and Ingle U M (1988) Production of quick-cooking moth bean (*Phaseolus aconitifolia* Jacq): IV Functional properties of flours. *Indian J. Nutr. Dietet.*, **25**: 18–25.

Pimentel D, Dristschilow W, Krummel J and Kutzman J (1975) Energy and land constraints in food protein production. *Science*, **190**: 754–761.

Pistoris R (1997) *Scientists, Plants and Politics – A history of the plant genetic resources movement*. IPGRI, Rome, Italy.

Prakash D and Misra P S (1987) Protein and amino acid profile of some wild leguminous seeds. *Plant Foods Hum. Nutr.*, **37**: 29–32.

Prakash D and Misra P S (1988) Protein content and amino acid profile of some wild leguminous seeds. *Plant Foods Hum. Nutr.*, **38**: 61–65.

Radhakrishnan K, Pandurangan A G and Pushpangadan P (1996) Ethnobotany of wild edible plants of Kerala, India. In: *Ethnobotany in Human Welfare* (Ed Jain S K), Deep Publications, New Delhi, India, pp. 48–51.

Ragunathan K (1978) Tribal pockets of Nilgiris. In: *Recording of the Field Study on Medicinal Flora and Health Practices* (Ed Ragunathan K), Central Council for Research in Indian Medicine and Homeopathy, New Delhi.

Rajaram N (1990) Studies on the Pulses of Tribal Utility and Their Wild Related Species in India. Ph.D. thesis, Bharathiar University, Coimbatore, India.

Rajaram N and Janardhanan K (1990) Chemical analysis and nutritional evaluation of certain under-exploited *Vigna* spp. *Food Sci. Nutr.*, **42**: 213–221.

Rajaram N and Janardhanan K (1991a) The biochemical composition and nutritional potential of the tribal pulse, *Mucuna gigantea* (Willd) DC. *Plant Foods Human Nutrn.*, **41**: 45–51.

Rajaram N and Janardhanan K (1991b) Studies on the underexploited tree pulses, *Acacia catechu* Willd., *Parkinsonia aculeata* L. and *Prosopis chilensis* (Molina) Stunz. chemical composition and antinutritional factors. *Food Chem.*, **42**: 265–273.

Rajaram N and Janardhanan K (1991c) Chemical composition and nutritional potential of the tribal pulses, *Bauhinia purpurea* L. *B. racemosa* Lamk. and *B. vahlii W. and A. J. Sci. Food Agricult.*, **55**: 423–431.

Rajaram N and Janardhanan K (1992a) Nutritional and chemical evaluation of raw seeds of *Canavalia gladiata* (Jacq) DC. and *C. ensiformis* DC.: The under utilized food and fodder crops in India. *Plant Foods Human Nutrn.*, **42**: 329–336.

Rajaram N and Janardhanan K (1992b) The chemical composition and nutritional potential of the tribal pulse, *Abrus precatorius* L. *Plant Foods Human Nutrn.*, **42**: 285–290.

Rajaram N and Janardhanan K (1992c) Agrobotanical and biochemical studies of some less known *Vigna* spp. *Acta Bot. Indica.*, **20**: 143–149.

Rajaram N and Janardhanan K (1993a) Biochemical composition of lima bean (*Phaseolus lunatus* L.) seeds. *Acta Bot. Indica.*, **21**: 39–43.

Rajaram N and Janardhanan K (1993b) *Ex situ* conservation of genetic resources of tribal pulses and their wild related species. *FAO/IBPGR Plant Genet. Resour. Newslett.*, **91/92**: 29–32.

Rajendran V, Joseph T and David J (1996) Reappraisal of dopamineric aspects *Mucuna pruriens* and comparative profile with L-DOPA on cardiovascular and central nervous system in animals. *Indian Drugs*, **33**: 465–472.

Rajyalakshmi P and Geervani P (1994) Nutritive value of the foods cultivated and consumed by the tribals of South India. *Plant Foods Hum. Nutr.*, **46**: 53–61.

Ramalingaswami V (1995) New global perspectives on overcoming malnutrition. *Am. J. Clin. Nutr.*, **61**: 259–263.

Rana R S and Arora R K (1991) Indian national plant genetic resource system – an overview. In: *Plant Genetic Resources Conservation and Management* (Eds Paroda R S and Arora R K), IBPGR Regional Office for South and Southeast Asia, New Delhi, India, pp. 329–349.

Rangalakshmi R (1998) Winged bean: a multipurpose, under-utilized legume for acidic soils. *Indian Farm.*, **47**: 13–15, 22.

Rao R R (1991) Ethnobotanical studies on some Adivasi tribes of Northeast India with special reference to Naga people. In: *Contribution to Ethnobotany of India* (Ed Jain SK), Scientific Publishers, Jodhpur, India, pp. 35–39.

Rao P S and Srivastava H C (1974) Tamarind. In: *Industrial Gums* (Ed Whistler R L), Academic Press, New York, USA, pp. 370–411.

Ravindran V and Ravindran G (1988) Nutritional and antinutritional characteristics of *Mucuna* (*Mucuna utilis*) bean seeds. *J. Sci. Food Agricult.*, **46**: 71–79.

Roberts L M (1970) *The Food Legumes: Recommendation for expansion and acceleration of research.* Rockefeller Foundation Report, USA.

Rodrigues B E and Torne S G (1991) A chemical study of seeds in three *Canavalia* species. *Trop. Sci.*, **31**: 101–103.

Sahu H S (1996) Life supporting promising food plants among aboriginals of Bastar (M.P.), India. In: *Ethnobotany in human welfare* (Ed Jain S K), Deep Publications, New Delhi, India, pp. 26–30.

Sakla A B, Ghali Y, El-Farra A and Rizk L F (1988) The effect of environmental conditions on the chemical composition of soybean seeds: Relationship between the protein, oil, carbohydrate and trypsin inhibitor content. *Food Chem.*, **29**: 221–231.

Salunke D M, Swamy M J, Islam Khan M, Mande S C, Surolia S and Vijayan M (1985) Arrangement of subunits in peanut lectin. Rotation function and chemical cross-linking studies. *J. Biol. Chem.*, **260**: 13576–13579.
Sastry M C S and Murray D R (1986) The tryptophan content of extractable seed proteins from cultivated legumes, sunflower and *Acacia*. *J. Sci. Food Agricult.*, **37**: 535–538.
Satya Sundaram I (2000) Pulses – Story of shortages. *Facts for You*. Market Survey, pp. 22–24.
Saxena S, Singh G and Mital B K (1994) Physical characteristics and composition of certain new varieties of soybean. *J. Food Sci. Technol.*, **31**: 145–147.
Schell J (1999) Plant Biotechnology: A Powerful Tool to Use Plant Resources and to Improve the Environmental Impact of Agriculture. In: *Global Biodiversity Conservation Measures* (Eds Khan T I and Al-Ajmi D N), Pointer Publishers, Jaipur, India.
Scrimshaw N S and Young V R (1977) Nutritional evaluation and the utilization of protein resources. In: *Evaluation of Proteins for Humans* (Eds Bodwell C E), The AVI Publishing Company, Inc. Westport, Connecticut, pp. 1–10.
Sekar M, Rajendra Prasad K J, Siddhuraju P and Janardhanan K (1999) New anthroquinones from *Cassia obtusa*. *Fitotherpia.*, **70**: 330–332.
Shah N C (1995) Main world centres and Workers of Ethnobotany. In: *A Manual of Ethnobotany* (Ed Jain S K), Scientific Publishers, Jodhpur, pp. 58–66.
Shankar R (1995) Tribal communities in India and PGR. In: *Farmer's Rights and Plant Genetic Resources Recognition and Reward: A Dialogue* (Ed Swaminathan M S), Macmillan India, Madras, India, pp. 106–111.
Sharma A and Sehgal S (1992) Chemical composition and physico-chemical properties of faba bean (*Vicia faba* L.) *Legume Res.*, **15**: 125–127.
Sharma A, Bakhshi A K and Verma M M (1991) Proximate composition and nutritional quality of some improved varieties of mung bean (*Vigna radiata*) *Legume Res.*, **14**: 197–200.
Shaw B P and Bera C H (1993) A preliminary clinical study to evaluate the effect of Vogorex-SF in sexual debility patients. *Indian J. Internal Medicine*, **3**: 165–169.
Shewry P R, Napier J A and Tatham A S (1995) Seed storage proteins: Structures and biosynthesis. *Plant Cell*, **7**: 945–956.
Shiv Raj A (1978) *An Introduction to Physiology of Field Crops*. Oxford and IBH Publishing Co., New Delhi, India.
Siddhuraju P (1994) Studies on the Seed Proteins of Tribal Pulses. Ph.D. thesis, Bharathiar University, Coimbatore, India.
Siddhuraju P, Vijayakumari K and Janardhanan K (1992a) Nutritional and chemical evaluation of raw seeds of the tribal pulse, *Vigna trilobata* (L.) *Verdc. Inter. J. Food Sci. Nutr.*, **43**: 97–103.
Siddhuraju P, Vijayakumari K and Janardhanan K (1992b) The biochemical composition and nutritional potential of the tribal pulse, *Alysicarpus rugosus* (Willd) DC. *Food Chem.*, **45**: 251–255.
Siddhuraju P, Vijayakumari K and Janardhanan K (1993) Genetic resources of tribal pulses. FAO/IBPGR *Plant Genet. Resour. Newslett.*, **96**: 47–49.
Siddhuraju P, Vijayakumari K and Janardhanan K (1994) Chemical analysis and nutritional assessment of the less known pulses, *Vigna aconitifolia* (Jacq.) Marechal and *V. vexillata* (L.) A. Rich. *Plant Foods Human Nutrn.*, **45**: 103–111.
Siddhuraju P, Vijayakumari K and Janardhanan K (1995a) Nutrient and chemical evaluation of raw seeds of *Xylia xylocarpa*: An under utilised food source. *Food Chem.*, **53**: 299–304.
Siddhuraju P, Vijayakumari K and Janardhanan K (1995b) Nutritional and antinutritional properties of the underexploited legumes, *Cassia laevigata* Willd. and *Tamarindus indica* L. *J. Food Comp. Analys.*, **8**: 351–363.
Siddhuraju P, Vijayakumari K and Janardhanan K (1995c) Studies on the underexploited legumes, *Indigofera linifolia* and *Sesbania bispinosa*: Nutrient composition and anti-nutritional factors. *Inter. J. Food Sci. Nutr.*, **46**: 195–203.
Siddhuraju P, Vijayakumari K and Janardhanan K (1996a) Chemical composition and nutritional evaluation of an underexploited legume, *Acacia nilotica* (L.) Del. *Food Chem.*, **57**: 385–391.
Siddhuraju P, Vijayakumari K and Janardhanan K (1996b) Chemical composition and protein quality of little known legume, velvet bean (*Mucuna pruriens* (L.) DC.) *J. Agricult. Food Chem.*, **44**: 2636–2641.
Siddhuraju P, Vijayakumari K and Janardhanan K (1997) Tryptophan contents of total (true) proteins and protein fractions, albumins and globulins of tribal/little known/under-exploited Indian legumes. *J. Food Sci.Technol.*, **34**: 140–142.

Singh S P, Misra B K, Chandel K P S and Pant K C (1985) Major food constituents of rice-bean (*Vigna umbellata*) *J. Food Sci. Technol.*, **17**: 238–240.
Singh U and Singh B (1992) Tropical grain legumes as important human foods. *Econ. Bot.*, **46**: 310–321.
Singh U, Kumar J and Gowda C L L (1983) Protein content of chickpea (*Cicer arietinum* L.) grown at different locations. *Qual. Plant- Plant Foods Hum. Nutr.*, **32**: 179–184.
Singh V and Premanath R K (1992) Some useful *Cassias* in India. *J. Econ. Tax. Bot.*, **10**: 29–54.
Sinha R K (1996) *Biodiversity Global Concerns*. Commonwealth Publishers, New Delhi, India.
Singhvi N R and Bhandari M M (1992) Non-conventional food plants of Rajasthan desert – an ethnobotanical study. *Adv. Plant Sci.*, **5**: 634–640.
Sprague G F (1980) Germplasm resources of plants: Their preservation and use. *Ann. Rev. Phytopathol.*, **18**: 147–165.
Subbulakshmi G and Chitra L (1996) Methods for determining nutrients in foods: A critical appraisal. *J. Food Sci. Technol.*, **33**: 267–284.
Sudha N, Mushtari Begum J, Shambulingappa K G and Babu C K (1995) Nutrients and anti-nutrients in horsegram (*Macrotyloma uniflorum* (Lam.) Verdc.). *Food Nutr. Bullet.*, **16**: 81–83.
Swaminathan M S (Ed) (1995) *Farmer's Rights and Plant Genetic Resources Recognition and Reward: A Dialogue*. Macmillan India Ltd., Madras, India.
Talpada P M, Pande M B, Shukla P C, Patel B M and Purohit L P (1980) Nutritive value of boiled *Cassia tora* seeds. *Indian Vet. J.*, **57**: 1022–1024.
Tarafder C R and Chaudhuri R H N (1991) Less known medicinal uses of plants among the tribals of Hazaribagh district of Bihar. In: *Contribution to Ethnobotany of India* (Ed Jain S K), Scientific Publishers, Jodhpur, India, pp. 91–100.
Thomson J A (1975) Genetic studies on the storage proteins of pea cotyledon. *CSIRO Division of Plant Industry Annual Report*, 26–30.
Udayasekhara Rao P and Deosthale Y G (1982) Tannin content of pulses; Varietal differences and effects of germination and cooking. *J. Sci. Food Agricult.*, **33**: 1013–1016.
Udayasekhara Rao P and Sharma R D (1987) An evaluation of protein quality of fenugreek seeds (*Trigonella foenumgraecum*) and their supplementary effects. *Food Chem.*, **24**: 1–9.
Vadivel V (1998) Nutritional and Antinutritional Properties of South Indian Tribal Pulses. Ph.D. thesis, Bharathiar University, Coimbatore, India.
Vadivel V and Janardhanan K (1997) Agrobotanical characters of different germplasm of velvet bean. *J. Swamy. Bot. Cl.*, **14**: 37–40.
Vadivel V and Janardhanan K (1998) Genetic resources of some South Indian tribal pulses. *IPGRI Newslett.*, **26**: 21–22.
Vadivel V and Janardhanan K (1999) Chemical composition of some little known/wild legumes of South India. *J. Swamy Bot. Cl.*, **16**: 25–29.
Vadivel V and Janardhanan K (2001) Nutritional and anti-nutritional composition of velvet bean: An underutilised food legume in South India. *Inter. J. Food. Sci. Nutr.* (in press).
Vadivel V, Vijayakumari K and Janardhanan K (1994) Agrobotanical characters of different germplasm of *Canavalia gladiata*. In: *Proceedings of the National Seminar on Biodiversity: Strategies for conservation and future challenges* (Eds Udaiyan K, Janardhanan K, Manian S and Reddy V R K), Bharathiar University, Coimbatore, India, pp. 21–24.
Vadivel V, Janardhanan K and Vijayakumari K (1998) Diversity in sword bean (*Canavalia gladiata* (Jacq.) DC.) collected from Tamil Nadu, India. *Genetic. Resour. Crop Evol.*, **45**: 63–68.
Van Sloten D H (1990a) IBPGR and the challenges of the 1990s: A personal point of view. *Diversity*, **6**: 36–39.
Van Sloten D H (1990b) *IBPGR – facing the future*. IBPGR in the 1990s: Presentation to International Centers week, Washington, DC, USA (mimeographed), p. 16.
Vasudeva Rao M K and Shanpru R (1991) Some plants in the life of the Garos of Meghalaya. In: *Contribution to Ethnobotany of India* (Ed Jain S K), Scientific Publishers, Jodhpur, India, pp. 183–190.
Vietmeyer N D (1986) Lesser-known plants for potential use in agriculture and forestry. *Science*, **232**: 1379–1384.
Vijaya Bhanu N, Ramachandra G and Monteiro P V (1991) Evaluation of protein isolate from *Cassia uniflora* as a source of plant protein. *J. Sci. Food Agric.*, **54**: 659–662.
Vijayakumari K (1994) Studies on the Antinutritional Factors in Tribal Pulses. Ph.D. thesis, Bharathiar University, Coimbatore, India.

Vijayakumari K, Sidduraju P and Janardhanan K (1993a) Chemical composition and nutritional potential of the tribal pulse, *Bauhinia malabarica* Roxb. *Plant Foods Human Nutr.*, **44**: 291–298.

Vijayakumari K, Sidduraju P and Janardhanan K (1993b) Nutritional and antinutritional properties of certain underexploited legume seeds. *Inter. J. Food Sci. Nutr.*, **44**: 181–189.

Vijakumari K, Sidduraju P and Janardhanan K (1994) Nutritional assessment and chemical composition of the less known tree legume, *Acacia leucophloea. Food Chem.*, **50**: 285–288.

Vijayakumari K, Siddhuraju P and Janardhanan K (1995) Effect of various water or hydrothermal treatments on certain antinutritional compounds in the seeds of the tribal pulse. *Dolichos lablab* var. *vulgaris* L. *Plant Foods Human Nutr.*, **48**: 17–29.

Vijayakumari K, Siddhuraju P and Janardhanan K (1997) Chemical composition, amino acid content and protein quality of little known legume, *Bauhinia purpurea* (L.) *J. Sci. Food Agricult.*, **73**: 279–286.

Waterlow J C (1994) Childhood malnutrition in developing nations: Looking back and looking forward. *Ann. Rev. Nutr.*, **14**: 1–19.

WHO (1993) *World Declaration on Nutrition. Plan of action for nutrition.* Adopted by International Conference on Nutrition, Jointly sponsored by Food and Agricultural Organization of the United Nations and World Health Organization, December 11, 1992. World Health Organization, Division of Food and Nutrition, Geneva.

WHO (1995) *Nutrition. Highlights of recent activities in the context of the world declaration and plan of action for nutrition.* Nutrition Programme, World Health Organization, Geneva.

Wijeratne S (1987) Poisonous plants, another's medicine. *Mirror Mag.* (Sri Lanka), **18**: 8–9.

Wilmot-Dear C M (1987) A revision of *Mucuna* (Leguminosae – *Phaseolae*) in the Indian sub-continent and Burma. *Kew Bulletin*, **42**: 23–46.

Yadav N R and Bharadwaj U R (1971) Biochemical studies on Indian wild legumes: (1) Chemical and nutritional constituents of some Indian wild legumes. *Indian J. Nutr. Dietet.*, **8**: 1–4.

Yohe J M and Poehlman J M (1972) Genetic variability in the mungbean (*Vigna radiata* (L.) Wilczek). *Crop Sci.*, **12**: 461–464.

Young V R and Pellett P L (1994) Plant proteins in relation to human protein and amino acid nutrition. *Amer. J. Clin. Nutr.*, **59**: 1203S–1212S.

Zeven A C and de Wet J M J (1982) *Dictionary of cultivated plants and their regions of diversity. Excluding most ornamentals, forest trees and lower plants.* (2nd edn). PUDOC, Wageningen, The Netherlands, p. 259.

SUBJECT INDEX

Abiotic stress 7
Abiotic stress signaling 223
Abiotic stress tolerance 223
Acacia catechu 88
A. koa 88
A. nilotica 88
Acetylene reduction assay 16, 211
Acidity 223
Advanced yield trial 14
Adventitious 66
Aeschynomene sensitiva 70
AFLP 9, 10, 140, 269, 271
Agrobacterium 25
A. rhizogenes 206, 237
A. tumefaciens 206, 237
Agrobacterium-mediated transformation 99
Albizia amara 88
A. lebbeck 70
A. lucida 88
A. richardiana 88
2S Albumin 286
Alfalfa 141, 204, 256
Alfalfa mosaic virus 149
Alien genes 17
Alkalinity 223
Alpha-amylase inhibitor 247, 249
Amino acid contents 282
γ-amino butyric acid 232
Ammonia assimilation 201, 204
Aneuploidy 142
Anoxia responsive element 229
Anthracnose 16
Anti-nutritional factors 11, 24, 291
Antisense RNA strategy 204
Antioxidant defense system 223, 224, 233
Aphanomyces 12
A. euteiches 16
Aphis craccivora 149
Arachis species 77, 157
A. batizocoi 158
A. cardenasii 158
A. chacoensis 160
A. duranensis 34, 158, 160
A. glabrata 158
A. hypogaea 24, 26, 34, 40, 70, 89, 156, 231
A. pintoi 70, 89
A. monticola 158
A. richardiana 70
A. villosa 70
Artificial seed 93, 117
Ascochyta blight 323, 324
Astragalus adsurgens 70, 89, 136
Atylosia platycarpa 41
A. lineata 43
Azorhizobium 175
Azuki bean 249

Bacillus thuringiensis 245
Bacterial resistance 16
Bean golden yellow mosaic virus 263, 264
Bean dwarf mosaic virus 264
Biodiversity 353, 354
Biological N_2 fixation 203
Biological yield 11, 13, 162
Biotic stress 7, 11, 24
Bioreactor 93
Biomass production 204, 211
Bloat safe legume 142
BOAA 291
Botrytis grey mould 16
Bradyrhizobium 175
Brazil nut 2S albumin 284
Breeding methodologies 3
Breeding program 7, 13
Breeding pyramid 13, 14, 15
Breeding strategies 13
Brome mosaic virus 149
Bt-transgenic legumes 250

C. albicans 165
Cajanus acutifolius 168
C. cajan 26, 31, 34, 40, 44, 46, 70, 77, 89, 153, 167, 169
C. lineatus 165
C. scaraboides 165, 168
C. sericius 165
C. platycarpus 155, 165, 167, 168, 169
Calcium 180
Calcium spiking 175, 182
Calmodulin 227
Callus 25
Clonal propagation 29
Carbohydrate metabolism 194

Cassia fistula 89
C. siamea 89
Cell membrane 144
Cell respiration 147
Cercis canadensis 89
Chamaecytisus austriacus 90
C. purpureus 90
Chitinase 206, 247, 271
Chickpea 28, 33, 40, 160, 303
Chickpea genome 303, 310, 321, 325
Cicer arietinum 16, 26, 28, 34, 40, 78, 90, 160, 162, 311, 312, 313, 319, 321
C. anatolicum 164, 311
C. bijugum 164
C. echinospermum 160
C. cuneatum 164
C. chorassanicum 164, 311
C. judaicum 28
C. oxyden 164
C. pinnatifidum 28, 163
C. microphyllum 164
C. monbretii 164
C. soongarium 164
Cladastris lutea 90
Coat protein 256
Comparative genomics 337
Compatible solutes 231
Cowpea mosaic comovirus 261
Crop genetic resource 155
Cross tolerance 225
Crotolaria juncea 70
Cyamopsis tetragonoloba 44
Cytoplasmic male sterility 140

Dalbergia sissoo 70, 90
D. latifolia 90
De novo regeneration 23, 28, 66
Desirable recombinants 8
Diazotrophic organisms 201
Disease diagnosis 3
Disease resistance 16, 24
DNA-DNA hybridization 10
DNA finger print 142
DNA marker 269, 270, 276, 306
DNA sequencing 10
DNA viruses 265
Drought 223
Drought tolerance 28

Electroporation 149, 284
ELISA 149
Elite germplasm 393
Embryo conversion 104
Embryo germination 104
Embryo rescue 25, 28, 43, 164
Endotoxin 245
ENOD 175, 184
Euploid 142
Ex vitro 135

FISH 310
Flooding 223
Fluorescein diacetate 141
Foliar disease 16, 142
Forage legumes 133
Functional perception 177
Fungal resistance 16, 269
Fusarium 12
F. oxysporum 16, 326, 327, 329

Gemini viruses 255, 263
Gene bank 164
Gene expression 145
Gene introgression 168
Gene pool 15, 163
Gene transfer 164
Gene pyramiding 164
Genista monosperma 90
Genome 303
Genotype 73, 107
Genetic distance 10
Genetic diversity 10, 393
Genetic erosion 393
Genetic manipulation 133
Genetic map 17
Genetic resource 3, 155, 355, 368, 369
Genetic transformation 23, 29, 43
Genetic uniformity 155
Germplasm 15, 65
Germplasm collection 368
Germplasm preservation 368
Germplasm resource management 380
Glucanase 271
Glutamate synthase 202
Glutamate synthetase 202
Glutathione reductase 224, 233, 236
Glycin betaine 224
Glycine species 76, 90
G. argyrea 70
G. canescens 27
G. clandestina 27, 70
G. latrobeana 70
G. falcata 70
G. max 24, 27, 35, 40, 44, 70, 90, 230, 260
G. tomentosa 70
G. tomentella 70
G-protein 176, 179, 180, 225

Grain legumes 3, 6, 17, 23, 26, 33, 39, 133, 134, 138
Groundnut 156
Growth regulators 196

Hardwickia binata 91
Harvest index 11, 13
Heat shock proteins 229, 230
Heavy metals 223
Helicoverpa armigera 166, 249
Herbicide 148
Heterokaryon 142, 143
High temperature 12
Histology 37
Holoenzyme 213
Hybridization 7, 160, 164
Hybrid embryo abortion 166

Ideotype 12
Immunofluorescence microscopy 149
Indigofera enneaphylla 70
Insect pests 246
Insect resistance 245, 248
Integrated pest management 245
Insecticidal proteins 247
Interspecific hybridization 15, 158, 161
In ovulo embryo culture 164
In vitro regeneration 11
Isogenic lines 11
Isozyme 10, 143, 203

Karyomorphology 307
Karyotype analysis 10, 308
Kidney bean 45

Lathyrus spp. 71
L. sativus 71, 141, 291
Lectins 5, 247, 249
Leghemoglobin promoter 208
Legume genomics 335
Lens spp. 10, 16
Lens culinaris 27, 44
Leucaena leucocephala 71
Linkage map 10, 325
Lipochito-oligosaccharides 175
Lotus species 204, 206
L. corniculatus 206
L. japonicus 6, 201, 206, 209, 212
Lupinus spp. 27, 78
L. augustifolius 44, 91
L. albus 71, 91
L. mutabilis 71, 91
Lysozymes 271

Macrophomina phaseolina 16
MAP-Kinases 180, 225
MAPMAKER 10
Marker assisted breeding 330
Marker assisted selection 3, 9
Medicago arborea 71, 139
M. borealis 71
M. intertexata 137
M. falcata 91
M. littoralis 137
M. lupulina 71
M. polymorpha 91, 137
M. sativa 71, 91, 139, 140, 182, 231
M. scutella 137
M. suffruticosa 91
M. truncatula 6, 7, 91, 137, 140, 177, 179
Membrane depolarization 177
Membrane transporters 175
Metabolic signals 211
Methionine deficient legumes 281
Methionine rich protein 284
Microprojectile-mediated transformation 26
Microsatellites 9, 312
Microsatellite based marker 314
Mimosa golois 91
M. pudica 71
Minerals 24
Mineral toxicity 12
Model diseases 7
Model legume 336
Molecular markers 9, 225, 313
Morphogenesis 11
Mothbean 39, 41, 42
Movement proteins 149, 256
Multiple shoots 29
Multiple stress resistance 17
Mungbean yellow mosaic virus 264
Mutagenesis 209
Mycosphaerella pinodes 11

Nematode resistance 269
Neurotoxin 291
Nitrate assimilation 191, 194
Nitrate reduction 192, 214
Nitrate reductase 193, 202
Nitrate suply 195
Nitrate reductase 202
Nitrogen 175, 191, 201
Nitrogen fixation 3, 15, 16, 211
Nitrogen utilization 201
Nitrogen use efficiency 203, 208
Nod O protein 183
Nod factor 175, 178
Nod factor perception 176

Nod factor signals 175
Nodulin 184
Nodulation 178
Nodule formation 185
Nodule primordium 178
Northern blot analysis 215
Novel genotypes 25
Nutritional value 24

ODAP 291
Oligonucleotide fingerprinting 314
Onobrychis spp. 136
Onobrychis viciaefolia 71, 136
Organogenesis 23, 28, 31, 65, 66
Osmosensor 225
Osmoprotectant 223, 231
Osmotic changes 144

Patch clamp technique 144
Parental selection 7
Pathogen-derived resistance 256, 261
Pathogenesis related proteins 269
PCR based techniques 9, 168
Pea 259, 262
Peanut 105, 260
PEG-mediated transformation 41
PEG-mediated protoplast fusion 141
Pest management 245
Phaseolus spp. 10, 75
P. aureus 27, 35, 91
P. acutifolius 9, 27, 35, 91
P. coccineus 4, 27, 35, 71, 91
P. lunatus 4
P. vulgaris 4, 5, 9, 16, 27, 35, 40, 45, 71
Phosphoinositide pathway 180, 225
Pigeonpea 31, 38, 41, 43, 164, 165, 261
Pisum spp. 141
Pisum sativum 5, 16, 27, 35, 40, 45, 71, 78, 92
Photorespiration 205
Phytohaemagglutinins 5
Phytophthora blight disease 155, 165
Phytohormones 175, 185
Phytohormone signaling 185
Phytoalexins 269
Plant growth regulators 111, 146
Plant-pathogen interaction 133, 149
Pod borer 166
Pod retention 12
Poinciana regia 92
Polyamines 224, 233
Polymorphic probes 9
Polyols 224, 233
Powdery mildew 16
Preliminary yield trial 13
Proline 224, 232
Proline dehydrogenase 232
Promoter 138, 228
Proteomics 339
Protease inhibitors 248
Protein kinases 176, 180, 225
Protein phosphatases 175, 225
Protein quality 11
Protein calorie malnutrition 353, 356
Protein energy malnutrition 353, 356
Protoplast 25, 133
Protoplast fusion 141
Protoplast regeneration 41
Protein quality improvement 283
Protoplast to plant systems 149
Pseudomans spp. 149
Psophocarpus tetragonolobus 35, 40, 71, 92
Pythium spp. 16

QTL 9, 10, 18, 211, 217

RAPD 9, 11, 17, 119, 140, 168, 169, 269, 270, 316
Reactive oxygen species 223, 224
Recalcitrance 66
Recombinant genotype 8
Recombinant DNA technology 25
Recurrent selection 15
Refugia 250
Replicase 264
Repetitive elements 311
Repetitive somatic embryogenesis 114
Resistance management 250
Restricted genetic base 10
Retrotransposons 311
RFLP 9, 10, 17, 119, 140, 168, 211, 270, 269
Rhizobium 175
Rhizoctonia spp. 16
Rhizosphere 175
Rhynchosia aurea 41
R. visicidia 41
Ribosome inactivating proteins 270
RNA viruses 255, 256, 257
Robinia pseudoacacia 92
RT-PCR 121
RUBISCO carboxylase 232
RUBISCO oxygenase 232
Rust 16

Salinity 223
Salt stress 12
Secondary messengers 223, 227
Secondary somatic embryogenesis 114

Scanning electron microscope 33
Seed composition 7
Seed yield 8, 13
Selection strategies 8
Sesbania rostrata 71, 92
Semi arid tropics 156
Sequence tagged microsattelite site 272, 317, 320, 331
STMS marker map 321
Simple sequence repeats 276
Somatic embryogenesis 23, 26, 33, 34, 87, 94, 107, 111, 119, 333
Somatic hybridization 25, 42, 138
Somaclonal variation 29, 41, 117, 118, 291, 333
Soil borne fungi 164
Soybean 99
Southern analysis 142
Spodoptera littoralis 248
Stem nematodes 12
Stringless pods 5
Storage proteins 24
Stress signal 225, 340
Stress tolerance 225, 340
Stress responsive genes 256
Stylosanthes guyanesis 71
Sunflower seed albumin 284
Superoxide dismutase 224, 233, 234, 235
Suspension culture 142
Symbiosis 204
Synteny 6
Synthetic seed 117

Tanins 142
Tetraploid chromosome 158
Transcriptomics 339
Transcription factors 223, 228, 229
Transformation systems 333
Tribal pulses 353, 365, 369, 375
Trifolium spp. 136
T. alpestre 72
T. armenium 72
T. arnense 72
T. batmanicum 72
T. cherleri 72
T. curvisepalum 72
T. fragiferum 92, 136
T. incarnatum 72
T. nanum 71
T. nigrescens 71
T. pratense 71, 92
T. repense 72, 92
T. resupinatum 72
T. rubens 72
T. striatum 72
T. subterraneum 72, 92
T. tridentatum 72
T. uniflorum 72

Underexploited pulses 365
UV-irradiation 223

Vicia faba 27, 35, 40, 72
V. narbonensis 27, 35, 40, 45, 92
Vigna spp. 75
V. aconitifolia 27, 32, 33, 39, 40, 45, 72, 92
V. angularis 138
V. glabrescens 92
V. mungo 13, 35, 72
V. radiata 27, 35, 72, 92
V. sinensis 27, 35, 93
V. sublobata 135
V. umbellata 135
V. unguiculata 27, 35, 45, 72, 228
Viral genome 149
Viral resistance 16, 259
Vitamins 24

Waterlogging 12
Wide hybridization 155, 164
Wild germplasm 3, 10
Wild relatives 377
Wild species 155, 168
Wollastonia biflora 146